ANNUAL REVIEW OF
EARTH AND
PLANETARY SCIENCES

EDITORIAL COMMITTEE (1993)

ANNUAL REVIEW OF EARTH AND PLANETARY SCIENCES

VOLUME 21, 1993

GEORGE W. WETHERILL, *Editor*
Carnegie Institution of Washington

ARDEN L. ALBEE, *Associate Editor*
California Institute of Technology

KEVIN C. BURKE, *Associate Editor*
National Research Council

ANNUAL REVIEWS INC   4139 EL CAMINO WAY   P.O. BOX 10139   PALO ALTO, CALIFORNIA 94303-0897

ANNUAL REVIEWS INC.
Palo Alto, California, USA

International Standard Serial Number: 0084-6597
International Standard Book Number: 0-8243-2021-2
Library of Congress Catalog Card Number: 72-82137

TYPESET BY BPCC-AUP GLASGOW LTD., SCOTLAND
PRINTED AND BOUND IN THE UNITED STATES OF AMERICA

Annual Review of Earth and Planetary Sciences
Volume 21, 1993

CONTENTS

SOME RELATED ARTICLES IN OTHER *ANNUAL REVIEWS*

From the *Annual Review of Astronomy and Astrophysics*, Volume 30 (1992):

From Steam to Stars to the Early Universe, William A. Fowler

Solar Flares and Coronal Mass Ejections, S. W. Kahler

The Pluto-Charon System, S. A. Stern

From the *Annual Review of Ecology and Systematics*, Volume 23 (1992):

Global Environmental Change: An Introduction, Peter M. Vitousek

The Potential for Applications of Individual-Based Simulation Models for Assessing the Effects of Global Change, H. H. Shugart, T. M. Smith, and W. M. Post

Global Change and Coral Reef Ecosystems, S. V. Smith and R. W. Buddemeier

Global Change and Freshwater Ecosystems, Stephen R. Carpenter, Stuart G. Fisher, Nancy B. Grimm, and James F. Kitchell

Global Changes During the Last Three Million Years: Climatic Controls and Biotic Responses, T. Webb III and P. J. Bartlein

Responses of Terrestrial Ecosystems to the Changing Atmosphere: A Resource-Based Approach, Christopher B. Field, F. Stuart Chapin III, Pamela A. Matson, and Harold A. Mooney

Trace Fossils and Bioturbidation: The Other Fossil Record, T. Peter Crimes and Mary L. Droser

From the *Annual Review of Entomology*, Volume 38 (1993):

Insects in Amber, G. O. Poinar Jr.

From the *Annual Review of Fluid Mechanics*, Volume 25 (1993):

Surface Waves and Coastal Dynamics, Chiang C. Mei and Phillip L.-F. Liu

Wave Breaking in Deep Water, M. L. Banner and D. H. Peregrine

Victor Vacquier

Annu. Rev. Earth Planet. Sci. 1992. 21:1–17

MANY JOBS

Victor Vacquier

Marine Physical Laboratory, Scripps Institution of Oceanography,
University of California San Diego

KEY WORDS: geomagnetism, heat flow, thermal conductivity, Earth's interior

Of speckled eggs the birdie sings
   And nests among the trees;
The sailor sings of ropes and things
   In ships upon the seas.

Robert Lewis Stevenson

In my long professional life in science and engineering, I often changed
my job, sometimes by choice, sometimes by the force of external circumstances. Often after a move, I would break out in a cold sweat from
fear of failing on the new job. This kept me on my toes, and prevented me
from becoming a fossil. Until 1953 I earned my living in industry and by
designing military instrumentation. During this first period I worked on
magnetic prospecting, first for oil and later for submerged submarines.
After the war, I was employed by Sperry Gyroscope Co., where my small
group of engineers and I developed new instruments for navigation and
stabilization of shipborne equipment. For four years after I left industry,
I prospected for groundwater in New Mexico, and finally joined the staff
of Scripps Institution of Oceanography where I was engaged in the magnetic and heat flow surveys of the oceans. When I retired in 1975, I was
trying to calculate heat flow from oil field data. I am now preoccupied
with the origin of terrestrial heat flow, half of which, I find, comes from
the release of gravitational potential energy via the absorption of iron
oxide of the mantle by the molten iron of the core.

   I was born on the third floor of No. 40, Tchaikovsky Street, St. Petersburg, Russia, on October 13, 1907. My first interest in science that I
remember made me dig a hole in search of a piece of lodestone in the
garden of the dacha the family rented on the shore of the Gulf of Finland.
I must have been seven years old. Two years later, I was winding a coil on

1

0084–6597/93/0515–0001$02.00

a magnetic compass, with wire cannibalized from an electric bell. By holding one end of the wire in one hand, and placing the other end on my tongue a sizable deflection of the compass needle confirmed the existence of electromagnetism and electrochemistry as described in my children's scientific encyclopedia. It was also fun to rig up the telescope from a surveying level, and watch from day to day the sunspots on the sun's projected image. Retired readers remember that in the early tungsten light bulbs the filament was strung in a zigzag between hooks held at the top and bottom of a central glass rod. My favorite test of electromagnetic interaction and Ohm's law was to move a magnet toward the bulb and watch the gradual increase in the amplitude of the filament's vibration, until a short with its neighboring section would brighten and whiten the light emission of the bulb. I also enjoyed mixing explosives from ingredients purchased at the drugstore across the street, where I had a charge account. Once, Grandfather really blew up when he found me reading the article on nitroglycerine in the Brockhaus & Effron encyclopedia (the equivalent of our Britannica).

This formative period of my life ended abruptly in February 1920, when Finnish smugglers carried our family, with one suitcase per head, from a Petersburg suburb to the shore of Finland, during the night in two one-horse open sleighs. They dumped us on the ice, in sight of shore. My father, a physician, kept us awake during the rest of the night to prevent us from lapsing into our last sleep. In the morning, we were duly arrested and taken to the police station. It was warm and cozy and they gave us white bread, which we had not eaten for two years. Sprawled on our luggage, I fell blissfully asleep with a piece of it in my hand.

In the summer of 1920 we moved from Finland to Paris, and as I had missed most of the 1919–1920 school year, two months of intensive tutoring by a bearded and nearly blind pedagogue, prepared me to enroll into the 4th class of Lycée Janson.

In the fall of 1921, to reduce living expenses the family bought a small farm near Pau in the foothills of the Pyrenees. I completed my third and second classes at the Lycée of Pau. The farming venture was a failure. In 1923, with the help of Charles R. Crane, of the Crane Plumbing Co. of Chicago, a Russophile and friend of the family, my mother and I emigrated to the United States.

My mother became a teaching assistant in the Romance Language Department of the University of Wisconsin, and I entered Madison Central High as a senior. At that time high school in America was not much different from the schools I went to in Europe. The classes numbered less than twenty pupils, and the required subjects were about the same. There was less homework and hardly any memorization of classic literature. The summer after graduation I worked as a farmhand for a tenant farmer on a small farm. The farming experience in France came in handy on this job. In the fall of 1924 I enrolled into the Electrical Engineering course at the University of Wisconsin. As I was going to school on borrowed money, I took heavy schedules and attended three summer schools, and thus graduated in three years.

I even took advantage of the professor in charge of teaching physics to undergraduate engineering students. He was a Francophile, and I succeeded in convincing him that I had taken the material in the sophomore physics class, which was a block of 12 credits in France, previously. It was, of course, a lie: More than half of the subjects in Duff's textbook were new to me. He quizzed me orally for 20 minutes, and passed me. Perhaps it was because, remembering how I used to destroy light bulbs with a magnet, I vehemently contradicted his probing suggestion that an electric current is increased by increasing the circuit's resistance. In my senior year he hired me to calculate currents in vacuum tube circuits.

I also translated Russian articles on gravity prospecting with the torsion balance for the firm of Mason, Slichter, and Hay. Mason was president of the University of Chicago, and L. B. (Uncle Louie) Slichter, the future founder of the Institute of Geophysics of the University of California, was the one who hired me. Later on, I corrected physics papers at the Extension Division for the father of Carl Jansky, the classmate who founded radio astronomy. Occasionally I substituted for a friend who clamped and unclamped the 12-inch refractor at the Washburn Observatory where the luminosity of Cepheid variable stars were measured.

After graduation I obtained an assistantship in the Physics Department. For a Master's thesis I built a fused quartz, high pressure mercury arc lamp for some photoelectric work of the department, where the photoelectric work function of various metals was being measured by PhD candidates. This was an exciting time in atomic physics. Debye, Brillouin, and Dirac gave short courses on their recent discoveries, and I was foolhardy enough to dream up a thesis subject in spectroscopy, in which my faculty sponsor had no expertise. When J. Frank, on a visit to my lab, heard of what I was trying to do, he shook his head and kindly advised me saying, "Young man, you should read my book." (*Excitation of Quantum Jumps by Collisions*; Jordan & Frank 1926).

This German book discouraged any further plans I had in spectroscopy, and in early summer of 1930, I was, without a PhD, helping my former professor of electric circuit theory, Leo J. Peters, interpret vertical magnetic profiles surveyed along the Orinoco river in Venezuela at the Gulf Research & Development Co. in Pittsburgh, Pennsylvania. A few months later, John Bardeen was put in charge of the theoretical work, while I was given the job of improving the precision of the Schmidt vertical intensity magnetic balance made by Askania in Germany. In this instrument, a horizontal magnet is pivoted about a knife edge of crystalline quartz resting on a pair of quartz cylinders. By moving weights on a pair of horizontal screws, a gravity couple is adjusted to balance the torque exerted on the magnet by the vertical component of the geomagnetic field. To eliminate short time fluctuations, a stationary instrument was read every 10 minutes by an operator who also reduced the observations of the previous day. In the U.S., the field operator drove along section fences and took readings a mile apart. In easy country, an average of 35 stations were observed per day. Nowadays geomagnetic surveys are flown at a speed of 100 knots. In the 1930s Gulf had about four or five magnetometer parties in the oil producing parts of the country. The base station records of the parties close to each other afforded an opportunity to study the spatial variability of the short time fluctuations. The management of Gulf R & D Co. had a generous policy on scientific publications by the staff. One of the five papers I published while I worked at Gulf, was "Short-Time Magnetic Fluctuations of Local Character" (1937) which demonstrated for the first time the existence of magnetic variations peculiar to a particular location. Some years later these variations were interpreted as caused by differences in electrical conductivity of the underlying geologic structure. This method of probing geologic structure is now called "deep magnetic sounding." The late J. Bartels of Göttingen became interested in this subject and sent U. Schmucker to my laboratory at Scripps in 1959, with a set of three component magnetic variographs built for this specific purpose. Schmucker covered the area from the Pacific Coast to a hundred miles east of the Texas–New Mexico border, and published a monograph in 1970.

He found that the highly conductive layer under the Basin and Range geologic province ends abruptly near the New Mexico–Texas border. This was later corroborated by a heat flow survey in specially drilled holes. In another paper (Vacquier & Affleck 1941) J. Affleck and I estimated the depth of the bottom of the magnetic layer of continental crust to be 23 kilometers. Recent work on this question finds about the same result.

In the middle 1930s there was a man called Lynton who claimed to be able to orient oil well cores from the horizontal component of their remanent magnetization. I designed the first spinner magnetometer to do this. It was a monstrous device that spun cores 4 inches in diameter at 11 rps. The coil pair rested on a pier disconnected from the floor of the building. Orienting cores this way was a failure, so I tried to measure the remanent magnetism of the rock formations near the laboratory. Inconsistency of the measurements made me abandon, alas too soon, this excursion into paleomagnetism. Fortunately, it stimulated the workers at the Department of Terrestrial Magnetism of the Carnegie Institution of Washington to design the air centrifuge spinner with which they extended the measurement of secular variation from the remanent magnetism of the annual lacustrine deposits of silt called varves.

Sometime in 1940, when Europe was already at war, I was experimenting with a bridge circuit with coils containing iron cores for the eventual purpose of detecting fatigue cracks in drill pipe. I noticed very sharp voltage signals sensitive to my moving a permanent magnet. I instantly realized that a sensitive electrical magnetometer could be made. The very first flux gate sensor I built was just as sensitive as the ones that followed it. I happened to have some permalloy wire about 1 mm in diameter from which I cut two lengths three inches long. Each one was wound with fine wire and the pair was inserted into a secondary coil. The principle on which the flux gate magnetometer works is very simple. The two cores are periodically magnetized beyond saturation in opposite direction. As long as the flux in a core increases, a voltage is induced in the secondary coil. When the core saturates the flux stops changing and the induced voltage drops suddenly to zero. The voltages induced in the secondary coil are subtracted. The ambient field advances the saturation of one core and retards the saturation of the other, producing a sharp voltage spike near saturation. The war on our minds, this first magnetometer was intended for an anti-tank land mine (U.S. Patent #2,406,870).

A condenser was charged through a high resistance leak. When its voltage reached the firing voltage of a cold cathode gas tube, it discharged through the primary windings of the flux gate. A change in the ambient magnetic field of 100 nT could fire another cold cathode tube through a dynamite cap. The army did not buy the fuze but "RESTRICTED ADMITTANCE" was stenciled onto the door of my lab. The cores in the next magnetometer were strips of mumetal 4 in long, 0.125 in wide and about 0.015 in thick. Strips of these dimensions were used in the magnetometers eventually built for the Navy. In the next magnetometer, the cold cathode tube was replaced by a hot cathode gas tube, the frequency of the condenser discharge raised to about 100 Hz, and an adjustable direct current was fed into the secondary coil, to cancel most of the ambient field. The sensitivity was about 5 nT per division on a needle milliammeter. The sensor was mounted on the telescope of a surveying transit. My chief, E. A. Eckhardt, and I took this instrument to New London, Connecticut. We set it up on the end of a pier, and recorded the signal from a passing submarine at progressively increasing distances.

A short time later, Eckhardt called me to his office: "Vic, I want you to put your magnetometer on an airplane." Thereupon he called the president of the Sperry Gyroscope Co. and made arrangements for collaboration. Frankly, I was saddened by seeing such a capable man lose touch with the world of reality. Luckily, Gary Muffly, our electronic and patent engineer, joined my lab. He realized that the errors from the change in orientation of the sensor caused by the roll and pitch of the aircraft would be reduced by directing it along the total magnetic intensity. The first airborne magnetometer to give useful results consisted of an air-driven horizon gyro carrying a light aluminum frame extending 15 cm to the right and left of the outer gimbal axis. At one end the frame carried a flux gate with its axis horizontal and normal to the plane of the frame. The anomaly detecting sensor was mounted at the other end of the frame so that it could be rotated and set in the E-W vertical plane by an angle equal to the average dip of the total intensity. The box containing this assembly was mounted on a turntable which was servo controlled by the orienting sensors so that the vertical plane containing the detecting sensor was normal to the magnetic meridian. The Pearl Harbor disaster found me experimenting with this instrument installed in a PBY flying boat at the Quonset Point Naval Air Station in Rhode Island. After an unsuccessful demonstration, the magnetometer was installed in a K-type blimp at Lakehurst, New Jersey. The Magnetic Anomaly Detector (MAD) was installed in the ballonet 15 feet forward of the gondola. It rested right on the outer skin of the blimp. A plywood frame prevented the inner bag from interfering with the rotation of the box containing the instrument when the helium bag was let to expand against the outer skin when more buoyancy was needed. To service the instrument in flight, I would take off my shoes so as not to tear the blimp's delicate skin and fall into the ocean far below. A sailor would then push me up into a canvas sleeve against the stream of compressed air. Once in the bag, the sleeve was tied shut, and I would find myself in the relative stillness of the big bag much as Jonah must have experienced in the belly of the whale. The blimps patrolled the eastern seaboard often within sight of land. At night ships were silhouetted against the continuous band of light, making perfect targets for German submarines. Blackout was very slow to come. The wrecks were numerous and the subs few. We depth-charged many wrecks, and President Roosevelt told the nation that the answer to the submarine was the blimp!

After several hundred hours on blimp patrols, I left Gulf Research & Development Co., to join the Airborne Instruments Laboratory (AIL) of Columbia University which was being created by D. G. C. Hare, under the auspices of the National Defense Research Committee. It occupied a small building in Mineola, New York. Hare decided to orient the flux gate along the total magnetic intensity by two other flux gates controlling servomotors. I witnessed his patent disclosure. It was a good idea if the servos could be made good enough. At that time, servomechanisms were a mysterious subject. Sperry Gyroscope Co. was pioneering this technique, and one of their engineers was assigned to consult part-time. In the meantime, in the magnetic quiet of the Department of Genetics of the Carnegie Institution at Cold Spring Harbor, NY, Walter Brattain, one of the future inventors of the transistor, devised a procedure for matching cores and coils for optimum sensitivity and identical characteristics in the manufacture of the flux gates. In about a year's time a beautiful prototype was produced and turned over to Texas Instruments Inc. (TI) for manufacture. It was the first manufacturing job undertaken by TI. In the meantime Rodney Simons and I were working nearby, at the Sperry Gyroscope Co., on controlling a air-driven gyro, so that its axis was along the total magnetic intensity. The gyro carried a flux gate parallel to the axis of the gyro, and a pair of flux gates mounted on a plate perpendicular to the gyro axis which maintained the gyro along the total magnetic intensity. This instrument had no servo errors, as the gyro kept the sensor assembly steady. The servomotor instrument developed at AIL was easier to manufacture, and the noise from servo errors was smaller than the noise from the maneuvers of the aircraft, so that the gyro-stabilized instrument was abandoned.

Operational problems and the actual use of the MAD occupied me until the end of the war. First I worked with W.E. Tolles on methods for compensating the noise produced by the maneuvers of the aircraft—as the Navy refused to tow the instrument behind the airplane. I also went on the next-to-the-last cruise of the escort carrier *Block Island* on which I had 3 TBFs equipped with an MAD. I went on most of the flights of my airplanes and although we sank more than three submarines, none of the kills could be credited to the MAD. This trip was during the last year of the war, after the German code was broken by an English cryptographer. The German submarines reported their position every night so that we knew exactly where to look for them. The MAD accounted for more than three kills by a squadron of PBY patrol aircraft continuously patrolling the Strait of Gibraltar. At ten freighters saved per submarine sunk, the MAD more than paid for itself.

As the magnetic work was being completed, the activity at AIL shifted to radar countermeasures. Many of the staff were radio engineers for whom this was a welcome transition, but I was beginning to feel like a fish out of water, and had to choose between Gulf and Sperry Gyroscope, where I had standing invitations. At Gulf I would be obviously doing more magnetic prospecting, but the work at Sperry was an unknown adventure, so I chose the latter. At Sperry, I joined the laboratory for the development of marine instruments. Our primary job was to build a new and better gyrocompass for the Bureau of Ships of the Navy.

The gyrocompass consists of a delicately suspended pendulous gyroscope which seeks to align its spin axis to the horizontal component of the Earth's rotation. The gyro spin axis tilts a few seconds of arc, just enough to keep up with the Earth's rotation, thus providing an accurate vertical reference about the E-W axis. By adding another horizontal gyro at right angles to the north-seeking one controlled by a damped pendulum about the N-S axis, we built an instrument that furnished an accurate reference to the vertical in addition to the direction of the ship's heading, for the stabilization of detection and ordnance devices. This instrument, the mk. 19 Gyrocompass was put into service about 1953, and is still in use (U.S. Patent #2,729,108). The mk. 23 Sperry gyrocompass was also designed in my laboratory. It was originally intended for army tanks so that they would not get lost in conditions like that of the battle of El Alamein, where landmarks for orientation were lacking. In this instrument a single gyro is floated in oil. It is now widely used. My lab worked on a variety of other projects such as a study of transistor amplifiers—at that time still made of germanium crystals, a far infrared detector which used Maksutov optics, etc.

During my last three years at Sperry, I often thought that I should not like to spend the rest of my professional life in a factory environment. To retain my status in geophysics, I asked Maurice Ewing if I could work with a graduate student in his department at Columbia University. I was still working full time at Sperry, so I became a professor without salary. At that time, Gulf duplicated the AIL airborne magnetometer thanks to E. Westrick, a Gulf employee who had been on leave as the civilian technician with the patrol squadron VP63 which had been sinking submarines in the Strait of Gibraltar. The Navy was slow in declassifying the equipment, so Gulf was in a privileged position. It was inevitable that airborne magnetic surveys would have a bright future. The principal application of mapping local geomagnetic anomalies in prospecting for oil is in estimating the total thickness of sediments. The greater the thickness of the sediments, the better chance there is of finding commercial oil deposits. Because the magnetization of sediments is very weak, the anomalies of the geomagnetic field come from magnetization contrasts within the crystalline basement rocks. The anomalies become sharper as the distance to the basement is decreased. Before the war, one method of interpreting the vertical magnetic anomalies, invented by John Bardeen, was to compare the sharpness of the magnetic anomaly to the inverse tangent curve which is the anomaly produced above two infinite quarter spaces separated by a vertical plane directed N-S and differing in their vertical magnetization.

For the inverse tangent the depth of burial is one half the distance under the steep slope of the curve. It was indeed amazing that this simplistic scheme worked as well as it did. Now, with the total intensity, instead of the vertical intensity, being mapped, some other method had to be used. I, therefore, put my student, Nelson Steenland, to work on computing model magnetic anomalies above vertical prisms of infinite depth extent at several latitudes. The depth to the basement was estimated by comparing anomalies in the surveys with the models. The only survey data available at that time belonged to the U.S. Geological Survey, so we acquired R.G. Henderson and I. Zietz as collaborators. They also contributed their method of calculating the second vertical derivative of the total magnetic intensity for delineating the contacts between the bodies of rock that had magnetization contrast. I was told that our system was widely used after Gulf licensed Aero Service Inc. and Fairchild to make aeromagnetic surveys and collected one dollar per mile flown from their customers, ruefully dubbed "the GULF dollar" by the industry. I have never actually practiced the method we developed, which was published as *Geological Society of America Memoir No.* 47 in 1951 (Vacquier et al 1951). It became obsolete when digital computers came into use.

In 1952 I used my vacation to look for a job in geophysics. It must have been at the suggestion of M. Tuve of the Carnegie Institution of Washington that I called on E.J. Workman, President of the New Mexico Institute of Mining and Technology (NMIMT) at Socorro. On this visit, a band of antelopes crossed the road 50 feet in front of my car, as I was touring the Plain of San Agustin. That, and E. J. Workman's small but modern research laboratory, made me take a job in geophysical prospecting for water at a 40% cut in salary. I moved to Socorro in 1953. At my first faculty meeting a heated discussion arose as to what should be done about students keeping dynamite under their bunks. On leaving the meeting, I noticed bullet holes in the window of the dormitory. For me, a denizen of New York suburbia, it was a stimulating new environment.

At the NMIMT my duties consisted of giving an elementary course in geophysical prospecting and supervising the research of graduate students. The research was to be directed toward increasing the water resources of the state. Prospecting for ores by induced electrical polarization had been invented a few years before by Arthur Brant, and I thought we should see if it could be adapted for finding fresh water in alluvium at the foot of mountains similar to the "kanats" that supply water for Teheran, Iran. The alluvium was chiefly argillaceous sand. We first studied the electrical response of mixtures of pure kaolin and montmorillonite with quartz sand. The mixtures were loaded into trays about 30 cm long, with current electrodes at the ends and nonpolarizing potential electrodes 10 cm from the ends of the tray. The sand was polarized by a direct current for a definite interval of time, like ten seconds, and the decaying voltage between the potential electrodes was recorded after the polarizing current was shut off and the area under the decay curve was integrated. When the tray was loaded with a slurry of sand and clay, no induced polarization (IP) appeared, but when the slurry was wetted without mixing after it was left to dry in the tray, it was polarizable. The effect was greatest with distilled water. Salinity destroyed it. In contrast with electrical resistivity prospecting, IP gave a specific test for fresh water. This is important because the groundwater in many deserts is a concentrated brine. Drying the slurry coated the sand grains with clay. When the sand was moistened after drying, the clay with its negatively charged surface remained stuck to the sand. When the water is fresh, most of the electrical conductivity resides in the clay which takes some time to return to equilibrium after the polarizing current is shut off. When the water is conducting, equilibrium is reestablished quickly after the polarizing current is turned off. In the field, equipment power was furnished by a permanent magnet alternator rotated from a power takeoff of a Jeep truck engine. After rectification by a selenium rectifier, it was led to current electrodes which consisted of a set of iron stakes in ground impregnated with brine. The potential electrodes were commercial porous pots containing copper sulphate solution. The amplifier had an extremely narrow bandwidth obtained by a tuning fork. The IP project caught the fancy of the Society of Exploration Geophysicists, and I was invited on the Distinguished Lecture tour to the local chapters of the Society. Shortly thereafter, I hosted the director of Egypt's Desert Research Institute and head of the Camel Corps—a veterinarian of high army rank, who invited us to work in Egypt. My PhD student, Paul Kintzinger, who had just gotten his degree, went to Cairo with the equipment. He was just ready to start work when the English bombed the Cairo airport runway near the building of the Desert Institute. This ended further field work on prospecting for water by IP. In the U.S. the method was a commercial failure. I competed unsuccessfully with a dowser with a forked stick who charged for his services—which thus became more valuable than the free advice of a State employee. Also, a small core drill was cheaper to use for exploration for water that had to be no deeper than 300 feet, because one ton of irrigation water had to cost no more than the postage of a first class letter (3 cents in those days). I was pretty much at loose ends, when Walter Munk came to Socorro to give the commencement address to our 1956 graduates. Soon I was invited to give a talk about our IP work at a seminar at Scripps Institution of Oceanography, where I got my last position in 1957.

At Scripps, I was given charge of magnetic surveys at sea. At that time,

R. Mason and A. Raff had just completed a magnetic survey of a strip of ocean along the west coast of the U.S. as guests on the U.S. Coast and Geodetic Survey ship *Pioneer* which revealed the lineated anomalies caused by seafloor spreading. The magnetometer they used was a modification of an MAD made by Bell Laboratories that was not used in wartime. The sensor was housed in a large fish that took a detail of six sailors to launch. The proton precession magnetometer was just appearing as a practical instrument. The Navy sent me one to experiment with. This instrument had a fish almost the same size as that of Bell Lab's. The fish contained water bottles plus tuning condensers connected by relays controlled from the ship by an 18-conductor tow cable. We spent six months building a proton magnetometer from scratch. Two coils at right angles containing bottles filled with water were housed in a bakelite cylinder about 6 inches in diameter and were connected in series to a low capacity, nonmicrophonic two-conductor cable 300 feet long. Only half of the signal was lost in the cable. The signal frequency was multiplied by 4, which gave a sensitivity of about 5 nT per count. The signal could be recorded on a digital printer or more commonly on a paper chart recorder. Several of these instruments were built at Scripps. They were used until Varian Associates came out with their commercial model.

The displacement of the magnetic anomaly pattern across the Murray Fracture Zone prompted us to look for similar displacements across other fracture zones. The next one studied was the Pioneer Fracture Zone. A few cruises later, we were able to map the displacements of magnetic anomalies across all the major faults of the eastern Pacific. Although I had offered a vague explanation for lineated magnetic anomalies parallel to midocean ridges, there was no reasonable explanation for the displacements of the magnetic patterns until J. T. Wilson conceived of the transform fault. Even though I was acquainted with the existence of reversely magnetized rocks, Néel had proposed mineralogical mechanisms that could explain thermoremanent magnetization contrary to the ambient field, such as was found in the dacite of Mount Haruna. This prevented me from believing in reversals of the field until the VEMA 19 magnetic profile in the southern Pacific showed exactly the same sequence of anomalies that we mapped north of the Mendocino Fracture Zone.

From magnetic and topographic surveys of a uniformly magnetized body, it is possible to compute the magnitude and direction of the magnetization of the body. This opens the possibility of measuring the change in latitude of the ocean floor by surveying seamounts. The technique was to plant an anchored buoy with a radar reflector on the summit of the seamount, and then survey a star pattern by dead reckoning. In this way we found that the floor of the Pacific Ocean had moved north about 30 degrees since the Cretaceous period.

In 1970 in Moscow, at the Geological Institute of the Academy of Science of the USSR, I gave a set of lectures on seafloor spreading and plate tectonics—subjects that were neglected in the USSR. These lectures later came out as a booklet entitled *Geomagnetism in Marine Geology* (Vacquier 1972).

When R. P. Vonherzen took a UNESCO job in Paris, I inherited his project for measuring heat flow at sea. The heat flow work at Scripps was begun by E. C. Bullard, who was wont to tell the story of the first lowering of the instrument. At that time, it recorded on photographic film. They were all standing in front of the darkroom door shouting questions, when the door opened and a lugubrious voice announced: "I put it into the hypo first." The Bullard instrument consists of a cylindrical pressure case housing a recorder. A hollow spear 150 cm long at the bottom of the case contained two thermistors 1 m apart connected in a bridge circuit. Before closing the case a timer is set to start the recorder a time ten minutes shorter than the time needed to reach bottom. The penetration of the sediment heats the spear, so that it is necessary to record the signal about 15 minutes in order to compute the asymptotic value of the temperature difference of the thermistors caused by the geothermal gradient in the sediment. The temperature difference is about 0.06°C measured to about 0.001°C. The thermal conductivity of the sediment was measured by recording the rise in temperature as a needle heated at a known number of watts per cm was inserted into a sample of the sediment extracted by a gravity corer. We combined the measurement of the gradient and conductivity by attaching a heated needle to a slider on the spear.

The high heat flow outboard of subducting oceanic plates prompted us to search for it on the continents where geothermal measurements are made principally in oil fields as part of the routine in the drilling procedure. We also made measurements in Lake Malawi with R. P. Vonherzen and in Lake Titicaca with J. G. Sclater. In 1973–1974, I spent nine months at the University of Bahia in Salvador, Brazil, working on determining the heat flow in the Reconcavo Basin fields from uncorrected bottom hole temperatures and conductivity measurements of water-saturated cores with a divided bar apparatus. The heat flow in Salvador was normal, but when we went to the fields of central Sumatra, the heat flow there was more than twice normal. In Sumatra, for the first time, we measured thermal conductivity by using a heated needle embedded in the surface of a block of poorly conducting plastic. For this measurement the specimen needs to have only a single flat surface to contact with the needle.

The rise of drilling costs are now restricting core drilling to oil producing formations. This does not give sufficient conductivity information for determining the thermal resistance needed to compute the heat flow. I, therefore, looked into the possibility of estimating thermal conductivity from well logging records. This had been attempted by others, without much success. In 1985, I spent six weeks in Paris at the Institut Français du Petrole making a careful laboratory study of two sets of cores from two exploratory wells in France. We found that a simple correlation of well logging parameters and conductivity was too weak without additional information from the lithologic log obtained from cuttings. While working on bottom hole temperatures, it became apparent that some oil fields over anticlinal structures had a heat flow high anomaly. This phenomenon can be caused only if the water under the hydrocarbon deposit is flowing from a recharge area on the Earth's surface. The water flows up over the structure bringing warmer water closer to the surface. In addition, the shale that caps the reservoir rock is hydrophilic, and lets the water rise, while preventing the oil from going through. This collects the oil dispersed in the water, according to the hydraulic theory of accumulation of oil. In a reasonably constant temperature environment—such as continental shelves—heat flow measured in shallow holes might indicate an oil deposit. When the reservoir formation is sealed off from meteoric waters, say, by an unconformity, for instance, there is no heat flow anomaly over the structures. In these fields the water is charged with salts, whereas fresh water underlies the fields where the producing formation is also an aquifer, like the Dakota sandstone in the Mountain States in the U.S.

My current interest is the origin of terrestrial heat flow. While working in 1980 on oil field heat flow data at the University of Bahia, Brazil, I read *Geodinamika* by Artiushkov (1983). The author adopts the notion that at the core mantle boundary (CMB) the molten iron of the core absorbs the metallized FeO which entered into the composition of the silicates of the mantle as proposed by Dubrovski & Pankov in 1972. The light material created in this way at the CMB rises in plumes and diapirs causing geotectonics on the surface of the Earth. Seven years later the international survey of helium of the world ocean, GEOSECS, found only 5% of radiogenic helium expected from the estimate of heat loss (Oxburgh & O'Nions 1987). Radioactive decay of U and Th generates both heat and helium, so that the rate of generation of He is an independent measure of the abundance of U and Th. It became obvious to me that the abundance of uranium and thorium in the mantle was negligible and that these elements had migrated into the continental crust early in the Earth's history because they are incompatible in the silicate lattice.

Table 1 shows the distribution of radiogenic and nonradiogenic sources of heat between continents and oceans. The radiogenic source is calculated

Table 1   Radiogenic and nonradiogenic heat production[a]

	Radiogenic ($W \times 10^{13}$)	Nonradiogenic ($W \times 10^{13}$)	Total heat prod. ($W \times 10^{13}$)	Heat flow ($mW\ m^{-2}$)
Potassium	0.21	—	0.21	4.2
U+Th	0.92	—	0.92	18.1
Oceans	0.27	2.67	2.94	98.0
Continents	0.87	0.39	1.26	60.0

[a] From: Sclater et al (1980) and Vacquier (1991, 1992).

from the abundances of U and Th in ordinary chondrites from which I assumed the Earth's mantle had accreted. The abundance of potassium is taken as the present amount in the mantle. It is apparent from Table 1 that more than 2/3 of the continental heat flow comes from radioactive nuclides, whereas in the oceans nearly all of the heat flow is of non-radioactive origin.

Laboratory experiments in the laser heated diamond cell by Knittle & Jeanloz (1989) have shown that below the 670 km seismic discontinuity, the silicates constituting the mantle transform into a silicate perovskite [Fe,Mg]SiO$_3$ and magnesiowuestite [Fe,Mg]O. These minerals remain stable down to the CMB, where the molten iron of the core absorbs the FeO from both of them, leaving the light oxides to form the mantle plumes. Frictional heat generated by the plumes is equal to the gravitational potential energy lost by the FeO absorbed into the core. By making many simplifying assumptions, I calculated that the energy released was 11,300 J/g of FeO, unaware that Birch had calculated it more accurately in 1965 by an elegant method and obtained 7,940 J/g.

To satisfy the Earth's moment of inertia and the velocity of elastic waves, there needs to be a light alloying element in the outer core. To release gravitational energy this element has to be oxygen. The needed amount of oxygen is given by the formula Fe$_2$O, or 12.5 wt% of oxygen. If all the oxygen got into the core by absorbing FeO from the silicates of the primordial mantle, the mass of FeO absorbed into the core would be 10^{27} g—more than half of the total mass of the core which is 1.89×10^{27}g. Some of the FeO must have gone into the core during protocore formation. If we choose to form the primordial mantle from the average of all ordinary chondrites, 17.5% of the FeO contained in the silicates had to be absorbed during protocore formation in order to leave 6.11% of Fe in the present mantle plus crust. This leaves 8.25×10^{26} g of FeO to be slowly absorbed during the life of the Earth.

This slow decay of the nonradiogenic heat source accounts for the remarkable persistence of geotectonic processes such as plate tectonics and the generation of the geomagnetic field. Using Birch's figure, the flux of FeO into the core is now 8.3×10^{16} g/yr, generating 2.1×10^{13} W. The 8.25×10^{26} g of FeO left in the primordial mantle at the beginning descended into the core during the life of the Earth, generating a total of 6.55×10^{30} J, or a little more than twice the heat generated at the present rate for the same elapsed time. Adding 4.2×10^{30} J contributed by cooling to the 6.6×10^{30} J due to the gravitational source, we get 11×10^{30} J for the total nonradiogenic heat lost since the Earth's formation. This is 2.4 times the nonradiogenic energy lost if the present rate were to be invoked during the last 4.6×10^9 yr, which should be sufficient to maintain single layer convection.

It is a logical necessity that the upper and lower mantles be chemically the same in the four major elements: Otherwise the upper mantle would have kept its original abundance of iron while the lower mantle would have lost up to three quarters of its iron to the core. Dividing the rate of absorption of FeO by the present surface of the core, gives about 1 kg m^{-2} yr^{-1}—a slow rate considering that the thickness of the boundary layer is about 200 km. The depleted material must be preventing the molten iron from reaching the iron-rich minerals until the flow exposes fresh material to the molten iron. A mechanism is also needed for segregating the plume material. Perhaps, the temperature at the core mantle boundary may be close enough to the melting point of the light minerals—consisting of stishovite (a high pressure form of SiO_2), MgO, and silicate perovskite devoid of iron—to allow segregation by melting in narrow magma chambers similar to those in the upper mantle which generate oceanic lithosphere at spreading ridges. The magma chambers may be strung out where the temperature is locally elevated as would be expected over rising convective flow in the core.

The production of oceanic lithosphere is 8.44×10^{17} g/yr, or 265 km^3/yr. If the average loss of nonradiogenic heat in the past was 2.4 times greater than now, and if the bottom of the upper mantle is 700 km down, then 9 times the volume of the upper mantle had been processed during the last 4.6×10^9 years. Although the flux of FeO into the core is smaller than the rate of creation of oceanic lithosphere, the production of lower mantle plumes is greater than the flux of FeO into the core because in the perovskite the molecular weight of FeO is much smaller than the sum of the molecular weights of the light oxides compounded with it. Furthermore, the plumes must pick up some of the country rock as they ascend. The flux of plumes is thus comparable to the rate of creation of oceanic lithosphere. It is therefore likely that the magma chambers which feed spreading ridges are located where the mantle plumes come up. This view is supported by the low abundance of incompatible elements in midocean ridge basalts (MORB), as compared with their abundance in both hotspot basalts of oceanic islands and flows of continental basalts such as Hawaii and the Deccan Traps. The hotspot volcanism is episodic, whereas the flow of MORB is continuous, as proved by the lineated magnetic patterns. Every diapir that creates an island or a flow of continental basalt has to break through the asthenosphere where the incompatible elements have accumulated during the early history of the Earth. On the other hand, the plumes that generate the oceanic lithosphere after the initial breakthrough flow continuously so that the MORB contains the smaller abundance of the incompatible elements characteristic of the deeper regions.

The isotopes of helium, just as the other incompatible elements, are concentrated in the enriched layer in the asthenosphere. The vertical gradient of the concentration of the primordial ^{3}He is steeper than the vertical concentration gradient of the radiogenic helium because primordial helium was more abundant when the incompatible elements migrated upward— say, when the magma ocean solidified, and also because a small amount of the radiogenic helium is still being produced. This accounts for the helium isotope ratio of hotspot volcanism. The difference in the helium isotope ratio of MORB and hotspot basalts reflects the relative concentration of the two isotopes in the regions where the barren magmas from the plumes arising from the CMB are contaminated by the incompatible elements.

Looking into the future, with the exception of the gyrocompasses I helped design, my work in instrumentation and engineering is only of historical interest. The cryogenic magnetometer has replaced the spinner and the flux gate in measurement of rock magnetism, and the proton precession magnetometer has largely replaced the flux gate anomaly detector in geophysical prospecting and in the measurement of the global geomagnetic field. In antisubmarine warfare, the only useful function of the MAD is to compel the submarine to carry degaussing gear and remain submerged when airplanes are about. The gyrocompass, on the other hand, will remain the mainstay of dead reckoning navigation, especially in foul weather.

Prospecting for groundwater by induced electrical polarization may be economically useful in countries where labor is cheap. Additional study of what causes the presence or absence of elevated heat flow over buried anticlinal structures may define the circumstances when hydraulic accumulation of petroleum occurs.

In science, the matching of magnetic anomalies across transform faults and the paleomagnetism of seamounts which began at the Scripps Institution of Oceanography will continue to provide data on plate motions in the geologic past. In my view, however, my lasting contribution is my theory of the origin of terrestrial heat flow. It consists of a synthesis of the work of many Earth scientists, who deserve most of the credit, even though many of them do not endorse it. In retrospect, it is aesthetically satisfying that the theory unifies—albeit unintentionally—my main professional endeavors, as it provides needed energy to plate tectonics in the oceans, along with the temperature needed for the maturation of kerogen into the petroleum I had been seeking in the deep sedimentary basins of the continents.

Literature Cited

Artiushkov, E. V. 1983. *Geodynamics. Developments in Geotectonics.* 18: 102. Amsterdam: Elsevier. (transl. of *Geodymamika*, 1979. Moscow: Nauka)

Dubrovsky, V. A., Pan'kov, V. L. 1972. On the composition of the Earth's core. *Acad. Sci., USSR, Phys. Solid Earth* 7: 452–55 (transl.)

Jordan, P., Frank, J. 1926. *Anregung von Quantensprungen durch Stössen.* Berlin: Springer-Verlag. 312 pp.

Knittle, E, Jeanloz, R. 1989. Simulating the core-mantle boundary—an experimental study of high-pressure reactions between silicates and liquid iron. *Geophys. Res. Lett.* 16: 609–12

Oxburgh, E. R., O'Nions, R. K. 1987. Helium loss, tectonics, and the terrestrial heat budget. *Science* 237: 1583–88

Schmucker, U. 1970. *Geomagnetic Variations in the Southwestern United States.* Berkeley: Univ. Calif. Press. 165 pp.

Sclater, J. G., Jaupart, C., Galson, D. 1980. The heat flow through oceanic and continental crust and the heat loss of the Earth. *Rev. Geophys. Space Phys.* 18: 269–311

Vacquier, V. 1937. Short-time magnetic fluctuations of local character. *Terr. Magn. Atmos. Electr.* 42: 17–28

Vacquier, V. 1972. *Geomagnetism in Marine Geology.* Elsevier Oceanogr. Ser., vol. 6. Amsterdam: Elsevier. 185 pp.

Vacquier, V. 1991. The origin of terrestrial heat flow. *Geophys. J. Int.* 106: 199–202

Vacquier, V. 1992. Corrigendum to "origin of terrestrial heat flow." *Geophys. J. Int.* In press

Vacquier, V., Affleck, J. 1941. A computation of the average depth of the bottom of the Earth's magnetic crust, based on a statistical study of local magnetic anomalies. *Trans. 1941 Am. Geophys. Union,* pp. 446–50

Vacquier, V., Steenland, N. C., Henderson, R. G., Zietz, I. 1951. Interpretation of aeromagnetic maps. *Geol. Soc. Am. Mem.* 47. New York: Geol. Soc. Am. 151 pp.

Annu. Rev. Earth Planet. Sci. 1993. 21:19–41

PETROLOGY OF THE MANTLE TRANSITION ZONE

Carl B. Agee

Department of Earth and Planetary Sciences, Harvard University, 20 Oxford St., Cambridge, Massachusetts 02138

KEY WORDS: high pressure, earth evolution

INTRODUCTION

The mantle transition zone is located between ~ 400 and ~ 670 km in the Earth (Dziewonski & Anderson 1981) and is characterized by rapidly increasing seismic wave velocities with depth. The velocity increases are of two sorts: discontinuous jumps over comparatively narrow distances (up to tens of km) at the bounds of the transition zone at 400 and 670 km and more or less continuous, relatively steep, gradients within the transition zone (Walck 1984, 1985; Grand & Helmberger 1984). The nature of the discontinuities and high velocity gradients have been the focus of an ongoing, lively debate. The debate has often been portrayed as a choice between two simple models of the mantle. One model proposes that the transition zone seismic velocities are a result of first-order phase transitions in an isochemical medium similar in composition to peridotite xenoliths (e.g. Ringwood 1979, Bina & Wood 1987). The alternative view has been that the transition zone is characterized by abrupt radial changes in mantle chemistry (e.g. Anderson & Bass 1986). These choices have implications not only for understanding the present state of the mantle transition zone but for the whole Earth and its formation. For instance, the isochemical model that proposes a peridotite mantle to depths of at least 670 km and possibly to the core-mantle boundary must be reconciled with chemical mass balance in an Earth that started out with chondritic Si/Mg and Ca/Al. Alternatively, a model that proposes abrupt radial changes in mantle chemistry associated with the 400 and 670 discontinuities implies a layered mantle and must be consistent with phase equilibria and phase density constraints.

19

0084–6597/93/0515–0019$02.00

Before embarking on a review of the petrology of the transition zone, a dose of reality is in order. Actually, the petrology of the transition zone is unknown. We have no evidence that pristine samples from below 400 km are ever exposed at the Earth's surface; to date, no geologist has ever held a proven, unadulterated, piece of the transition zone in their hand. To characterize the transition zone and its formation we cannot rely on rock specimens as do volcanologists, lunar petrologists, or meteoriticists. Instead we must seek indirect information from sources such as high pressure phase equilibria, mineral physics, and cosmochemistry. Phase equilibria experiments can establish which mineral assemblages are stable in proposed compositions at mantle *P-T* conditions. Mineral physics provides elastic data on mantle minerals so that the experimentally produced assemblages may be tested against observed seismic velocities. Such tests can distinguish between mineral assemblages that are good matches for the mantle at a particular depth and ones that are poor matches.

Even though mantle models often present peridotite as a "given" for the transition zone composition, this assumption should be carefully scrutinized. It is possible that peridotite xenoliths may only be representative samples of the lower lithosphere; deeper parts of the Earth could be significantly different in composition. In a later section we explore a petrologic model for the transition zone that is based on an initial Earth similar in composition to carbonaceous chondrites. The model accommodates both phase changes and chemical variations while satisfying seismic observations. First however, an isochemical peridotite mantle is examined.

PHASE EQUILIBRIA IN A PERIDOTITE TRANSITION ZONE

Multi-Anvil Experiments

Over the past decade there have been dramatic advances made in the application of large volume, solid media, high-pressure technology to experimental petrology. Previously, experiments designed to simulate equilibrium conditions in the Earth's mantle were usually performed in a piston-cylinder apparatus. The piston-cylinder apparatus is an excellent choice for experimental petrology in the range 0.5 to 4.0 GPa (~ 20 to ~ 120 km depth in the Earth), but beyond 4.0, possibly at 6.0 GPa, the apparatus reaches its practical limit. Creative applications of the original octahedron-within-cubes multi-anvil design of Kawai & Endo (1970) have opened the door to experimental examination of the pressure temperature regime of the transition zone. Akaogi & Akimoto (1979), Irifune et al (1986), Irifune (1987), Ito & Takahashi (1987), Irifune & Ringwood (1987),

Akaogi et al (1987), and Takahashi & Ito (1987) have investigated the mineralogy of a peridotite mantle along model geotherms up to 25 GPa (~ 700 km). From these data an inventory of the phase proportions of a peridotitic mantle have been estimated for a 1400°C adiabat and is presented in Figure 1. We emphasize that Figure 1 is for an isochemical

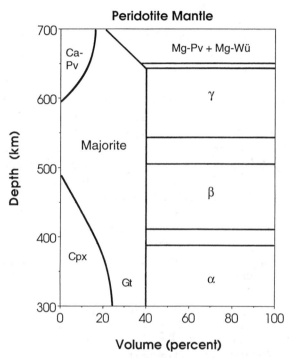

Figure 1 Volumetric mineral content of a model peridotite mantle $(MgO/MgO + FeO = 0.88)$ along a 1400°C adiabat from experimental studies of Akaogi & Akimoto (1979), Irifune et al (1986), Irifune (1987), Ito & Takahashi (1987), Irifune & Ringwood (1987), Akaogi et al (1987), and Takahashi & Ito (1987). $\alpha = (Mg,Fe)_2SiO_4$ olivine, $\beta = (Mg,Fe)_2SiO_4$ β-phase, $\gamma = (Mg,Fe)_2SiO_4$ γ-phase, Gt $= (Mg,Fe,Ca)_3Al_2Si_3O_{12}$ garnet, Cpx $= (Mg,Fe,Ca)_4(Al,Si)_4O_{12}$ clinopyroxene, Majorite $=$ garnet-pyroxene $(Mg,Fe,Ca)_3$ $Al_2Si_3O_{12}-(Mg,Fe,Ca)_4(Al,Si)_4O_{12}$ solid solution, Ca-Pv $=$ CaSiO₃ perovskite, Mg-Pv $=$ $(Mg,Fe,Ca)(Al,Si)O_3$ perovskite, Mg-Wü $= (Mg,Fe)O$ magnesiowüstite (elements shown in italics are minor constituents). Horizontal segments between α and β, β and γ, and γ and Mg-Pv + Mg-Wü denote the approximate kilometer widths of corresponding divariant phase loops where both low pressure and high pressure phases coexist. α, β, and γ are shown as a constant 60 volume percent for simplicity. Variations in Fe-Mg partitioning between $(Mg,Fe)_2SiO_4$ polymorphs and majorite could cause the volume for $(Mg,Fe)_2SiO_4$ to be either slightly less than or greater than 60 volume percent at different depths. Possible minor mantle constituents such as a post-spinel Al_2O_3-rich phase or SiO_2-stishovite have been omitted.

mantle and all changes in phase proportions shown are due to effects of pressure and temperature. Olivine (60%) is the dominant mineral in a shallow peridotitic mantle while orthopyroxene (opx) and clinopyroxene (cpx) together account for 25% of the volume, with pyropic garnet at approximately 15%. With increasing pressure, the first change in phase proportions is caused by the gradual increasing solubility of opx into cpx (not shown). At a depth of 275 km opx is no longer stable and the entire pyroxene volume is represented by cpx. In the depth range 300 to 500 km, cpx undergoes a continuous solid solution reaction with pyropic garnet to form majorite garnet; cpx is no longer stable at ~ 500 km. Because majorite accommodates most essential element constituents of both pyroxenes and garnets, its volume expands in the mid transition zone (500–600 km depth) to nearly $\sim 40\%$. At a depth near 600 km $CaSiO_3$-perovskite becomes stable and the majorite volume share of the transition zone shrinks to $\sim 30\%$ at 650 km. Majorite stability extends to depths below the 670 km transition zone boundary, but a reaction is initiated at 650 km that forms $(Mg,Fe)SiO_3$-perovskite at the expense of majorite. This reaction is spread out over a depth range of ~ 100 km and is completed at ~ 750 km with majorite being entirely consumed in the process.

Much attention has been given to the olivine to β-phase transformation since it closely coincides with the discontinuity in seismic wave velocities near 400 km. Recent work on the Mg_2SiO_4-Fe_2SiO_4 phase diagram appropriate to the 400 km discontinuity comes from multi-anvil experiments by Katsura & Ito (1989). Assuming a temperature range of 1400–1750°C in a peridotite mantle they calculate, from the shape of the olivine-β phase divariant phase loop, a discontinuity thickness of 11–19 km. The minimum possible thickness, assuming no dry melting at 400 km, is constrained by the peridotite solidus to be 8.8 km. The peridotite model mantle contains olivines with MgO/MgO + FeO (molar) = 0.88. If the mantle near 400 km is more depleted in FeO than the model value, then the transformation will be narrower still and at a slightly deeper level. On the other hand if that region contains more FeO than the model, then the transformation will be broader (> 20 km) and at a shallower level. The experimental results of Katsura & Ito also predict the conversion of β-phase to γ-spinel to commence at ~ 510 km and be completed by ~ 540 km.

Ito & Takahashi (1989) have investigated the post-spinel transformations in the system Mg_2SiO_4-Fe_2SiO_4 with the multi-anvil device. They concluded that the shape of the divariant phase loop for the transformation of γ-spinel to $(Mg,Fe)SiO_3$-perovskite + magnesiowüstite is vanishingly narrow for MgO/MgO + FeO values in the range 1.0 to 0.8. The narrowness of their experimental transformation determined at 23.1 GPa and 1600°C would translate into a very thin (< 4 km) discontinuity located at a depth of ~ 650 km. In contrast to the positive slope of the P-T phase boundary associated with the olivine to β-phase and the β-phase to γ-spinel transformations, the γ-spinel to $(Mg,Fe)SiO_3$-perovskite + magnesiowüstite transformation is characterized by a negative dT/dP. Hence, if the mantle temperature is higher than 1600°C in this region, then the transformation will occur at a shallower depth than 650 km. Assuming no dry partial melting, the minimum depth of the transformation is constrained by the peridotite dry solidus to be ~ 625 km. In a mantle colder than 1600°C the transformation occurs at a depth > 650 km.

To summarize, the important minerals in a model peridotite transition zone are olivine/β-phase, cpx, and garnet at shallow levels, β-phase/γ-spinel and majorite at intermediate levels, and γ-spinel, majorite, and $CaSiO_3$-perovskite at the deepest levels. With the experimental phases identified, and their proportions established, a test of the isochemical peridotite mantle hypothesis can be performed using elastic properties of mantle minerals.

MINERAL PHYSICS CONSTRAINTS

Elastic Properties of Mantle Minerals at High Pressure and Temperature

Elastic properties of minerals can place important constraints on the composition of the mantle transition zone. For isotropic bodies we have

$$V_P^2 = \frac{K + \frac{4G}{3}}{\rho}$$

$$V_S^2 = \frac{G}{\rho}$$

where V_P and V_S are the velocities of compressional and shear waves, K is the bulk modulus, G is the shear modulus, and ρ is the density. If random orientation of crystals is assumed, and K, G, and ρ are known, then P- and S-wave velocities for mineral assemblages such as those present in peridotite at high pressure and temperature can be calculated and compared with seismic velocities observed in the transition zone. Unfortunately, knowledge of the elastic properties of minerals at mantle conditions is, at present, far from complete. A catalog of known and unknown properties is summarized in Weidner & Ito (1987) and although recent progress has increased the number of "measured" values and reduced the "estimated" or "guessed" categories, uncertainties, especially in pressure and temperature derivatives of K and G, still make it difficult to categorically rule out either a mantle that is homogeneous or one that contains chemical layering. The best that is possible at present is to distinguish between compositions that are better or poorer matches for observed seismic profiles.

It is also important to understand the meaning of "measured" in the present context of mineral elastic properties. Most published measurements on mantle minerals have not been performed at simultaneous high pressure and temperature. For instance the bulk modulus of a mineral is often determined at high pressure and room temperature by x-ray diffraction in a diamond anvil cell or at 1 atm and room temperature by acoustic velocity methods. The effect of thermal expansion on unit cell volume is then ascertained separately, in a different experiment, most commonly at 1 atm. The shortcomings of this approach have been recognized for sometime and recent successful efforts to apply a synchrotron energy source to simultaneous high P-T measurements (e.g. Mao et al 1991) are an important contribution to the current incomplete data set. Another source of uncertainty comes from applying single crystal data to multiphase aggregates. Care must also be taken to account for compositional effects on elastic moduli. Often, reported values are from measurements on simple synthetic endmember crystals that are not necessarily valid for solid solution minerals stable in natural rocks. For an overview of mineral physics experimental techniques and lists of preferred elastic constants the interested reader is referred to Manghnani & Syono (1987), Anderson (1989), and Duffy & Anderson (1989). For present purposes elastic properties and calculated seismic velocities of mantle minerals are adopted from Duffy & Anderson (1989). The data include yet unmeasured temperature and pressure derivatives of elastic moduli that are consistent with structural and chemical trends present in well characterized phases. Obviously, the Duffy & Anderson data set is not the final word on high P-T elastic properties. However, it forms a convenient, systematic, and internally consistent basis for applying recent mineral physics results to the region of the mantle transition zone; future revisions and improvements are expected. Figures 2a and 2b are graphical inventories of the seismic velocity-with-depth trajectories for minerals present in peridotite phase equilibria experiments. It should be noted that majorite is assigned the same elastic constants as pyropic garnet, which is in accord with the measurements of Yagi et al (1987) and Yeganeh-Haeri et al (1990). The trajectories for $(Mg,Fe)SiO_3$-perovskite may be revised to higher values (especially S waves) based on the larger than expected shear modulus from measurements by Yeganeh-Haeri et al (1989). However for the minerals of the transition zone in particular, all velocity trajectories shown are in good agreement with the most current elastic properties measurements. The minerals with the lowest seismic velocities are olivine and clinopyroxene; $CaSiO_3$-perovskite, β-phase, and γ-spinel have the highest velocities, while garnet and majorite possess intermediate velocities.

Seismic Velocities of a Peridotite Transition Zone

Figures 3a and 3b present a comparison of the seismic velocity profiles for the model peridotite mantle of Figure 1 and the observationally-based mantle models of Walck (1985) and Grand & Helmberger (1984). It can be seen that the broad features of peridotite and the observed mantle are similar; both contain abrupt discontinuities around 400 and 670 km. However there are also significant differences between the two profiles. Firstly, the velocity jump at 400 km is significantly greater for the peridotite model: about 6.6 versus 3.9% for P waves and 8.4 versus 4.6% for S waves. Secondly, the gradient of the model peridotite profile within the transition zone (between 400 and 670 km) is less steep than that observed; this is especially evident with respect to shear waves.

Anderson & Bass (1986) noted that the poor match between peridotite and the observed mantle at 400 km was due to the very large seismic velocity jump required for the olivine to β-spinel transformation. They hypothesized that the only probable composition immediately below 400 km must consist of comparatively olivine-poor rocks, while the shallow mantle above could resemble olivine-rich peridotite. Duffy & Anderson (1989) estimated that the observed P-wave jump could be explained by a mantle at 400 km with 40–53% olivine, while S waves suggest an olivine content of 35%. Alternatively, Weidner & Ito (1987) noted that the then, unmeasured, pressure derivative of the shear modulus (dG/dP) for β-phase might be significantly lower than that adopted from olivine data. This would in turn reduce the large difference in S-wave velocity between olivine and β-phase and make peridotite a better match for the observed mantle. More recently however, Gwanmesia et al (1990) measured the dG/dP of β-phase at room temperature and found it to be indistinguishable from that of olivine, in accord with previous estimates and table 1 in Duffy & Anderson (1989). The temperature derivatives of elastic moduli for β-phase have yet to be measured, but assuming they are also similar to olivine, an isochemical peridotite mantle extending across the 400 km remains a less than ideal fit to seismic Earth models.

The comparatively shallow P- and S-wave gradients calculated for model peridotite between 400 and 670 km are also difficult to reconcile with the gradients from the data of Walck (1984, 1985) and Grand & Helmberger (1984). The increasing amount of high velocity material with depth associated with the conversion of cpx and garnet to the majorite structure had, at one time, been considered a possible candidate for

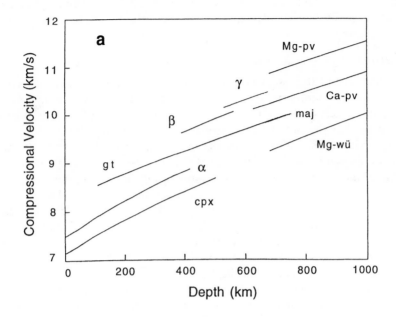

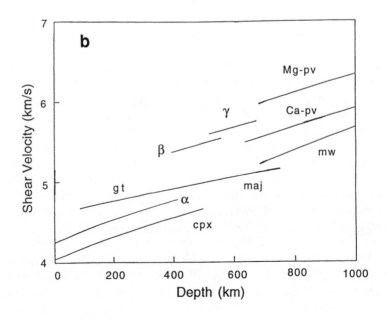

improving the gradient match. Experimental studies by Irifune et al (1986) and Gasparik (1990) showed that this reaction is completed at depths shallower than 500 km and hence could not be called upon to produce the observed gradient throughout the transition zone. Also ruled out for explaining the gradient are the velocity increases associated with the β-phase to γ-spinel transformation which are now known to be confined within the range of ~ 505 to ~ 545 km (Katsura & Ito 1989). Duffy & Anderson (1989) speculated that the observed velocity increases with depth could be caused by a chemical gradient in the transition zone. Rigden et al (1991) suggested that anelastic relaxation in the shallow transition zone might produce the observed steeper S-wave gradient, though they noted that the compositional heterogeneity proposal of Duffy & Anderson (1989) could also explain the steep gradient.

The 670 km discontinuity is characterized by a sharp jump in seismic velocities, occurring over a narrow range—perhaps less than 5 km (Lees et al 1983). The experiments of Ito & Takahashi (1989) indicate that the abrupt discontinuity can be explained by an isochemical phase transformation of γ-spinel to $(Mg,Fe)SiO_3$-perovskite + magnesiowüstite. However because the same transformation can occur in peridotite, chondrite, komatiite, and a number of other hypothetical mantle compositions, a chemical boundary at 670 km cannot be ruled out. The proportions of participating phases would be different for each composition, thus very accurate pressure and temperature derivatives of elastic moduli for $(Mg,Fe)SiO_3$-perovskite and magnesiowüstite, if they were available, might allow a determination. Finally, between 650 and 750 km there is a relatively high velocity gradient (Grand & Helmberger 1984) that is consistent with the "smeared out" reaction observed for the conversion of majorite to $(Mg,Fe)SiO_3$-perovskite. This reaction is also expected to occur in a wide range of possible mantle compositions.

The mineral physics test shows that though an isochemical peridotite mantle cannot be ruled out, it is not a particularly good match for the seismically observed transition zone. It appears, at least between 400 and 600 km depth, that the transition zone is olivine-poor relative to peridotite. A more complete view of transition zone composition can be obtained by considering models of formation. We approach this subject by assuming

---

Figure 2   (a) Compressional velocity versus depth diagram (after Duffy & Anderson 1989). (b) Shear velocity versus depth diagram (after Duffy & Anderson 1989). Trajectories are for mantle minerals produced in peridotite multi-anvil experiments and are initiated along a 1400°C adiabat. Elastic constants for calculated trajectories are taken from table 1 in Duffy & Anderson (1989). Horizontal components of the trajectories are approximately equivalent to the thermodynamic stability range of the mantle phases.

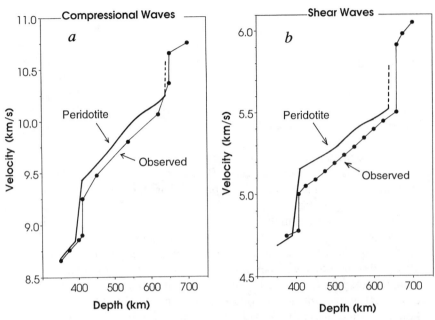

Figure 3 (*a*) Compressional velocity versus depth diagram comparing a model peridotite mantle with the Earth model GCA of Walck (1985). Velocities for peridotite were calculated at various depths using the phase proportions of Figure 1 and the elastic data represented in Figure 2a. Peridotite is shown as a solid bold line; the Earth model GCA is shown as dots connected by thin lines and is labelled "Observed." For peridotite the "400" km discontinuity starts at 390 km and is complete at 410 km (Katsura & Ito 1989). The "670" km discontinuity in peridotite starts at 640 km and is assumed to be completed over a distance of less than 5 km (Ito & Takahashi 1989). The size of the velocity jump is not shown (*dashed line*) because of the present uncertainty in elastic constants for $(Mg,Fe)SiO_3$-perovskite. (*b*) Shear velocity versus depth diagram comparing a model peridotite mantle with the Earth models TNA and SNA of Grand & Helmberger (1984). Velocities for peridotite were calculated at various depths using the phase proportions of Figure 1 and the elastic data represented in Figure 2b. Peridotite is shown as a solid bold line; the Earth models TNA and SNA are shown as dots connected by lines and are labelled "Observed." For peridotite the "400" km discontinuity starts at 390 km and is complete at 410 km (Katsura & Ito 1989). The "670" km discontinuity in peridotite starts at 640 km and is assumed to be completed over a distance of less than 5 km (Ito & Takahashi 1989). The size of the velocity jump is not shown (*dashed line*) because of the present uncertainty in elastic constants for $(Mg,Fe)SiO_3$-perovskite.

that the broad features of the modern transition zone were formed during primordial differentiation of the Earth. Cosmochemical constraints are employed to elucidate Earth formation; the starting point is carbonaceous chondrite.

COSMOCHEMICAL CONSTRAINTS

Carbonaceous Chondrite

There are a number of compelling reasons to choose carbonaceous chondrite as an initial starting composition for a model of the accreting Earth. Carbonaceous chondrites contain approximately solar abundances of the nonvolatile elements (Anders & Grevesse 1989). They are also the most primitive of the solar system objects that have been sampled (Dodd 1981). Ordinary chondrites are more abundant in terms of recorded terrestrial falls, but this may reflect a temporal sampling bias since carbonaceous-like bodies are dominant among the undifferentiated objects in the asteroid belt (Chapman 1976). Accordingly, carbonaceous chondrites may be good analogues for the solar nebula material that accreted to form early planetesimals and ultimately the terrestrial planets. In contrast, peridotite xenoliths from the lithospheric mantle may be derivative.

Carbonaceous chondrites (CC) differ chemically from modern upper mantle peridotites in that they are much more FeO-rich (Jarosewich 1990)—in fact CC contain more than 3.5 times the FeO concentration of average peridotite [for average peridotite values see Maaloe & Aoki (1977), Jagoutz et al (1979), and Irving (1981)]. FeO in this case implies iron in the ferrous state, for the most part, as an essential structural component in silicate minerals. If subjected to crustal or shallow mantle conditions the stable ferro-silicates in a carbonaceous chondrite are primarily olivines and pyroxenes. At higher pressures in the deeper mantle it is expected that ferrous iron resides in garnet, silicate spinel, silicate perovskite, and magnesiowüstite. Carbonaceous chondrites are also the most oxidized of the chondrites and this is reflected in the very low concentration levels of iron-metal present in bulk chemical analyses. In short, an Earth with bulk composition of CC has an apparent excess of ferrous iron. To reconcile the composition of the present peridotite lithospheric mantle with a CC-bulk Earth, a mechanism must have existed that depleted the early chondritic silicates of iron (+nickel) and segregated it to form the core. Relevant to the petrology of the transition zone is whether the mechanism produced a mantle that is chemically homogeneous with depth or whether it produced one with chemical layering or gradients.

FORMATION OF AN ISOCHEMICAL PERIDOTITE MANTLE

One proposed formation mechanism is that the accreting Earth was subjected to highly reducing conditions, ferrous iron was extracted from silicates at subsolidus temperatures as Fe(Ni)-metal, and the metal, in liquid state, segregated to the core by percolation through a solid silicate matrix (Elsasser 1963). This mechanism assumes the Earth's interior was not hot enough for silicate melting to be important, and that subsequent to metal removal, the resulting mantle was essentially a homogeneous undepleted peridotite (undepleted in the sense that it had not yet experienced significant basalt extraction). But, such a mechanism could not have operated at the oxygen fugacities present in today's upper mantle or in an oxidized body similar to carbonaceous chondrite. For example, had the present mantle undergone a single stage Fe-reduction reaction to form the core in this way, then a geochemical signature in the moderately siderophile elements in the samples of upper mantle peridotites should be evident. These hypothetical peridotites would contain olivines with forsterite contents of 98 mol% versus the ~ 90 mol% actually observed, and both Ni^{+2} and Co^{+2} would be barely detectable in the mantle (O'Neill 1991), rather than the $\sim 1000s$ (Ni) and 100s (Co) ppm trace levels often recorded in xenoliths. To correct the very low oxidation state produced in the Elasser-type models, a late-stage bombardment of fresh oxidized meteoritical component is commonly included in the scenario (Wänke et al 1984, O'Neill 1991). This oxidized component is intended to penetrate and mix with the mantle in order to boost the oxygen fugacity up to observed levels. The result is a silicate mantle with lower than observed FeO/MgO, and also one with a SiO_2/MgO that is significantly higher than that measured in peridotite xenoliths. Hence this excess Fe^{+2} remedy for the mantle not only overshoots the target with respect to the moderately siderophile elements, but it also creates an excess SiO_2 problem. Ringwood (1979) proposed that Si would be selectively volatilized during the high temperature stage of accretion and that this accounts for the observed, lower than chondritic, SiO_2/MgO in the upper mantle. Si-volatilization, combined with the correct amount of late-stage oxidized component added to an Fe-reduced mantle, can indeed produce the prescribed FeO-MgO-SiO_2 value of average peridotite. An Earth formation model that produces a homogeneous peridotite mantle between crust and core cannot be ruled out. However, a formation model that produces an isochemical peridotite mantle may not be the best explanation for Earth differentiation. The special set of conditions— Si-volatilization, Fe-reduction, followed by Fe-oxidation—even though consistent with the state of peridotite xenoliths, is an ad hoc combination. The Earth may have formed and evolved by this particular path, but simpler and testable explanations, if they exist, should be preferred.

FORMATION OF A MANTLE WITH CHEMICAL LAYERING

Melting the Allende Meteorite at High Pressure

As an alternative to the isochemical peridotite model, recent high pressure melting experiments on the Allende CV3 carbonaceous chondrite (Agee 1990) suggest that core formation does not require highly reducing conditions in the mantle of the early Earth. Instead, a large portion of the iron(nickel) that reaches core depths during earliest differentiation may be incorporated there not as Fe(Ni)-metal but as Fe(Ni)-sulfide and Fe-oxide. In contrast to the isochemical peridotite model, this mode of core formation requires that the early Earth was extensively molten—in other words, it passed through a high temperature magma ocean stage in which crystal fractionation was important. As a consequence of crystal fractionation, the mantle solidified with chemical layering or zonation. The nature of the proposed layering can be deduced from experimentally determined high-pressure phase equilibria and phase density constraints.

Solidification of the Lower Mantle

Before focusing on the formation of a carbonaceous chondrite-based transition zone it is appropriate to set the stage by considering the solidification of the lower mantle. High pressure Allende experiments show that FeO-rich magnesiowüstite, majorite, and $(Mg,Fe)SiO_3$-perovskite (hereafter "perovskite") are crystallizing phases at temperatures near the silicate liquidus in carbonaceous chondrite at ~ 26 GPa (~ 720 km depth in the Earth). At pressures above 26 GPa only magnesiowüstite and perovskite are expected to be stable in the presence of CC-silicate melt. Miller et al (1991a,b) proposed that equilibrium perovskite would be neutrally buoyant in ultrabasic silicate liquid at ~ 1300 km depth in the mantle. FeO-rich magnesiowüstite on the other hand is predicted to be denser than Allende silicate liquid at all depths within the Earth (Agee 1990). During lower mantle solidification perovskite and magnesiowüstite fractionation may occur simultaneously, but perovskite flotation will transport crystals upward within the magma ocean. Perovskite ascent is along an adiabat that ultimately coincides with the CC liquidus where crystals dissolve. Opposite signs of buoyancy for magnesiowüstite and perovskite may result in crystal sorting that concentrates magnesiowüstite into a basal cumulate and perovskite back into the magma ocean liquid.

As magnesiowüstite is fractionated from the CC silicate magma ocean, the initial FeO content decreases while the $MgSiO_3$ component increases. This change in composition boosts the topology of the magma ocean liquidus and solidus to higher temperatures and rapid crystallization of the lower mantle ensues. The physical details of a rapidly crystallizing lower mantle are obscure. For instance, would a large amount of melt be trapped in the solidifying perovskite-rich matrix or could segregation by melt percolation, either up or down, still occur? Agee & Walker (1988a) have shown that the composition of average upper mantle peridotites can be explained, in part, by approximately 30% perovskite fractionation. Hence some perovskite/melt segregation, even during rapid lower mantle solidification, may be feasible.

Solidification of the Shallow Mantle·

As the lower mantle solidifies at depth, the shallow mantle may receive descending blocks of foundered olivine-rich quench crust, which according to the olivine flotation experiments of Agee & Walker (1988b) will be neutrally buoyant at ~250 km depth and form a solid layer of dunite there. Growth and thickening of the dunite layer by olivine flotation from below promotes isolation of the upper magma ocean from the lower magma ocean. Hence the shallow mantle experiences efficient heat loss to the surface and solidifies rapidly, while the trapped melt below the dunite layer cools slowly. The dunite layer, being gravitationally unstable with respect to the solidifying mantle above is mixed convectively upward. Olivine addition by buoyancy into the shallow mantle completes the major element mass balance of producing average upper mantle peridotites from an FeO-depleted chondritic magma ocean. In the mass balance and physical model of Agee & Walker (1988a), the proto-transition zone (400–670 km) is a dense partial melt of komatiitic composition and is the last major deep-Earth reservoir to solidify. Below we speculate on the formation of the transition zone in a chondritic Earth.

CHONDRITE-BASED TRANSITION ZONE

The Proto-Transition Zone

Phase equilibria studies suggest (Wei et al 1990) that majorite is on the liquidus in komatiite at transition zone pressures (13 to 23 GPa). Hence an Agee & Walker-style proto-transition zone, upon cooling and solidification, should be dominated by majorite crystallization.

MAJORITE DENSITY CROSSOVER   The distribution of chemical components during transition zone solidification is expected to be controlled by phase density relations. Miller et al (1991a,b) were the first to note that majorite would be neutrally buoyant in komatiite at pressures of ~19 GPa. Agee & Walker (1992) concurred by proposing that their experimentally produced, liquidus pyropic garnet, observed at 8.4 GPa, would float in komatiite at

19.5 GPa. The estimate of 19.5 GPa for the density crossover is conservative because garnet chemistry and structure are pressure dependent, changing from a denser pyrope-rich to a less dense majorite-rich form between 10 and 16 GPa (Akaogi & Akimoto 1977, Irifune 1987, Yagi et al 1987). Takahashi (1990) has also noted that the Fe-Mg garnet/liquid distribution coefficient

$$K_D = \frac{(X_{FeO}^{Gt})\,(X_{MgO}^{Liq})}{(X_{FeO}^{Liq})\,(X_{MgO}^{Gt})}$$

is sensitive to pressure and decreases from 0.5–0.8 in lower pressure pyrope/liquid pairs to 0.3–0.4 in higher pressure majorite/liquid pairs. Increasing pressure will therefore produce liquidus garnets with decreasing $Fe_3Al_2Si_3O_{12}$ and $Fe_4Si_4O_{12}$ components and hence will also contribute to producing a garnet that is less dense than the experimental 0.84 pyrope/0.082 almandine/0.078 grossular garnet of Agee & Walker (1992). Hence, the density crossover between komatiite and majorite garnet could be as low as 18 GPa (Figure 4) or ~ 540 km depth in the transition zone.

SOLIDIFICATION SEQUENCE   The chemically evolved, dense, komatiite magma ocean that occupies the proto-transition zone, is sandwiched

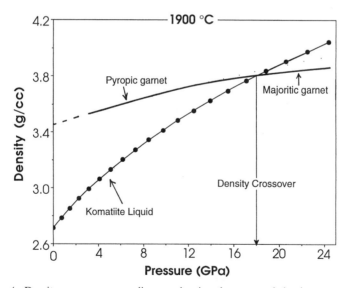

Figure 4 Density versus pressure diagram showing the proposed density crossover for majorite in the partially molten proto-transition zone. One effect of the gradual transition of pyropic garnet to majorite with pressure is to translate the density crossover to lower pressure (~ 18 GPa). Compression curves are calculated using the third-order Birch-Murnaghan equation; elastic parameters are from Agee & Walker (1992).

between an olivine-rich peridotitic solid shallow mantle lid and a perovskite-rich solid lower mantle. As majorite fractionation proceeds during further cooling, a solid layer at the level of neutral buoyancy ($\sim$ 540 km) is formed. Residual melts above and below the majorite-rich layer become increasingly enriched in MgO, FeO, and CaO. When sufficient majorite component has been removed from these melts, they will precipitate liquidus phases rich in MgO, FeO, and CaO. In the region from 400 to 500 km, liquidus precipitates will consist primarily of β-phase and clinopyroxene. At these depths cpx will float, possibly to the roof of the buried magma ocean or be dissolved back into the melt in a fashion analogous to floating perovskite in the lower mantle. β-phase will be near neutral buoyancy in this region and may sink or float depending on changes in melt chemistry. From 600 to 670 km the predominant crystalline products will be γ-spinel and $CaSiO_3$-perovskite. Both γ-spinel and $CaSiO_3$-perovskite are expected to sink in coexisting liquid to form cumulates. If the top of the lower mantle is not completely solidified and they sink below $\sim$650 km, the γ-spinel will convert to $(Mg,Fe)SiO_3$-perovskite + magnesiowüstite. If the lower mantle has completely solidified, sinking γ-spinel and $CaSiO_3$-perovskite will build a cumulate at the base of the proto-transition zone.

The shift in fractionation from majorite-dominated crystallization to phases rich in MgO, FeO, and CaO is gradual and may undergo several reversal iterations as liquids evolve. Density relations act to sort phases; however as the melt fraction shrinks, liquid/solid separation becomes increasingly inefficient. Thus solidification of the proto-transition zone is not expected to form abrupt discontinuities, but rather, smooth major element chemical gradients that are roughly symmetric about the majorite maximum at $\sim$ 540 km. Figure 5 illustrates one possible distribution of phases consistent with this proposed chemical gradient. The majorite volume occupies a larger portion of the transition zone than in the isochemical peridotite model. This element distribution is in agreement with mass balance and also obeys phase equilibria constraints.

Finally, it is noted that the proposed gradient is also expected to be gravitationally stable. The densest phases, $CaSiO_3$-perovskite + γ-spinel, will be concentrated at the base of the transition zone, intermediate density β-phase + majorite, β-phase + γ-spinel + majorite, or γ-spinel + majorite in the mid-transition zone, and the least dense phases cpx + olivine + β-phase or cpx + β-phase will reside at higher levels.

A Match for the Seismic Velocities in the Transition Zone

In Figures 6a and 6b we compare the observed seismic gradients between 400 and 670 km (Walck 1985, Grand & Helmberger 1984) with those of

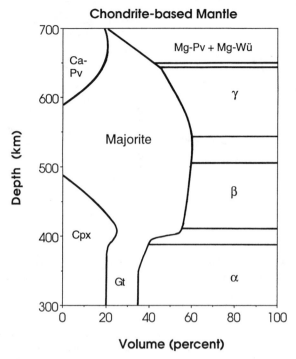

Figure 5 Depth versus volume percent diagram for a transition zone solidified from an initially molten chondritic Earth ("Chondrite-based mantle"). Symbols are the same as in Figure 1. A majorite maximum of 60% at 540 km which corresponds to the level of neutral buoyancy in komatiite (Figure 4) is proposed. This is in contrast to a model peridotite composition that has 40% majorite and 60% γ-spinel at this depth. The phase proportions at 540 km of 60% majorite and 40% γ-spinel are an excellent approximation for the corresponding observed shear and compressional wave velocities (see Figures 6a,b). Phases rich in MgO, FeO, and CaO are concentrated at levels above and below the majorite maximum. A maximum in clinopyroxene content exists at 410 km owing to crystal fractionation and buoyant accumulation at the roof of the komatiite magma ocean. The rapid increase in olivine content above 410 is in accord with the olivine flotation hypothesis of Stolper et al (1981) and the experiments of Agee & Walker (1988b) and Miller et al (1991a,b). Calcium sequestered below 580 km residues in $CaSiO_3$-perovskite. Horizontal segments between α and β, β and γ, and γ and Mg-Pv + Mg-Wü denote the approximate kilometer widths of corresponding divariant phase loops where both low pressure and high pressure phases coexist. The deepest region in the transition zone (600–670 km) is increasingly rich in olivine(γ-spinel) component due to crystal fractionation. Thermally equilibrated, γ-spinel-rich, subducted slabs (80% harzburgite + 20% basalt) residing in this region may be difficult to distinguish from a primordial magma ocean, γ-spinel-rich, cumulate layer. In the chondrite-based model the lower mantle is enriched in $(Mg,Fe)SiO_3$-perovskite owing to 30% fractionation during solidification.

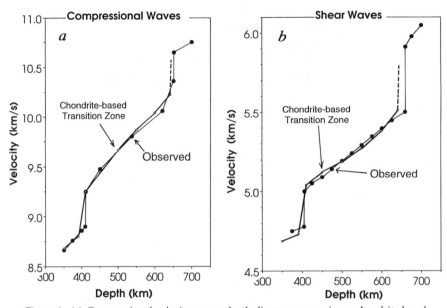

Figure 6 (*a*) Compressional velocity versus depth diagram comparing a chondrite-based mantle with the Earth model GCA of Walck (1985). Velocities for the mantle were calculated at various depths using the phase proportions of Figure 5 and the elastic data represented in Figure 2*a*. The chondrite-based mantle is shown as small open circles connected by a solid bold line; the Earth model GCA is shown as large dots connected by thin lines and is labelled "Observed." For the chondrite-based model the "400" km discontinuity starts at 390 km and is complete at 410 km (Katsura & Ito 1989). The "670" km discontinuity in peridotite starts at 640 km and is assumed to be completed over a distance of less than 5 km (Ito & Takahashi 1989). The size of the velocity jump is not shown (*dashed line*) because of the present uncertainty in elastic constants for $(Mg,Fe)SiO_3$-perovskite. (*b*) Shear velocity versus depth diagram comparing a chondrite-based mantle with the Earth models TNA and SNA of Grand & Helmberger (1984). Velocities for peridotite were calculated at various depths using the phase proportions of Figure 5 and the elastic data represented in Figure 2*b*. The chondrite-based mantle is shown as small open circles connected by a solid bold line; the Earth models TNA and SNA are shown as large dots connected by thin lines and are labelled "Observed." For the chondrite-based model the "400" km discontinuity starts at 390 km and is complete at 410 km (Katsura & Ito 1989). The "670" km discontinuity in peridotite starts at 640 km and is assumed to be completed over a distance of less than 5 km (Ito & Takahashi 1989). The size of the velocity jump is not shown (*dashed line*) because of the present uncertainty in elastic constants for $(Mg,Fe)SiO_3$perovskite.

the mineral assemblages produced by majorite flotation and crystal/liquid fractionation in the chondrite-based proto-transition zone (Figure 5).

The chondrite-based mantle provides a better match for the 400 km discontinuity than isochemical peridotite. The velocity jumps are smaller— 5.1% for *P* waves and 6.6% for *S* waves—and the calculated velocity values just below 400 km are nearly identical to the seismic data. Mantle chemistry must change near 400 km in order to match to the size of the velocity jumps. Above 400 km it must be olivine-rich, garnet-poor, and contain some 20–25% cpx. The abrupt increase in olivine component above 400 km is in agreement with olivine flotation and concentration in the shallow mantle proposed by Agee & Walker (1988a,b). Below the 400 km level the mantle must be olivine(β-phase)-poor and garnet(majorite)- and cpx-rich. This conclusion is in accord with Anderson & Bass (1986) and Duffy & Anderson (1989).

The velocity profiles within the transition zone show that the chondrite-based model produces values that are indistinguishable from seismic observations. The phase proportions in Figure 5 are not always unique solutions for the compressional and shear seismic gradient at every depth. A unique solution for P- and S-wave velocities *is* achievable in regions where only two phases (i.e. majorite and γ-spinel) coexist. The proposed majorite maximum at 540 km is such a location. As an example, the proposed 60 : 40 majorite-γ-spinel ratio for this depth, is the best fit to equations

$$(X_{maj}) \times 9.59 \text{ km/s} + (X_{sp}) \times 10.21 \text{ km/s} = 9.84 \text{ km/s } [P \text{ waves}]$$

$$(X_{maj}) \times 5.02 \text{ km/s} + (X_{sp}) \times 5.62 \text{ km/s} = 5.27 \text{ km/s } [S \text{ waves}],$$

where X_{maj} and X_{sp} are the volume fractions of majorite and γ-spinel and values in km/s are the P- or S-wave velocities of majorite, γ-spinel, and the Grand & Helmberger or Walck models at 540 km. If mantle density could be measured independently of seismic velocity, then a unique solution for the mineralogy of the transition zone, where three major phases coexist, would be available. Then the proposed volumes of β-phase, cpx, and $CaSiO_3$-perovskite at shallower and deeper levels respectively, could be tested. On the other hand, the velocity solutions of Figures 6a and 6b are not selected at random: They are constrained by phase equilibria and density relations in a chondrite-based, solidifying transition zone.

A transition zone containing a chemical gradient is a better match for the observed mantle than isochemical peridotite. At shallow and intermediate levels cpx and majorite maxima respectively account for the good match. At deeper levels in the transition zone majorite concentration wanes and the bulk composition approaches that of model peridotite. High seismic velocities at depths greater than 540 km are well explained by the transition of β-phase to γ-spinel, and a gradual increase in γ-spinel and $CaSiO_3$-perovskite at the expense of majorite garnet.

SUBDUCTED SLABS IN THE TRANSITION ZONE

The formation models presented here have assumed that primordial differentiation was the primary mechanism in producing the structure of the present transition zone. This may be an oversimplification. Modern processes and processes that have operated over most of geologic time may also contribute to transition zone structure. Known features of the transition zone, which have been omitted from the discussion up to now, are of course, subducted slabs. A central question concerning slabs is their fate in the transition zone. Are they deposited near the 670 km discontinuity to contribute to the gross structure of that region or are they merely transient features that ultimately pass into the lower mantle?

Conditions in slabs, both thermal and chemical, are expected to be different than in surrounding mantle, and these factors will determine relative buoyancy. For example the existence of positive and negative dT/dP phase boundaries can translate into shallower and deeper phase transitions in a cold slab either enhancing or inhibiting descent relative to surrounding mantle. The chemical nature of the mantle below 670 km will also affect the relative buoyancy of slabs. Jeanloz (1991) has pointed out that a small, but in terms of density, significant increase in ferrous iron content at 670 km would be difficult if not impossible to detect from seismic velocity data. Such a chemically induced density increase combined with the negative dT/dP associated with γ-spinel to perovskite + magnesiowüstite transformation might be sufficient to prohibit slab penetration into the lower mantle.

Ringwood & Irifune (1988) proposed an idealized slab with a 6.5 km top layer of basalt (cpx-rich) underlain by thicker layers of residual harzburgite (82% olivine, 18% opx) and pyrolite (62% olivine, 38% opx, cpx, and garnet). During descent the "pyrolite" layer is resorbed into the surrounding upper mantle by convective circulation and the slab enters the transition zone as a 30–40 km thick package of 20% basalt and 80% harzburgite. The model proposes that young, thin oceanic plates undergo rapid thermal equilibration and can be gravitationally trapped between 600 and 700 km. Thick, cold, mature slabs may pass through the 670 km discontinuity; however if they first encounter a layer of ancient trapped oceanic lithosphere between 600 and 700 km they will experience buoyant resistance. The cold slab buckles and thickens eventually forming a "megalith" which protrudes into and becomes entrained in a lower mantle, downward moving, convection current.

The Ringwood & Irifune hypothesis is well supported by the most recent tomographic imaging of subduction zones. Van der Hilst et al (1991) have shown that slabs beneath Japan and Izu Bonin arcs are deflected at the 670 km discontinuity while the slabs beneath Kurile and Mariana descend into the lower mantle. Fukao et al (1992) found that the subducting slab beneath Southern Kurile to Bonin arcs becomes subhorizontal near 670 km, and extends in this orientation for more than 1000 km, where it connects to a high velocity blob at $\sim$800 km depth. Global topography near 670 km, reported by Shearer & Masters (1992), is also consistent with the "megalith" of Ringwood & Irifune. They propose that the broad regional depressions observed in the 670 km discontinuity are correlated with the locations of subduction zones, though whether deeper penetration into the lower mantle occurs or not, remains equivocal. Woodward et al (1992) interpret the 670 km discontinuity as a boundary that may, under some circumstances, be breached by upwelling or downwelling convection currents, and under other circumstances, be an impenetrable barrier that sustains layering. Hence, primordial structure as envisaged in mantle formation models may in principle be well preserved in some regions of the Earth's interior, while in other locations it may have been altered by later stage convective mixing. If slabs are gravitationally trapped between 600–700 km, then chemical layering, regardless of primordial differentiation model, must exist in the transition zone.

CONCLUSION

The mantle transition zone is not well matched by an isochemical model peridotite composition. The region between 400 and 600 km must be depleted in olivine-component relative to peridotite in order to be consistent with seismic velocity profiles, though deeper levels (600–670 km) and the shallow mantle ($<$400 km) may resemble peridotite. It is the author's view that the upper and mid transition zone are clinopyroxene- and majorite-rich as a result of primordial differentiation that included a deep magma ocean stage. The lower transition zone (600–670) may possess high seismic velocities because of γ-spinel and $CaSiO_3$-perovskite fractionation and accumulation associated with magma ocean solidification or because of subducted slab deposition over geologic time or because of both. The transition zone boundaries at 400 km and 670 km may be sites of chemical change, but the seismic velocity jumps are primarily due to mineralogic phase transformations that can occur in a wide range of natural and hypothetical mantle compositions.

ACKNOWLEDGMENTS

The experimental work on olivine and majorite flotation was performed in collaboration with Dave Walker at Lamont-Doherty Geological Observatory, Palisades, NY and at the Bayerisches Geoinstitut, Bayreuth, Germany. I thank Tom Duffy for discussions on mineral physics data and Bob Woodward for recent references on mantle tomography.

Literature Cited

Agee, C. B. 1990. A new look at differentiation of the Earth from melting experiments on the Allende meteorite. *Nature* 346: 834–37

Agee, C. B., Walker, D. 1988a. Mass balance and phase density constraints on early differentiation of chondritic mantle. *Earth Planet. Sci. Lett.* 90: 144–56

Agee, C. B., Walker, D. 1988b. Static compression and olivine flotation in ultrabasic silicate liquid. *J. Geophys. Res.* 93(B4): 3437–49

Agee, C. B., Walker, D. 1992. Olivine flotation in mantle melt. *Earth Planet. Sci. Lett.* Submitted

Akaogi, M., Akimoto, S. 1977. Pyroxene-garnet solid solution equilibrium. *Phys. Earth Planet. Inter.* 15: 90–106

Akaogi, M., Akimoto, S. 1979. High-pressure phase equilibria in a garnet lherzolite, with special reference to Mg2+-Fe2+ partitioning among constituent minerals. *Phys. Earth Planet. Inter.* 15: 31–51

Akaogi, M., Navrotsky, A., Yagi, T., Akimoto, S. 1987. Pyroxene-Garnet transformation: thermochemistry and elasticity of garnet solid solutions, and application to a pyrolite mantle. See Manghnani & Syono, 1987, pp. 251–60

Anders, E., Grevesse, N. 1989. Solar system abundances of the elements. *Geochim. Cosmochim. Acta* 53: 197–224

Anderson, D. L. 1989. *Theory of the Earth.* Boston: Blackwell. 366 pp.

Anderson, D. L., Bass, J. D. 1986. Transition region of the Earth's upper mantle. *Nature* 320: 321–28

Bina, C. R., Wood, B. J. 1987. Olivine-spinel transitions: experimental and thermodynamic constraints and implications for the nature of the 400-km discontinuity. *J. Geophys. Res.* 92: 4853–66

Chapman, C. R. 1976. Asteroids as meteorite parent bodies: the astronomical perspective. *Geochim. Cosmochim. Acta* 40: 701–19

Dodd, R. T. 1981. *Meteorites.* Cambridge: Cambridge Univ. Press. 368 pp.

Duffy, T. S., Anderson, D. L. 1989. Seismic velocities in mantle minerals and the mineralogy of the upper mantle. *J. Geophys. Res.* 94: 1895–1912

Dziewonski, A. M., Anderson, D. L. 1981. Preliminary reference Earth model. *Phys. Earth Planet. Inter.* 25: 297–356

Elsasser, W. M. 1963. Early history of the Earth. In *Earth Science and Meteorites*, ed. J. Geiss, E. Goldberg, pp. 1–30. Amsterdam: North-Holland

Fukao, Y., Obayashi, M., Inoue, H., Nenbai, N. 1992. Subducting slabs stagnant in the mantle transition zone. *J. Geophys. Res.* 97(B4): 4809–22

Gasparik, T. 1990. Phase relations in the transition zone. *J. Geophys. Res.* 95(B10): 15,751–69

Grand, S. P., Helmberger, D. V. 1984. Upper mantle shear structure of North America. *Geophys. J. R. Astron. Soc.* 76: 399–438

Gwanmesia, G. D., Rigden, S., Jackson, I., Liebermann, R. C. 1990. Pressure dependence of elastic wave velocity for β-Mg_2SiO_4 and the composition of the Earth's mantle. *Science* 250: 794–97

Irifune, T. 1987. An experimental investigation of pyroxene-garnet transformation in a pyrolite composition and its bearing on the constitution of the mantle. *Phys. Earth Planet. Inter.* 45: 324–36

Irifune, T., Ringwood, A. E. 1987. Phase transformation in primitive MORB and pyrolite compositions to 25 GPa and some geophysical implications. See Manghnani & Syono 1987, pp. 231–42

Irifune, T., Sekine, T., Ringwood, A. E., Hibberson, W. O. 1986. The eclogite-garnetite transformation at high pressure and some geophysical implications. *Earth Planet. Sci. Lett.* 77: 245–56

Irving, A. J. 1981. Ultramafic Xenoliths in terrestrial volcanics and mantle processes. In *Basaltic Volcanism on the Terrestrial Planets*, ed. Members of the Basaltic Volcanism Study Project. New York: Pergamon. pp. 282–307

Ito, E., Takahashi, E. 1987. Ultrahigh-pressure phase transformations and the constitution of the deep mantle. See Manghnani & Syono 1987, pp. 221–29

Ito, E., Takahashi, E. 1989. Postspinel transformations in the system Mg_2SiO_4-Fe_2SiO_4 and some geophysical implications. *J. Geophys. Res.* 94(B8): 10,637–46

Jagoutz, E. H., Palme, H., Baddenhausen, K., Blum, M., Cendales, G., et al. 1979. The abundances of major, minor, and trace elements in the Earth's mantle as derived from primitive ultramafic nodules. *Proc. Lunar Planet. Sci. Conf. X* 2031–50

Jarosewich, E. 1990. Chemical analyses of meteorites: a compilation of stony and iron meteorite analyses. *Meteoritics* 25: 323–37

Jeanloz, R. 1991. Effects of phase transitions and possible compositional changes on the seismological structure near 650 km depth. *Geophys. Res. Lett.* 18: 1743–46

Katsura, T., Ito, E. 1989. The system Mg_2SiO_4-Fe_2SiO_4 at high pressures and temperatures: precise determination of stabilities of olivine, modified spinel, and spinel. *J. Geophys. Res.* 94: 15,663–70

Kawai, N., Endo, S. 1970. The generation of ultrahigh hydrostatic pressure by a split sphere apparatus. *Rev. Sci. Instrum.* 41: 425–28

Lees, A. C., Bukowinski, M. S. T., Jeanloz, R. 1983. Reflection properties of phase transition and compositional change models of the 670 km discontinuity. *J. Geophys. Res.* 88: 8145–59

Maaloe, S., Aoki, K. 1977. The major element composition of the upper mantle estimated from the composition of lherzolites. *Contrib. Mineral. Petrol.* 63: 161–73

Manghnani, M. H., Syono, Y., eds. 1987. *High-Pressure Research in Mineral Physics.* Tokyo: Terra Sci. Washington, DC: Am. Geophys. Union. 486 pp.

Mao, H. K., Hemley, R. J., Fei, Y., Shu, J. F., Chen, L. C., et al. 1991. Effect of pressure, temperature, and composition on lattice parameters and density of (Fe,Mg)SiO$_3$-perovskite to 30 GPa. *J. Geophys. Res.* 96(B5): 8069–79

Miller, G. H., Stolper, E. M., Ahrens, T. J. 1991a. The equation of state of molten komatiite. 1. Shock wave compression to 36 GPa. *J. Geophys. Res.* 96: 11,831–48

Miller, G. H., Stolper, E. M., Ahrens, T. J. 1991b. The equation of state of molten komatiite. 2. Application to komatiite petrogenesis and the Hadean mantle. *J. Geophys. Res.* 96: 11,849–64

O'Neill, H. St. C. 1991. The origin of the moon and the early history of the Earth—a chemical model. *Geochim. Cosmochim. Acta* 55: 1135–58

Rigden, S. M., Gwanmesia, G. D., FitzGerald, J. D., Jackson, I., Liebermann, R. C. 1991. Spinel elasticity and seismic structure of the transition zone of the mantle. *Nature* 354: 143–45

Ringwood, A. E. 1979. *Origin of the Earth and Moon.* New York: Springer-Verlag. 295 pp.

Ringwood, A. E., Irifune, T. 1988. Nature of the 650-km seismic discontinuity: implications for mantle dynamics. *Nature* 331: 131–36

Shearer, P. M., Masters, G. T. 1992. Global mapping of topography on the 660-km discontinuity. *Nature* 355: 791–96

Stolper, E. M., Walker, D., Hager, B. H.,

Hays, J. F. 1981. Melt segregation from partially molten source region: the importance of melt density and source region size. *J. Geophys. Res.* 86: 6161–6271

Takahashi, E. 1990. Speculations on the Archean mantle: missing link between komatiite and garnet peridotite. *J. Geophys. Res.* 95: 15,941–54

Takahashi, E., Ito, E. 1987. Mineralogy of mantle peridotite along a model geotherm up to 700 km depth. See Manghnani & Syono 1987, pp. 427–37

van der Hilst, R., Engdahl, R., Spakman, W., Nolet, G. 1991. Tomographic imaging of subducted lithosphere below northwest Pacific island arcs. *Nature* 353: 37–43

Walck, M. C. 1984. The P-wave upper mantle structure beneath an active spreading center: the Gulf of California. *Geophys. J. R. Astron. Soc.* 76: 697–723

Walck, M. C. 1985. The upper mantle beneath the north-east Pacific rim: a comparison with the Gulf of California. *Geophys. J. R. Astron. Soc.* 81: 243–76

Wänke, H., Dreibus, G., Jagoutz, E. 1984. Mantle chemistry and accretion history of the Earth. In *Archean Geochemistry*, ed. A. Kröner, G. N. Hanson, A. W. Goodwin, pp. 1–24. Berlin: Springer-Verlag

Wei, K., Trønnes, R. G., Scarfe, C. M. 1990. Phase relations of aluminum-undepleted and aluminum-depleted komatiites at pressures of 4–12 GPa. *J. Geophys. Res.* 95: 15,817–27

Weidner, D. J., Ito, E. 1987. Mineral physics constraints on a uniform mantle composition. See Manghnani & Syono 1987, pp. 439–46

Woodward, R. L., Dziewonski, A. M., Peltier, W. R. 1992. Comparison of seismic heterogeneity models and convective flow simulations. *Nature*. Submitted

Yagi, T., Akaogi, M., Shimomura, O., Tamai, H., Akimoto, S. 1987. High pressure high temperature equations of state of majorite. See Manghnani & Syono 1987, pp. 141–47

Yeganeh-Haeri, Weidner, D. J., Ito, E. 1989. Elasticity of MgSiO$_3$ in the perovskite structure. *Science* 243: 787–89

Yeganeh-Haeri, Weidner, D. J., Ito, E. 1990. Elastic properties of the pyrope-majorite solid solution series. *Geophys. Res. Lett.* 17: 2453–56

Annu. Rev. Earth Planet. Sci. 1993. 21:43–87

PLANETARY LIGHTNING

C. T. Russell

Institute of Geophysics and Planetary Physics, University of California, Los Angeles, California 90024–1567

KEY WORDS: atmospheric electricity, planetary atmospheres

INTRODUCTION

Lightning from a planet-wide perspective is a relatively rare phenomenon on Earth. Averaged over the entire globe a square kilometer is hit by lightning only once every 50 days. Since the cross section of a lightning bolt is only about 100 cm^2, only 1 air molecule out of every 100 million will experience the heating of that lightning flash and only for a fraction of a second. Nevertheless, lightning discharges are important for the terrestrial atmosphere. Because the conditions inside the lightning stroke are so extreme, chemical reactions can occur that are otherwise impossible to achieve in the terrestrial atmosphere. Such reactions may have been important in the Earth's early atmosphere for the evolution of life and are important now for the production of nitrous oxide which affects the chemistry of the upper atmosphere and is becoming a major concern due to the worsening ozone depletion.

Lightning is also not so rare that it is unfamiliar to the inhabitants of this planet. Since the sound and flash from a lightning stroke can be heard and seen for tens of kilometers, the average inhabitant of this planet will sense one or more lightning strokes per day. Since lightning occurs during particular weather conditions and is not uniformly distributed geographically, the above statement is true only in an average sense. Lightning may not be detected for a long period and then there may be many flashes in a short period of time. Some inhabitants see many lightning strokes a day. Others see none. So too our planetary spacecraft may only infrequently detect lightning with their optical detectors when these detectors view only a portion of the planet for a limited view period, nor can they peer deeply into the atmosphere.

Since human beings are usually quite familiar with terrestrial lightning,

43

0084–6597/93/0515–0043$02.00

and usually in awe of lightning because of its energy density, they are often quite interested in the existence of planetary lightning regardless of the relative importance of lightning to the physics and chemistry of the planet. As a result the topic of planetary lightning has received frequent attention in the press and has been a popular subject of planetary artists. Lightning bolts in Venusian volcanic plumes and intracloud discharges in the Jovian atmosphere add dramatic touches to artists' renderings of planetary settings.

We do not know for certain from our terrestrial experience that lightning occurs on the other planets, but lighthning does appear to be a robust phenomenon on Earth. While it most commonly occurs in water/ice clouds, it can also be found in volcanic plumes, snowstorms, and dust storms. Thus we should not be surprised if we find lightning in the sulfuric acid clouds of Venus or the ammonia clouds of Jupiter, nor should we take the observation of lightning on a planet to imply the presence of water/ice clouds.

Scientifically we are interested in planetary processes such as lightning both to understand the workings of the atmosphere of the particular planet being studied as well as to learn more about the generic processes through comparison with the terrestrial processes. This latter objective is called comparative planetology. Knowledge usually flows both ways. In the early stage of a planetary investigation, the understanding of the terrestrial process is used to interpret the planetary process. As the study matures the differences in the planetary process become understood more clearly and that new understanding can be used to refine our knowledge of the original terrestrial process. For example, if we find planetary lightning in a situation where the clouds are not thought to be convective we will first question whether our evidence for lightning is correct. Having verified that it is, we then question whether our knowledge of the convective behavior of the cloud is correct. If we were to then find that the cloud was not convective, we would put that understanding back into our original paradigm and adjust it so that we no longer believe lightning had a one-to-one correlation with convection.

When a phenomenon is being studied, the first phase of that investigation is to be certain that the phenomenon exists. Once existence is proven, the significance is tested by, for example, measuring the rate of occurrence. If the process is significant we study it further first attempting to obtain a qualitative understanding of how it works, then a quantitative understanding, and finally a predictive understanding. At present we have just completed the existence test for planetary lightning for Venus and the Jovian planets, and have some constraints on the significance of the lightning discharge on these planets, but we do not yet have the definitive answer because of the limitations of our observations. We certainly do not yet have much qualitative understanding of the processes involved. Likewise, our understanding of terrestrial lightning is no more than semi-quantitative.

THE LIGHTNING DISCHARGE

Lightning on Earth occurs when charge accumulated in a portion of a cloud creates an electric field so strong that the neutral gas becomes ionized and an electrically conducting channel forms which allows the accumulated charge to be neutralized. On Earth this ionized path leads to the ground about one third of the time and to another cloud or another part of the cloud about two thirds of the time. The current surge through the ionized channel heats the air until it glows over much of the channel (see, for example, Braginskii 1958). The heated channel sends a shock wave into the surrounding air and it is this shock wave that generates the acoustic waves known as thunder. The electric current flowing in the ionized channel generates electromagnetic waves which propagate around the world and deep into space. On Earth it is these radio waves that are most often used to detect lightning rather than the optical flash.

Over the last few years many reviews of the lightning discharge process have been written (Rinnert 1982, Vonnegut 1982, Levin et al 1983, Williams et al 1983, Uman 1987). Most of the lightning on Earth is produced in strongly convective cumulus clouds but as noted above lightning can also be produced in volcanic clouds, dust storms, and snowstorms. While lightning associated with terrestrial volcanoes is rare in comparison with weather-related lightning, volcanic eruptions are probably the most active of all lightning generators (Vonnegut 1982). These clouds vary in altitude from $\frac{1}{2}$ km to over 30 km. While there was some early speculation that Venusian lightning was associated with volcanism (Scarf et al 1980b, Scarf & Russell 1983), such an association is now deemed very unlikely (Russell et al 1988c, 1989a). On the Jovian planets, which are principally gas spheres, a volcanic source is obviously ruled out, but a volcanic source could occur on the Jovian moon Io. Likewise dust and snowstorm sources of lightning are unlikely on all the planets except for Mars—a planet for which we have surprisingly little data on electromagnetic phenomena despite the many Martian missions.

Terrestrial lightning occurs in two quite distinct forms: cloud-to-ground discharge and intracloud discharges. Such discharges occur when the electric field in a cloud exceeds the breakdown value which ranges from about 10^6 Vm^{-1} in wet air to about 3×10^6 Vm^{-1} in dry air. Usually cloud-to-ground discharges begin in the cloud with a stepped leader carrying about

200–300 A and moving about 50–100 m per step. The stepped leader travels with an average velocity of about 1.5×10^5 ms^{-1}. This is followed by a return stroke in the same channel at a velocity of about 6×10^7 ms^{-1} carrying 10–20 kA which decays in about 20 to 50 μs. There are usually 3 or more of these strokes in any one flash separated by about 40 ms and lasting about 0.2 s.

Discharges within a cloud are 2 to 4 times more frequent than cloud-to-ground discharges. There is no return stroke, per se, in an intracloud discharge but there is a recoil streamer which propagates about 2×10^6 ms^{-1} over channel lengths of 1 to 3 km with peak currents of 1 to 4 kA. Figure 1 shows the magnetic field pulses during eleven intracloud discharges (Rinnert et al 1989). The pulses are narrower than those of the cloud-to-ground return strokes, and they frequently occur in sequences or trains of pulses. The data shown here were obtained with the same instrument that is being flown to Jupiter on the *Galileo* probe (Lanzerotti et al 1992).

A moderate cloud-to-ground flash will generate about 4.5×10^8 J and a large one about 2×10^{10} J of which only about 0.1% of the energy goes into optical radiation. Most of the energy released in a lightning flash is expended in ionizing the air and heating it. Although intracloud discharges generate less energy per stroke than cloud-to-ground discharges, such discharges are more numerous.

Terrestrial lightning has both geographic and local time correlations (Bliokh et al 1980). There are three major centers of activity as a function of longitude: the Americas, Africa, and Indonesia. Figure 2 shows the Universal Time distribution of lightning in each of these regions and the corresponding local time distribution. Each region has a similar local time profile which peaks around 1600 LT when the surface temperature due to solar heating and convection is at a maximum. However, in some localized areas the maximum frequency of thunderstorms and precipitation occurs late at night (Vonnegut 1982).

In order to create lightning there must be an abundance of a substance that can be readily electrified, a process to electrify these particles, and a large-scale charge-separation process. The electrification is proportional to the polarizability of a molecule which in turn is proportional to its dielectric constant. The dielectric constant for water is 80, and for H_2SO_4—which is present in the clouds of Venus—110. The largest charges in terrestrial clouds occur at altitudes where the water has become supercooled to temperatures from -10 to $-40°C$ and where ice crystals are present. Most of the charge resides in a layer hundreds of meters thick centered around the altitude at which the temperature is approximately $-15°C$. H_2SO_4, the main constituent of Venus clouds, freezes at tem-

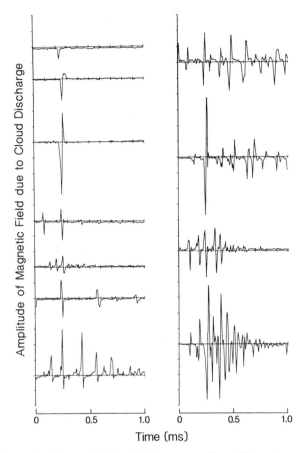

Figure 1 Examples of terrestrial cloud discharge waveforms that were recorded by the *Galileo* lightning detector during ground testing. Each amplitude division corresponds to 20 nT for frequencies below 50 kHz (Rinnert et al 1989).

peratures similar to that of water. Water clouds are expected to be present in the atmospheres of the Jovian planets. However, the upper layer of clouds that we see at these planets is not water but rather ammonia.

The terrestrial clouds in which lightning discharges arise vary in height from 4 to 20 km. The distance between the effective centers of charge in a cloud is similar to the distance of the lower charge center to the ground. In contrast, the clouds on Venus occur at about 55 km which is much greater than the cloud layer thickness of about 10 km. Thus, it is much more difficult for a discharge to occur from the Venus cloud layers to ground. However, if a Venus discharge were due to vertical charge sep-

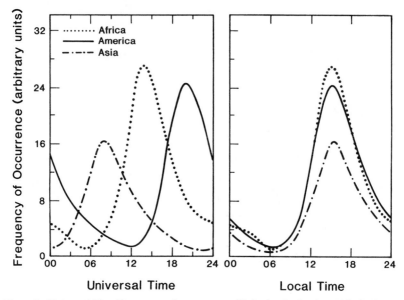

Figure 2 Universal Time/frequency of occurrence of lightning in the three principal zones of activity and their corresponding local time distributions (after Bliokh et al 1980).

aration within a cloud one might expect it to have a discharge length comparable to that on Earth. On the Jovian planets there is no surface for cloud-to-ground discharges, and the water clouds are thought to form at atmospheric pressures greater than those on Earth.

A recent terrestrial "discovery" made possible by the development of very sensitive low-light-level television systems is the upward going flash from the tops of clouds (Franz et al 1990, Boeck et al 1991). Both from the ground and space, flashes of great length have been observed which extend well into the stratosphere. This may be much more important at both Venus and the Jovian planets. At Venus the clouds are much closer to the ionosphere than on Earth. On the Jovian planets, as mentioned above, there is no planetary surface for discharge.

The charge separation process is not well understood. One possible mechanism is that particles become charged differently according to size and then separate according to size as large particles fall more rapidly in updrafts under the action of gravity. One mechanism that has been proposed for the production of the charge gathered by these falling particles is ion production by cosmic rays in the atmosphere (Wilson 1929). While this process may be too weak on the Earth, it may be much more productive on Venus which has no shielding magnetic field (Russell et al 1980, Phillips

& Russell 1987) and which is much closer to the Sun. It is also difficult to compare the expected magnitude of updrafts on Earth and Venus. While strong updrafts were observed on the *Vega* balloons (Sagdeev et al 1986), no balloon or probe data were available at the local times where we would expect from our terrestrial experience that lightning would be present. At Jupiter, and to a lesser extent Saturn, the radiation belts could provide such an ionizing source. Another possible electrification process would be charge exchange due to the collision of different sized ice particles. This explains several of the observed properties of the charge distribution in terrestrial clouds and could also work at Venus if Venusian clouds contain particles with properties similar to ice. The total mass loading of the Venusian atmosphere is much lower than in terrestrial cumulonimbus clouds. While the average particle number density in the Venus clouds is comparable to that in the terrestrial clouds, the particle size is smaller (Levin et al 1983), and the existence of particles large enough to fall in strong electric fields (~ 400 μm) has been questioned (Knollenberg & Hunten 1980).

While optical data on lightning exist for both Venus and Jupiter most of our constraints on the rate of lightning occurrence and the proof of the existence of planetary lightning rests on observations of the electromagnetic energy released by lightning. Thus we should examine how that energy propagates in the Earth's environment. On Earth the lightning discharge creates electromagnetic fluctuations over a wide range of frequencies from ELF frequencies which resonate in the Earth-ionosphere waveguide as the Schumann resonances to gigahertz waves which propagate unaffected by the ionosphere. At VLF frequencies, where the wave power is the most intense, the waves are principally guided by the Earth-ionospheric waveguide. Waves which enter the ionosphere are refracted vertically by the increasing density of the lower ionosphere. Thus on Earth the angle of propagation of the waves to the magnetic field is a strong function of magnetic latitude. This affects both the access of waves to the magnetosphere and their damping as they propagate through the ionosphere. Waves can enter the ionosphere up to the local electron gyrofrequency (about 1 MHz at low altitudes) but only much lower frequencies can propagate far into the magnetosphere because the electron gyrofrequency drops rapidly with altitude. Thus most studies in the ionosphere of the waves generated by lightning are performed at VLF frequencies, near the peak energy of the lightning-generated electromagnetic spectrum.

The higher frequencies can penetrate the ionosphere above the electron plasma frequency, which is of the order of 10 MHz. Thus it is possible to detect lightning with sensitive radio wave receivers from 10–100 MHz and at higher frequencies, but the power rapidly decreases with increasing frequency. These waves are not trapped in the Earth-ionosphere waveguide and can be attenuated as they propagate through the D-region of the ionosphere where electrons accelerated by the wave can collide with atoms of the neutral atmosphere and remove energy from the wave.

Between the electron gyrofrequency and the electron plasma frequency electromagnetic waves do not freely propagate. However, on Venus strong VLF signals apparently generated in the atmosphere are observed at these frequencies in the ionosphere. Thus, if signals at higher frequencies are observed to penetrate into Venus's ionosphere, one must find a separate mechanism to explain that observation. That such a mechanism exists has been demonstrated by the rocket experiments of Kelly et al (1985, 1990), but it is as yet poorly understood.

We now examine the observations of planetary lightning by space probes. We begin with the Jovian planets—Jupiter, Saturn, Uranus, and Neptune—for which the evidence was obtained entirely by the *Voyager* mission.

JUPITER

The possibility of lightning on Jupiter was anticipated by Bar-Nun (1975) who had suggested that lightning could account for the observed abundance of acetylene in the Jovian atmosphere (Ridgway 1974) but, unlike Jovian magnetospheric radio emissions, neither the optical emissions nor the radio emissions from Jovian lightning could be detected from the Earth. Despite the inability to detect Jovian lightning from 1 AU, the massive convecting clouds of Jupiter seemed ripe for the generation of lightning. *Voyager* scientists were hoping to detect evidence for lightning when the spacecraft arrived in 1979 and they were not to be disappointed. Images from both *Voyagers* 1 and 2 revealed lightning storms on the dark side of the planet (Cook et al 1979; Smith et al 1979a,b; Magalhaes & Borucki 1991), and the *Voyager* 1 plasma wave instrument detected whistlers (Scarf et al 1979). While the existence of lightning on Jupiter seems not to have been controversial, the occurrence rate and the resulting significance of this phenomenon were.

Both the optical and the VLF data were affected by the differing trajectories of *Voyagers* 1 and 2. *Voyager* 1 passed inside the orbit of Io, crossing the Io torus and approaching within 350,000 km of the center of Jupiter. This produced a closer view, a shorter observing period, and a smaller observing area than for *Voyager* 2, which passed between the orbits of Eurpoa and Ganymede, approaching within 725,000 km of Jupiter. Since *Voyager* 1 flew closer to Jupiter it could observe weaker lightning flashes than *Voyager* 2 but *Voyager* 2 was able to survey a larger portion of the nightside. In fact, *Voyager* 1 seems to have been able to detect individual (superbolt) flashes, whereas the *Voyager* 2 images seem to record only the integrated activity of storm centers (W. J. Borucki 1991, personal communication). Similarly at VLF frequencies the closer passage of *Voyager* 1 allowed the detection of whistlers generated by lightning (Gurnett et al 1979, Scarf et al 1979, Kurth et al 1985) while the greater distance of *Voyager* 2 prevented this detection.

Optical Observations

Figure 3 shows *Voyager* photo 16396.39, a multiple exposure of the night-side of Jupiter (Borucki et al 1982). This photo shows both aurora, as an arc from upper left to the lower right, and lightning, as bright patches. After each of the exposures the camera was moved slightly to the lower right. If a lightning storm was continuously active this would produce three separate spots. As indicated by the three arrows this did occur, but only in one region. The differing intensities of the three spots correspond to the differing exposure times of 35, 40, and 85 s. The other active areas seem not to be triplicated.

The intensity of these flashes is comparable to superbolts on Earth. Borucki et al (1982) have calculated the optical and total energy of the brightest 16 *Voyager* flashes. These range in optical brightness from 4.3×10^8 to 6.6×10^9 J. To calculate the total energy of the flash one must know what fraction is converted into light. Borucki et al (1982) use an efficiency of 0.0038 deduced by Krider et al (1968) for the Earth. A later study by Borucki & McKay (1987) for a realistic Jovian atmosphere lowered this efficiency by a factor of 3.8 so the estimates of the total energy of these flashes must be raised by this amount. The total energy dissipated per flash during the *Voyager* 1 flyby ranged from 1.1×10^{12} to 1.7×10^{13} J. This calculation does not account for atmospheric scattering and attenuation. This attenuation depends on the source altitude of the lightning and could be as high as a factor of 10.

Figure 4 shows a schematic of the Jovian cloud layers contrasted with terrestrial clouds (Lanzerotti et al 1989). Clouds at the 1 bar level on Jupiter consist of ammonia, not water. Below the ammonia clouds and above the 2 bar level are thought to be clouds of ammonium disulfide, and below these clouds at levels down to 5 bars are thought to be water clouds. At one time it was not certain whether the apparent dryness of the upper atmosphere was due to an atmosphere wide depletion of water (Bjoraker 1985, West et al 1986) or whether the lower atmosphere was water rich (Weidenschilling & Lewis 1973). However, recent studies indicate that the water content of the lower atmosphere is about twice solar abundance and hence is wet (Carlson et al 1992). The optical data of the *Voyager* 1 flyby

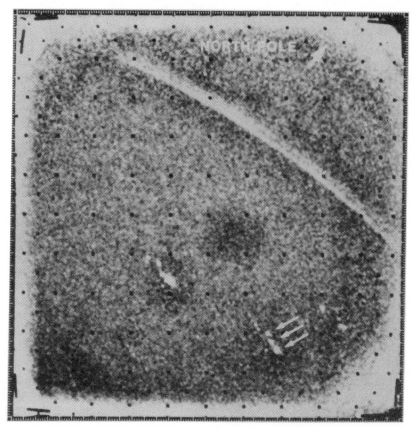

Figure 3 *Voyager* 1 photo 16396.39, showing lightning on the nightside of Jupiter. The north pole is at the top right. The arrows point to repeated images of the same storm center. This multiple exposure consists of three individual exposures of 35, 40, and 85 seconds. Between exposures the camera slew nearly parallel to the limb and the auroral arc in the upper right (Borucki et al 1982).

permit a test of these two atmospheric models because the spot size of the lightning activity is determined by the depth of the source below the upper cloud layer. Borucki & Williams (1986) find that the source altitude is most consistent with a cloud layer in the 2–5 bar level, thus supporting the Weidenschilling & Lewis (1973) model and the Carlson et al (1992) analysis.

The *Voyager* 2 imager data were not as ideally suited for the lightning study as the *Voyager* 1 data but recently have been carefully analyzed by Magalhaes & Borucki (1991) and Borucki & Magalhaes (1992). They find

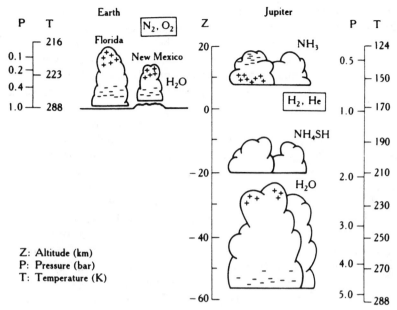

Figure 4 Altitude profiles of terrestrial and expected Jovian clouds. Altitudes in kilometers are measured from the 1 bar level. Charge separation of water clouds chosen to illustrate typical conditions found in terrestrial clouds. Charge separation in the Jovian ammonia cloud is arbitrarily reversed from water clouds. We do not even know if we should expect charging in ammonia clouds (Lanzerotti et al 1989).

that, because of its greater distance, *Voyager* 2 could detect only storms larger than those seen by *Voyager* 1. Correspondingly, *Voyager* 2 detected fewer storms per unit time and area, but the dissipation in these storms was similar to that of *Voyager* 1, about 700 W/km^2 averaged over the two missions. The corresponding number for Earth is about 10% of this value (Borucki & Chameides 1984). Thus Jovian lightning is much more intense than terrestrial lightning. Since the available convective energy on Jupiter is much less than on Earth (about 14 W/m^2) the ratio of energy dissipated by lightning to the convective energy is much greater than on Earth, by a factor of about 1000. This factor agrees with an estimate of Ingersoll (1990) who used Jovian images to infer that Jupiter was 150 times as efficient as the Earth in converting thermal flux to mechanical energy.

These estimates run directly counter to the calculations of the energy dissipation rates of Lewis (1980). However, Lewis did not include the nonoptical energy in his calculation which is by far the bulk of the energy. The rate of Jovian flashes was also controversial. Evidence based on the occurrence of whistlers suggests a much higher rate (Scarf et al 1981) than from otical data. The resolution of this dichotomy is simply that the optical data are recording the largest events while the VLF data are recording the much more frequent smaller flashes. Furthermore, Scarf et al (1981) also assumed ducted propagation rather than unducted which may have also increased their estimate above what we now believe is the true rate of discharge. Thus these optical data appear to be only the tip of the iceberg. The VLF data are discussed below.

A most interesting result of the reanalysis of the *Voyager* 2 lightning images by Borucki & Magalhaes (1992) is the limited latitude range observed for the lightning flashes. Cook et al (1979) had shown that the events were at high northerly latitudes and that many of the events were concentrated at particular altitudes, but the addition of the *Voyager* 2 data showed how truly remarkable the distribution was. Figure 5 shows a map of the lightning events from the two missions (Magalhaes & Borucki 1991). Most of the lightning is concentrated in a single band at 49°N latitude with a small number of events at slightly greater latitudes. There are two closely spaced storms at 13.5°N in a region described by Smith et al (1979a) as disturbed, and there are no events in the southern hemisphere. The lack of southern hemisphere events is explicable by the lack of high latitude southern observations but the absence of events in the large midlatitude range surveyed also emphasizes the rarity of the two storms at 13.5°N. Borucki & Magalhaes (1991) attribute the lightning observations at high latitudes, rather than at equatorial latitudes as on Earth, to the strength of the internal heat source at Jupiter together with the ability of the polar regions to better radiate that energy to space.

VLF Observations

The *Voyagers* 1 and 2 spacecraft included a plasma wave instrument that recorded Very Low Frequency (VLF) electric field measurements on both flyby trajectories (Scarf et al 1979, Gurnett et al 1979) but only the *Voyager* 1 spacecraft came close enough to the planet to detect whistlers generated by lightning. The waveform of the signals can be captured by recording 4-bit digital words at a rate of $28,800 \text{ s}^{-1}$ for a period of up to 48 s. During the *Voyager* 1 flyby a total of 141 48-s waveform segments were obtained over a 9-hour period surrounding closest approach. A total of 167 whistlers were identified of which 90 were sufficiently intense to be analyzed (Kurth et al 1985).

Figure 6 shows frequency-time spectrograms of whistlers at three different times near closest approach. (Kurth et al 1985). The descending tone is produced by the slowing down of the electromagnetic waves produced over a broad band of frequencies by the lightning discharge. The lowest frequencies are delayed the most by the plasma. The more slowly falling

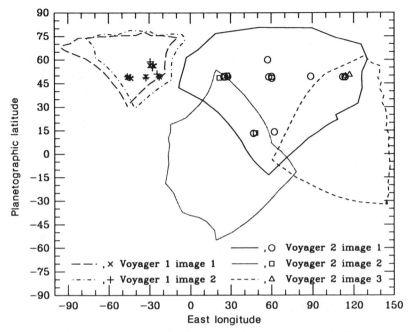

Figure 5 Latitude and longitude of Jovian lightning flashes observed on the *Voyager* 1 and 2 missions (Magalhaes & Borucki 1991).

tones have encountered the greatest density—in this case the plasma in the Io torus (Menietti & Gurnett 1980). During the *Voyager* 1 encounter whistlers are observed only over a limited range of frequencies from about 2 to 7 kHz. The lower cutoff may be due to damping by the magnetospheric electrons and the upper cutoff due to propagation effects.

The whistler waves were observable only at distinct times during the flyby but these times were controlled principally by the receiver gain of the instrument. Thus, the rates of occurrence at the times when other sources of noise were less intense and not the average rate of occurrence is the appropriate number to quote for the whistler generation rate. This number is close to 1 s^{-1} (Kurth et al 1985), which implies a much greater planetary flash rate than the optical results would imply. Thus, these results must be due to weaker events than the optical flashes. Since there is no equivalent of the Earth-ionosphere waveguide on Jupiter (Rinnert et al 1979) we do not expect the lightning-generated whistlers to travel far from their source before entering the ionosphere. Thus, the energy of a whistler will be strongest right above the discharge and fall off with the square of the distance from the source. Since the maximum ionospheric density occurs about 2000 km above the cloud tops (Eshleman et al 1979) the effective area illuminated by a discharge is about 10^7 km^2. The 1 s^{-1} rate at *Voyager* therefore implies a flash rate of about 3 km^{-2}yr^{-1}. Scarf's rate of ~ 40 km^{-2}yr^{-1} seems to be somewhat too high, again possibly because he assumed ducted propagation which lowers the size of the possible source region.

There is no reason to believe that the Jovian whistlers were ducted by field-aligned density enhancements nor restricted to a source region at the foot of the Io flux tube as assumed by Tokar et al (1982). Jovian whistlers probably traveled in the unducted mode, as "magnetospherically reflected" whistlers. If so then multiple paths are possible, yet none were seen. Also

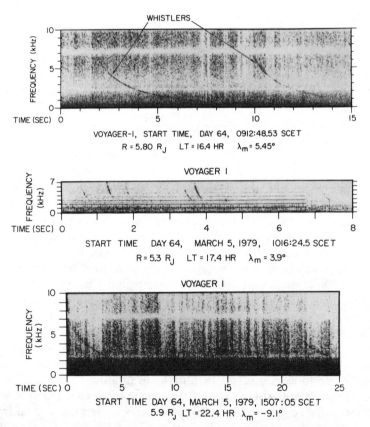

Figure 6 Frequency-time diagrams of Jovian whistlers at three times during the Io torus passage. The dispersions of these whistlers are about 300, 50, and 500 Hz$^{1/2}$ respectively in each of the three panels (Kurth et al 1985).

multiple-hop whistlers were not seen. The observed waves at most made one crossing of the torus (Kurth et al 1985), but contrary to the optical results, whistler sources in both hemispheres were seen. These observations suggest that damping is important for these waves and that only the least damped waves are observed. Menietti & Gurnett (1980) have looked at the damping rate of magnetospherically reflected whistlers in the Jovian magnetosphere and concluded that the electron temperature in the torus must be less than a few times 10^5 K.

It is clear from our comparison of the whistler and optical data that the optical data refer only to superbolts and not to the majority of lightning discharges on Jupiter. The rate of occurrence of weaker flashes on Jupiter (those comparable to normal terrestrial flashes) is clearly large, probably similar to terrestrial values. The *Galileo* mission may shed more light on this activity but like *Voyager* will spend only a brief time at radial distances inside of Io.

SATURN

When *Voyagers* 1 and 2 reached Saturn on August 22, 1980 and August 26, 1981 respectively, they did not detect the optical signature of lightning (Smith et al 1981, 1982)—despite attempts to do so. These attempts were thwarted by sunlight reflected by the rings and other factors, such as greater attenuation by the thicker ammonia cloud deck on Saturn than on Jupiter (Burns et al 1983) and by hazes. As at Jupiter we expect significant attenuation of lightning flashes from low levels in the cloud deck. Moreover, because of Saturn's colder temperatures and lower gravity, atmospheric layers are thicker at Saturn than at Jupiter, and haze layers are more extensive than at Jupiter. This can be seen in images of Saturn which show much less distinct features than images of Jupiter.

Whistlers were also not detected at Saturn. While some impulsive signals were seen, they did not have the expected dispersion of whistler mode signals generated by discharges in the atmosphere (Gurnett et al 1981). To understand why whistlers may not have reached *Voyager* let us consider the Jovian whistlers shown in Figure 6. The frequency band of these whistlers is restricted to a narrow band of 2–7 kHz. The lower cutoff is probably controlled by magnetospheric thermal electrons through damping of these nonducted waves whose wave normals are at a large angle to the magnetic field. The upper cutoff at about 10% of the local electron gyrofrequency is probably controlled by propagation effects. If we have the same allowed bandwidth for whistlers at Saturn as at Jupiter, then the *Voyager* spacecraft did not come close enough to Saturn to detect these waves. *Voyager* 1 came only as close as 3.1 R_S where the expected upper cutoff at 10% of the local electron gyrofrequency was slightly less than 2 kHz and *Voyager* 2 to 2.7 R$_S$ where the expected upper cutoff was somewhat greater than 2 kHz. If any whistler energy were present, the bandwidth of these signals may have been simply too narrow to allow the distinctive dispersive behavior of whistlers to be recognized. Furthermore, fewer wideband measurements were made at Saturn than at Jupiter so there were fewer detection opportunities (Burns et al 1983).

Although lightning was detected at Jupiter only through its optical and VLF signatures, it is also possible to detect lightning at frequencies above the maximum electron plasma frequency in the ionosphere. These waves can freely propagate and be little affected (in general) by the ionospheric plasma. Such waves were observed at Saturn and dubbed Saturn Electrostatic Discharges or SEDs. Figure 7 shows an example of these SEDs (Desch 1992). These waves are similar in many ways to lightning-generated terrestrial emissions (Warwick et al 1981, Evans et al 1981). They are broadband, impulsive, and unpolarized. The waves appear as brief bursts over the entire frequency range of the *Voyager* radio wave receiver because the broadband SED pulses are far shorter than the sweep time of the receiver. The source region appears to be quite small, ≤ 50 km. However, because the source appeared to rotate significantly faster than the body of the planet, the *Voyager* radio wave team preferred to interpret the source of the bursts as some unidentified mechanism operating in the rings.

Burns et al (1983) were the first to point out that the objections to an atmospheric source were easily answered. The reason that the waves recurred faster than the period of the planet as a whole was that the "storm" generating the lightning was traveling at the velocity of the equatorial atmosphere (which superrotates). The region of the atmosphere with winds fast enough to explain the observed recurrence period is about 20° wide centered on the equator (Burns et al 1983). The reason for the extended frequency range of the emissions, to frequencies below the usual ionospheric plasma frequency of about 1 MHz, is due to leakage from the night ionosphere and the fact that the shadow of Saturn's ring on the equatorial ionosphere creates a hole in ionospheric density allowing lower frequencies to escape. At the time of the *Voyager* 1 encounter the ring shadow covered from 1 to 5°S latitude; during the *Voyager* 2 encounter it covered latitudes 2–10°S.

Kaiser & Desch (1984) have carefully analyzed the geometry of the *Voyager* 1 encounter with Saturn and concur with the conclusions of Burns et al (1983). They find that one extended storm in the equatorial region produced detectable bursts at a rate of up to 0.2 s^{-1}, and that more than one storm must be present to explain the observations. Desch (1992) estimates that a typical SED discharge dissipates 10^{12}–10^{13} J—larger than

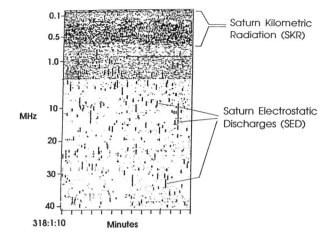

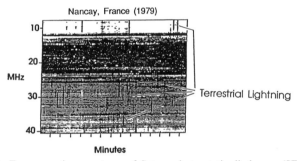

Figure 7   Frequency-time spectrum of Saturn electrostatic discharges (SEDs) in the top panel compared with simlar measurements of terrestrial lightning in the bottom panel using the *Voyager* 1 planetary radio astronomy instrument. The strong similarity of the two sets of data argues that the Saturn electrostatic discharges are also due to lightning (Desch et al 1991). (Note: In contrast to earlier frequency-time spectra, here lowest frequencies are at the top of the diagram.)

even terrestrial superbolts. The most extensive study of SEDs was led by P. Zarka who studied 23,000 SED discharges (Zarka & Pedersen 1983). Figure 8 shows the spectrum deduced from the data. The pre-encounter data (*heavy line*) clearly shows the absorption by the dayside ionosphere below about 5 Mhz. The post encounter data (*light line*) whose ray paths from the atmosphere passed through the night ionosphere show no such absorption. Zarka (1984) has used the variation in this absorption feature throughout the encounter period to deduce the peak ionospheric density.

The observed SED rate of 0.2 s^{-1} observed by *Voyager* combined with

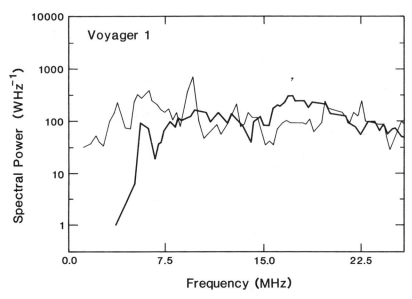

Figure 8   Pre-encounter (*dark line*) and post-encounter (*light line*) spectra constructed from 23,000 SED discharges. Pre-encounter data show the effects of absorption by the ionosphere below about 5 MHz (Zarka & Pedersen 1983).

an estimate of the source area of 2.6×10^8 km^2 leads to an overall estimate of 0.02 km^{-2}yr^{-1} (Desch 1992). However, this rate is the rate for super-bolts, not the weakest of signals. Moreover, these waves all appear to be generated in a narrow region of latitude and longitude. Thus it is difficult to extend this rate to the planet as a whole. However, Desch (1992) has done this extrapolation and estimates that perhaps up to several hundred flashes per square kilometer per year are generated at Saturn.

One remaining puzzle about the SED events is why they were not detected at Jupiter. Kaiser & Desch (1984) point out that the *Voyager* 1 encounter was twice as far from the Jovian cloud top as it was at Saturn. Moreover, 15 db attenuators were used by both spacecraft during their closest approach to Jupiter. Furthermore, the plasma frequency cutoff is much higher at Jupiter than at Saturn, thus preventing much of the signal greater than 1 MHz from leaving the Jovian atmosphere. Zarka (1984) has suggested yet another possible reason: that the D-layer of the Jovian ionosphere where collisions take place is simply too attenuating to allow the Jovian waves to escape. No matter which explanation is correct for why analogues of the SED events were not seen at Jupiter (and in fact all could contribute), the case for lightning at Saturn was further strengthened by observations at Uranus.

TITAN

Before moving on to the next planet on *Voyager's* tour, it is worth discussing Titan which has the most massive atmosphere of any planetary moon. Titan's atmosphere is approximately 96.8% nitrogen, 3% methane, and 0.2% hydrogen, with trace amounts of hydrogen cyanide, ethane, propane, acetylene and other hydrocarbons (Lindal et al 1983, Samuelson et al 1981). It is thought that the atmosphere may have once been principally NH_3 possibly converted to N_2 and H_2 through meteorite-impact generated shocks (McKay et al 1988), and it has been proposed that lightning played a role in producing the hydrocarbons (Borucki et al 1987, 1988). Borucki et al (1984) studied the expected electrical conductivity and charging of aerosols in Titan's atmosphere and later simulated Titan lightning in the laboratory (Borucki et al 1988). Borucki et al's calculations indicated a rate of lightning energy dissipation much less than on Earth or Jupiter but one that could be significant over geologic time. The *Voyager* 1 passage at an altitude of 4400 km above the Titan cloud tops on Nov 12, 1980, however, detected no radio emissions that could be identified with Titan lightning (Desch & Kaiser 1990), implying that the present rate of lightning on Titan is much less than that on Earth.

URANUS

Uranus has the most featureless atmosphere at optical wavelengths of any of the gas giants as a result of its extensive haze. Thus, it is not surprising that no optical signature of lightning was detected (Smith et al 1986). Furthermore, *Voyager* 2 passed no closer than 4.1 R_U so that the maximum electron gyrofrequency was only about 9 kHz. If whistlers could propagate only up to 10% of the local electron gyrofrequency as was seen at Jupiter, and they had the same (2 kHz) lower cutoff, then they would not have been expected to be seen by *Voyager* 2 at Uranus. Impulsive broadband emissions from 0.9 to 40 MHz were observed as at Saturn and dubbed Uranian Electrostatic Discharges or UEDs (Zarka & Pedersen 1986). Figure 9 shows the frequency and time of occurrence of the 140 UED events. The analysis of Zarka & Pedersen of this admittedly sparse sample shows that the waves are consistent with their generation by lightning that is much more powerful than terrestrial bursts (about 10^7 J per flash) but comparable in intensity to Jovian flashes and much less intense than Saturnian SED events. Since Uranus does not have an extensive massive ring system like that at Saturn, support for the ring generation hypothesis faded away.

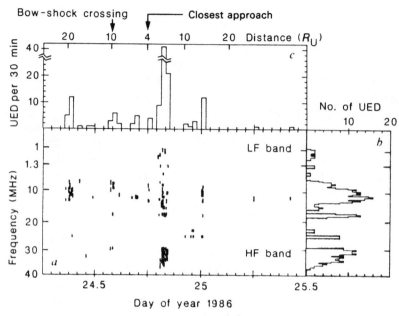

Figure 9 Frequency and time of occurrence of the 140 Uranian electrostatic discharges observed by *Voyager* 2 (Zarka & Pedersen 1986).

NEPTUNE

By the time *Voyager* 2 arrived at Neptune in August 1989, the lightning observations of the previous decade at Jupiter were sufficiently well understood that Borucki (1989) attempted to extrapolate them to Neptune. He based his estimate on the postulate of Lewis (1976) that the energy dissipation rate in lightning would be a small fraction of the dissipation rate of the convective energy in the atmosphere. On Earth this ratio is 4×10^{-7} (Borucki & Chameides 1984) while on Jupiter it is 4×10^{-5} (Borucki et al 1982) so clearly is not a constant of planetary atmospheres. Assuming that this ratio was Jupiter-like on Neptune and that the distribution of flash energy was Jupiter-like, Borucki (1989) predicted that the imager would see approximately 60 events for the planned look directions and observing time.

As of this writing no optical observations of lightning have been reported at Neptune. There could be several reasons for the lack of optical signatures, none of which is that lightning does not occur at Neptune. The plasma wave instrument clearly detected whistlers near closest approach (Gurnett et al 1990). It is likely that the lightning generating region is simply so deep in the atmosphere that the optical radiation from the flash is too attenuated by scattering and absorption to be detected by the *Voyager* cameras. Only four possible lightning-generated radio emissions were seen by the *Voyager* radio astronomy receiver. This rules out activity of the level of Saturn or Uranus but allows the presence of lightning with energies slightly stronger than terrestrial lightning but with a much lower occurrence rate (Kaiser et al 1991).

The whistler identification is unambiguous at Neptune. There were 16 whistlers captured by the wideband system. Figure 10 shows a frequency-time diagram for each of the whistlers for which the onset time could be estimated. A common characteristic of all these events is that they have large dispersion. The low frequencies are greatly delayed implying that the waves have traveled through a lot of plasma. This could mean that they traveled a great distance or that the plasma density is high. Initially Gurnett et al (1991) interpreted these events in terms of multiple-hop ducted whistlers originating on the dayside. However, Menietti et al (1991) have shown that direct propagation along a nonducted path provides a simple explanation for the observed signals. The plasma density implied by these observations is of the order of $100 \, \text{cm}^{-3}$ at the location of the *Voyager* spacecraft.

Part of the reason for *Voyager 2's* success in finding whistlers at Neptune in the face of the absence of such phenomena at Saturn and Uranus is not that Neptune is in some way different than these other two bodies but that the *Voyager* trajectory was quite different. As shown in Figure 11 *Voyager* approached within 1.2 R_N of the planet and all whistlers were observed

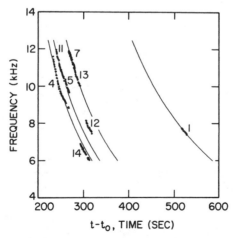

Figure 10   Frequency versus time delay for each whistler detected by *Voyager* 2 at Neptune for which it was possible to estimate an initiation time.

within 2 R_N and at latitudes less than 33°. Such data were not available at Saturn and Uranus. As with the Jovian whistlers the lower frequency cutoff is probably due to damping by the cold magnetospheric and ionospheric electrons. The high frequency cutoff, here up to 20% of the local electron gyrofrequency, is probably due to a propagation effect so that only this range of frequencies can reach the spacecraft.

OUTER PLANET SUMMARY

The combined optical, radio wave, and plasma wave data from *Voyager* has clearly established the presence of lightning discharges at each of the four Jovian planets, but at no planet were data available from all three techniques. At Jupiter optical flashes were detected as were whistlers, but radio waves were not detected due possibly to the lack of sensitivity of the receiver at its gain setting during the Jovian encounter together with some D-region absorption. At Saturn and Uranus only radio wave data detected the lightning discharges and at Neptune only the plasma wave instrument

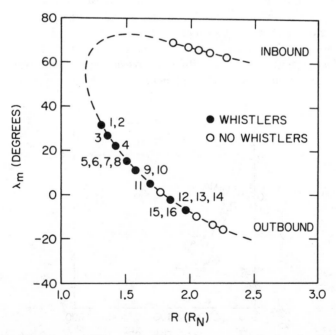

Figure 11 The trajectory of *Voyager* 2 at Neptune in magnetic latitude and radial distance coordinates using an offset tilted dipole magnetic field model. No whistlers were observed during the wideband frames denoted by the open circles.

clearly detected signals originating in lightning. Optical signals appear to have been absorbed by the clouds and haze in the upper atmospheres of Saturn, Uranus, and Neptune.

VENUS

Lightning on Venus has been the subject of speculation for many years since Meinel & Hoxie (1962) attributed the faint glows, on the nightside of Venus, known as Ashen Light, to lightning (see also Krasnopol'sky 1980). Recent studies have dismissed the Ashen Light as an artifact and claim that lightning on Venus would be too weak to cause the reported effect (Borucki et al 1981, Williams et al 1982), and a recent Ashen Light campaign produced no evidence that the Ashen Light was a phenomenon intrinsic to Venus (Phillips & Russell 1991). Nevertheless, the existence of the necessary (but not sufficient) condition of a well developed cloud system as sketched in Figure 12 combined with the early speculation led to the installation on the *Veneras* 11, 12, 13, and 14 landers of instrumentation capable of detecting lightning (Ksanformaliti 1979). Lightning detection was also one of the objectives of the *Pioneer Venus* plasma wave investigation (Colin & Hunten 1977) and a photometer capable of detecting optical flashes was included in the *Vega* balloon mission (Sagdeev et al 1986). Lightning also became the subject of investigation with the *Venera* 9 orbiting spectrometer (Krasnopol'sky 1980), the *Pioneer Venus* star scanner (Borucki et al 1991), the *Galileo* imager, the *Galileo* plasma wave spectrometer (Gurnett et al 1991), and possibly in future with the *Magellan* radiometer. Some of these studies have successfully detected lightning and others have not, in the process engendering some controversy.

The importance of lightning on Venus depends, as it does on Earth and on the Jovian planets, on the rate of lightning occurrence and its energy per flash. On Venus lightning could produce an amount of CO and O_2 that could play a key role in maintaining the clouds by fueling the conversion of S and SO_2 to H_2SO_4 and H_2O and SO_2 to S if its occurrence rate were much greater than that on Earth (Chameides et al 1979). It is also important to determine the electrical activity of clouds because we would like to obtain measurements in the Venusian clouds and electrical activity would present a hazard to these measuring devices.

Because of the large number of lightning investigations undertaken at Venus, there is a much larger literature on Venus lightning than on lightning at the four Jovian planets combined. Therefore we have left the discussion of Venus lightning until now. The evidence for lightning on Venus is not any less compelling than on the other planets, although

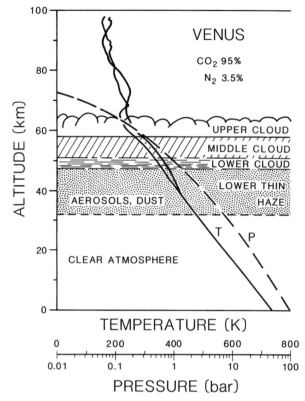

Figure 12   The height profile of the Venus cloud system vs atmospheric temperature and pressure (after Rinnert 1982).

the topic has at times been even more controversial. This controversy is surprising since evidence for Venus lightning can be found in all three sets of optical, VLF, and radio emission data. Furthermore, Venus is the only planet, other than the Earth, for which evidence for lightning was obtained within the planetary atmosphere. In the sections below we discuss first these in situ, atmospheric data from the *Venera* lander, and then we discuss the various optical observations. Next we treat the *Pioneer Venus* VLF data and finally the *Galileo* radio observations. For those wishing a more detailed review of Venus lightning (prior to the *Galileo* observations), see the review by Russell (1991a).

Venera *Landers*

The *Venera* 11 and 12 landers carried both a high sensitivity loop antenna and an acoustic sensor. While the acoustic sensor was saturated during descent, the VLF measurements from the loop antenna provided the first unambiguous evidence of lightning (Ksanfomaliti 1979). The instrument returned amplitudes in four narrow-band channels centered at 10, 18, 36, and 89 kHz and in a wideband channel. The amplitude profiles obtained on the four missions looked very different from each other even though the descent trajectories were very similar. Thus, the observed waves were not due to the interaction of the probe with the Venus atmosphere. Rather the signal variation was consistent with a temporally varying source like lightning. Figure 13 shows high resolution VLF data obtained during weak activity at high altitudes by the *Venera* 12 lander. Here the pulses occur infrequently enough that they can be counted by eye. In this interval there is a burst about every 12 s on the average. Later during the descent the rate became too great to count. *Veneras* 13 and 14 carried electromagnetic sensors similar to those on *Veneras* 11 and 12 but with no acoustic sensors. Instead they monitored the coronal discharge current from the spacecraft. No discharge currents were detected (Ksanfomaliti 1983). The activity on *Veneras* 13 and 14 was similar to that on *Venera* 12 and significantly less than on *Venera* 11.

The landing area for the probes was close to the subsolar point and far away (∼ 8000 km) from the equivalent region in which terrestrial lightning is most frequent. On the surface the instruments continued to operate but only *Venera* 12 saw a burst of VLF noise for the brief period shown in Figure 14. No acoustic signals other than those associated with spacecraft operations were observed after landing. This suggests that the lightning was distant.

The *Venera* VLF instruments included an impulse counter. The mean rate was 16.5 s^{-1} overall. Below 20 km altitude the rate dropped to 13 s^{-1} and below 5 km it dropped to 10 s^{-1}. These rates are much greater than terrestrial rates. The rates obtained during a test run on a clear day in the terrestrial atmosphere were similar to the lowest rates obtained on the *Venera* 12 entry from 25 to 12 km altitude and while the instrument was on the surface of the planet. On the *Venera* 11 descent the rates were an order of magnitude higher than the terrestrial rate until the vehicle landed,

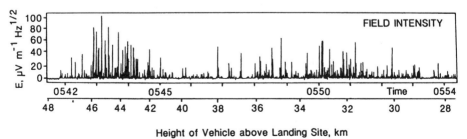

Figure 13 High resolution measurements of the wideband field intensity on *Venera* 12 (Ksanfomaliti et al 1983).

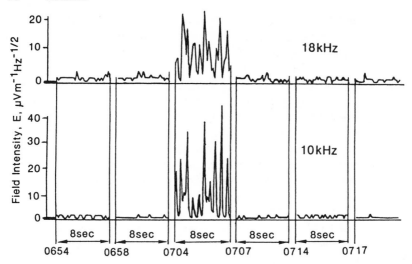

Figure 14   A burst of VLF noise while *Venera* 12 was sitting on the surface (Ksanfomaliti et al 1983).

Estimates of the source location are difficult. The detection of spin modulation during descent (Ksanfomaliti el al 1983) indicates a small angular source as opposed to an extended source and perhaps therefore a distant one. Ksanfomaliti et al (1983) estimated that *Veneras* 11 and 12 were able to monitor lightning discharges from 7.5% of the planet. By extrapolating the number of discrete burst sites, they suggested that there were about 50 sites over the entire planet compared to the 2000 that one would detect on Earth. However, their extrapolation depends on the assumption that storms are uniformly distributed, and thus ignores the possibility that they could be more frequent on Venus beyond the radio horizon of the *Venera* landers. It is very difficult to estimate the strength of an individual stroke from these measurements or a rate of occurrence because we do not know the location of the flashes responsible for the observed VLF pulses. Thus the *Venera* data provide almost no constraints on the strength and occurrence of Venus lightning. The differences in the rates of occurrence and the temporal pattern of occurrence during each descent through the atmosphere are reminiscent of the variability expected of lightning associated with terrestrial weather patterns.

Optical Studies

On October 26, 1975, the *Venera* 9 visible spectrometer detected a period of apparent optical flashes on the nightside of Venus at 1930 LT at 9°S

(Krasnopol'sky 1983a). The flashes were seen over a period of 70 seconds corresponding to 7 sweeps of the instrument while the field of view moved 450 km. No such flashes were seen when the instrument pointed away from the planet. Figure 15 shows the observations obtained during one such sweep (Krasnopol'sky 1983a). We have corrected this plot from the original by applying the factor of 8 gain to the points indicated as being at reduced gain in the original figure. In doing so we assumed that the baseline represented zero signal and that the scales were linear (V. Krasnopol'sky 1991, personal communication). The dashed and dotted line shows the instrument response to a 1 $W/Å/km^2$ signal. The dashed line shows the radiation background. The narrow pulses at either end show calibration marks at 3000 and 8000Å. The observations shown by the solid lines clearly show impulsive signals of about $\frac{1}{4}$ s duration, the amplitudes of which mimic the instrument response indicating that they arise outside the instrument not inside such as would occur due to bit errors. The durations of each peak, lasting several samples, also indicate that the signals are real and not telemetry noise. If we now correct the peak amplitudes for the instrument response we obtain the peak responses indicated by the asterisks. These peak amplitudes seem to be much more uniform with wavelength. The slight weakening of amplitudes seen at shorter wavelengths is consistent with the greater atmospheric absorption expected in the UV. Both Krasnopol'sky (1983b) and Borucki (1982) point out that the weakness of the absorption with wavelength suggests that the source must be at high altitudes, and not near the surface.

The expected spectrum should consist of a series of emission lines. While temporal variability should cause the relative strengths of these emission features to differ from those anticipated, we would expect the position of the lines to be invariant. Borucki et al (1985) have simulated Venus lightning spectra using an atmosphere of 96% CO_2 and 4% N_2. These spectra show only weak spectral features between 500 and 650 nm where there are several strong features in Figure 15. A possible resolution of this discrepancy is that the "lines" are due to constituents absent in the simulated Venus atmosphere, such as helium and sulfur. However, this question does not seem to have been addressed by either Krasnopol'sky or Borucki.

The optical energy in each flash was about 2.5×10^7 J corresponding to a total energy of about 7×10^9 J (Borucki & McKay 1987). Although the Venusian strokes dissipate more energy than terrestrial strokes, their longer pulse duration causes the peak optical power to be much less (Levin et al 1983). Since only one storm was found over the 10^7 km^2 surveyed by *Venera* 9 the average rate was 45 flashes $km^{-2}yr^{-1}$—a rate 6 times that of Earth but one that is very uncertain. If the flash energy is 20 times that

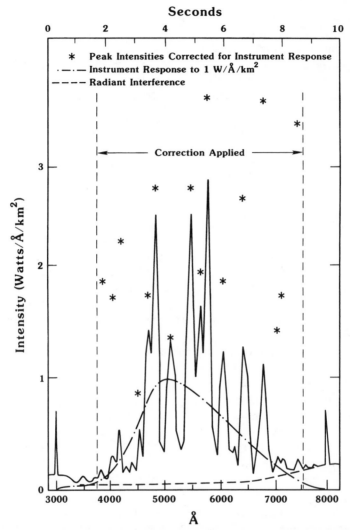

Seconds

Figure 15 One sweep of 7 of the *Venera* 9 spectrometer during which time flashes were observed. No such flashes were observed when the instrument viewed space. The short pulses at the start and end of the sweep are calibration pulses. The lower dashed line is the radiative background noise observed by the instrument. The dash-dot line is the response of the instrument to a 1 Watt/Å/km^2 energy source. The instrument was operating in gain states 2 and 3 (of 4) at this time and we have accounted for the switching between gains in drawing this plot. The asterisks show the peak amplitudes after correcting for the spectral response function (*dot-dash curve*) of the instrument. (After Kransnopol'sky 1983; V. A. Krasnopol'sky 1991, personal communication.)

on Earth, as noted above, than the energy dissipated by lightning on Venus could be much greater than that on Earth.

The *Pioneer Venus* star sensor has also been used in an attempt to detect lightning. The star sensor cannot be directly used to search for lightning on Venus because the visible night airglow saturates the sensor. Thus, scattered light when the sensor points off the planet has to be used. Two studies of the star sensor data have been completed (Borucki et al 1981, 1991). The latter study supercedes the first due to a recalibration of the star sensor axial response function. No optical pulses clearly due to lightning were seen in this survey. However, the sensitivity of the survey depends on the duration of the pulse. The data during only one of the two seasons studied was sensitive to long pulses as seen by the *Venera* 9 spectrometer. Also very few data were obtained over the local time sector identified as the "lightning source" region by the VLF data which will be discussed below. The *Venera* 9 optical data on the other hand were obtained over the "active" sector.

Two other optical results should be mentioned. The first is the report of a very bright, nitrogen band on two successive short wavelength scans of the instrument in which only a few adjacent lines of the band were excited. (T. Slanger 1991, personal communication). While these two "flashes" need further study, their duration appears similar to the flashes shown in Figure 15. Finally we note that the *Vega* balloon photometers saw no flashes as they moved from midnight to dawn in the Venus clouds. However, we note that it is harder to see distant lightning while in a cloud than above it and that the *Vega* balloons did not fly through the region thought to be the "active" local time sector according to the VLF data.

Orbital VLF Studies

The in situ VLF measurements with the *Venera* landers provided evidence for Venus lightning but provided little information on the source location of the lightning. The one storm observed by *Venera* 9 established some of the optical properties of lightning in the Venus atmosphere but by no means provided a statistical sample. The *Pioneer Venus Orbiter* (*PVO*) included a VLF receiver with four narrow-band channels. Its long lifetime permitted statistical sampling of both the source location (with some important constraints to be discussed below) and the rate of occurrence of flashes. As discussed in the introduction, VLF signals on Earth enter the ionosphere in two quite different ways. VLF waves in the Earth-ionosphere waveguide, with frequencies below the local electron gyro-frequency, are refracted so that their wave normals are oriented upwards when they enter the ionosphere. As long as their wave normals are within a certain angle of the magnetic field direction (inside the resonance cone) the waves can propagate with little damping in the whistler mode. In the Earth's magnetosphere these waves can propagate many 10,000's of km roughly guided by the magnetic field. A second mode of entry also seems to be allowed in the Earth's lower ionosphere. VLF waves which should not propagate are found immediately above thunderstorms (Kelley et al 1985). On Venus, the magnetic field magnitude is such that the three higher frequency channels are almost always above the local electron gyrofrequency. Thus waves at these frequencies do not propagate and if detected should be diagnostic of lightning source regions. At 100 Hz, waves can propagate if the magnetic field is sufficiently inclined to the horizontal direction. However, when the ionospheric magnetic field is nearly horizontal the vertically propagating waves lie outside the resonance cone and, as with the higher frequency waves, these 100 Hz waves should not propagate. In short, in studying the 100 Hz signals care must be taken to note both the strength and orientation of the local magnetic field. This was not always done in initial studies.

INITIAL RESULTS   The *Pioneer Venus* spacecraft includes a simple plasma wave instrument which monitors the wave power in four narrow (15%) frequency bands centered at 100 Hz, 730 Hz, 5.4 kHz, and 30 kHz (Scarf et al 1980a). The antenna consists of two wire cages about 12 cm across deployed on 1 m booms about 0.8 m apart. The instrument has a fast response with a rise time of several tens of ms but a much slower decay time of about $\frac{1}{2}$ s. In sunlight the noise level of the instrument exceeds all but the strongest natural emissions. However, in the shadow of the planet the instrument appears to be quite sensitive. At orbit injection *Pioneer Venus* periapsis was in sunlight and not until a month later did the spacecraft begin to be in shadow as it passed through closest approach. During this period the spacecraft observed the first electric field signal that appeared to be due to lightning (Taylor et al 1979). The signals had all the characteristics expected for lightning-generated signals. They were intense, impulsive, and broadband. They occurred on all four frequency channels and lasted less than 0.5 s because the signals were shorter than the decay constant of the instrument. The signals appeared to be more frequent at lower altitudes suggesting a low altitude or atmospheric source.

After analyzing a larger sample of data, Scarf et al (1980b) decided to use a more conservative definition of a possible lightning-generated signal. Since signals above the electron gyrofrequency should not propagate to the spacecraft in a collisionless plasma, Scarf et al (1980b) defined lightning events as impulsive signals at 100 Hz that were unaccompanied by signals at other frequencies. Furthermore, the magnetic field had to be strong enough so that the electron gyrofrequency well exceeded 100 Hz. However, he did not check the orientation of the magnetic field. Since Venus clouds were 50 km or more above the Venus surface, it is unlikely that cloud-to-ground strokes occur. However, since volcanoes also generate clouds and lightning on the Earth, Scarf et al (1980b) looked for a possible correlation of the emissions with volcanoes seen in the radar-derived topography.

A larger sample of data was analyzed by Scarf & Russell (1983) who examined records for the first 1185 orbits of *Pioneer Venus*. These orbits covered five complete traversals of the night ionosphere (or observing seasons) plus part of a sixth. This survey covered 14% of Venus' surface. In this study 65% of the observed signals came from regions near Phoebe, Beta, and Atla which were topographically high regions where volcanism was suspected. The data seemed to be consistent with a small number of surface sources, strengthening in the minds of the authors a possible link between volcanism and the lightning events. The study reported a rate of 2.4 bursts $km^{-2}yr^{-1}$. However, this burst rate cannot be considered to be a flash rate both because there are many flashes in a burst and because many bursts were not counted since they failed one of the conservative selection criteria. Furthermore, the area of surface monitored at any one time was not known, nor was the size of the discharge that could cause a burst at *Pioneer Venus*. As before there was no attempt to normalize the distribution for observing time.

Scarf and coworkers later attempted to revise the definition of an event to make it more nearly equal to a flash by counting each impulse in a group of impulses. However, the time resolution of the instrument still limited this approach so that the flash rate was still underestimated. In this study they also chose a threshold just above that of the instrument of $20 \mu Vm^{-1}Hz^{-1/2}$ at 100 Hz. They required that the electron gyrofrequency be greater than 100 Hz and that the magnetic field line through the spacecraft intersect the planet. Examining data from the first 2124 orbits (9.5 observing seasons), Scarf et al (1987) found 4240 bursts covering the altitude range of 150–2900 km. After the first three seasons, the periapsis altitude of *Pioneer Venus* was allowed to follow the rise caused by solar gravitational perturbations. As periapsis rose, the seasonal total number of bursts fell, indicating that the event rate was greatest at low altitudes. An attempt to show this altitude dependence using the rate versus altitude, independent of season, was flawed because of the nonuniform sampling of altitudes within any season. As before no attempt was made to normalize the occurrence rates.

Despite the evidence from the *Venera* probes which they did not challenge, Taylor et al (1985) took issue with the lightning interpretation of the *Pioneer Venus* VLF data and the speculation that the lightning was associated with volcanism. Taylor et al (1985) noted the frequent occurrence of the VLF bursts when the magnetic field was strong and the plasma density was low and suggested that these were the conditions that facilitated local plasma wave formation and not propagation from below the ionosphere. This debate continued for several years (Scarf 1986, Taylor et al 1986, Taylor & Cloutier 1986, Taylor et al 1987a, Scarf & Russell 1988). While Taylor and coworkers as of this writing appear still to be unconvinced that the events identified by Scarf and coworkers were due to lightning, their criticisms did encourage Scarf and coworkers to establish that the waves were electromagnetic as would be expected for lightning-generated waves (Scarf & Russell 1988). However, these later studies also showed that the apparent correlation with geographical location seemed to have been influenced by the coverage available from *Pioneer Venus*, so that a volcanic source was not needed to explain the observations.

QUANTITATIVE STUDIES   As noted above VLF waves appear to be able to penetrate the ionosphere from below by two different mechanisms: a propagating whistler mode wave and a nonpropagating disturbance. Taylor et al (1979) ascribed all signals in their initial study to lightning but the following studies by Scarf and coworkers were restricted to 100 Hz signals which they felt could propagate from the atmosphere. Singh & Russell (1986) took issue with this conservative approach since the higher frequency signals had many of the properties expected for lightning-generated signals. A new controversy arose over this interpretation (Taylor et al 1987b, Singh & Russell 1987, Taylor et al 1988, Russell & Singh 1989, Russell 1991b, Taylor & Cloutier 1991). Thus a new approach to the study was adopted.

QUANTITATIVE OCCURRENCE RATES: HIGH FREQUENCIES   In order to derive quantitative occurrence rates Russell et al (1988a) noted both where signals are detected and where they are not detected. All data were examined when *Pioneer Venus* was in the unilluminated ionosphere at a solar zenith angle greater than $90°$ and within 1.05 Venus radii of the extended Sun-Venus line. To be characterized as impulsive or bursty the magnitude of the signal had to vary in 30 s by an amount comparable to the mean of the signal and to exceed thresholds of 1×10^{-5} $Vm^{-1}Hz^{-1/2}$ at 0.73 kHz, 3×10^{-6} $Vm^{-1}Hz^{1/2}$ at 5.4 kHz and 9×10^{-7} $Vm^{-1}Hz^{1/2}$ at 30 kHz. Each 30-s interval was classified. In addition to the four electric field measurements, the magnetometer data were examined to eliminate periods in which telemetry errors occurred. Figure 16 shows typical signals classified in this study. The signals labeled *b* are interference associated with the motion of the antenna into the spacecraft wake. The signals labeled *a* and *c* are typical of those thought to be associated with lightning. As we discuss below, the broadband nature of the *a* signals implies an origin probably

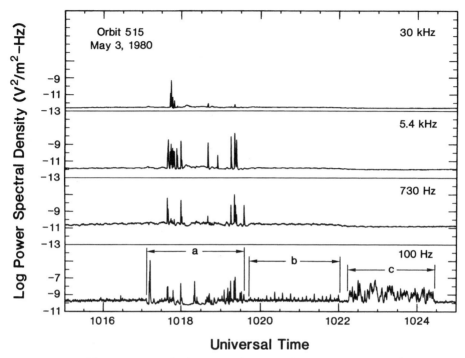

Figure 16 Plasma wave amplitudes versus time on orbit 515 showing types of signals observed. Type *b* is due to the interaction of the antenna or the spacecraft with the local plasma environment (Russell & Scarf 1990).

from a source immediately below the satellite. Such signals are seen at the lowest altitudes. Since the *c* signals occur only on the 100 Hz channel, the spacecraft does not appear to be in the near field of the lightning stroke. Probably these signals have propagated a large distance to the spacecraft and come from an extended region. The initial examination of these data showed that the occurrence rate decreased with increasing altitude at all frequencies, suggesting that at least some of the signals present were generated below the ionosphere. It also showed that the occurrence rates varied from year to year (Russell et al 1988a). The magnetic field strength affects the occurrence rate of these signals but the direction of the magnetic field has only a small or negligible effect (Russell et al 1988b). The signals above the electron gyrofrequency are useful for mapping the source locations precisely because they attenuate rapidly. Figure 17 shows the occurrence rate at 730 Hz as a function of latitude and local time for each of the first three seasons (Russell et al 1989a). During the first half of the second season Venus passed behind the Sun preventing telemetry transmission to Earth. Thus only $2\frac{1}{2}$ seasons of low altitude data are available. In the middle panel the region covered by the first star sensor survey (Borucki et al 1981) is also indicated. The first and third seasons show a very strong local time dependence with a region of highest occurrence rate on the dusk side which is confirmed during the second season by the relative absence of signals on the dawn side. The star sensor search was clearly performed over a region of infrequent plasma wave activity as was the *Vega* balloon mission which covered the sector from midnight to dawn. In contrast the *Venera* 9 observation was at 1930 LT. We note that if we correct for the altitude dependence of the recurrence rate of these signals they maximize in the equatorial regions as on Earth (Russell et al 1989a). We believe that the local time variation seen on the dawn side of the peak activity in Figure 17 is due to variations in the source because the ionosphere does not vary much across midnight to dawn but the decrease in occurrence rate from 2100 LT to 1800 LT is due to attenuation in the increasingly dense ionosphere (Russell et al 1989a). The 100 Hz signals which generally occur below the electron gyrofrequency behave in some ways the same and in someways different than those above 100 Hz (Russell et al 1990). Figure 18 shows the local time of occurrence of the 100 Hz signal when *B* is greater than 15 nT, together with the occurrence rate of fields greater than 15 nT and with the occurrence rate of emissions above the electron gryofrequency. This latter quantity was inferred earlier to point to the source region of the lightning emissions because of the rapid decrease in occurrence (and amplitude) with altitude. The 100 Hz occurrence rate does not resemble this curve nor does it resemble the occurrence of strong fields. If it did we would infer that the occurrence of 100 Hz waves was controlled solely by the magnetic field. Rather the 100 Hz waves peak over the so-called source region but extend far across the nightside, gradually decreasing in occurrence. We deduce from this that the 100 Hz signals may be generated in the "source region" but that they can propagate a long way in the Venus-ionosphere waveguide before they enter the ionosphere. We recall that we know nothing about the possible extension of the source region into the afternoon sector because the ionospheric density appears to control the access of signals to the spacecraft.

The studies by Russell and coworkers of the occurrence of VLF bursts treated only 30-second intervals of data. The study did not distinguish if an interval had a single impulse or was continually active. Thus these results should be viewed more as an activity study rather than an occurrence rate study. In order to obtain a measure that more nearly reflected the flash rate, Ho et al (1991) developed a method to count individual

0.73 kHz

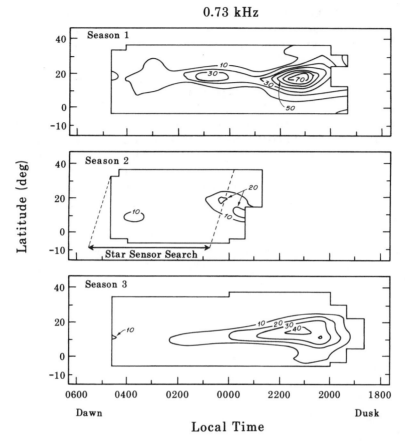

Figure 17   Contour maps of the occurrence rate of bursts at 730 Hz in local time and latitude for each of the three observing seasons. The interval during the second observing season over which the star sensor search was conducted is also shown (Russell et al 1989a).

VLF bursts or impulses on each of the four channels of the *PVO* plasma wave instrument. Qualitatively the results of the study were the same as in the earlier "activity" study. The burst rate decreased rapidly with altitude above the electron gyrofrequency (with a scale length of about 20 km) and less rapidly below. The local time distribution of the high frequencies was also the same as before. However, the post midnight sector had higher 100 Hz occurrence rates, possibly in part due to the magnetic geometry in this region being more favorable to the leakage of the 100 Hz noise from the sub-ionospheric waveguide. Most importantly the authors were able to estimate a global flash rate of 250 s^{-1}, a rate $2\frac{1}{2}$ times larger than the terrestrial rate. However, this estimate depends on the size of the lightning generation region which extends an unknown distance into the dayside of the planet.

Another attempt to determine a quantitative measure of the VLF energy propagating out of the atmosphere into the ionosphere was the attempt to measure the Poynting flux of 100 Hz energy using the 100 Hz electric field intensity and the ionospheric electron density and magnetic field strength (Russell et al 1989b). The study showed that there was about 10^{-7} W/m^2 of 100 Hz energy over the active region. This was not inconsistent with expectations from terrestrial lightning activity. However, the estimate depends on the assumption that the waves are propagating along the magnetic field direction. We expect the waves to be refracted vertically by the ionosphere at low altitudes but we do not expect the magnetic field always to be vertical. Thus, this estimate needs to be refined. The expected vertical refraction of waves, however, provides a very powerful test of the

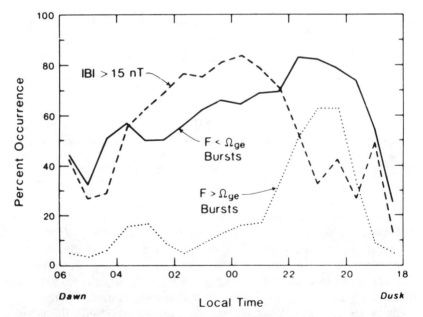

Figure 18   Local time dependence of the low altitude VLF bursts observed by *Pioneer Venus* when the magnetic field strength at the satellite was greater than 15 nT. Solid line shows percent active 30-s intervals for 100 Hz channel; dotted line for higher 3 channels. (Russell et al 1990.)

lightning generation of the observed VLF signals. We discuss this test and others in the follow section.

TESTING THE LIGHTNING HYPOTHESIS FOR THE VLF IMPULSES   Despite the *Venera* observations of electromagnetic VLF bursts in the atmosphere and the *Venera* 9 observations of flashes from the night atmosphere, there were some who sought alternate explanations for the *PVO* VLF bursts. It was proposed that the waves were not electromagnetic and were subject to Doppler shifting. Scarf & Russell (1988) countered this with examples of waves that could not be due to Doppler shifting and waves that were electromagnetic in their polarization relative to the field. Later Ho et al (1992) showed statistically that the waves were not caused by Doppler shifting of lower frequency signals and Strangeway (1992a) showed statistically that the waves had electromagnetic polarization. Since electromagnetic waves can be generated by a plasma as well as by lightning, Maeda & Grebowsky (1989) and Huba (1992) have proposed alternates to the lightning mechanism. These suggestions have been tested by Strangeway (1992b) and found not to be consistent with the *PVO* data. Thus, at present each of the proposed in situ source mechanisms for the VLF impulses in the night ionosphere of Venus has been found to be inconsistent with at least some properties of the observed signals.

In addition to testing whether there are any viable alternates to the lightning generation hypothesis, it is possible to test predictions of the lightning hypothesis itself. A valuable test was proposed by Sonwalkar et al (1991) who noted that at frequencies below the electron gyrofrequency whistler mode waves were restricted to a cone of propagation, or resonance cone, around the field. Since waves would be refracted vertically by the ionospheric density increase as the waves propagated upward from a lightning discharge, these waves could not enter the ionosphere if the magnetic field were horizontal but could freely penetrate if it were vertical. Upon testing this hypothesis, Sonwalkar et al (1991) found that a significant number of the VLF bursts were consistent with the lightning hypothesis. Ho et al (1992) used this same test to make an even more powerful statement about the source of the waves. They examined the altitude distribution of the 100 Hz waves (the ones below the electron gyrofrequency) in the allowed and forbidden range of angles based on the sole assumption that the waves were propagating vertically upward out of the atmosphere. Figure 19 shows the resulting altitude distribution. Inside the resonance cone, at allowed whistler mode angles, the waves have little altitude dependence as one would expect for whistler mode propagation. Outside the resonance cone the 100 Hz waves decrease rapidly in occurrence in a manner similar to the nonpropagating waves at 730 Hz, 5.4

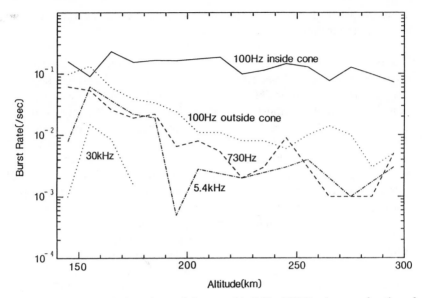

Figure 19 The altitude dependence of the rate of individual VLF pulses as a function of altitude for signals above the local gyrofrequency at 730 Hz, 5.4 kHz, and 30 kHz, together with the rate of occurrence of 100 Hz bursts inside the resonance cone and outside the resonance cone. The only assumption used in deriving the direction of propagation of these 100 Hz waves is that the source of the waves was in the atmosphere. This resonance cone appears to be quite successful in separating propagating waves (*top curve*) from non-propagating waves (*dotted line*). (Ho et al 1992.)

kHz, and 30 kHz. The only known explanation of this separation is that the waves came out of the atmosphere from below the ionosphere.

Radio Wave Observations

Lightning generates a broad spectrum of electromagnetic energy, extending well above the megahertz range. At Saturn and Uranus these radio waves were very important for the verification of lightning activity because neither optical flashes nor whistler mode signals could be detected at these planets. Neither the *Venera* spacecraft nor the *Pioneer Venus Orbiter* were instrumented to receive MHz signals so could not detect lightning this way. The *Galileo* spacecraft, however, was detoured past Venus on its way to Jupiter and its plasma wave receiver did extend well past 1 MHz. This receiver was turned on for only 1 hr as the spacecraft flew by Venus but in that period found nine radio wave bursts that had the expected characteristics of lightning (Gurnett et al 1991). The spectrum of these emissions is shown in Figure 20. The intensities of these signals is similar

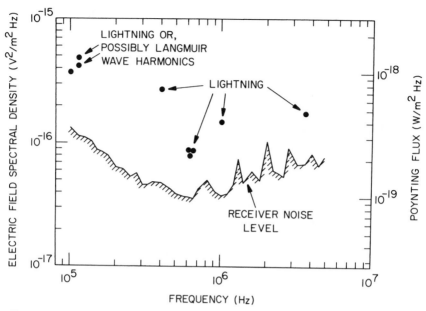

Figure 20   Amplitude of radio wave impulses seen by the plasma wave instrument on *Galileo* as it passed Venus on Feb. 10, 1990.

to what one would expect from terrestrial lightning at a similar radial distance, and except for possibly the lowest frequency bursts in this figure lightning is the only known source that is consistent with the observed properties of the waves

VENUS SUMMARY

Venus is the only planet besides the Earth for which we have in situ atmospheric measurements as well as ionospheric data on lightning. The *Venera* landers clearly registered electromagnetic pulses with the properties of lightning generation. The mission to mission differences were reminiscent of the variability due to weather differences. The similar design and operation of the landers would have led to similar behavior of any artifacts. The *Venera* orbiter data too showed examples of optical flashes. Unfortunately these data have usually been dismissed because they are few in number and were poorly presented in the literature. However as we have seen above, close examination of these data adds credence to their veracity. The optical data are sparse because even the flashes associated with high rates of lighting activity are hard to find optically: A detector has to be looking at the source to observe it. Radio signals can be detected over a much larger solid angle.

Pioneer Venus' nearly 14 years in orbit have allowed ionospheric VLF signals to be extensively probed. This probing is effective only over the nightside of the planet but within this limitation the observations reveal a flash rate similar or greater than the terrestrial one. Moreover, the properties of the observed signals are those expected for waves originating in the atmosphere and propagating upward into the ionosphere. They do not have the properties expected of waves generated in the ionosphere itself.

The *Galileo* data while brief in time are totally consistent with the *Venera* and *PVO* data. No optical observations were obtained but the geometry was not optimum and scattered light set too high a background level, while the radio wave instrument was successful in detecting lightning, at a rate and at intensities reminiscent of Earth.

The exploration of Venus lightning is not complete. Perhaps more could be done with the *PVO* ultraviolet data. Even if we could get a more reliable flash rate from the *PVO* ultraviolet spectrometer data it would still pertain only to the nightside. In order to probe the dayside where we would expect the highest flash rate, we need radio techniques or an optical detector designed to operate in the presence of reflected sunlight. The only possibility at the present time is to use the *Magellan* radiometer at 2.65 GHz. However, this frequency is quite a bit higher than that which would be most desirable. A new mission with optimized optical and radio sensors would seem to be necessary to provide a satisfactory measure of the rate, strength, and atmospheric significance of Venus lightning. It is possible that the lightning rate in the Venus atmosphere is much greater on the dayside than on the nightside.

CONCLUSIONS

The study of planetary lightning is still in its infancy. The existence of planetary lightning at Venus, Jupiter, Saturn, Uranus, and Neptune is no longer the subject of serious debate but the rate of occurrence and the energy dissipation rate due to lightning remain to be determined. We have no satisfactory measure of the dayside rate of occurrence of lightning except perhaps at Saturn and Uranus. Our optical data are poor even at Jupiter where *Voyagers* 1 and 2 provided several nightside images of lightning. At VLF frequencies where lightning emissions are strongest no signals at all were detected by the *Voyager* spacecraft. Clearly we need new and improved observations. Some of these will be provided by the *Galileo* mission which will drop a probe into the atmosphere in 1995 and then orbit the planet for nearly two years with radio, VLF, and optical sensors. *Cassini*, in some respects similar to *Galileo*, will provide a similar set of data for Saturn. However, at present there are no future missions being planned for Venus, Uranus, or Neptune. This is particularly unfortunate for Venus where the rate of lightning flashes may be a significant factor in the generation of some chemical species in the atmosphere.

ACKNOWLEDGMENTS

The author would like to thank W. J. Borucki, W. S. Kurth, and R. J. Strangeway for their helpful comments and careful reading of an early draft of this manuscript. The preparation of this review was supported in part by the National Aeronautics and Space Administration under research grant NAG2–501.

Literature Cited

Bar-Nun, A. 1975. Thunderstorms on Jupiter. *Icarus* 24: 86–94

Bjoraker G. L. 1985. *The gas composition and vertical cloud structure of Jupiter's troposphere derived from five-micron spectroscopic observations*. PhD thesis. U. Ariz., Tucson

Bliokh, P. V., Nicholaenko, A. P., Fillippov, Yu. F. 1980. *Schumann Resonances in the Earth-Ionosphere Wave Guide*. New York: Peregrinns. 165pp.

Boeck, W. L., Vaughan, O. H. Jr., Blakelee, R. J. 1991. Low light level TV images of terrestrial lightning viewed from space. *Eos, Trans. Am. Geophys. Union* 72: 171–72 (Abstr.)

Borucki, W. J. 1982. Comparison of Venusian lightning observations. *Icarus* 52: 354–62

Borucki, W. J. 1989. Predictions of lightning activity at Neptune. *Geophys. Res. Lett.* 16: 937–39

Borucki, W. J., Bar-Nun, A., Scarf, F. L., Cook , A. F. II, Hunt, G. E. 1982. Lightning activity on Jupiter. *Icarus* 52: 492–502

Borucki, W. J., Chameides, W. L. 1984. Lightning: estimates of the rates of energy dissipation and nitrogen fixation. *Rev. Geophys.* 22: 363–72

Borucki, W. J., Dyer, J. W., Thomas, G. Z., Jordan, J. C., Comstock, D. A. 1981. Optical search for lightning on Venus. *Geophys. Res. Lett.* 8: 233–36

Borucki, W. J., Dyer, J. W., Phillips, J. R., Pham, P. 1991. Pioneer Venus Orbiter search for Venusian lightning. *J. Geophys. Res.* 96: 11,033–43

Borucki, W. J., Giver, L. P., McKay, C. P.,

Scattergood, T., Parris, J. E. 1988. Lightning production of hydrocarbons and HCN on Titan: laboratory measurements. *Icarus* 76: 125–34

Borucki, W. J., Levin, Z., Whitten, R. C., Keese, R. G., Capone, L. A., et al. 1987. Predictions of the electrical conductivity and charging of aerosols in Titan's atmosphere. *Icarus* 72: 604–22

Borucki, W. J., Magalhaes, J. A. 1992. Analysis of Voyager 2 images of Jovian lightning. *Icarus* 96: 1–14

Borucki, W. J., McKay, C. P., Whitten, R. C. 1984. Possible production by lightning of aerosols and trace gases in Titan's atmosphere. *Icarus* 60: 260–73

Borucki, W. J., McKay, C. P. 1987. Optical efficiencies of lightning in planetary atmospheres. *Nature* 328: 509–10

Borucki, W. J., McKenzie, R. L., McKay, C. P., Duong, N. D., Boac, D. S. 1985. Spectra of simulated lightning on Venus, Jupiter and Titan. *Icarus* 64: 221–32

Borucki, W. J., Williams, M. A. 1986. Lightning in the Jovian water cloud. *J. Geophys. Res.* 91: 9893–903

Braginskii, S. I. 1958. Theory of the development of a spark channel. *Sov. Phys. JETP* 34: 1068–74

Burns, J. A., Sonwalkar, M. R., Cuzzi, J. N., Durisen, R. H. 1983. Saturn's electrostatic discharges: Could lightning be the cause? *Icarus* 54: 280–95

Carlson, B., Lacis, A. A., Rossow, W. B. 1992. The abundance and distribution of water vapor in the Jovian troposphere as inferred from Voyager IRIS data. *Astrophys. J.* 388: 648–68

Chameides, W. L., Walker, J. C. G., Nagy,

A. F. 1979. Possible chemical impact of planetary lightning. *Nature* 280: 820–22

Colin, L., Hunten, D. M. 1977. Pioneer Venus experiment descriptions. *Space Sci. Rev.* 20: 451–525

Cook, A. F. II, Duxbury, T. C., Hunt, G. E. 1979. First results on Jovian lightning. *Nature* 280: 794

Desch, M. D. 1992. Lightning at planets in the outer solar system. In *Planetary Radio Emissions III*, ed. H. O. Rucker, S. J. Bauer. Vienna: Austrian Acad. Sci. In press

Desch, M. D., Kaiser, M. L. 1990. Upper limit set for level of lightning activity on Titan. *Nature* 343: 442–44

Eshleman, V. R., et al. 1979. Radio science with Voyager 1 at Jupiter: preliminary profiles of the atmosphere and ionosphere. *Science* 204: 976–78

Evans, D. R., Warwick, J. W., Pearce, J. B., Carr, T. D., Schauble, J. J. 1981. Impulsive radio discharges near Saturn. *Nature* 292: 716–18

Franz, R. C., Nemzek, R. J., Winckler, J. R. 1990. Television image of a large upward electrical discharge above a thunderstorm system. *Science* 249: 48–51

Gurnett, D. A., Kurth, W. S., Scarf, F. L. 1981. Plasma waves near Saturn: initial result. *Science* 212: 235–39

Gurnett, D. A., Kurth, W. S., Cairns, I. H., Granroth, L. J. 1990. Whistlers in Neptune's magnetosphere: evidence of atmospheric lightning. *J. Geophys. Res.* 95: 20,967–76

Gurnett, D. A., Kurth, W. S., Roux, A., Gendrin, R., Kennel, C. F., Bolton, S. J. 1991. Lightning and plasma wave observations from the Galileo flyby of Venus. *Science* 253: 1522–25

Gurnett, D. A., Shaw, R. R., Anderson, R. R., Kurth, W. S. 1979. Whistlers observed on Voyager 1: detection of lightning on Jupiter. *Geophys. Res. Lett.* 6: 511–14

Ho, C.-M., Strangeway, R. J., Russell, C. T. 1991. Occurrence characteristics of VLF bursts in the night ionosphere of Venus. *J.Geophys. Res.* 96: 21,361–69

Ho, C.-M., Strangeway, R. J., Russell, C. T. 1992. Control of VLF burst activity in the nightside ionosphere of Venus by the magnetic field orientation. *J. Geophys. Res.* In press

Huba, J. D. 1992. Theory of small scale density and electric field fluctuations in the nightside Venus ionosphere. *J. Geophys. Res.* 97: 43–50

Ingersoll, A. P. 1990. Atmospheric dynamics of the outer planets. *Science* 248: 308–15

Kaiser, M. L., Connerney, J. E. P., Desch, M. D. 1984. Atmospheric storm explanation of Saturnian electrostatic discharges. *Nature* 303: 50–53

Kaiser, M. L., Desch, M. D. 1984. Voyager detection of lightning on Saturn. *Proc. 7th Int. Conf. of Atmosph. Electr.*, p. 472. Boston: Am. Meteorol. Soc.

Kaiser, M. L., Zarka, P., Desch, M. D., Farrell, W. M. 1991. Restriction on the characteristics of Neptunian lightning. *J. Geophys. Res.* 96: 19,043–47

Kelley, M. C., et al. 1985. Electrical measurements in the atmosphere and the ionosphere over an active thunderstorm. 1. Campaign overview and initial ionospheric results. *J. Geophys. Res.* 90: 9815–23

Kelley, M. C., Ding, J. G., Holzworth, R. H. 1990. Intense ionosphere electric and magnetic field pulses generated by lightning. *Geophys. Res. Lett.* 17: 2,221–24

Knollenberg, R. G., Hunten, D. M. 1980. The microphysics of the clouds of Venus: results of the Pioneer Venus particle size spectrometer experiment. *J. Geophys. Res.* 85: 8039–58

Krasnopol'sky, V. A. 1980. Lightning on Venus according to information obtained by the satellites Venera 9 and 10. *Kosm. Issled.* 18: 429–34

Krasnopol'sky, V. A. 1983a. Venus spectroscopy in the 3000–8000Å region by Veneras 9 and 10. In *Venus*, ed. D. M. Hunten, L. Colin, T. M. Donahue, V. I. Moroz, pp. 459–83. Tucson: Univ. Ariz. Press

Krasnopol'sky, V. A. 1983b. Lightnings and nitric oxide on Venus. *Planet. Space Sci.* 31: 1363–69

Krider, E. P., Dawson, G. A., Uman. M. A. 1968. Peak power and energy dissipation in a single-stroke lightning flash. *J. Geophys. Res.* 73: 3335–39

Ksanfomaliti, L. V. 1979. Lightning in the cloud layers of Venus. *Kosm. Issled.* 17: 747–62

Ksanfomaliti, L. V. 1983. Electrical activity in the atmosphere of Venus. I. Measurements on descending probes. *Kosm. Issled.* 21: 279–96

Ksanfomaliti, L. V., Scarf, F. L., Taylor, W. W. L. 1983. The electrical activity of the atmosphere of Venus. In *Venus*, ed. D. M. Hunten, L. Colin, T. M. Donahue, V. I. Moroz, pp. 565–603. Tucson: Univ. Ariz. Press

Kurth, W. S., Strayer, B. D., Gurnett, D. S., Scarf, F. L. 1985. A summary of whistlers observed by Voyager 1 at Jupiter. *Icarus* 61: 497–507

Lanzerotti, L. J., et al. 1992. The lightning and radio emission detector (LRD) instrument. *Space Sci. Rev.* 60: 91–109

Lanzerotti, L. J., Rinnert, K.E., Krider, P., Uman, M. A. 1989. Jovian lightning. In *Time Variable Phenomena in the Jovian*

System, ed. M. J. S. Belton, R. A. West, J. Rahe, pp. 374–83. *NASA SP–494*, Washington, DC

Levin, Z., Borucki, W. J., Toon, O. B. 1983. Lightning generation in planetary atmospheres. *Icarus* 56: 80–115

Lewis, J. S. 1976. Equilibrium and disequilibrium chemistry of adiabatic, solar-composition planetary atmospheres. In *Chemical Evolution of the Giant Planets*, ed. C. Ponnaperuma, pp. 13–25. New York: Academic

Lewis, J. S. 1980. Lightning on Jupiter: rate, energetics and effects. *Science* 210: 1351–52

Lindal, G. G., Wood, G. E., Hotz, H. B., Sweetnam, D. N., Eshleman, V. R., Tyler, G. L. 1983. The atmosphere of Titan: an analysis of the Voyager 1 radio occultation measurements. *Icarus* 53: 348–68

Maeda, K., Grebowsky, J. M. 1989. VLF emission bursts in the terrestrial and Venusian nightside troughs. *Nature* 341: 219

Magalhaes, J. A., Borucki, W. J. 1991. Spatial distribution of visible lightning on Jupiter. *Nature* 349: 311–13

McKay, C. P., Scattergood, T. W., Pollack, J. B., Borucki, W. J., Van Ghysegham, H. T. 1988. High temperature shock formation of N_2 and organics on primordial Titan. *Nature* 332: 520–22

Meinel, A. B., Hoxie, D. T. 1962. On the spectrum of lightning in the atmosphere of Venus. *Comm. Lunar Planet. Lab.* 1: 35–38

Menietti, J. D., Gurnett, D. A. 1980. Whistler propagation in the Jovian magnetosphere. *Geophys. Res. Lett.* 7: 49–52

Menietti, J. D., Tsintikidis, D., Gurnett, D. A., Curran, D. B. 1991. Modeling of whistler ray paths in the magnetosphere of Neptune. *J. Geophys. Res.* 96: 19,117–22

Phillips, J. L., Russell, C. T. 1987. Upper limit on the intrinsic magnetic field of Venus. *J. Geophys. Res.* 92: 2253–63

Phillips, J. L., Russell, C. T. 1991. The Venus Ashen Light: results of the 1988 observing campaign. *Adv. Space Res.* 12(9): 51–56

Ridgway, S. T. 1974. Jupiter: identification of ethane and acetylene. *Astrophys. J.* 187: L41–L43

Rinnert, K. 1982. Lightning within planetary atmospheres. In *Handbook of Atmospheres*, Vol. 2, ed. H. Volland, pp. 1–22. Boca Raton, Florida: CRC Press

Rinnert, K., Lanzerotti, L. J., Krider, E. P., Uman, M. A., Dehmel, G., et al. 1979. Electromagnetic noise and radiowave propagation below 100 Hz in the Jovian atmosphere. 1. Equatorial region. *J. Geophys. Res.* 84: 5181–88

Rinnert, K., Lauderdale, R. II, Lanzerotti,

L. J., Krider, E. P., Uman, M. A. 1989. Characteristics of magnetic field pulses in Earth lightning measured by the Galileo probe instrument. *J. Geophys. Res.* 94: 13,229–35

Russell, C. T. 1991a. Venus lightning. *Space Sci. Rev.* 55: 317–56

Russell, C. T. 1991b. Reply to Taylor and Cloutier. *Geophys. Res. Lett.* 18: 755–58

Russell, C. T., Elphic, R. C., Luhmann, J. G., Slavin, J. A. 1980. On the search for an intrinsic magnetic field at Venus. 11*th Proc. Lunar Planet. Sci.*, pp. 1894–906

Russell, C. T., Singh, R. N. 1989. A re-examination of impulsive VLF signals in the night ionosphere of Venus. *Geophys. Res. Lett.* 16: 1481–484

Russell, C. T., Scarf, F. L. 1990. Evidence for lightning on Venus. *Adv. Space Res.* 10(5): 137–41

Russell, C. T., von Dornum, M., Scarf, F. L. 1988a. The altitude distribution of impulsive signals in the night ionosphere of Venus. *J. Geophys. Res.* 93: 5915–21

Russell, C. T., von Dornum, M., Scarf, F. L. 1988b. VLF bursts in the night ionosphere of Venus. *Planet. Space Sci.* 36: 1211–18

Russell, C. T., von Dornum, M., Scarf, F. L. 1988c. Planetocentric clustering of low altitude impulsive electric signals in the night ionosphere of Venus. *Nature* 331: 591–94

Russell, C. T., von Dornum, M., Scarf, F. L. 1989a. Source locations for impulsive electric signals seen in the night ionosphere of Venus. *Icarus* 80: 390–415

Russell, C. T., von Dornum, M., Strangeway, R. J. 1989b. VLF bursts in the night ionosphere of Venus: estimates of the Poynting flux. *Geophys. Res. Lett.* 16: 579–82

Russell, C. T., von Dornum, M., Scarf, F. L. 1990. Impulsive signals in the night ionosphere of Venus: comparison of results obtained below the local electron gyrofrequency with those above. *Adv. Space Res.* 10(5): 37–40

Sagdeev, R. Z., et al. 1986. Overview of VEGA Venus balloon in-situ meteorological measurements. *Science* 231: 1411–14

Samuelson, R. E., Hanel, R. A., Kunde, V. G., Maguire, W. C. 1981. Mean molecular weight and hydrogen abundance of Titan's atmosphere. *Nature* 292: 688–93

Scarf, F. L. 1986. Comment on "Venus nightside ionospheric troughs: implications for evidence of lightning and volcanism" by H. A. Taylor, Jr., J. M. Grebowsky and P. A. Cloutier. *J. Geophys. Res.* 91: 4594–98

Scarf, F. L., Gurnett, D. A., Kurth, W. S. 1979. Jupiter plasma wave observations:

an initial Voyager 1 overview. *Science* 204: 991–95

Scarf, F. L., Gurnett, D. A., Kurth, W. S., Anderson, R. R., Shaw, R. R. 1981. An upper bound to the lightning flash rate in Jupiter's atmosphere. *Science* 213: 684–85

Scarf, F. L., Jordan, K. F., Russell, C. T. 1987. Distribution of whistler mode bursts at Venus. *J. Geophys. Res.* 92: 12,407–11

Scarf, F. L., Russell, C. T. 1983. Lightning measurements from the Pioneer Venus Orbiter. *Geophys. Res. Lett.* 10: 1192–95

Scarf, F. L., Russell, C. T. 1988. Evidence of lightning and volcanic activity on Venus. *Science* 240: 222–24

Scarf, F. L., Taylor, W. W. L., Russell, C. T., Brace, L. H. 1980b. Lightning on Venus: orbiter detection of whistler signals. *J. Geophys. Res.* 85: 8158–66

Scarf, F. L., Taylor, W. W. L., Virobik, P. F. 1980a. The Pioneer Venus orbiter plasma wave investigation. *IEEE Trans. Geosci. Remote Sensing* GE18: 36–38

Singh, R. N., Russell, C. T. 1986. Further evidence for lightning on Venus. *Geophys. Res. Lett.* 13: 1071–74

Singh, R. N., Russell, C. T. 1987. Reply to Taylor and Cloutier. *Geophys. Res. Lett.* 14: 571–72

Smith, B. A., Soderblom, L. A., Johnson, T. V., Ingersoll, A. P., Collins, S. A., et al. 1979a. The Jupiter system through the eyes of Voyager 1. *Science* 204: 951–72

Smith, B. A., Soderblom, L. A., Beebe, R., Boyce, J., Briggs, G., et al. 1979b. The Galilean satellites and Jupiter: Voyager 2 imaging science results. *Science* 206: 927–50

Smith, B. A., Soderblom, L. A., Beebe, R., Boyce, J., Briggs, G., et al. 1981. Encounter with Saturn: Voyager 1 imaging science results. *Science* 212: 163–91

Smith, B. A., Soderblom, L. A., Batson, R., Bridges, P., Inge, J., et al. 1982. A new look at the Saturn system: the Voyager 2 images. *Science* 215: 504–37

Smith, B. A., Soderblom, L. A., Beebe, R., Bliss, D., Boyce, J. M., et al. 1986. Voyager 2 in the Uranian system: imaging science results. *Science* 233: 43–64

Sonwalkar, V. S., Carpenter, D. L., Strangeway, R. J. 1991. Testing radio bursts observed on the nightside of Venus for evidence of whistler-mode propagation from lightning. *J. Geophys. Res.* 96: 17,763–78

Strangeway, R. J. 1992a. Polarization of the impulsive signals observed in the nightside ionosphere of Venus. *J. Geophys. Res.* 97: In press

Strangeway, R. J. 1992b. An assessment of lightning or in situ instabilities as a source for whistler mode waves in the night iono-

sphere of Venus. *J. Geophys. Res.* 97: In press

Taylor, H. A. Jr., Cloutier, P. A. 1986. Venus: dead or alive. *Science* 234: 1087–93

Taylor, H. A. Jr., Cloutier, P. A., Zheng, Z. 1987a. Venus lightning signals reinterpreted as in situ plasma noise. *J. Geophys. Res.* 92: 9907–19

Taylor, H. A. Jr., Cloutier, P. A., Zheng, Z. 1987b. Comment on "Further evidence for lightning at Venus." *Geophys. Res. Lett.* 14: 568–70

Taylor, H. A. Jr., Cloutier, P. A., Zheng, Z. 1988. Telemetry interference incorrectly interpreted as evidence for lightning and present-day volcanism at Venus. *Geophys. Res. Lett.* 15: 729–32

Taylor, H. A. Jr., Cloutier, P. A. 1991. Comment on "A re-examination of impulsive VLF signals in the night ionosphere of Venus." *Geophys. Res. Lett.* 18: 753–54

Taylor, H. A. Jr., Grebowsky, J. M., Cloutier, P. A. 1985. Venus nightside ionospheric troughs: implications for evidence of lightning and volcanism. *J. Geophys. Res.* 90: 7415–26

Taylor, H. A. Jr., Grebowsky, J. M., Cloutier, P. A. 1986. Reply. *J. Geophys. Res.* 91: 4599–605

Taylor, W. W. L., Scarf, F. L., Russell, C. T., Brace, L. H. 1979. Evidence for lightning on Venus. *Nature* 282: 614–16

Tokar, R. L., Gurnett, D. A., Bagenal, F., Shaw R. R. 1982. Light ion concentrations in Jupiter's inner magnetosphere. *J. Geophys. Res.* 87: 2241

Uman, M. 1987. *The Lightning Discharge.* Orlando: Academic

Vonnegut, B. 1982. The physics of thunderclouds. In *Handbook of Atmospheric Electricity*, Vol. 1, ed. H. Volland, pp. 1–22. Boca Raton, Florida: CRC Press

Warwick, J. W., Evans, D. R., Romig, J. H., Alexander, J. K., Desch, M. D., et al. 1981. Planetary radio astronomy observations from Voyager 1 near Saturn. *Science* 212: 239–44

Weidenschilling, S. J., Lewis, J. S. 1973. Atmosphere and cloud structures of the Jovian planets. *Icarus* 20: 465–76

West, R. A., Strobel, D. F., Tomasko, M. G. 1986. Clouds aerosols and photochemistry in the Jovian atmosphere. *Icarus* 65: 161–217

Williams, M. A., Dyer, J. W., Thomas, G. Z., Jordan, J. C., Comstock, D. A. 1982. The transmission to space of the light produced by lightning in the clouds of Venus. *Icarus* 52: 166–70

Williams, M. A., Krider, E. P., Hunten, D. M. 1983. Planetary lightning: Earth, Jupiter, and Venus. *Rev. Geophys.* 21: 892–902

Wilson, C. T. R. 1929. Some thundercloud problems. *J. Franklin Inst.* 208: 1–12

Zarka, P. 1984. Saturn electrostatic discharges. In *Planetary Radio Emissions,* ed. H. O. Rucker, S. J. Bauer, pp. 237–70. Vienna: Austrian Acad. Sci.

Zarka P., Pedersen, B. M. 1983. Statistical study of Saturn electrostatic discharges. *J. Geophys. Res.* 88: 9007–18

Zarka, P., Pedersen, B. M. 1986. Radio detection of Uranian lightning by Voyager 2. *Nature* 323: 605–8

Annu. Rev. Earth Planet. Sci. 1993. 21:89–114

SEDIMENT DEPOSITION FROM TURBIDITY CURRENTS

Gerard V. Middleton

Department of Geology, McMaster University, Hamilton, Ontario L8S 4M1, Canada

KEY WORDS: autosuspension, gravity flow, hydraulics, turbidite

1.   INTRODUCTION

Turbidity currents are *gravity currents* (Simpson 1987; also called *density currents* by many authors, following Bell 1942) in which the excess density or unit weight is due to suspended sediment. Gravity (or density) currents are a general class of flows (also know as stratified flows) in which flow takes place because of relatively small differences in unit weight between two fluids: The gravity flow may move below, above, or between ambient fluid of different unit weight (or weights). The difference in unit weight may be due to differences in composition (for example, the flow of oil over water), in salinity, or in temperature. The flow of rivers is not normally considered a type of gravity current because the unit weight of water is several hundred times larger than that of air. The flow of cold air beneath warm air, however, is considered to be a gravity or density current. The term turbidity current is generally applied only to flows of sediment suspended in water, though strictly it applies also to flows of suspended solids in air—for example, to some atmospheric duststorms, to powder snow avalanches (Hopfinger 1983, and see *Annals of Glaciology* 13, 1989), and to phenomena such as pyroclastic flows or base surges.

Gravity currents that are driven by gravity acting on dispersed sediment in the flow were called *sediment gravity flows* by Middleton & Hampton (1973, 1976). Turbidity currents are one type of sediment gravity flow in which the sediment is held in suspension by fluid turbulence. They can be distinguished from other types (e.g. debris flows) in which sediment is

89

dispersed by other mechanisms (e.g. matrix strength, grain collisions, buoyancy, etc).

This review is concerned with the mechanics of sediment deposition from turbidity currents. Beds deposited from turbidity currents (called *turbidites*) are one of the commonest types of sedimentary rocks: They include both sands and muds, and may be of siliciclastic or other composition (e.g. carbonates, cherts). For example, the majority of sandstones in the geologic record were deposited either from rivers or from turbidity currents, and the largest sedimentary features on the modern earth are submarine fans and abyssal plains, both of which result from turbidity currents. This review will not attempt to summarize the large number of published descriptions of turbidite facies (for a recent summary see Pickering et al 1986, 1989).

The early history of the turbidity current concept has been reviewed by Walker (1973). Though Daly (1936) and Bell (1942) were pioneers, it was the experimental and field observations of Philip Kuenen that convinced geologists that many sandstones previously believed to have been deposited in shallow water were in fact deposited by turbidity currents in water hundreds or thousands of meters deep. This notion had far reaching consequences for stratigraphy, and indeed for the whole of geology. This review is dedicated to Kuenen's memory—an appreciation of his work is now available in the *Dictionary of Scientific Biography* (Bourgeois 1990).

2. HYDRAULICS

2.1 *Introduction*

Many aspects of turbidity current hydraulics are similar to those of other types of gravity flows. These have been reviewed by Turner (1979), Yih (1980), Simpson (1982, 1987), and Hopfinger (1983). The dynamics of granular flows—a class of sediment gravity flows that probably grades into turbidity currents—has been reviewed by Campbell (1990). In this section, therefore, we do not attempt to describe all aspects of gravity flow hydraulics, but discuss only the few topics that are necessary to the discussion of sediment deposition given later in the review.

2.2 *Complexity of Gravity Flows*

It must first be emphasized that gravity flows are a very complex class of flows, more complex for example than open-channel flows such as rivers (which might be considered to be merely one extreme end member of the gravity flow spectrum). No geologist would expect that rivers could deposit only a single type of sediment bed, characterized by a single suite of sedimentary structures and textures—yet many geologists have such a notion about turbidites. Turbidity currents are generally thought to be unsteady, indeed catastrophic events, generated by sediment slumping on an oversteepened subqueous slope. They flow down a subqueous channel, erosional in its upper part, becoming depositional, with prominent subqueous levees, near the base of slope. Further from the base of slope, the channel may disappear and the turbidity current spreads out as a flow that is unconfined, except by the topography of the sedimentary basin. As the location of the upper part of the channel (or "canyon") is fixed, and the lower depositional channels are free to migrate by avulsion, lateral migration, or braiding, such idealized flows give rise to subqueous fans and basin-plain deposits. A single bed, whether deposited on the fan or on the plain, is expected to be size graded, and to show a sequence of structures known as the *Bouma sequence* (after the pioneering work of Bouma 1962; see Figure 1).

Such a simple model of turbidity current behavior has been surprisingly successful in accounting for many observed features of ancient turbidites, but it is now becoming clear that a better understanding is required to account for the many diverse phenomena reported from modern environments and ancient turbidite systems (Shanmugam & Moiola 1991, Hesse & Rakofsky 1992). The mechanisms for turbidity current generation are

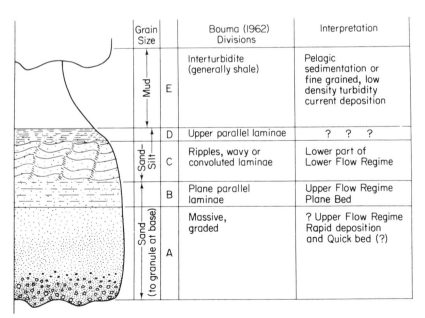

Grain Size		Bouma (1962) Divisions	Interpretation
Mud	E	Interturbidite (generally shale)	Pelagic sedimentation or fine grained, low density turbidity current deposition
Sand–Silt	D	Upper parallel laminae	? ? ?
Sand–Silt	C	Ripples, wavy or convoluted laminae	Lower part of Lower Flow Regime
Sand–Silt	B	Plane parallel laminae	Upper Flow Regime Plane Bed
Sand (to granule at base)	A	Massive, graded	? Upper Flow Regime Rapid deposition and Quick bed (?)

Figure 1   The Bouma sequence.

certainly diverse. Though most turbidity currents are unsteady surges, others are long-lived, more or less steady flows. Turbidity currents are capable of erosion or deposition. They span an immense range of scales from flows that are more than a hundred meters thick, carry cubic kms of sediment, and deposit the "megaturbidites" of Labaume et al (1983; Mutti et al 1984), to thin, dilute flows. They probably vary greatly in sediment concentration from "hyperconcentrations" of more than 40% by weight (Pierson & Costa 1987, Smith 1986), to "high density" turbidity currents (defined, by Kuenen 1966 and Middleton 1970, as those with more than 10% of sediment by weight) to flows with only a few parts per thousand of sediment (see below).

Like other gravity flows turbidity currents display supercritical and subcritical flow regimes, and the transition from the former to the latter can take place only through a submerged hydraulic jump (Middleton 1970, Komar 1971, Hand 1974, Weirich 1988). In rivers, such jumps take place only in small, steep tributaries: Though important in the design of hydraulic structures they have little geomorphological importance. In turbidity current systems, hydraulic jumps must take place at the base of slope, and should produce important—though presently largely unrecognized—geomorphological and sedimentological effects.

Because of the effects of buoyancy, turbidity currents are generally thicker than river flows of corresponding power: This is shown by the large size of turbidity current channels (generally over 10 m deep, even for small systems such as those in modern fjords). Channel levees are meters to tens of meters or more in thickness, and extend for hundreds of meters or more from the channels. Frequent overflow may be partly a consequence of spill from the head which forms at the front of turbidity current surges and is generally at least twice the thickness of the succeeding flow. It must also result in part from the stratification in sediment concentration and size that results from turbulent suspension of sediment, and from mixing of the finer sediment fraction from the current proper to a zone of ambient fluid that is entrained by the flow. Yet another factor is superelevation of the current surface in channel bends—transverse slopes in meandering turbidity currents are much larger than those in rivers.

Turbidity current channels seem to display the full range of channel patterns shown by fluvial channels. Meandering is well developed on some submarine fans (e.g. the Amazon fan channels described by Flood & Damuth 1987). Other large meandering, levee-bounded channels consititute a sedimentary system in themselves, which can develop independent from any submarine fan (Hesse & Rakofsky 1992). Large meandering patterns of flow are also recorded by sole marks and grain orientation in thin, apparently unchannelized turbidites (Parkash & Middleton 1970,

Middleton 1970). Braiding seems to be common on the lower parts of some submarine fans, and within large channels. Though cross-bedding formed by migration of dunes is uncommon in most ancient turbidites, some modern turbidite channels are floored by large dunes (Piper et al 1988, Malinverno et al 1988) and some ancient submarine channels contain trough-cross bedded sandstones, presumably deposited by turbidity currents (Hein 1982).

Pickering & Hiscott (1985) were the first to demonstrate conclusively that some turbidite beds were deposited by a single current that reversed direction within the same basin. This phenomenon, essentially unknown in fluvial systems, has been studied experimentally by Pantin & Leeder (1987) and reported in modern basins by Muck & Underwood (1990) and Kneller et al (1991).

2.3   *Simple Hydraulic Equations*

Gravity flows, like other fluid flows, exhibit laminar and turbulent regimes. Work on laminar flows has been summarized by Harleman (1961) and Huppert (1991). The transition to the turbulent regime takes place at Reynolds numbers of less than 1000 (where the characteristic length is taken as the flow thickness). In nature, therefore, laminar flows are confined to magma chambers. An exception might be small gravity currents with very high concentrations of mud (McCave & Jones 1988). In these, the "fluid" certainly shows non-Newtonian properties, and if it is truly nonturbulent it can no longer be classified as a turbidity current. In this review, it is assumed that the flows are fully turbulent—a condition achieved in most, though not all, experimental investigations.

Based on gravity surge experiments, Middleton (1966b, 1967) divided a typical gravity current into three parts: the head, body, and tail. As first documented by Keulegan (1958) the head of a density surge has a distinct shape and hydraulics, which differ from the region behind the head. The heads of saline gravity flows were subject to very thorough experimental investigation during the 1970s and 1980s, and the results are reported by Simpson (1982, 1987). Recent numerical simulations include those by Crook & Miller (1985), Droegemeier & Wilhelmson (1985), Haase & Smith (1989a,b), and Xu et al (1992). The implications of these observations for sediment deposition from turbidity currents have been discussed by Middleton (1967), Allen (1971a), Komar (1972), and Middleton & Southard (1984).

The mass and momentum balance of the head of a gravity flow differ significantly from those of the body and tail.(See Figures 2 and 3.) In order for the head to advance it must displace the ambient fluid, which is generally at rest (for the case of a gravity flow advancing into a moving

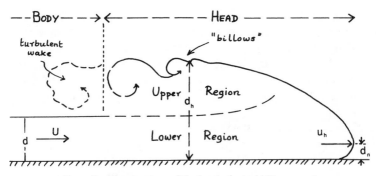

Figure 2   The structure of the head of a turbidity current.

fluid see Bühler et al 1991; for the effect of turbulence in the ambient fluid, see Linden & Simpson 1986; Noh & Fernando 1991b, 1992). Accelerating the ambient fluid produces a resistance to the flow which is larger than friction at the bed or the upper interface. Therefore the head of the current must be thicker (have more gravitational potential energy) than the current behind the head, where only frictional resistance is important. Not that friction is unimportant at the head: The bed friction produces an overhanging "nose," so that the head overrides lighter ambient fluid. This is one way that ambient fluid is mixed into the flow, and results in a gravitational instability that in turn produces the three-dimensional, lobate structure seen at the front of the flow. Though this effect should be less important for large-scale natural flows than it is in the laboratory, Simpson & Britter (1980) showed that a 10% overhang was present even in very large scale atmospheric fronts. Viscous shear at the upper surface of the head leads to the formation of Kelvin-Helmholz "billows" and ultimately to large-scale turbulent mixing at the back of the head.

The back of the head is marked by a sharp discontinuity in flow thickness. The flow pattern is similar to the flow separation and turbulent wake observed in the lee of a blunt (solid) body moving through a stationary fluid. In this case, however, the moving body is fluid, not solid, and mass is constantly lost from it to large eddies that break away from the back of the head. To achieve a constant rate of advance, this fluid must be replaced from the body behind the head, and this implies that the average speed of the flow in the body must be larger than the speed of advance of the head. How much larger depends on the rate of loss of fluid from the head, which is determined mainly by the densiometric Froude number and therefore by the slope (Middleton 1967; and see discussion below).

Figure 3   Experimental turbidity currents. The top photo, taken by Roger Walker in 1964, shows the author admiring one of the flows he produced in the W. M. Keck laboratory at the California Institute of Technology. The bottom photo (courtesy of Roger Walker) shows details of the head of another turbidity current.

The *densiometric Froude number* for a flow of thickness d and depth-averaged velocity U is defined as

$$Fr = \frac{U}{\sqrt{g'd}},\tag{1}$$

where g' is the buoyancy-reduced gravitational acceleration,

$$g' = \frac{(\rho_f - \rho)}{\rho}g,\tag{2}$$

ρ_f is the bulk density of the flow, and ρ is the density of the ambient fluid. The densiometric Froude number of the head itself can be defined as

$$Fr_h = \frac{u_h}{\sqrt{g'd_h}},\tag{3}$$

where u_h is the speed of the head, and d_h is its thickness. Keulegan showed that, for two-dimensional saline heads advancing into a horizontal channel filled with freshwater, this number was a constant with a value of about 0.7–0.8. Subsequent experiments have shown (e.g. Simpson 1982, Hay 1983) that this value is only slightly altered by large increases in scale (Reynolds number), several degrees of bed slope, or by an increase in the depth of ambient fluid, or by replacing the saline solution with a sediment suspension (even a suspension with as much as 40% sediment by volume).

Provided that there is a sufficient volume of suspension released by the initiating event, a region will develop behind the head of a gravity surge in which there is a fairly close approximation to steady, uniform flow. Such flows are also expected if there is a steady discharge of denser fluid supplied from the source (e.g. from a river entering a lake). For the two-dimensional case, a simple force balance applied to a control volume leads to the Chézy-type equation:

$$U = \sqrt{\frac{8g'}{f_o + f_i}}\sqrt{dS},\tag{4}$$

where U is the velocity averaged over the depth d, and f_o and f_i are the friction factors for the bottom and upper interface respectively, defined by the equations

$$\tau_o = \frac{f_o}{4}\frac{\rho_f U^2}{2},\tag{5}$$

$$\tau_i = \frac{f_i}{4}\frac{\rho_f U^2}{2},\tag{6}$$

where τ_o and τ_i are the shear stresses at the bottom and upper interface respectively. Notice that the ratio between the friction coefficients is equal to the ratio between the shear stresses at the bottom and upper interface. Assuming, as a first approximation, that the bulk density is uniform within the flow so that the shear stress varies linearly from the top to the bottom, the ratio between the two friction coefficients is also equal to the ratio of the thicknesses of the part of the flow below the velocity maximum (at which the average shear stress is zero) d_o and the thickness of the part of the flow above the velocity maximum d_i (Figure 4). The influence of slope on the densiometric Froude number is shown by using the Chézy equation for U in the definition

$$Fr = \sqrt{\frac{8S}{f_o + f_i}}.$$ (7)

This illustrates one of the problems in estimating f_i, since experimental evidence indicates that it, in turn, depends on Fr.

Hydraulic equations of the type given above have been widely used to

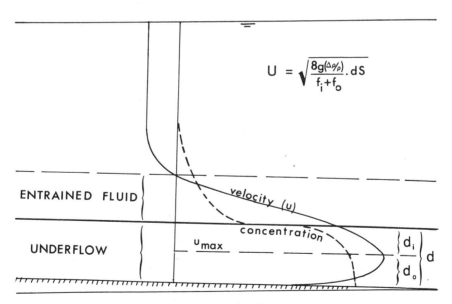

$$U = \sqrt{\frac{8g(\Delta\%)}{f_i + f_o}} \cdot dS$$

Figure 4   Uniform flow of a turbidity current. In this diagram the top of the underflow is considered to be defined by the inflection point in the sediment concentration profile. See text for discussion.

analyze the flow of turbidity currents (e.g. Kersey & Hsü 1976; Komar 1977, 1985; van Tassel 1981; Bowen et al 1984; Zeng et al 1991). Their use is, however, limited by the necessity to make educated guesses not only about the bulk density of the flow (which has generally not been measured) but also about the friction coefficients f_o and f_i. It may be expected that f_o is a function mainly of the bottom roughness, so it can be roughly estimated—though the estimate is more difficult than it is for rivers because of high sediment concentrations (see below). But f_i depends on mixing at the upper interface, and therefore on the hydraulics of the flow. It is certainly some function of the densiometric Froude number (Middleton 1966c, Fernando 1991) and is fairly small for low Fr and becomes large as the critical value is approached. But it is not easy to provide realistic estimates for large-scale flows. It certainly cannot be assumed that it is simply a fraction of the bottom friction, since the resistances in the two cases arise from totally different physical mechanisms and are subject to different hydraulic controls (Parker et al 1987, Normark 1989).

3. MODERN TURBIDITY CURRENTS

From the beginning, geologists have sought information about modern turbidity currents in oceans and lakes in order to understand how ancient turbidites were deposited. At first, observations were mainly of currents in reservoirs such as Lake Mead. They have been followed by a large number of other observations in reservoirs (Fan 1986; Fan & Morris 1992; Chikita 1989, 1990) and lakes (Normark 1989, Lambert et al 1976, Lambert & Giovanoli 1988, Gilbert & Shaw 1981, Smith et al 1982, Weirich 1986).

The existence of catastrophic, large-scale turbidity currents in the oceans was inferred mainly from cable-breaks, submarine topography, and the presence of sand beds with displaced shallow water fauna in deep-sea environments. Attempts to directly measure the properties of turbidity currents by monitoring currents in submarine channels were begun by Shepard in 1968 (see Shepard et al 1979, Inman et al 1976) and have continued to the present (Dengler et al 1984, Reynolds 1987). Most attempts have yielded only imperfect records, due largely to the unpredictability of the larger events and their ability to destroy installed monitoring instruments. A relatively successful program has been carried out in some fjords, where currents are, however, weaker and more continuous than those of most interest to geologists (Hay et al 1982, 1987a,b; Prior et al 1987; Zeng et al 1991; Phillips & Smith 1992).

It is impossible to give a complete summary of the observations in this review. In lakes and reservoirs, most recorded currents have been relatively continuous, extending over several days, relatively slow (less than 1 ms^{-1}), relatively thin (a few meters), and relatively low density (less than 1% by weight of sediment). They carry mainly silt and clay: The sand delivered by the source river is generally deposited in a delta at the head of the lake or reservoir. Because of the low sediment concentration, temperature may be as important in forming gravity flows as suspended sediment.

The turbidity currents in Chinese reservoirs described by Fan (1986) and Fan & Morris (1992) represent an extreme example. Suspended sediment consisted mainly of silt; concentrations were over 10% by weight in the source river, and reached 5% or more in gravity flows. Currents were up to 15 m thick and maximum velocities over 1 ms^{-1} were recorded. Fan (1986, figure 5) presents velocity and density profiles for a flow which traveled 50 km down Sanmenxia Reservoir in 1961. The slope was about 0.00025, the thickness about 2–3 m, the concentration about 6%—and fairly uniform over most of the depth, and the maximum velocity of about 0.6 ms^{-1} was located near the upper interface. This indicates there was little mixing and resistance at the upper interface ($Fr \approx 0.5$.)

Gravity flows measured in fjords are similar to those in reservoirs and lakes. They are pulsed, but may be relatively long-lived flows, though they are mostly generated by slumping at the fjord-head delta, rather than by the direct plunge of muddy inflow below the surface as in most lakes and reservoirs. In Queen Inlet (Phillips & Smith 1992) concentrations and velocities measured one meter above the bed in a 20 m deep channel were low (less than 5 ppt, and 0.1–0.2 ms^{-1} respectively), though there were a few surges with velocities of more than 1 ms^{-1}. Flow thickness is difficult to estimate, but in the July 27 flows it probably reached about 10 m, with the upper part of the flow having lower sediment concentrations (about 1 ppt) than the lowest meter (about 3 ppt). Channel slope was high (0.01), and Fr probably quite high (0.7)—though the significance of the number becomes dubious where there are large variations in concentration with depth. In Bute Inlet (Zeng et al 1991), current velocities were measured 4 m above the bottom, and flow thicknesses were measured roughly by the use of vane deflectors, but sediment concentrations were not measured directly. Speeds measured reached more than 3 ms^{-1} in short surges more than 30 m thick near the source but declined rapidly to less than 1 ms^{-1} 30 km down fjord. From these measurements and estimates of velocity, depth, and slope derived from the morphology of the channels, Zeng et al (1991) estimated average sediment concentrations in the range from nearly 2% by weight to 5 ppt (declining rapidly downflow).

Evidence for more powerful gravity surges in the oceans is provided by displacement of heavy instrument package anchors and a few surviving current meter records of speeds above 3 ms^{-1} (e.g. Dengler at al 1984).

But there are essentially no direct measurements of flow thicknesses, or velocity and sediment concentration profiles from such strong flows. Cable-breaks give evidence of head velocities reaching at least 19 ms^{-1} (Piper et al 1988).

4.   INFERENCES FROM TURBIDITES

An alternative to measuring the properties of turbidity currents in models or in nature is to infer their behavior from the beds they deposit (*turbidites*). This has been attempted since the concept of turbidity currents was first used to explain ancient sandstone beds. A review of the early studies was given by Walker (1973), and more recent work is summarized by Allen (1982, vol. 2) and Pickering et al (1989).

An early problem was to distingish turbidites from beds deposited from other types of flows, including other types of gravity flows. This problem is still not completely solved: Observations on ancient deep-sea deposits indicate that some must have been deposited by other mechanisms, such as submarine debris flows, but that there are transitions between deposits that can safely be ascribed to deposition from turbidity currents and those possibly deposited by other types of flow. Transitions in sedimentary facies are often assumed to correspond to transitions in flow types. Middleton & Hampton (1973, 1976) and Lowe (1979, 1982) have stressed that the textures and structures observed in the deposit, formed during the final stages of flow or even shortly after deposition, do not necessarily tell us much about the nature of the flow while it was actively transporting sediment. The same problems exist for subaerial flows (Pierson & Costa 1987, Smith 1986).

The existence of a single model of the vertical sequence of textures and structures found in an "ideal" turbidite bed (the *Bouma sequence*) has tempted many geologists to propose an "ideal" sequence of hydraulic events leading to the deposition of that sequence. Very few turbidites show the complete Bouma sequence (Figure 1), however, and most fall into one of at least two categories: *proximal* or *distal* turbidites. Proximal turbidites show an erosional base, followed by a "massive" (Bouma A) division. The A division may pass up gradually into faint parallel lamination (the Bouma B division), which in turn may be capped by a weak development of current ripples (the Bouma C division). Distal turbidites show an erosional base, followed by development of distinct plane lamination (B division) and/or ripple cross-lamination (see Pickering et al 1989, for the many possible variations on this theme). Convolute lamination, originally recognized by Bouma (1962) as an alternate criterion for the C division, is formed by a gravitational instability acting on rapidly deposited sediment, and may be developed from either a plane bed or ripples. It should no longer be used to define the structural divisions.

The importance of not lumping proximal and distal turbidite types together is readily revealed by making inferences from the structures about rates of erosion or deposition. The structures were formed as different parts of a surge (head, body, tail) pass over a bed location. Many turbidites, both proximal and distal, have a sharp erosive base indicating that the head of the flow is not generally a region of sediment deposition. In some turbidites it can be inferred, either from the erosive marks themselves (Allen 1971d) or from the evidence of trace fossils preserved on the sole (Seilacher 1962, Wetzel & Aigner 1986), that several centimeters of erosion was involved, in others that there was time only for the removal of the least consolidated muddy material.

In proximal turbidites deposition of a massive interval that passes up gradationally into poorly developed plane lamination is most plausibly interpreted as due to a sudden change from erosion, to deposition from suspension which was so rapid that it did not permit sufficient grain movement by traction on the bed to form visible lamination. Strong experimental evidence for this interpretation has been described by Arnott & Hand (1989) and discussed by Allen (1991), and includes the experimental production not only of massive beds, but also of the high-angle grain imbrication characteristic of the A division of proximal turbidites.

In many distal turbidites, however, there is strong evidence that, after a period of bed erosion, deposition began slowly, with a relatively extended period of grain traction, and then accelerated with an increasing rate of sedimentation from suspension towards the upper part of the bed. The evidence consists in an upward change in the types of structures (well-developed plane lamination, to ripple cross-lamination showing stoss-side erosion, to ripple cross-lamination showing preservation of stoss-side lamination) and in a common upward increase in the rate of climb of ripple drift cross-lamination (Walker 1969; Allen 1971b,c; Ashley et al 1982). Flutes at the base of distal turbidites frequently trap grains that are much coarser than those in the rest of the bed. This indicates that the head of the surge was not only capable of erosion but also able to carry much coarser sediment than the body of the current. As recognized by Kuenen long ago, such evidence indicates that far-traveled turbidity currents have a well-developed longitudinal hydraulic structure. The coarsest sediment is either deposited upstream, or transfered from the body into the head of the flow. It is then recycled, either within the head itself, or by dropping out of eddies torn off the rear of the head, and falling into the faster moving body of the flow, where it will be carried once again into the head. It is thus possible for some coarse grains to bypass more proximal regions.

It is surely to be expected that different types of turbidity currents will deposit sediment in different ways in different parts of the same sedimentary basin. A single hydraulic model can have only limited utility.

5. EXPERIMENTAL STUDIES

In contrast to the large number of experimental studies of gravity flow hydraulics, and to the rapidly increasing number of theoretical studies of sediment deposition (see next section) there have been relatively few experimental studies of sediment deposition from turbidity currents. Since the work of Middleton (1967), the following constitutes a complete list: Riddell (1969), Tesaker (1969), Fan (1980), Lüthi (1980a,b, 1981), Pallesen (1983), Ravenne & Beghin (1983), Siegenthaler et al (1984), Siegenthaler & Bühler (1985, 1986), Bühler & Siegenthaler (1986), Parker et al (1987), Garcia & Parker (1988,1989), Garcia (1990), Scheiwiller (1986), Scheiwiller et al (1987), Laval et al (1988), Middleton & Neal (1989), Norem et al (1990), Altinakar et al (1990), Duringer et al (1991), and Kerr (1991). To this list should be added some studies of sediment deposition from fluid gravity flows that were made specifically to mimic deposition of sediment from turbidity currents. The most important are those by Ashley et al (1982) and Southard & Mackintosh (1981).

The scaling of laboratory experiments was discussed by Middleton (1966a) and Middleton & Neal (1989). Besides ensuring that the current is fully turbulent, the most important dimensionless numbers are the densiometric Froude number, and the ratio between the sediment settling velocity w and some characteristic velocity of the flow (generally U or u_h). As large turbidity currents generally have speeds in the range of 1–10 ms^{-1} and laboratory flows only reach about 0.1 ms^{-1}, this means that settling velocity should be reduced in the model by a factor of 10 to 100. To model a sand with a settling velocity of 5 cms^{-1} one should use a silt—but this raises several problems: Such sediments may exhibit cohesion, beds composed of silt are hydraulically smooth—not rough like beds of sand, and deposits produced by the experimental flows are very thin and difficult to analyze. To overcome this problem, Middleton (1967), Middleton & Neal (1989), Riddell (1969), and Garcia (1990) used sediment of sand size but reduced density (plastic or coal). Unfortunately, this also reduces the bulk density of the flow, an effect that can only be offset by raising the volume concentration. Laval et al (1988) tried mixed experiments where the density contrast was produced using dissolved salt, but sand was added to the flow to act as a tracer. Garcia & Parker (1988) studied erosion of a sand or silt bed by a saline flow.

Middleton (1967), Riddell (1969), and Middleton & Neal (1989) studied deposition of sand from surges released from a lock into a horizontal channel. One surprising aspect of these flows is that they deposit a bed of sediment that is at first almost uniform in thickness, and wedges out only where the flow nears the end of its run out. Middleton & Neal (1989) found that increasing the volume concentration increased both thickness and length of the bed in about the same proportion, and the data for several grain sizes and two different densities could be collapsed onto a single line by using the ratio between the bed thickness and the cube root of the initial volume of suspension. For unit width and 20% volume concentration the equation is:

$$\log(w/u_h) = 1.75 + 2.0 \log[t/(LH)^{1/2}], \tag{8}$$

where L and H are the length and height of the lock from which the suspension was released. In words, the result is that the dimensionless bed thickness is proportional to the square root of the dimensionless settling velocity. Extrapolation of this relationship to much larger scales seems possible: Turbidite beds on abyssal plains (which have slopes of less than one in a thousand) have remarkably constant thickness, and application of the experimental equation yields estimates of current speed in the right range. But the result needs much more testing and is poorly understood theoretically.

The studies by Ravenne & Beghin (1983), Laval et al (1988), Lüthi (1980a,b, 1981), Garcia & Parker (1989), Duringer et al (1991), and Kerr (1991) were exploratory in nature. Laval et al (1988) studied "pure" surges (produced by using a large H/L ratio), which decelerate rapidly. Only Hauenstein & Dracos (1984), Beghin & Olagne (1991), and Lüthi have made large-scale experiments on any kind of gravity flow in which the flow was allowed to expand in the transverse direction as it flowed down a slope (in other words, experiments on a negatively buoyant wall-jet.) Hauenstein & Dracos (1984) reported that steady discharge saline inflows, on slopes less than 15°, did not show a constant angle of spreading, though a constant angle of 30° was observed by Beghin & Olagne (1991) for saline surges released onto slopes greater than 15°. In experiments using quartz silts, Lüthi (1991) found that dilution accounted for most of the density decrease of the flow, with sediment deposition being only of secondary importance. Spreading, combined with dilution by mixing, led to rapid deceleration of the flows even down relatively steep slopes and to a nearly exponential decrease in bed thickness with distance from the source.

The experiments by Altinakar et al (1990) and Garcia (1990) were concerned mainly with near-uniform flow of turbidity currents down slopes. Together with the older studies of Michon et al (1955) they provide most of the measurements of velocity and sediment concentration profiles available for such flows. Profiles were also measured by Scheiwiller (1986), but only for flows on slopes higher than 30°. Discussion of these results is deferred to the next section.

6. THEORY

6.1 *Introduction*

Most of the theoretical discussion of turbidity currents has addressed two-dimensional, subcritical, nearly-uniform flow down a slope. The basic problem is how the suspended sediment interacts with the flow, changing the turbulent structure of the flow, and therefore the mixing with ambient fluid at the upper interface and the velocity and sediment concentration profiles in the flow. These in turn determine whether the flow will grow in volume, due to incorporation of ambient fluid, and whether the sediment concentration will increase or decrease, due to erosion from or deposition on the bed. If an initial cloud of suspended sediment on a slope loses sediment and becomes more dilute, then it may either reach some quasi-equilibrium, or it may be dissipated completely (flows in the "subsiding" field of Pantin 1979; flows that fail to "ignite" in the analysis by Parker 1982). Alternatively, the flow may accelerate and erode more sediment from the bed (i.e. it may be in the "exploding" field of Pantin 1979, or it may "ignite" in the analysis of Parker 1982) until it reaches some quasi-equilibrium state where flow volume and sediment concentration change only very slowly, if at all, so long as the slope remains constant.

Models of such flows fall into two major categories: those that use vertical averaging techniques, so that at each position along the travel path, the flow is characterized by a single velocity and density (sediment concentration), and those that attempt to derive the velocity and density profiles from a turbulence model.

The first category consists of models which expand the simple Chézy-type equation (Section 2.3) to flows where the slope, depth, and other parameters vary slowly down flow. The way in which the parameters vary is a consequence either of conservation equations or of the assumptions of the model. There are numerous examples including Johnson (1962), Hopfinger & Tochon-Danguy (1977; applied to the Oahu turbidity current by Dengler & Wilde 1987), Chu et al (1979), Fan (1980), Siegenthaler & Bühler (1985), Siegenthaler & Bühler (1986), Kirwan et al (1986), Caserta et al (1990), Norem et al (1990), and the models of Pantin and Parker referenced above. The models can be classified by the number of equations involved: for example the Kirwan et al model uses a single, conservation of momentum equation, and the Johnson and Parker et al (1987) models use four equations, for conservation of momentum, sediment mass, fluid mass, and energy. Also some models assume strongly non-Newtonian materials (Dengler & Wilde 1987, Norem et al 1990) so are more appropriate for submarine debris flows than for turbidity currents.

The second category uses models of turbulent suspension and mixing to predict the evolution of velocity and density profiles down the flow path, from some assumed initial condition. These models require use of turbulence models, either implicitly or explicitly. Turbulence models make use of the time-averaged equations of motion of a turbulent fluid. These include the Reynolds stress terms, which cannot be predicted from the Navier-Stokes equations (thus giving rise to the turbulence "closure problem"). A large variety of possible models has been developed that attempt to predict these terms for practical use. The hydraulics applications have been reviewed by Rodi (1984). One simple model, proposed by Gelfenbaum (1988), is discussed below. Another model was developed by Stacey & Bowen (1988a,b). A somewhat more complex model, the k-ε model, uses equations for the total turbulent energy k, and its rate of dissipation ε to obtain closure. This model has been modified to allow for density stratification and applied to turbidity currents by Scheiwiller et al (1987), Kupusović (1989), and Eidsvik & Brørs (1989).

All of these are two-dimensional *forward models*, i.e. they predict the evolution of a turbidity current as it flows in a channel down a varying slope, assuming a (generally constant) specified source of suspension. They do not directly address the main problem of interest to geologists: how to reconstruct the properties of the current from the observed properties of the bed that it deposits. It is possible to calculate some properties of the bed produced from most of the models, though none of the authors has actually gone through this exercise, but to simulate a given bed would require multiple iterations of the model, and the result would probably not be unique. One advantage of the equation of Middleton & Neal (1989) is that it can be used to infer the current speed directly from the measured volume, size distribution, and thickness of a given bed (provided it was deposited on a flat, confined basin floor).

For discussion, three aspects of the problem may be distinguished: (*a*) the autosuspension criterion; (*b*) velocity and concentration profiles in the lower part of the flow, and sediment exchange with the bed; and (*c*) velocity and concentration profiles in the upper part of the flow, and mixing at the upper interface.

6.2 *Autosuspension*

Bagnold (1962) argued that there should exist a condition in which the input of gravitational energy is just sufficient to sustain the turbulence necessary to keep the sediment in suspension and to overcome friction at the upper and lower interface. As it is the suspended sediment that supplies the gravitational energy to drive the flow and create the turbulence, it may be said that the sediment "suspends itself," hence the term *autosuspension*. From a simple energy balance Bagnold (1962) derived the necessary condition to be

$$w \leqslant SU. \tag{9}$$

Pantin (1979), Parker (1982), Seymour (1986), and Stacey & Bowen (1988b, 1990) proposed modifying this criterion to

$$w \leqslant eSU, \tag{10}$$

where e is an efficiency coefficient, less than unity, calculated in various ways by different authors. Southard & Mackintosh (1981) devised an experimental test for Bagnold's original criterion and could find no evidence to support it: There is a disagreement between them and other authors whether this was because their test was not sensitive enough to detect the reduced effect of a small value of e in Equation (10) or because the whole concept of autosuspension is based on a faulty concept of suspension energetics (Paola & Southard 1983).

If an autosuspension criterion does exist, field evidence examined by Seymour (1986) suggests $e < 0.1$. Unfortunately the interpretation of the scanty field data is not at all clear. In the oceans, large turbidity currents certainly travel on low slopes for hundreds of kilometers without depositing all their sediment. One may argue that this is evidence for at least a near-approach to the autosuspension condition. On the other hand, there is evidence that even large flows deposit sediment along their complete travel-path (e.g. Pilkey et al 1980, Piper et al 1988). For this to be consistent with autosuspension, one must suppose that, in regions of near constant slope, sediment deposition took place only from the tail of the flow.

6.3   *Lower Part of the Flow*

The simplest theory of sediment suspension is the diffusion theory of Rouse, based on the mixing-length theory for velocity distribution in a turbulent boundary layer (for a review see Middleton & Southard 1984). This theory assumes that the velocity profile is independent of the sediment concentration—an assumption long known to be incorrect. The problem is bad enough in rivers, where suspended sediment concentrations are generally low, but becomes critical for turbidity currents. There is a large literature which cannot be discussed here: Some recent contributors are Smith & McLean (1977), Noh & Fernando (1991a), Villaret & Trowbridge (1991), McLean (1992), and Umeyama & Gerritsen (1992).

Perhaps the simplest solution is that proposed by Smith & McLean

(1977) and tested by Gelfenbaum & Smith (1986) and again by Villaret & Trowbridge (1991) using a larger data set. These authors use a depth-dependent turbulent diffusion coefficient, as does the classical theory, but modified by the stable stratification induced by the suspended sediment profile and calculated by iteration (Villaret & Trowbridge provide direct approximation methods.)

In a slug of suspended sediment starting to move down a slope some particles will settle to the bed. The critical question is then whether or not they will remain on the bed or be eroded and taken back into suspension. The problem of entrainment of bed material into suspension (rather than into tractive motion, which is the usual "initiation of motion" problem in sediment transport) was studied by Akiyama & Fukushima (1985) and their results were used to construct a theory of turbidity currents by Akiyama & Stefan (1985) and Parker et al (1987). In these theories, as in many other theories of turbidity currents, the velocity and sediment concentration in the model is averaged over the depth, rather than derived explicitly. Parker et al (1987; see also Akiyama & Stefan 1985) argued that three equation models, based on depth-averaged conservation equations for fluid mass, sediment mass, and momentum, were inadequate because they did not take into account the turbulent energy balance. They proposed adding a fourth equation to link the sediment entrainment to the turbulent energy k rather than to the mean velocity of the flow.

6.4   Upper Part of the Flow

A steady gravity current flowing down a uniform slope mixes with the ambient fluid at its upper interface. This raises problems in defining exactly what constitutes the current itself: It cannot be all the fluid moving down the slope, because some of the overlying fluid is dragged along by the flow, even though it has nearly the ambient density. Mixing at the interface increases the fluid discharge downslope, so that the volume (and thus the depth) of the flow gradually increases in that direction—therefore it is not possible to have a strictly uniform gravity flow (Stacey & Bowen 1988a), and the flow depth cannot reasonably be calculated from the upslope discharge and the velocity averaged up from the bed. Most (but not all) authors have measured the depth from the bed to the inflection point of the concentration profile (Figure 4). A few authors prefer to include all of the profile that differs significantly in density from the ambient fluid: This implies by definition that mixing goes only one way. Ambient fluid may be mixed into the gravity current, but denser fluid cannot be mixed into the ambient fluid.

Factors governing the rate of mixing have been reviewed by Fernando (1991). In geophysical flows, there is always mixing, therefore no sharp interface between the denser and ambient fluid. Mixing across the zone where both the density ρ and the velocity u are functions of elevation z is therefore better related to the *gradient Richardson number*

$$Ri = \frac{-g}{\rho} \frac{(\partial\rho/\partial z)}{(\partial u/\partial z)^2} \tag{11}$$

than to the densiometric Froude number [in the case of a sharp interface, the Richardson number can be defined as $Ri = (1/Fr)^2$ (Turner 1979)]. The critical condition for stability is $Ri > 0.25$ (Miles 1990). Gelfenbaum (1988), using the results of Geyer & Smith (1987), proposed that mixing processes maintain a constant Richardson number only slightly larger than critical, and therefore a linear velocity and concentration profile in the upper part of the flow. The thickness of the upper layer can then be calculated from the critical value of the Richardson number and the velocity and sediment concentration at the velocity maximum (the top of the lower layer). The entrainment of ambient fluid into the flow, and its downslope evolution (including erosion or deposition of sediment), can be calculated by using the equations of motion, and the velocity and sediment concentrations for the lower part of the flow derived from the stratification-corrected diffusion theory, There is, however, still no generally accepted theory of entrainment in stratified shear flows (Fernando 1991).

7.　CONCLUSIONS

Turbidity currents are very complex hydraulic phenomena. Those of greatest interest to geologists are large-scale catastrophic events which are extremely difficult to observe and measure in nature. Though small-scale models have been used successfully to explore some aspects of gravity flow hydraulics, there are serious limitations on their use to reproduce sedimentation from such flows directly. The complexity of the interaction between sediment suspension, turbulence, and mixing at the top of the flow extends present turbulence and numerical models up to and beyond their known limits. It seems that progress in the future will depend upon more detailed monitoring of natural flows at intermediate scales (in lakes and fjords), renewed experimentation with sediment-bearing flows at larger laboratory scales than generally attempted in the past, and further application of forward models which attempt to predict velocity and sediment concentration profiles, rather than use depth-averaging techniques. Verification of model assumptions is generally attempted only piece-by-piece, and it will be very difficult to obtain field data that is sufficiently complete to allow testing of the whole model. These models must not only be verified in large-scale experiments, but also used to predict turbidite size distribution and bed-thickness. Such predictions, though not yet attempted, could be made using several existing models.

A fundamental problem that still remains concerns the general nature of turbidity currents: Can a cloud of suspended sediment achieve some sort of equilibrium state (autosuspension), where it moves downslope without either eroding or depositing sediment? Or, are suspension clouds, once created, doomed to rapid dissipation by spreading, mixing with ambient fluid, and sediment deposition?

Though large uncertainties remain, remarkable progress has been made in the tracing and interpretation of ancient turbidites, and in understanding the basic hydraulics of gravity currents. The problems that remain, though difficult, have been vigorously attacked in the past ten years by workers trained in several different disciplines (geology, oceanography, meteorology, fluid mechanics) using techniques unknown to an earlier generation. The future seems hopeful.

ACKNOWLEDGMENTS

I thank Marcelo Garcia, Donald Lowe, Gary Parker, and Jianjun Zeng for their comments on an earlier version of the manuscript. This review was written while I was on sabbatical leave at the Department of Geological Sciences, University of Washington. My research on turbidity currents has been supported by grants from the National Science and Engineering Research Council of Canada.

Literature Cited

Akiyama. J., Fukushima, Y. 1985. Entrainment of noncohesive bed sediment into suspension. *Third Int. Symp. on River Sedimentation, Univ. Miss.*, pp. 804–13

Akiyama, J., Stefan, H. 1985. Turbidity current with erosion and deposition. *J. Hydraul. Eng.* 111: 1473–96

Allen, J. R. L. 1971a. Mixing at turbidity current heads, and its geological implications. *J. Sediment. Petrol.* 41: 97–113

Allen, J. R. L. 1971b. A theoretical and experimental study of climbing ripple cross-stratification, with a field application to the Uppsala Esker. *Geogr. Ann.* A53: 157–87

Allen, J. R. L. 1971c. Instantaneous sediment deposition rates deduced from climbing-ripple cross-lamination. *J. Geol. Soc. London* 127: 553–61

Allen, J. R. L. 1971d. Transverse erosional marks of mud and rock: their physical basis and geological significance. *Sediment. Geol.* 5: 167–384

Allen, J. R. L. 1982. *Sedimentary Structures: Their Character and Physical Basis.* New York: Elsevier Co. 2 Vol., 593 pp. and 663 pp.

Allen, J. R. L. 1991. The Bouma division A and the possible duration of turbidity currents. *J. Sediment. Petrol.* 61: 291–95

Altinakar, S., Graf, W. H., Hopfinger, E. J. 1990. Weakly depositing turbidity current on a small slope. *J. Hydraul. Res.* 28: 55–80

Arnott, R. W. C., Hand, B. M. 1989. Bedforms, primary structures and grain fabric in the presence of sediment rain. *J. Sediment. Petrol.* 59: 1062–69

Ashley, G. M., Southard, J. B., Boothroyd, J. C. 1982. Deposition of climbing-ripple beds: a flume simulation. *Sedimentology* 29: 67–79

110 MIDDLETON

Bagnold, R. A. 1962. Auto-suspension of transported sediment: turbidity currents. *Proc. R. Soc. London* 265A: 315–19

Beghin, P., Olagne, X. 1991. Experimental and theoretical study of the dynamics of powder snow avalanches. *Cold Reg. Sci. Technol.* 19: 317–26

Bell, H. S. 1942. Density currents as agents for transporting sediments. *J. Geol.* 50: 512–47

Bouma, A. H. 1962. *Sedimentology of Some Flysch Deposits: A Graphic Approach to Facies Interpretation.* New York: Elsevier. 168 pp.

Bourgeois, J., 1990. Kuenen, Philip Henry. In *Dictionary of Scientific Biography*, ed. F. L. Holmes, 17(Suppl. II): 509–14. New York: Scribener's

Bowen, A. J., Normark, W. R., Piper, D. J. P. 1984. Modelling of turbidity currents on Navy Submarine Fan, California Continental Borderland. *Sedimentology* 31: 169–85

Bühler, J., Wright, S. J., Kim, Y. 1991. Gravity current advancing into a flowing fluid. *J. Hydraul. Res.* 29: 243–57

Campbell, C. S. 1990. Rapid granular flows. *Annu. Rev. Fluid Mech.* 22: 57–92

Caserta, A., Mieli, E., Salusti, E. 1990. On a model of bottom erosion by dense water steady veins. *Geophys. Astrophys. Fluid Dyn.* 55: 117–35

Chikita, K. 1989. A field study on turbidity currents initiated from spring runoffs. *Water Resources Res.* 25: 257–71

Chikita, K. 1990. Sedimentation by river-induced turbidity currents: field measurements and interpretation. *Sedimentology* 37: 891–905

Chu, F. H., Pilkey, W. D., Pilkey, O. H. 1979. An analytical study of turbidity current steady flow. *Mar. Geol.* 33: 205–20

Crook, N. A., Miller, M. J. 1985. A numerical and analytical study of atmospheric undular bores. *Q. J. R. Meteorol. Soc.* 111: 225–42

Daly, R. A. 1936. Origin of submarine canyons. *Am. J. Sci.* 31: 410–20

Dengler, A. T., Wilde, P., Noda, E. K., Normark, W. R. 1984. Turbidity currents generated by Hurricane Iwa. *Geo-Mar. Lett.* 4: 5–11

Dengler, A. T., Wilde, P. 1987. Turbidity currents on steep slopes: application of an avalanche-type numeric model for ocean thermal energy conversion design. *Ocean Eng.* 14: 409–33

Droegemeier, K. K., Wilhelmson, R. B. 1985. Kelvin-Helmholtz instability in a numerically simulated thunderstorm outflow. *14th Conf. on Severe Local Storms*, Indianapolis, Ind., pp. 151–54

Duringer, P., Paicheler, J. C., Schneider, J. L. 1991. Un courant d'eau continu peut-il générer des turbidites? Résultats d'expérimentations analogiques. *Mar. Geol.* 99: 231–46

Eidsvik, K. J., Brørs, B. 1989. Self-accelerating turbidity current prediction based upon $(k-\varepsilon)$ turbulence. *Cont. Shelf Res.* 9: 617–27

Fan, J. 1980. Analysis of the sediment deposition in density currents. *Scientia Sinica* 23(4): 526–38

Fan, J. 1986. Turbid density currents in reservoirs. *Water Int.* 11: 107–16

Fan, J., Morris, G. L. 1992. Reservoir sedimentation. I: Delta and density current deposits. *J. Hydraul. Eng.* 118(3): 354–69.

Fernando, Harindra J. S. 1991. Turbulent mixing in stratified fluids. *Annu. Rev. Fluid Mech.* 23: 455–91

Flood, R. D., Damuth, J. E. 1987. Quantitative characteristics of sinuous distributary channels on the Amazon deep-sea fan. *Geol. Soc. Am. Bull.* 98: 728–38

Garcia, M. H. 1990. Depositing and eroding sediment-driven flows: turbidity currents. *Univ. Minn., St. Anthony Falls Hydraul. Lab. Proj. Rep. No.* 306. 179 pp.+2 Append.

Garcia, M., Parker, G. 1988. Entrainment of bed sediment by density underflows. In *Hydraulic Engineering, Proc. 1988 Natl. Conf., Am. Soc. Civil Eng.*, ed. S. R. Abt, J. Gessler, pp. 270–75

Garcia, M., Parker, G. 1989. Experiments on hydraulic jumps in turbidity currents near a canyon-fan transition. *Science* 245: 393–96

Gelfenbaum, G., Smith, J. D. 1986. Experimental evaluation of a generalized suspended-sediment transport theory. In *Shelf Sand and Sandstones, Can. Soc. Petroleum Geol. Mem.*, ed. J. R. Knight, 2: 133–44

Gelfenbaum, G. R., 1988. *Mechanics of Steady Turbidity Currents.* PhD thesis. Univ. Wash., Seattle. 137 pp.

Geyer, W. R., Smith, J. D. 1987. Shear instability in a highly stratified estuary. *J. Phys. Oceanogr.* 17: 1668–79

Gilbert, R., Shaw, J. 1981. Sedimentation in proglacial Sunwapta lake, Alberta. *Can. J. Earth Sci.* 18: 81–93

Haase, S. P., Smith, R. K. 1989a. The numerical simulation of atmospheric gravity current environments. Part I: Neutrally-stable environments. *Geophys. Astrophys. Fluid Dyn.* 46: 1–33

Haase, Sabine P., Smith, Roger K. 1989b. The numerical simulation of atmospheric gravity currents. Part II: Environments with stable layers. *Geophys. Astrophys. Fluid Dyn.* 46: 35–51

Hand, B. M. 1974. Supercritical flow in turbidites. *J. Sediment. Petrol.* 44: 637–48

Harleman, D. R. F. 1961. Stratified flow. In

Handbook of Fluid Dynamics, ed. V. L. Streeter, Chap. 26. New York: McGraw Hill

Hauenstein, W., Dracos, Th. 1984. Investigation of plunging density currents generated by inflows in lakes. *J. Hydraul. Res.* 22: 157–79

Hay, A. E. 1983. On the frontal speeds of internal gravity surges on sloping boundaries. *J. Geophys. Res.* 88(C1): 751–54

Hay, A. E. 1987a. Turbidity currents and submarine channel formation in Rupert Inlet, British Columbia, 1. Surge observations. *J. Geophys. Res.* 92(C3): 2875–81

Hay, A. E. 1987b. Turbidity currents and submarine channel formation in Rupert Inlet, British Columbia, 2. The roles of continuous and surge-type flow. *J. Geophys. Res.* 92(C3): 2883–900

Hay, A. E., Burling, R. W., Murray, J. W. 1982. Remote acoustic detection of a turbidity current surge. *Science* 217: 833–35

Hein, F. J. 1982. Depositional mechanisms of deep-sea coarse clastic sediments, Cap Enragé Formation, Quebec. *Can. J. Earth Sci.* 19: 267–87

Hesse, R., Rakofsky, A. 1992. Deep-sea channel/submarine-yazoo system of the Labrador Sea: a new deep-water facies model. *Am. Assoc. Petroleum Geol. Bull.* 76: 680–707

Hopfinger, E. J. 1983. Snow avalanche motion and related phenomena. *Annu. Rev. Fluid Mech.* 15: 47–76

Hopfinger, E. J., Tochon-Danguy, J.-C. 1977. A model study of powder-snow avalanches. *J. Glaciol.* 19: 343–56

Huppert, H. E. 1991. Buoyancy-driven motions in particle-laden fluids. In *Of Fluid Mechanics and Related Matters*, Proc. of a Symp. Honouring John Miles on his Seventieth Birthday, Scripps Inst. Oceanogr. Ref. Ser. 91–24, ed. R. Salmon, D. Betts, pp. 141–59.

Inman, D. L., Nordstrom, C. E., Flick, R. E. 1976. Currents in submarine canyons: an air-sea-land interaction. *Annu. Rev. Fluid Mech.* 8: 275–310

Johnson, M. A. 1962. Turbidity currents. *Science Prog. (London)* 50: 257–73

Kerr, R. C. 1991. Erosion of a stable density gradient by sedimentation-driven convection. *Nature* 353: 423–25

Kersey, D. G., Hsü, K. J. 1976. Energy relations of density-current flow: an experimental investigation. *Sedimentology* 23: 761–89

Keulegan, Garbis H. 1958. Twelfth progress report on model laws for density currents: the motion of saline fronts in still water. *U.S. Natl. Bur. Stand. Rep.* 5831, 29 pp.

Kirwan, A. D. Jr., Doyle, L. J., Bowles, W. D., Brooks, G. R. 1986. Time-dependent hydrodynamic models of turbidity currents analysed with data from the Grand Banks and Orleansville events. *J. Sediment. Petrol.* 56: 379–86

Kneller, B., Edwards, D., McCaffrey, W., Moore, R. 1991. Oblique reflection of turbidity currents. *Geology* 14: 250–52

Komar, P. D. 1971. Hydraulic jumps in turbidity currents. *Geol. Soc. Am. Bull.* 82: 1477–88

Komar, P. D. 1972. Relative significance of head and body spill from a channellized turbidity current. *Geol. Soc. Am. Bull.* 83: 1151–56

Komar, P. D. 1977. Computer simulation of turbidity current flow and the study of deep-sea channels and fan sedimentation. In *The Sea, Ideas and Observation on Progress in the Study of the Seas*: 6, *Marine Modelling*, ed. E. D. Goldberg, I. N. McCave, J. J. O'Brien, J. H. Steele, pp. 603–21

Komar, P. D. 1985. The hydraulic interpretation of turbidites from their grain sizes and sedimentary structures. *Sedimentology* 32: 395–407

Kuenen, P. H. 1966. Matrix of turbidites: experimental approach. *Sedimentology* 7: 267–97

Kupusović, T. 1989. A two dimensional model of turbulent flow applied to density currents. In *Computational Modelling and Experimental Methods in Hydraulics* (*Hydrocomp '89*), ed. Ć. Maksimović, M. Radojković, pp. 169–78. New York: Elsevier

Labaume, P. , Mutti, E., Séguret, M., Roselle, J. 1983. Mégaturbidites carbonatées du bassin turbiditiques de l'Eocène inférieur de moyen sud-pyrénéen. *Bull. Soc. Géol. France Ser.* 7 25: 927–41

Lambert, A. M., Kelts, K. R., Marshall, N. F. 1976. Measurements of density underflows from Walensee, Switzerland. *Sedimentology* 23: 87–105

Lambert, A. M., Giovanoli, F. 1988. Records of riverborne turbidity currents and indications of slope failures in the Rhone delta of Lake Geneva. *Limnol. Oceanogr.* 33: 458–68

Laval, A., Cremer, M., Beghin, P., Ravenne, C. 1988. Density surges: two dimensional experiments. *Sedimentology* 35: 73–84

Linden, P. F., Simpson, J. E. 1986. Gravity-driven flows in a turbulent fluid. *J. Fluid Mech.* 172: 481–97

Lowe, D. R. 1979. Sediment gravity flows: their classification and some problems of application to natural flows and deposits. In *Geology of Continental Slopes*, SEPM Spec. Publ. 27, ed. L. J. Doyle, O. H. Pilkey Jr., pp. 75–82

Lowe, D. R. 1982. Sediment gravity flows, II. Depositional models with special reference to the deposits of high-density turbidity currents. *J. Sediment. Petrol.* 52: 279–97

Lüthi, S. 1980a. Die Eigenschaften nicht-kanalisierter· Trübeströme: Eine experimentelle Untersuchung. *Eclogae Geol. Helv.* 73: 881–904

Lüthi, S. 1980b. Some new aspects of two-dimensional turbidity currents. *Sedimentology* 28: 97–105

Lüthi, S. 1981. Experiments on non-channelized turbidity currents and their deposits. *Mar. Geol.* 40: M59–M68

Malinverno, A. , Ryan, W. B. F., Auffret, G., Pautot, G. 1988. Sonar images of the path of recent failure events on the continental margin off Nice, France. In *Sedimentologic Consequences of Convulsive Geologic Events*, ed. H. E. Clifton, *Geol. Soc. Am. Spec. Pap.* 229: 59–75

McCave, I. N., Jones, K. P. N. 1988. Deposition of ungraded muds from high-density non-turbulent turbidity currents. *Nature* 333: 250–52

McLean, S. R. 1992. On the calculation of suspended load for noncohesive sediments. *J. Geophys. Res.* 97(C4): 5759–70

Michon, X., Goddet, J., Bonnefille, R. 1955. *Etude Théoritique et Expérimentale des Courants de Densité.* Chatou, France: Lab. Natl. Hydraul. 2 vol.

Middleton, G. V. 1966a. Small scale models of turbidity currents and the criterion for auto-suspension. *J. Sediment. Petrol.* 36: 202–8

Middleton, G. V. 1966b. Experiments on density and turbidity currents, I. Motion of the head. *Can. J. Earth Sci.* 3: 523–46

Middleton, G. V. 1966c. Experiments on density and turbidity currents, II. Uniform flow of density currents. *Can. J. Earth Sci.* 3: 627–37

Middleton, G. V. 1967. Experiments on density and turbidity currents, III. Deposition of sediment. *Can. J. Earth Sci.* 4: 475–505

Middleton, G. V. 1970. Experimental studies related to problems of flysch sedimentation. In *Flysch Sedimentology in North America*, ed. J. Lajoie, *Geol. Assoc. Can. Spec. Pap.* 7: 253–72

Middleton, G. V., Hampton, M. A. 1973. Part I. Sediment gravity flows: mechanics of flow and deposition. In *Turbidites and Deep Water Sedimentation*, ed. G. V. Middleton, A. H. Bouma, *SEPM Pac. Sec. Short Course Notes.* Anaheim, Calif: SEPM. 38p.

Middleton, G. V., Hampton, M. A. 1976. Subaqueous sediment transport and deposition by sediment gravity flows. In *Marine Sediment Transport and Environmental Management*, ed. D. J. Stanley, D. J. P. Swift, pp. 197–218. New York: Wiley

Middleton, G. V., Southard, J. B. 1984. *Mechanics of Sediment Movement.* Soc. Econ. Paleontol. Mineral. Short Course Notes 3. 401 pp. 2nd Ed.

Middleton, G. V., Neal, W. J. 1989. Experiments on the thickness of beds deposited by turbidity currents. *J. Sediment. Petrol.* 59: 297–307

Miles, J. 1990. Richardson's number revisited. In *Stratified Flows, Proc. Third Int. Symp. on Stratified Flows, Feb. 3–5, 1987, Pasadena CA*, Am. Soc. Civil Eng., ed. E. J. List, G. H. Jirka, pp. 1–7

Muck, M. T., Underwood, M. B. 1990. Upslope flow of turbidity currents: a comparison among field observations, theory, and laboratory models. *Geology* 18: 54–57

Mutti, E. , Ricci Lucchi, F., Séguret, M., Zanzucchi, G. 1984. Seismoturbidites: a new group of resedimented deposits. *Mar. Geol.* 55: 103–16

Noh, Y., Fernando, H. J. S. 1991a. Dispersion of suspended particles in turbulent flow. *Phys. Fluids* A3: 1730–40

Noh, Y., Fernando, H. J. S. 1991b. Gravity current propagation along an incline in the presence of boundary mixing. *J. Geophys. Res.* 96(C7): 12,586–92

Noh, Y., Fernando, H. J. S. 1992. The motion of a buoyant cloud along an incline in the presence of boundary mixing. *J. Fluid Mech.* 235: 557–77

Norem, H. , Locat, J. , Schieldrop, B. 1990. An approach to the physics and modeling of submarine flowslides. *Mar. Geotechnol.* 9: 93–111

Normark, W. R. 1989. Observed parameters for turbidity-current flow in channels, Reserve Fan, Lake Superior. *J. Sediment. Petrol.* 59: 423–31

Pallesen, T. R. 1983. *Turbidity Currents.* Tech. Univ. Denmark, Inst. Hydrodyn. Hydraul. Eng., Ser. Pap. 32. 115 pp.

Pantin, H. M. 1979. Interaction between velocity and effective density in turbidity flow: phase-plane analysis, with criteria for autosuspension. *Mar. Geol.* 31: 59–99

Pantin, H. M., Leeder, M. R. 1987. Reverse flow in turbidity currents: the role of internal solitons. *Sedimentology* 34: 1143–55

Paola, C. , Southard, J. B. 1983. Auto-suspension and the energetics of two-phase flows: reply to comments on "Experimental test of autosuspension" by J. B. Southard and M. E. Mackintosh. *Earth Surf. Processes Landf.* 8: 273–79

Parkash, B., Middleton, G. V. 1970. Down-current textural changes in Ordovician turbidite greywackes. *Sedimentology* 14: 259–93

Parker, G. 1982. Conditions for the ignition of catastrophically erosive turbidity currents. *Mar. Geol.* 46: 307–27

Parker, G., Garcia, M., Fukushima, Y., Yu, W. 1987. Experiments on turbidity currents over an erodible bed. *J. Hydraul. Res.* 25(1): 123–47

Phillips, A. C., Smith, N. D. 1992. Delta slope processes and turbidity currents in prodeltaic submarine channels, Queen Inlet, Glacier Bay, Alaska. *Can. J. Earth Sci.* 29: 93–101

Pickering, K. T., Hiscott, R. N. 1985. Contained (reflected) turbidity currents from the Middle Ordovician Cloridorme Formation, Quebec, Canada: an alternative to the antidune hypothesis. *Sedimentology* 32: 373–94

Pickering, K., Stow, D., Watson, M., Hiscott, R. 1986. Deep-water facies, processes and models: a review and classification scheme for modern and ancient sediments. *Earth Sci. Rev.* 23: 75–174

Pickering, K. T., Hiscott, R. N., Hein, F. J. 1989. *Deep-Marine Environments: Clastic Sedimentation and Tectonics.* London: Unwin Hyman. 416 pp.

Pierson, T. C., Costa, J. E. 1987. A rheological classification of subaerial sediment-water flows. *Geol. Soc. Am. Rev. Eng. Geol.* 7: 1–12

Pilkey, O. H., Locker, S. D., Cleary, W. J. 1980. Comparison of sand-layer geometry on flat floors of 10 modern depositional basins. *Am. Assoc. Petrol. Geol. Bull.* 64: 841–56

Piper, D. J. W., Shor A. N., Hughes Clarke, J. E. 1988. The 1929 "Grand Banks" earthquake, slump, and turbidity current. In *Sedimentologic Consequences of Convulsive Geologic Events,* ed. H. E. Clifton, *Geol. Soc. Am. Spec. Pap.* 229: 77–92

Prior, D. B., Bornhold, B. D., Wiseman, W. J. Jr., Lowe, D. R. 1987. Turbidity current activity in a British Columbia fjord. *Science* 237: 1330–33

Ravenne, C., Beghin, P. 1983. Apport des expériences en canal à l'interprétation sédimentologique des dépots de cones détritiques sous-marins. *Rev. Inst. Fr. Pétr.* 38: 279–97

Reynolds, S. 1987. A recent turbidity current event, Hueneme Fan, California: reconstruction of flow properties. *Sedimentology* 34: 129–37

Riddell, J. F. 1969. A laboratory study of suspension-effect density currents. *Can. J. Earth Sci.* 6: 231–46

Rodi, W. 1984. *Turbulence Models and their Application in Hydraulics—A State of the Art Review.* Amsterdam: Int. Assoc. Hydraul. Res. 104 pp. 2nd ed.

Scheiwiller, T. 1986. *Dynamics of powdersnow avalanches.* D. Nat. Sci. Diss. Swiss Fed. Inst. Technol., Zurich (Diss. ETH 7951). 116 pp.

Scheiwiller, T., Hutter, K., Hermann, F. 1987. Dynamics of powder snow avalanches. *Ann. Geophysicae* 5B: 569–88

Seilacher, A. 1962. Paleontological studies on turbidite sedimentation and erosion. *J. Geol.* 70: 227–34

Seymour, R. J. 1986. Nearshore auto-suspending turbidity currents. *Ocean Eng.* 13: 435–47

Shanmugam, G., Moiola, R. J. 1991. Types of submarine fan lobes: models and implications. *Am. Assoc. Petrol. Geol. Bull.* 75: 156–79

Shepard, F. P., Marshall, N. F., McLoughlin, P. A., Sullivan, G. G. 1979. *Currents in Submarine Canyons and Other Seavalleys, Am. Assoc. Petrol. Geol. Stud. Geol. no.* 8.. 173 pp.

Siegenthaler, C., Hsü, K. J., Kleboth, P. 1984. Longitudinal transport of turbidity currents—a model study of the Horgen events. *Sedimentology* 31: 187–93

Siegenthaler, C., Buehler, J. 1986. The reconstruction of the paleo-slope of turbidity currents, based on simple hydromechanical parameters of the deposit. *Acta Mech.* 63: 235–44

Siegenthaler, C., Bühler, J. 1985. The kinematics of turbulent suspension currents (turbidity currents) on inclined boundaries. *Mar. Geol.* 64: 19–40

Simpson, J. E. 1982. Gravity currents in the laboratory, atmosphere, and ocean. *Annu. Rev. Fluid Mech.* 14: 213–34

Simpson, J. E., 1987. *Gravity Currents in the Environment and the Laboratory.* New York: Wiley. 244 pp.

Simpson, J. E., Britter, R. E. 1980. A laboratory model of an atmospheric mesofront. *Q. J. R. Meteorol. Soc.* 106: 485–

Smith, G. A. 1986. Coarse-grained nonmarine volcaniclastic sediment: terminology and depositional process. *Geol. Soc. Am. Bull.* 97: 1–10

Smith, J. D., McLean, S. R. 1977. Spatially averaged flow over a wavy surface. *J. Geophys. Res.* 82: 1735–46

Smith, N. D., Vendl, M. A., Kennedy, S. K. 1982. Comparison of sedimentation regimes in four glacier-fed lakes of western Alberta. In *Research in Glacial, Glaciofluvial, and Glaciolacustrine Systems,* ed. R. Davidson-Arnott, W. Nickling, B. D. Fahey, pp. 203–38. Norwich, UK: Geobooks

Southard, J. B., Mackintosh, M. E. 1981. Experimental test of autosuspension. *Earth Surf. Processes Landf.* 6: 103–11

Stacey, M. W., Bowen, A. J. 1988a. The vertical structure of density and turbidity currents: theory and observations. *J. Geophys. Res.* 93: 3528–42

Stacey, M. W., Bowen, A. J. 1988b. The vertical structure of turbidity currents and a necessary condition for self-maintenance. *J. Geophys. Res.* 93: 3543–53

Stacey, M. W., Bowen, A. J. 1990. A comparison of an autosuspension criterion to field observations of five turbidity currents. *Sedimentology* 37: 1–5

Tesaker, E. 1969. *Uniform turbidity current experiments.* Thesis for Licentiatus Tech. Civ. Eng. Inst. Vassbygging. Norges Tekniske Hogskole, Trondheim, Norway

Turner, J. S. 1979. *Buoyancy Effects in Fluids.* Cambridge: Cambridge Univ. Press. 368 pp. 2nd ed.

Umeyama, M. , Gerritsen, F. 1992. Velocity distribution in unform sediment-laden flow. *J. Hydraul. Eng.* 118: 229–45

Van Tassell, J. 1981. Silver abyssal plain carbonate turbidite: flow characteristics. *J. Geol.* 89: 317–33

Villaret, C. , Trowbridge, J. H. 1991. Effects of stratification by suspended sediments on turbulent shear flows. *J. Geophys. Res.* 96(C6): 10,659–80

Walker, R. G. 1969. Geometrical analysis of ripple-drift cross-lamination. *Can. J. Earth Sci.* 6: 683–91

Walker, R. G. 1973. Mopping-up the turbidite mess. In *Evolving Concepts in Sedimentology,* ed. R. N. Ginsburg, pp. 1–37. Baltimore: Johns Hopkins Press

Weirich, F. H. 1986. The record of density-induced underflows in a glacial lake. *Sedimentology* 33: 261–77

Weirich, F. H. 1988. Field evidence for hydraulic jumps in subaqueous sediment gravity flows. *Nature* 332: 626–29

Wetzel, A., Aigner, T. 1986. Stratigraphic completeness: teired trace fossils provide a measuring stick. *Geology* 14: 234–37

Xu, Q., Zhang, F. S., Lou, G. P. 1992. Finite element solutions of free-interface density currents. *Mon. Weather Rev.* 110: 230–31

Yih, C.-S., 1980. *Stratified Flows.* New York: Academic. 418 pp.

Zeng, J. , Lowe, D. R., Prior, D. B., Wiseman, W. J. Jr., Bornhold, B. D. 1991. Flow properties of turbidity currents in Bute Inlet, British Columbia. *Sedimentology* 38: 975–96

Annu. Rev. Earth Planet. Sci. 1993. 21:115–49

OXYGEN ISOTOPES IN METEORITES

Robert N. Clayton

Enrico Fermi Institute, Department of Chemistry, and Department of the Geophysical Sciences, University of Chicago, 5640 S. Ellis Avenue, Chicago, Illinois 60637

KEY WORDS: cosmochemistry, solar nebula, isotopic anomalies

INTRODUCTION

Among all the chemical elements, oxygen has a combination of properties that makes it uniquely important in cosmochemistry. It is an abundant element, and is the principal constituent of most minerals and rocks. It is a light element, so that its three stable isotopes are subject to large mass-dependent fractionation effects in both equilibrium and kinetically-controlled processes. Its isotopes are formed in different nucleosynthetic processes, so that stellar nucleosynthesis can produce isotopic heterogeneity. It can undergo chemical reactions in which non-mass-dependent isotope fractionations occur. Its cosmic abundance, relative to that of carbon, on the one hand, and metallic elements on the other, causes it to occur simultaneously in two cosmochemical reservoirs: a gas (mostly CO and H_2O) and a solid (oxides and silicates of the metallic elements). This last property is especially important, as it allows oxygen to avoid isotopic homogenization in the interstellar medium and in the early solar system, and thus to serve as a natural tracer for interactions of different reservoirs.

The abundant isotope, ^{16}O (99.76%), is produced in stellar nucleosynthesis by helium burning, and is returned to the interstellar medium in supernova explosions. The rare isotopes, ^{17}O (0.04%) and ^{18}O (0.20%), are produced by hot CNO cycles in zones rich in H and He, respectively, in both novae and supernovae (Audouze & Vauclair 1980). For all three isotopes, rapid cooling of stellar ejecta can lead to the formation of molecules and/or refractory solids, which bear the signature of the particular nucleosynthetic processes (Lattimer et al 1978). It might, therefore, be

115

0084–6597/93/0515–0115$02.00

expected that the interstellar medium should contain molecules and grains that have anomalous abundances of the isotopes of oxygen *and* of the elements with which it is combined. These molecules and particles might be preserved in primitive meteorites in some form. Recognition and characterization of such "presolar carriers" would provide direct experimental information on the processes of nucleosynthesis and on the nature of interstellar materials.

An analogy of the expected presolar carriers of oxygen anomalies has already been found for other elements in the form of interstellar diamond, graphite, and SiC in meteorites (Lewis et al 1987, Tang et al 1989). The astrophysical setting is different, however, in that these carbon-rich grains require an oxygen-deficient stellar environment for their formation.

At the present state of research, the existence of oxygen isotopic anomalies in meteorites has been well established (Clayton et al 1973), but specific presolar carrier phases have not been identified. Nevertheless, a great deal can be learned about chemical processes in the early solar system by using the isotopic anomalies as natural tracers, i.e. as "fingerprints," for different parts of an originally heterogeneous solar nebula.

ORIGIN OF THE ISOTOPIC ANOMALIES IN OXYGEN

If the raw materials of the solar system had been homogenized so as to remove all memory of the different nucleosynthetic origins of the three stable isotopes, and if only mass-dependent isotopic fractionations occurred during chemical processing in the solar nebula, then all samples of the solar system (outside the Sun) would have isotopic compositions, expressed as $\delta^{17}O$ and $\delta^{18}O$,[1] which fall on a single line on the three-isotope graph. This curve, approximated by the equation $\delta^{17}O = 0.52\delta^{18}O$ for small values of δ, describes the effect of the mass dependence of equilibrium constants and rate constants for most chemical processes. Deviations from such a relationship might have been caused by nuclear reactions involving energetic particles from the Sun or from the galactic cosmic rays. The possibility of significant perturbation of the oxygen isotope ratios by nuclear reactions *within* the solar nebula (by bombardment with solar protons) has been considered by Lee (1978). Although such effects are energetically feasible, the necessary large proton fluence would have left specific records in other elements, which are not found. It

---

[1] Isotopic compositions are given in the conventional δ notation: $\delta^{18}O = (R_{sam.}^{18}/R_{std}^{18} - 1)1000$, where $R^{18} = {}^{18}O/{}^{16}O$. Departures from the terrestrial fractionation line are given as ${}^{17}O$-excesses, defined by $\Delta^{17}O = \delta^{17}O - 0.52\delta^{18}O$.

is likely, therefore, that oxygen isotopic compositions of the meteorites and planets were not significantly affected by solar system nuclear processes.

The observation that the oxygen isotopes of various solar system samples do not follow a simple mass-dependent pattern (Clayton et al 1973) implies either: 1. that the nebular material was not initially homogeneous, or 2. that non-mass-dependent fractionation processes played an important role (Thiemens & Heidenreich 1983). Neither of these two possibilities has been ruled out. It is characteristic of both mechanisms that they produce isotopic variations such that $^{17}O/^{18}O$ ratios are nearly constant, whereas ^{16}O abundances vary by several percent relative to the other isotopes (Figure 1). In the nucleosynthetic scenario, the ^{16}O variations are attributed to contributions from explosive stellar nucleosynthesis with the likelihood that the dust and gas components of the nebula contained different proportions of the ^{16}O-rich component. In the chemical fractionation scenario, differences in reaction rates of symmetrical molecules, such as $^{16}O^{16}O$, compared to asymmetrical molecules, such as $^{16}O^{17}O$, can lead to the observed isotopic pattern, since the abundant isotope (^{16}O) is almost all contained in symmetrical molecules, while both rare isotopes (^{17}O and ^{18}O) are almost all contained in asymmetrical molecules (Heidenreich & Thiemens 1982).

For some elements, discrimination between nuclear and chemical sources of isotopic variation can be made on the basis of the size of the effects. For example, variations of $^{13}C/^{12}C$ or $^{15}N/^{14}N$ by more than a factor of two in meteorites are usually considered too large to have been caused by chemical fractionation (Lewis et al 1983) and are thus attributed to nuclear processes. However, the observed variations in $^{17}O/^{16}O$ and $^{18}O/^{16}O$ in meteorites are less than 15%, and are thus within the range accessible by chemical fractionations. Another kind of evidence used to identify nuclear origins of isotopic anomalies in meteorites is the association of effects in different elements in the same samples. For example, excesses and deficits of neutron-rich isotopes of calcium, titanium, chromium, iron, and nickel are all correlated (see Clayton et al 1988 for a review), clearly implying a common nucleosynthetic environment. In the case of oxygen, however, no simple correlation has been found with nuclear isotopic anomalies in other elements. Since there does not yet appear to be any unambiguous basis on which to make a distinction between nuclear and chemical origins for the oxygen isotope anomalies, both possibilities must be considered, and evaluated in terms of the observed distribution of the effects in parent bodies, rocks, and minerals.

Another characteristic held in common by both the nuclear and the chemical origins is that they require nonequilibrium chemical processes. In either case, if reactions such as evaporation, condensation, gas-solid exchange, etc were governed by thermodynamic equilibrium, all anomalies

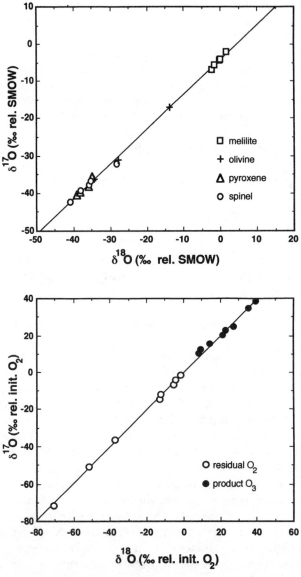

Figure 1 (*Top*) Three-isotope plot for separated minerals from calcium-aluminum-rich inclusions (CAI) from the Allende meteorite. Slope-1 line corresponds to constant $^{17}O/^{18}O$ and variable ^{16}O. Pure ^{16}O plots at $-1000‰$ on both axes. Note the mineralogical control on isotopic compositions. (*Bottom*) Three-isotope plot for ozone product and residual O_2 from the reaction: $3O_2 \rightarrow 2O_3$, carried out in the laboratory by Heidenreich & Thiemens (1982). Molecular symmetry kinetically favors production of $^{17}O^{16}O^{16}O$ and $^{18}O^{16}O^{16}O$ relative to $^{16}O^{16}O^{16}O$. Ordinary mass-dependent isotopic fractionation would produce a line with slope = 1/2 on this diagram.

would disappear. Hence some widespread form of disequilibrium chemistry is required to account for the ubiquitous anomalous oxygen.

Yet another observation which sets oxygen apart from other elements with isotopic anomalies is the absence of any known reservoir which could be called "normal." The situation for most elements is illustrated by magnesium, calcium, iron, and most metallic elements: Their mass-dependent isotopic fractionations are negligibly small in parent-body (including terrestrial) processes, so that their isotopic compositions are uniform in almost all terrestrial and meteoritic materials. The few meteorites with exceptional isotopic compositions are then easily recognized and their anomalous abundances can usually be subdivided into mass-dependent fractionation effects and nuclear effects. For some other elements, such as silicon and sulfur, fractionation effects within parent bodies are not negligible, but their systematic behavior allows assignment of "whole-Earth" and "whole-solar system" isotopic compositions which are essentially identical, and which provide a baseline for measurement of meteoritic anomalies. In the case of oxygen isotopes, each planet or parent body can be described by a characteristic value of $\Delta^{17}O$ (the excess of $\delta^{17}O$ relative to the terrestrial fractionation line). However, there is no convergence to some mean solar system value, and the terrestrial reference line has no special position relative to the various meteoritic groups. The isotopic composition of the Sun would provide a good baseline, but it is not known with enough precision to be useful. As a consequence, we have no sample or definition of "normal solar system oxygen" in the sense of "normal magnesium," etc, or even of "planetary oxygen," in the sense of "planetary neon, or xenon."

The two models (nuclear and chemical) under consideration here have different roles for "normal solar system oxygen." In its pure form, the chemical model would begin with some such "normal" composition, and then fractionate it into chemically different forms by some non-mass-dependent process(es). Unless all the oxygen in the solar system went through such fractionation, a residue (probably large) of the "normal" oxygen should remain. The nuclear model, in its pure form, always has at least two isotopically different reservoirs, in different chemical forms and with different nucleosynthetic histories, so that a "normal" composition could only be defined for the whole solar system, or for the whole Sun.

Further discussion of the origin of oxygen isotope anomalies will be presented below, under "Refractory Inclusions."

PARENT-BODY PROCESSES

Oxygen isotopic fractionations among mineral phases in terrestrial rocks have been used extensively for geothermometry and for study of fluid-rock interactions (for reviews, see Valley et al 1986 and Kyser 1987). Similar applications can be made for processes occurring within other solar system bodies: planets and meteorite parent bodies.

Ordinary Chondrites

Thermal metamorphism of the ordinary chondrites can be studied by means of the isotopic fractionation among their major oxygen-bearing minerals: olivine, pyroxene, and plagioclase. Results for nine equilibrated ordinary chondrites are shown in Table 1. Data for three minerals allow calculation of two independent isotopic temperatures for each meteorite and provide a test of concordancy. The analytical uncertainty of about $\pm 0.1‰$ in each measured fractionation factor contributes an uncertainty of $\pm 60°$ to the plagioclase-pyroxene temperatures, and an uncertainty of similar magnitude is associated with the calibration curves for these mineral pairs. The calibration error may be significant for this application, since the curves were based on experiments with diopside (Chiba et al 1989), whereas the meteorites contain mostly Ca-poor orthopyroxene. A decrease of the $\delta^{18}O$ value of pyroxene in Table 1 by $0.1‰$ removes almost all of the systematic discrepancies. With these caveats, it is seen that reasonably concordant temperatures are measured. For the five meteorites of petrographic grade 6, the plagioclase-olivine temperatures average $900 \pm 50°C$ ($\pm 2\sigma_m$), in very good agreement with element-partitioning geothermometry (Olsen & Bunch 1984).

Two chondrites in Table 1 have quite different isotopic temperatures: Bjurböle (L4), with a value of about 600°C, and Shaw (L7), with a value

Table 1 Oxygen isotope thermometry of ordinary chondrites

		$\delta^{18}O^a$			Temperature °C[b]		
Class	Meteorite	Pc (An 12)	Px	Ol	Pc-Px	Pc-Ol	Px-Ol
L4	Bjurböle	7.70	5.87	4.44	670	600	490
H5	Allegan	4.90	4.09	3.37	1180	1030	820
LL5	Olivenza	6.27	5.16	4.33	960	870	740
L6	Bruderheim	6.20	5.10	4.24	970	870	730
H6	Estacado	5.27	4.16	3.49	960	930	870
L6	Mocs	5.86	4.88	4.10	1050	940	780
L6	Modoc	5.87	4.85	4.17	1020	960	860
LL6	St. Severin	6.33	4.99	4.19	850	820	760
L7	Shaw	5.73	5.03	4.70	1290	1320	1370

[a] Isotopic data modified after Onuma et al (1972a) and Clayton et al (1976).
[b] Temperatures based on calibrations of Clayton & Kieffer (1991).

near 1300°C. The temperature for Bjurböle is probably a reasonable value for its petrographic grade. Shaw is an exceptional meteorite, probably produced by impact melting (Taylor et al 1979), so that its very high temperature is plausible.

In comparison with geothermometry of terrestrial metamorphic rocks, the application to chondrites has the advantage of minimal retrograde exchange, due to their anhydrous character. On the other hand, their mineralogy restricts the range of sensitivity of isotopic thermometers, so that temperatures are inherently not very precise.

Achondrites

Oxygen isotopic thermometry can also be applied to achondrites, most of which are igneous rocks. Results for 11 achondrites and a lunar average are presented in Table 2. On the basis of experimental petrology, Stolper (1977) suggested that eucrites were quenched from liquids at temperatures between 1150° and 1190°C. The oxygen isotope temperatures are significantly lower than this, and probably reflect the post-crystallization "annealing" described by Steele & Smith (1977), which produced ordering in plagioclase and exsolution in pigeonite. Ishii et al (1976) inferred a temperature of 910–1000°C for pigeonite exsolution in a eucrite clast in the howardite Y-7308. Another long-standing and unexplained observation is the existence of very ^{18}O-rich silica minerals in eucrites (Taylor et al 1965).

Table 2 Oxygen isotope thermometry of achondrites

Class	Meteorite	$\delta^{18}O^a$			Temperature °C[b]		
		Pc (An)	Px	Ol	Pc-Px	Pc-Ol	Px-Ol
Eucrite	Ibitira	3.97 (95)	3.47		960		
	Juvinas	3.54 (90)	2.90		850		
Howardite	Yurtuk	3.94 (80)	3.30		930		
Mesosiderite	Crab Orchard	3.50 (95)	3.18		1290		
	Estherville	3.74 (92)	3.36		1190		
Shergottite	Shergotty	4.82 (50)	4.17		1130		
Nakhlite	Lafayette		4.58	4.03			990
	Nakhla		4.59	4.10			1070
Ureilite	ALH 82130		5.95	4.85			610
	EET 83309		6.97	6.06			720
	Kenna		8.16	7.51			890
Moon	Apollo 12 mean	5.98 (90)	5.59	5.15	1140	1140	1140

[a] Isotopic data from Clayton et al (1976), Clayton & Mayeda (1983, 1988).
[b] Temperatures based on calibrations of Clayton & Kieffer (1991).

These are 7–10‰ higher than coexisting plagioclase and pyroxene, and would require temperatures on the order of 300°C if they had formed in equilibrium with the major minerals.

The other exceptionally low isotopic temperatures in Table 2 are those for three ureilites. Of the several published petrogenetic models for ureilites, those in which igneous differentiation plays a major role are ruled out by the oxygen isotope heterogeneity of this group (see Figure 5) (Clayton & Mayeda 1988). The oxygen data favor some form of catastrophic disruption of a heterogeneous, carbonaceous-chondrite-like parent body as proposed by Takeda (1987). The coarse-grained olivine-pigeonite assemblage represents the pre-impact state, which was briefly heated then quenched on impact, allowing preservation of pigeonite. The oxygen isotopic compositions probably preserve their pre-impact values, and provide estimates of temperatures in the source regions of individual meteorites. The isotopic temperatures increase in order of increasing $\delta^{18}O$, which in turn is correlated with increasing Fa content of the olivine (Clayton & Mayeda 1988). Additional sampling would be needed to test whether this trend reflects a radial chemical zoning in the parent body.

NEBULAR PROCESSES

^{16}O Mixing Lines

The scales over which oxygen isotope heterogeneity has been observed range from micrometers to thousands of kilometers, i.e. from mineral grains to planets (Clayton et al 1977). The appropriate sampling scale to reveal the underlying regularities is the scale of individual chondrules and inclusions in primitive meteorites. Figure 2 shows oxygen isotopic compositions for four groups of such samples: 1. refractory inclusions from C3 carbonaceous chondrites, 2. chondrules from C3 and C2 carbonaceous chondrites, 3. chondrules from ordinary chondrites, and 4. chondrules from enstatite chondrites. Although each of these groups occupies a distinct region of the diagram, it is clear that a single overall pattern applies to all the data: a straight line with slope near unity. Such a line is produced by mixtures of two components: one near the terrestrial composition at the upper right, the other enriched in ^{16}O at the lower left. In the following sections each of the four groups will be discussed in more detail, in order to determine their similarities and differences, and thereby to deduce the ultimate origin of the mixing line.

Refractory Inclusions

Historically, the first evidence for isotopic anomalies in oxygen in meteorites was found in calcium-aluminum-rich refractory inclusions (CAI) in

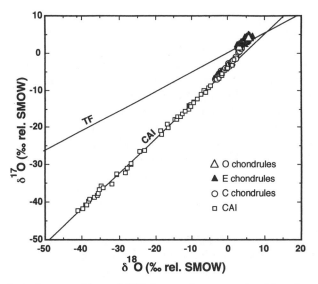

Figure 2 Isotopic compositions of CAI from carbonaceous chondrites, ferromagnesian chondrules from carbonaceous chondrites, and chondrules from ordinary and enstatite chondrites. All of these follow ^{16}O-mixing lines, with small differences which are shown in Figure 3. Terrestrial fractionation line ($\delta^{17}O = 0.52\delta^{18}O$) (marked TF) is the locus of terrestrial rock data; the line marked CAI is a least-squares fit to all the data from refractory inclusions. These same lines are shown for reference in subsequent figures.

Allende (Clayton et al 1973). The CAI are typically mm- to cm-sized objects in carbonaceous chondrites, composed mostly of melilite, pyroxene, and spinel. It was shown that the isotopic compositions of an array of such objects had the property that their $^{17}O/^{18}O$ ratio was nearly constant, whereas the $^{17}O/^{16}O$ and $^{18}O/^{16}O$ ratios varied by several percent. It was recognized that such a pattern could result from mixing of two components, one of which was a pure-^{16}O compound. To account for the observed range of isotopic abundances, the putative ^{16}O-bearing "nuggets" would have to make up several percent of each CAI, and should therefore be detectable and identifiable as presolar grains. Searches for such "nuggets" have been carried out and have been fruitless.

Clayton et al (1977) analyzed various mineral fractions of CAI and found a systematic relationship between isotopic composition and mineralogy, illustrated in Figure 1. Spinel was always found to be ^{16}O-rich, with the most extreme compositions being: $\delta^{18}O \approx -40\text{‰}$, $\delta^{17}O \approx -42\text{‰}$. Clinopyroxene and forsterite often have similarly ^{16}O-rich compositions, whereas melilite, plagioclase, and iron-bearing olivine are much less ^{16}O-rich. These analyses were done on milligram quantities of material, containing millions of grains, and thus gave no indication of the possibility of much larger ^{16}O-excesses in individual crystals. More recent measurements, using ion microprobe techniques for analysis of single grains a few microns in size, have shown that spinel grains in a CAI vary in isotope ratios by a percent or so, but no extreme compositions ($\delta < -90‰$; Anders et al 1991) have been found. Also, a class of spinels with no ^{16}O-excess has been found (Grossman et al 1988).

Many CAI, particularly those classified as coarse-grained Type A and coarse-grained types B1 and B2, show textural and chemical evidence of having crystallized from a molten state as isolated droplets. The existence of well-defined ^{26}Al/^{26}Mg isochrons in some of them indicates that the isotopic composition of magnesium was initially uniform from mineral to mineral in these inclusions (Hutcheon 1982). Nevertheless, these same minerals are found to have widely different oxygen isotopic compositions, falling along the usual ^{16}O-mixing line. This fundamentally different behavior between two major elements in the principal minerals of the rock is a key piece of information in understanding isotopic cosmochemistry. No process is known that could give rise to such an isotopic distribution of oxygen by closed-system crystallization of a melt. Thus an interaction (isotopic exchange) of the CAI with some external oxygen-bearing reservoir has been called upon (Clayton et al 1977, Clayton & Mayeda 1977). In order to produce the observed correlation between mineralogy and isotopic composition, two alternatives might be considered: 1. the CAI and external reservoir exchanged *during* crystallization, so that early-crystallizing minerals (e.g. spinel) have isotopic compositions close to that of the initial liquid (^{16}O-rich), whereas late-crystallizing minerals (e.g. plagioclase) have compositions close to that of the external reservoir (^{16}O-poor); this model has difficulty in that crystallization of melilite (^{16}O-poor) precedes that of pyroxene (^{16}O-rich) (Stolper 1982); 2. exchange took place *after* solidification, with rates differing from mineral to mineral according to their diffusion coefficients. Thus spinel, with a very small oxygen self-diffusion coefficient (Ando & Oishi 1974) remained little altered from its initial ^{16}O-rich composition, while melilite, with a large diffusion coefficient, exchanged extensively with the ^{16}O-poor external reservoir. Both of these alternatives require that the initial CAI liquid be ^{16}O-rich, and that the external reservoir, presumably the nebular gas, be ^{16}O-poor. At present, the relevant oxygen diffusion coefficients are not all available for a detailed modeling of the proposed exchange processes, and some of the published diffusion data are in serious disagreement, particularly for diffusion in melilite (Hayashi & Muehlenbachs 1986, Yurimoto et al 1989). In any case, a major conclusion is that a nebular gas reservoir was present during, or immediately after, the melting of the CAI.

A closely related line of evidence concerning the thermal history of CAI comes from the observation of mass-dependent fractionation of some of the major metallic elements: magnesium, silicon, calcium, and titanium. These elements undergo negligible isotopic fractionation in igneous processes of melting and crystallization, yet their isotopic compositions are highly variable in CAI (see Clayton et al 1988 for a review). The variations are attributable to large kinetic isotope effects during evaporation and condensation (Niederer & Papanastassiou 1984, Clayton et al 1988). Isotopic fractionation of magnesium, silicon, and oxygen during evaporation has been studied in the laboratory using Mg_2SiO_4 as a simple model system (Hashimoto 1990, Davis et al 1990). They found that evaporation produced negligible isotopic change in the residue when sublimation occurred from the solid state. Wang et al (1991) showed that heavy-isotope enrichment occurs only in a steady-state surface layer <100 μm thick. However, evaporation at higher temperature, from liquid Mg_2SiO_4, produced large heavy-isotope enrichment in the residue for all three elements. The process follows a Rayleigh law, with the single-stage fractionation factors closely approximated by the inverse square-root of the ratio of masses of the evaporating atomic or molecular species. Comparison of the laboratory results with analyses of natural CAI shows that the magnitudes of the fractionations for magnesium and silicon, and their ratios in individual CAI, are the same for the natural and laboratory systems. The implication is that heavy-isotope enrichment of magnesium and silicon in coarse-grained CAI is the result of evaporation of wholly or partly molten precursor material. The heavy-isotope depletion commonly observed in fine-grained CAI (Niederer & Papanstassiou 1984) suggests that they formed by condensation of the vapors produced by the evaporation process.

In the laboratory experiments, oxygen isotope fractionation effects are proportional to the fractionation effects in magnesium and silicon. In natural CAI, however, the oxygen variations, although large, are not correlated with fractionations in silicon (Clayton et al 1983a, 1988). Thus the oxygen variations are dominated by the process of exchange with an external reservoir, as discussed above, rather than by the kinetics of evaporation.

One subset of the CAI, known as FUN inclusions, does show evidence of both mass-dependent fractionation of oxygen and later exchange with an external reservoir (Clayton & Mayeda 1977, Wasserburg et al 1977). These inclusions also have exceptionally large fractionations in magnesium and silicon. They were probably evaporated from completely molten droplets at a rate fast enough to minimize oxygen exchange with the nebular gas until after solidification. It is in these FUN inclusions that nucleosynthetic isotopic anomalies have been found for many heavy elements [e.g. barium, neodymium, and samarium (McCulloch & Wasserburg 1978a,b)]. The reason for the association between large fractionation (F) and unidentified nuclear effects (UN) remains unclear. It may be that the precursor material contained "normal" mixtures of r-process, s-process, and p-process isotopes of the heavy elements, and that this mixture was fractionated by preferential evaporation of s-process-containing compounds.

The cosmochemical significance of the large oxygen isotope anomalies in CAI depends on our understanding of the ultimate origin of the anomalies: whether they are residual nucleosynthetic effects, or whether they are chemically-generated solar system effects. If the ^{16}O excesses in CAI are of nuclear origin, the arguments given above imply the existence of two distinct reservoirs which differ in their relative abundances of ^{16}O. If these reservoirs are identified as a dust component and a gas component of the nebula, then the CAI data require that the dust was the ^{16}O-rich component, and that it had $\delta^{18}O$ and $\delta^{17}O$ similar to that now observed in spinels in carbonaceous chondrites. It could be asked whether the spinel crystals in CAI are actually presolar grains. Other than oxygen, the only elements for which isotopic data in spinels are available are magnesium and aluminum. Magnesium isotopic compositions in Al-poor phases are the same as terrestrial, except for mass-fractionation effects. However, since the three magnesium isotopes are produced together in the same process, nucleosynthetic variations may be very small (D. Clayton 1988). They have not been recognized in any sample of meteoritic magnesium. Spinel from Murchison shows evidence of both mass-dependent isotope fractionation and a ^{26}Mg excess attributable to ^{26}Al decay (Ireland et al 1986). The inferred initial $^{26}Al/^{27}Al$ is $5.8(\pm 1.4) \times 10^{-5}$, a value consistent with that found in many CAI. The concentrations of other elements are very low in spinel, so that it is not known whether or not other anomalous compositions are present. On the basis of the available evidence, a presolar origin for spinel in CAI cannot be ruled out, but seems unlikely.

The large ^{16}O-excesses in minerals in CAI are not limited to spinel and other very refractory phases such as hibonite and corundum. They are also sometimes found in clinopyroxene and forsterite (Clayton et al 1977). It can be concluded that the materials now constituting the CAI began as ^{16}O-enriched solids, which underwent heating, melting, partial evaporation, and exchange with an oxygen-bearing gaseous reservoir. The important question for solar system studies is whether the ^{16}O-enrichment of solids was a special attribute of the CAI or a general property of the dust component of the nebula.

If the ^{16}O-excesses in CAI originated through the action of a non-mass-dependent chemical isotope effect, one should seek a process which could produce ^{16}O-enrichment selectively in CAI. The only chemical reaction yet identified which produces a slope-1 array of products on the oxygen three-isotope graph (Figure 1) is the reaction:

$$O + O_2 \rightarrow O_3 \qquad (1)$$

(Thiemens & Heidenreich 1983). Bhattacharya & Thiemens (1989) have found non-mass-dependent effects in the reaction:

$$O + CO \rightarrow CO_2, \qquad (2)$$

which might be expected to play some role in a CO-rich nebula. They found that the product CO_2 was enriched in ^{17}O, and either enriched or depleted in ^{18}O, with respect to the starting materials ($O_2 + CO$). They interpreted their results in terms of a multi-stage process involving both mass-dependent isotopic exchange and a non-mass-dependent recombination reaction. The proposed non-mass-dependent step involves heavy-isotope enrichment in the CO_2 product. There is no step in their proposed mechanism that might generate a ^{16}O-rich species which could be trapped in a solid compound, as is required for a CAI precursor.

The various types of non-mass-dependent isotopic fractionation processes which have been observed (molecular symmetry—Thiemens & Heidenreich 1983; vibrational pumping—Basov et al 1974, Bergman et al 1983; photochemical self-shielding—Sander et al 1977) all depend on gas-phase reactions operating under non-Boltzmann conditions. In order for such reactions to leave a record in meteoritic materials, the nonequilibrium products must be trapped and effectively removed. The difficulty of doing this under nebular conditions has been discussed by Navon & Wasserburg (1985). No mechanism has yet been proposed to connect non-mass-dependent chemical isotope effects with the anomalies found in meteorites.

It is also in CAI that isotopic anomalies of undisputed nucleosynthetic origin have been found in many heavy elements (for a review, see Begemann 1980). No proposal has been advanced that would couple a chemically-produced isotopic anomaly in oxygen with nucleosynthetic anomalies in other elements in the same mineral. The present state of knowledge does not favor a chemical origin for oxygen isotope anomalies; the remainder of this review is written with the assumption that the anomalies are of nucleosynthetic origin.

Chondrules

CHONDRULES IN CARBONACEOUS CHONDRITES   Like CAI, chondrules are rounded objects of previously molten matter: silicates, usually with sulfides and metal. They differ from CAI in three major respects: 1. their elemental abundances are close to the solar values except for atmophile elements; 2.

they are typically an order of magnitude smaller in size (100–1000 μm); and 3. they are very much more abundant. The processes by which the stony part of the proto-solar cloud, presumably in the form of dust, was converted into chondrules are not known. However, it will be argued below that these processes were so efficient that almost all of the material of the meteorites and inner planets passed through a chondrule stage.

Oxygen isotopic compositions of individual chondrules have several features in common with compositions of CAI, implying some common aspects of their origin. Chondrules from all classes of carbonaceous chondrite (CC) form a slope-1 mixing line on the three-isotope graph (Figures 2 and 3), covering a range of about 10‰ in $\delta^{18}O$ and $\delta^{17}O$ (Clayton et al 1983a, Rubin et al 1990). As in the case of CAI, this ^{16}O-mixing line probably results from interaction of a ^{16}O-rich reservoir and a ^{16}O-poor reservoir. Also as in the case of CAI, oxygen isotope variations are decoupled from silicon isotopic fractionation in the same chondrules (Clayton et al 1983a). These observations suggest interaction between a solid precursor and a nebular gas during some heating event, probably the chondrule-forming event. There is no strong chemical, mineralogical, or textural evidence to allow assignment of one or the other end of the mixing line to the solid component, as there is for CAI. However, the chondrule mixing line for carbonaceous chondrites is only slightly displaced parallel to the CAI mixing line, and overlaps it, so that the same process operating in the same direction seems most likely.

An additional observation of oxygen isotope systematics in Allende chondrules is that the common rimmed chondrules have rims that are more ^{16}O-depleted than their cores (Rubin et al 1990), which further supports the interpretation that the direction of isotopic evolution of the chondrules is from ^{16}O-rich to ^{16}O-poor (i.e. toward the upper right on the three-isotope graph). A similar isotopic zoning in a single olivine grain from Allende was observed by Weinbruch et al (1989) using microbeam techniques.

The CC chondrule data in Figure 3 include a cluster of points close to the ranges of ordinary chondrite and enstatite chondrite chondrules. These are all from barred olivine chondrules from Allende (Clayton et al 1983a). McSween (1985a) showed that these chondrules are FeO-rich. It is inferred that their chemical compositions led to more complete melting and more extensive exchange than other CC chondrules.

CHONDRULES IN ORDINARY CHONDRITES   Although chondrules from ordinary chondrites (OC) have chemical, mineralogical, textural, and size characteristics which overlap those of chondrules from carbonaceous chondrites, their oxygen isotopic compositions form a cluster which does

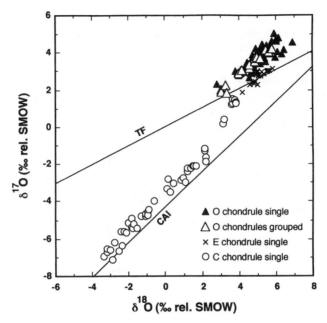

Figure 3 Oxygen isotopic composition of chondrules from different chondrite groups: carbonaceous chondrites, enstatite chondrites, and ordinary chondrites (▲—individual chondrules, △—composites of size-sorted chondrules). Terrestrial fractionation (TF) and refractory inclusion (CAI) lines shown for reference. Each group follows a ^{16}O-mixing line, but populations of different chondrite groups do not overlap.

not overlap the carbonaceous chondrite array (Clayton et al 1991, Figure 3). These two sets of chondrules must have sampled the solar nebula in different regions or at different times. On the three-isotope plot, OC chondrules fall along their own ^{16}O-mixing line with about the same amount of scatter as CC chondrules, but covering a three-fold smaller range. Clayton et al (1991) discussed the implications of the differences in range of variation, and concluded that the short mixing line for OC chondrules is a consequence of repeated melting and exchange events for each chondrule. The "piling up" of data points also suggests that the OC chondrules reached compositions not far from equilibrium with the ambient gas reservoir.

Various explanations have been proposed for the origin of the OC chondrule mixing line. One possibility is the contribution of two different solid precursor materials. If these had different chemical compositions, there should be a correlation between oxygen isotopes and chemistry. Such correlations were sought for a small sample of OC chondrules by Gooding et al (1983). No strong correlations were found between oxygen isotopes and any of the following: chondrule size, texture, mineralogy, major elements, lithophile trace elements, siderophile trace elements. Thus mixtures of different solid precursors do not account for the isotopic variability. The same conclusion was drawn by McSween (1985a) for CC chondrules. Another possible origin of the mixing line is the same solid/gas or melt/gas exchange as was suggested for CAI and CC chondrules. In the case of the OC chondrules, it is less clear whether the solid precursor occupies the lower or upper end of the mixing line (Clayton et al 1991). Based on the observation of a correlation between isotopic composition and chondrule size for the Dhajala (H3.7) chondrite, Clayton et al (1986a) proposed that the solid precursor lies at the upper end of the mixing line. However, internal variations in Mezö-Madaras (L3) are best interpreted in terms of a precursor at the lower end of the mixing line (Mayeda et al 1989). The postulate of a separate ^{16}O-poor solid component for the OC chondrules, while maintaining an ^{16}O-rich solid component for the CC chondrules, is unattractive due to its complication, but cannot be ruled out by the available data.

The simplest model for the oxygen isotopic compositions of CAI and all chondrules is one involving the fewest components and the fewest processes. In broad terms, this implies a common ^{16}O-rich dust component as the solid precursor, and a ^{16}O-poor gas component with which the solid exchanges during one or more heating events. Based on solar system abundances of the elements, only about 17% of the total oxygen could be bound as non-icy solids (i.e. compounds of Mg, Si, Fe, Na, Ca, Al, etc), whereas the principal oxygen reservoir was in the gas (or icy) phase—presumably mostly as H_2O and CO. The differences among the different mixing lines can be produced by variations of the isotopic composition of the gas phase, possibly in space, but more likely in time, due to its interaction with the solid components during nebular heating events.

A very important characteristic of the OC chondrule population is the absence of any correlation between the isotopic compositions of individual chondrules and the iron group (H, L, or LL) of its host meteorite (Clayton et al 1991). This is remarkable, since the bulk meteorites do show a clear correlation between oxygen isotopes and iron group. This question is discussed further under "Ordinary Chondrites."

CHONDRULES IN ENSTATITE CHONDRITES   Oxygen isotope data are available for chondrules from Parsa and Qingzhen (both EH3) and Indarch (EH4) (Clayton & Mayeda 1985). They occupy a fourth distinct area on the three-isotope diagram, falling between the arrays for CC and OC chondrules (Figure 3). The range of variation is smaller than that for ordinary chondrites. The slope of the trend is about 0.7 ± 0.1, significantly steeper than a mass-dependent trend of 0.5. Thus interaction between reservoirs is implied, as in the case of other chondrule groups. The small spread in the data probably indicates close approach to equilibrium between chondrules and the ambient nebular gas.

CHONDRITES

Chondrite Groups

The ten well-recognized chondrite groups—CO, CV, CR, CM, CI, H, L, LL, EH, and EL—have distinctive oxygen isotopic compositions or ranges, with the exceptions that CO and CV overlap, as do EH and EL. The isotopic groups are shown in Figure 4. The isotopic distinctions among classes arise from three fundamentally different causes: 1. primary differences in chondrule origins, 2. addition of a distinct matrix component, and 3. secondary processing on the parent body. The ordinary chondrites

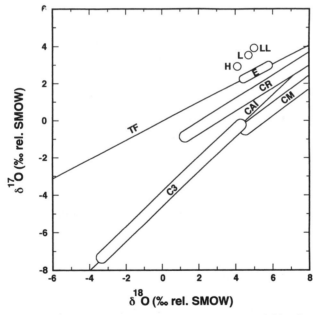

Figure 4 Oxygen isotopic compositions of chondrite groups. Terrestrial fractionation (TF) and refractory inclusion (CAI) lines shown for reference. Fields are shown for H, L, and LL ordinary chondrites, for enstatite chondrites (EH and EL together), and for C3, CR, and CM chondrites. CI chondrites are off scale to the right. For more detail on the carbonaceous chondrite groups, see Figure 5.

are composed primarily of chondrules from the OC group (Figure 3); the isotopic differences among iron groups appear to result primarily from a size-sorting effect (discussed below). The EH and EL groups have isotopic compositions characteristic of their chondrules, and fall in a narrow range. The CO and CV chondrites derive their chondrules from the more heterogeneous CC group (Figure 2), and also have isotopically variable anhydrous matrices. The isotopic compositions of CR, CM, and CI chondrites are dominated by the effects of low-temperature aqueous alteration, which introduced large mass-dependent isotopic fractionation in the formation of phyllosilicate matrix. Each of these groups is discussed in more detail below.

ORDINARY CHONDRITES   Oxygen isotopic compositions of ordinary chondrites and their constituent chondrules were recently discussed by Clayton et al (1991). The detailed data and arguments will not be repeated here, but the following is a summary of their conclusions, as they bear on the origin and history of the ordinary chondrites, the most abundant class of stony meteorites.

1. Ordinary chondrite chondrules formed in an environment spatially or temporally separated from the environment of formation of chondrules in carbonaceous chondrites or enstatite chondrites.
2. Chondrules in H, L, and LL chondrites are all derived from a single chondrule population.
3. Differences in oxygen isotopic compositions of chondrules within the ordinary chondrite population are due primarily to brief high-temperature exchange processes between chondrule precursors and nebular gas. For ordinary chondrite chondrules, multiple heating events are likely for each chondrule. Chondrules evolved from ^{16}O-rich to less ^{16}O-rich in this process.
4. The principal cause for differences in composition (chemical and isotopic) of the different iron groups is a size-sorting effect, with H-material being closest to mean solar composition, and L and LL material being progressively depleted in small chondrules and metal.
5. The fine-grained fraction of H chondrites is an important component in their isotopic material balance. However, isotopic identification of a specific "matrix" component has not yet been achieved.
6. For each of the three iron groups, the material of the type 3 chondrites (UOC) is a satisfactory precursor for formation of types 4, 5, and 6 (EOC) by thermal metamorphism. Small systematic differences between UOC and EOC may result from a combination of two open-system processes: (*a*) aqueous alteration, and (*b*) reduction by carbon.
7. The EOC meteorites are derived from three distinct reservoirs, cor-

responding to the H, L, and LL groups. Each reservoir could be one parent body or a large number of very similar parent bodies.

8. No known stony achondrites were derived from the same reservoirs identified in item 7. However, some genetic associations between ordinary chondrites and iron meteorites are possible, such as between H chondrites and IIE irons (Clayton et al 1983b).

9. Other chondritic reservoirs are implied by the existence of a few isotopically unusual meteorites (Weisberg et al 1991).

10. Xenoliths of material from one iron group in a host from another iron group are not rare, implying that collision partners of ordinary chondrite parent bodies are likely to be made of ordinary chondrite material. Oxygen isotopic compositions provide the best "fingerprint" for the origin of xenoliths, since this property is preserved after other chemical criteria have disappeared by metamorphic interaction with the host.

ENSTATITE CHONDRITES   The enstatite chondrites (EC) occupy a very restricted range in the oxygen isotope diagram, falling on the fractionation line defined by terrestrial samples, which lies between the territories of the CC and the OC. The isotopic difference between EH and EL is barely resolvable (0.26‰ in $\delta^{18}O$) (Clayton et al 1984a), consistent with the proposal of Baedecker & Wasson (1975) that the major chemical difference between EH and EL is due to their degree of accretion of a metallic phase (analogous to the differences among H, L, and LL ordinary chondrites). The isotopic compositions of the enstatite chondrites are determined by the compositions of their constituent chondrules: There is no evidence either of additional oxygen-bearing components or of parent-body modification of the isotopic composition.

There is a recurrent theme in the meteorite literature which maintains that the enstatite chondrites were formed in the inner part of the solar system. The argument is based on the progression of oxidation states from reduced to oxidized in the order: EC–OC–CC, combined with the assumption that the carbonaceous chondrites formed at large distances from the proto-sun (Baedecker & Wasson 1975). However, other properties of these meteorite groups produce different sequences: The order of increasing volatile-element abundances is OC < EC < CC (Palme et al 1988); the order of oxygen isotope abundances is OC > EC = Earth-Moon > CC. These different sequences suggest that a single radial distance parameter is inadequate to explain the major differences among chondrite groups. A time coordinate must also be important, as is shown by the stratigraphic relationships in rimmed chondrules discussed earlier (Chondrules in Carbonaceous Chondrites). Chondrule materials appear to have evolved toward more oxidizing and more ^{16}O-poor with time (at least at the location of the CC chondrules).

As will be discussed below under "Achondrites," the EC group is the only known example of a chondrite group which has an isotopically equivalent achondrite group—the enstatite achondrites (aubrites). Their chemical similarities also make it possible that these are different parts of the same parent body.

CARBONACEOUS CHONDRITES—CO AND CV   Oxygen isotopic compositions of CO and CV whole-rocks are shown in Figure 5, along with data for other carbonaceous chondrite groups (Clayton & Mayeda 1984, 1989). These meteorites are internally isotopically heterogeneous on a chondrule-to-chondrule scale (Figure 2), so that different chips of the same meteorite may give very different isotopic compositions. Two samples of Leoville were found to have δ^{18}O which differed by 7‰. Thus, although there is a general tendency for CO to have a greater ^{16}O-enrichment than CV, there is overlap between the two groups. Only four analyses are available for chondrules from a CO chondrite, Ornans. These are included in Figure 3, and fall within the range of CV chondrules. The range of oxygen isotopic compositions of bulk CO chondrites lies within the range of CC chondrules, implying that the chondrule- and non-chondrule-portions of CO have similar isotopic compositions. For CV chondrites, on the other

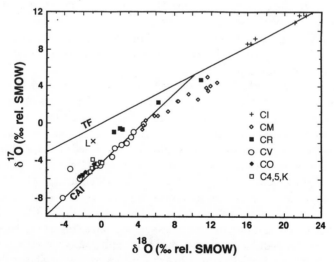

Figure 5 Oxygen isotopic compositions of carbonaceous chondrites. Terrestrial fractionation (TF) and refractory inclusion (CAI) lines shown for reference. Point L is LEW 85332, potential precursor material for CR and/or CI.

hand, the bulk meteorite compositions mostly lie higher on the mixing line than their chondrules (Figure 5), and thus require the presence of a component more ^{16}O-poor than the chondrules. This must be the CV matrix, which has δ^{18}O ranging from $+2‰$ to $+6‰$ (Clayton et al 1977). Similar isotopic compositions are found for dark inclusions in CV chondrites (δ^{18}O from $+3‰$ to $+12‰$) (Johnson et al 1990), which appear to have undergone gas-solid oxidation processes before accretion to the parent body. The greater isotopic heterogeneity of CV relative to CO chondrites arises from the presence in the former of mm-sized dark inclusions (^{16}O-poor) and CAI (^{16}O-rich).

The metamorphosed carbonaceous chondrites (C4, C5, or CK) have isotopic compositions in the CO field (Figure 5). Their chemical affinities with CO and/or CV are ambiguous (Dodd 1981, Kallemeyn 1987). Closed-system metamorphism of Karoonda has produced internal isotopic equilibration among plagioclase, olivine, and magnetite corresponding to a temperature of $590 \pm 50°C$ (Clayton et al 1977).

CARBONACEOUS CHONDRITES—CM   A detailed study of various mineral fractions of the Murchison (CM2) chondrite revealed a range of oxygen isotopic composition as large as that for all other meteorites together (Clayton & Mayeda 1984). A spinel fraction, presumably derived from CAI, has δ^{18}O $= -40‰$, δ^{17}O $= -41‰$, like spinel from Allende CAI. Chondrules and olivine-pyroxene fractions fall along the ^{16}O-mixing line for CAI and CC chondrules (Figure 2). Phyllosilicate matrix is heavy-isotope-enriched, with δ^{18}O $= +12.6‰$, δ^{17}O $= +4.7‰$, and a carbonate fraction has δ^{18}O $= +35.1‰$, δ^{17}O $= +16.7‰$. These large isotopic variations are attributable to two processes: 1. formation of CAI and chondrules, accompanied by exchange with nebular gas at high temperatures, and 2. reaction of anhydrous silicates with water at low temperature (near $0°C$) to form phyllosilicates and carbonates. The low-temperature processes occur with large mass-dependent fractionation factors, comparable in size with the isotopic variations due to nuclear anomalies. The compositions of the low-temperature products are therefore dependent on the parameters of the alteration process, particularly temperature and water/rock ratio.

Oxygen isotope ratios of bulk CM chondrites are highly variable, due to differences in composition of the anhydrous and hydrous end-members, and to their different proportions from one meteorite to another (Clayton & Mayeda 1989). Compositions of 15 CM2 meteorites are shown in Figure 5. The mixing line has a slope of 0.6 ± 0.04. At its lower end, it passes through the data points for two of the CV chondrites, Bali and Mokoia, which would serve as satisfactory precursors for the aqueous alteration of the CM group. Alteration of iron-rich olivine to saponite in the matrix of Mokoia has been reported by Tomeoka & Buseck (1990). These observations suggest a gradational transition from CV to CM chondrites. The upper end of the CM mixing line is defined by the composition of the phyllosilicate matrix.

It has not been established with certainty whether all aqueous alteration of CM matrix occurred on the parent body, although such a view is widely held (e.g. Grimm & McSween 1989 and references therein). The model used by Clayton & Mayeda (1984) for interpretation of the isotopic record in Murchison *assumed* that hydration took place by action of liquid water on the parent body. However, the oxygen isotope data could probably also fit a model involving reaction with nebular gas, with appropriate changes of model parameters.

CARBONACEOUS CHONDRITES—CR   The CR group, consisting of Renazzo, Al Rais, and several Antarctic and Sahara meteorites, shares many features with the CM group, although the two groups form distinct trends on the oxygen isotope graph (Figures 4 and 5) (Weisberg et al 1992, Bischoff et al 1992). These meteorites contain abundant large chondrules which have undergone various degrees of aqueous alteration. The chondrules and the bulk meteorites together form an ^{16}O-mixing line, which lies above the CM bulk meteorite line. However, the first stage of evolution of CM components, following the CAI mixing line, seems to be absent in the CR chondrites. The CR chondrules with the least aqueous alteration fall near an extrapolation of the CC chondrule trend toward more ^{16}O-poor compositions. Other features which distinguish CR matter from CM matter are the presence of abundant Fe-Ni metal in CR chondrules (Weisberg et al 1992) and different isotopic compositions of nitrogen: typically $\sim +40‰$ in CM and $+180‰$ in CR (Kung & Clayton 1978, Kerridge 1985).

The conditions under which aqueous alteration took place in CR chondrites have not yet been established. A detailed isotopic study of magnetite and carbonate may help to constrain temperature and water/rock ratios.

CARBONACEOUS CHONDRITES—CI   Although CI chondrites are usually taken to represent the least chemically fractionated material among all meteorite types, it is clear from their mineralogy and textures that they are not pristine samples of the solar nebula. This is also evident in their oxygen isotopic abundances, which are determined by the low-temperature aqueous reactions which produced their phyllosilicates, carbonates, sulfates, magnetite, etc. Although small amounts of residual olivine are present in CI, no oxygen isotope data are yet available to provide an estimate of the isotopic composition of anhydrous precursors to the present mineral assemblage. A possible candidate for precursor material for CI chondrites is represented by an exceptional carbonaceous chondrite, LEW 85332 (Rubin & Kallemeyn 1990, Brearley 1992). Its oxygen isotopic composition is similar to that of unaltered minerals in CR chondrites (Figure 5). Based on the assumption that Orgueil was produced by alteration of the same precursor as that for Murchison, Clayton & Mayeda (1984) deduced hydration conditions of about 150°C, with a water/rock ratio greater than that for CM.

The oxygen isotopic compositions of CI chondrites fall very near the terrestrial fractionation line. On the assumption that the low-temperature minerals of CI formed by interaction with an aqueous reservoir within the parent body, it can be inferred that the reservoir also had an isotopic composition near the terrestrial fractionation line, displaced to the left of the phyllosilicate composition by an amount equal to the mass-dependent fractionation at the temperature of chemical interaction. At the estimated temperature of 150°C for Orgueil, the phyllosilicate-water fractionation in $^{18}O/^{16}O$ is about 7‰ (Onuma et al 1972b), which leads to an estimate of about +9‰ for $\delta^{18}O$ of the water. The isotopic composition of the solar nebular gases will be discussed further below.

Although the isotopic compositions of Alais, Ivuna and Orgueil (three CI1 meteorites) are very similar to one another and are clearly separated from the most ^{16}O-poor members of the CM2 group, there is another cluster of carbonaceous chondrites of uncertain classification which are even more ^{16}O-depleted: B-7904, Y-82162, and Y-86720 (Figure 5) (Clayton & Mayeda 1989). B-7904 contains chondrules and has other attributes of a CM2 chondrite, but has an exceptionally "heavy" matrix, implying aqueous alteration under conditions different from those of most CM meteorites. Y-82162 and Y-86720 both appear to have undergone thermal metamorphism (Akai 1988, Tomeoka et al 1989a,b). They might be classified as CI2. It is likely that their oxygen isotopic compositions have not been significantly perturbed by the modest degree of metamorphism.

No comprehensive study of the oxygen isotopic composition of organic matter in meteorites has been reported. A single measurement of kerogen from Orgueil gave $\delta^{18}O = +6.0$, $\delta^{17}O = +3.4$, $\Delta^{17}O = +0.3 \pm 0.8$‰ (Halbout et al 1990). This composition lies on the fractionation line through Orgueil silicates, and is thus compatible with origination of the organic matter on the parent body.

UNGROUPED CHONDRITES Several chondritic meteorites have isotopic and chemical compositions unlike those of any of the groups discussed above, and will no doubt be classified as new groups as additional representatives are found.

The Carlisle Lakes-type chondrites (Carlisle Lakes, ALH-85151, and Y-

75302) are characterized by a higher state of oxidation than any of the ordinary chondrites (olivine, typically Fa_{38}), and have $\Delta^{17}O$ greater than any other known meteoritic material (Weisberg et al 1991). Too few individual chondrules of this group have been analyzed to determine whether they are derived from a chondrule population completely separate from the OC chondrules.

Kakangari is a unique chondrite (Graham & Hutchison 1974, Davis et al 1977) with olivine and pyroxene compositions unlike those of other chondrite groups. Oxygen isotopic compositions of its chondrules and matrix define a mixing line lying between the CR chondrites and the E chondrites (Prinz et al 1989). Its matrix has an oxygen isotopic composition similar to the anhydrous minerals of the CR group, but the chondrules follow a steeper ^{16}O-depletion trend. Lea County 002 has mineralogical and isotopic similarities to Kakangari (Zolensky et al 1989).

ACHONDRITES

Eight achondrite groups are recognized on the basis of their oxygen isotopic compositions: 1. HED (howardites, eucrites, diogenites), mesosiderites, and main-group pallasites; 2. SNC (shergottites, nakhlites, and Chassigny); 3. aubrites; 4. winonaites and IAB irons; 5. ureilites; 6. lodranites and acapulcoites; 7. brachinites; and 8. lunar meteorites. Each of these groups occupies a unique field in the oxygen isotope graph (Figure 6) with the exception that the aubrites and lunar meteorites are indistinguishable

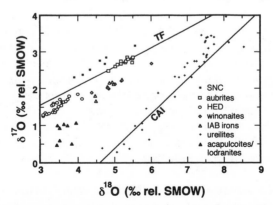

Figure 6   Oxygen isotopic compositions of achondrites. Terrestrial fractionation (TF) and refractory inclusion (CAI) lines shown for reference. Differentiated parent bodies form slope-1/2 groups, as in SNC, HED, aubrites. Heterogeneous parent bodies do not, as in ureilites, acapulcoites/lodranites.

from the Earth (Clayton et al 1976, 1984a, 1992; Clayton & Mayeda 1978a,b, 1983, 1988; Mayeda & Clayton 1980; Mayeda et al 1983, 1984, 1987). The isotopic compositions will be discussed first in terms of the variations within each group, then in terms of the variations from one group to another.

Two of the achondrite groups, HED and SNC, are composed of several different meteorite types probably related to one another by igneous differentiation on a parent body which had undergone extensive melting (Hewins & Newsom 1988, McSween 1985b). By analogy with igneous processes on the Earth and Moon, we anticipate that samples from such a parent body should have a constant $\Delta^{17}O$ (i.e. be related to one another by mass-dependent fractionation only) and ranges of $\delta^{18}O$ of 1–2‰, associated with the small mineral-mineral fractionation factors at igneous temperatures. All lithologies of the SNC group follow such a pattern, so that oxygen isotope data support the origin of all the SNC meteorites in the same parent body. [It was recognized on the basis of oxygen isotopes (Clayton & Mayeda 1983) that the dunite Brachina was *not* a member of the SNC group.] Karlsson et al (1992) have shown that water extracted from some SNC meteorites has even higher $\Delta^{17}O$ than the meteorites themselves, implying an extraplanetary source for the Martian atmosphere.

Many lines of evidence imply that eucrites and diogenites are related to one another through igneous differentiation processes (see Hewins & Newsome 1988, and references therein). Howardites are breccias consisting primarily of eucritic and diogenitic clasts. Hence, there is general agreement that eucrites, diogenites, and howardites (HED) are from the same parent body. The association between HED and mesosiderites is less direct: bulk chemical compositions of mesosiderite silicates and howardites are very similar (Mittlefehldt et al 1979) and eucritic clasts are observed in some mesosiderites (Rubin & Jerde 1987). Similarities between mesosiderites and main-group pallasites are found in the chemical compositions of olivine (Mittlefehldt 1980). Oxygen isotopic compositions of chromite suggest a genetic connection between the main-group pallasites and the IIIAB irons (Clayton et al 1986b). Chemical evidence for this association has been presented by Scott (1977) and Buseck (1977). Chemical similarity between metal in mesosiderites and the IIIAB iron meteorites was noted by Hassanzadeh et al (1990). These and other authors propose that the metal and silicate parts of the mesosiderites represent the two partners of a collision. However, the oxygen isotopic composition of the IIIAB chromite suggests that it was derived from the same oxygen reservoir as the silicates in the pallasites, mesosiderites, and HED meteorites, implying that all of these meteorites may come from a single parent body. If all these rock-types are exposed at the surface of one parent body, it must have been previously disrupted and reassembled, Miranda-like. The perennial problem of under-representation of dunites remains.

It should be noted that two non-main-group pallasites, Eagle Station and Itzawisis, have oxygen isotopic compositions far from the HED field, and must represent samples of a different parent body (Clayton & Mayeda 1978b). Thus pallasites were formed in at least two bodies, which opens the possibility that the main-group pallasites and other members of the HED group may also come from several parent bodies with very similar materials and processes.

The aubrites (enstatite achondrites) are the only achondrites with oxygen isotopic compositions the same as a chondrite group (the enstatite chondrites) (Mayeda & Clayton 1980, Clayton et al 1984a). Because of the strong chemical similarity between these two groups, a genetic association has generally been accepted (Watters & Prinz 1979). Relative to the enstatite chondrites, the aubrites are depleted in a feldspathic component. Wilson & Keil (1991) have proposed an explosive volcanic mechanism for the removal of a basaltic liquid from the parent body, leaving an aubrite residue.

The ureilites have an oxygen isotope pattern totally unlike any other achondrite group (Clayton & Mayeda 1988). Instead of being constant, the $\Delta^{17}O$ value ranges from -0.4 to -2.5‰. The data fall along an ^{16}O-mixing line which overlaps and extends the mixing line defined by CAI and dark inclusions from C3 chondrites. The mixing line must represent either the sampling of a very heterogeneous parent body or a series of intimate mixtures of two different bodies. The latter hypothesis would require that both bodies originally lay on the carbonaceous chondrite mixing line. Some of the ureilites are polymict breccias: Nilpena, for example, contains clasts spanning 8‰ in $\delta^{18}O$, which seems easier to reconcile with a single heterogeneous body (as CV3 parents are known to be), rather than mixing on a millimeter scale of two asteroidal bodies.

However the oxygen isotopic heterogeneity originated, it was clearly a nebular process rather than one of planetary differentiation. The same must be true of the principal chemical variation of the ureilites: the content of ferrous iron in the silicates, which correlates well with the oxygen isotopic composition. Thus both the chemical and isotopic compositions of the ureilites result from pre-parent-body nebular processes, with little disturbance resulting from subsequent igneous and metamorphic processes. Within this framework, the mineralogical and textural features of the ureilites are best accounted for by a catastrophic disruption of the hot interior of a parent body with overall carbonaceous chondrite composition (Takeda 1987).

The lodranites and acapulcoites appear to belong to a distinct oxygen isotope group. The observed range of $\Delta^{17}O$, however, is about three times as great as the estimated analytical uncertainty, which implies an origin either in an isotopically heterogeneous parent body or in several different parent bodies. McCoy et al (1992) proposed that the principal difference between lodranites and acapulcoites is in their thermal history, lodranites having melted and undergone some removal of metal, troilite, and feldspathic liquid, while acapulcoites were very little melted. The iron meteorite Sombrerete has albite + orthopyroxene inclusions with oxygen isotopic ratios at the lower end of this group, and might be complementary to the lodranites (Mayeda & Clayton 1980).

The winonaites (Winona, Mt. Morris WI, Pontlyfni, Tierra Blanca, and Y-75300) have oxygen isotope ratios and chemical compositions indistinguishable from the silicate inclusions in IAB iron meteorites, and are therefore classified in the same group (Mayeda & Clayton 1980, Clayton et al 1984b). Their $\Delta^{17}O$ value of -0.46 ± 0.05 is resolved from the value of -0.25 ± 0.08 for the HED group.

The brachinites (Brachina, Eagles Nest, Window Butte, and ALH 84025) are a group of primitive achondrites (dunites) which may form a separate oxygen isotope class with $\Delta^{17}O = -0.28 \pm 0.08$. This group is not resolvable from the HED group on the basis of $\Delta^{17}O$, but is consistently higher in $\delta^{18}O$ than the HED group, whereas the olivine-rich members of HED (the pallasites) are at the low-^{18}O end of that group.

IRON METEORITES

Many iron meteorites contain oxygen-bearing phases in which oxygen isotopes can be measured. The mesosiderites and IAB irons are breccias containing angular fragments of polymineralic silicates; pallasites and IIE irons contain large monomineralic silicate crystals in a metallic matrix; some IVA irons contain angular monomineralic silicate fragments; some IIIAB irons contain chromite and/or phosphates which may have crystallized or exsolved directly from the metal. Oxygen isotopic compositions may be useful in recognizing genetic associations between irons and stones, and in identifying the partners in some asteroidal collisions.

Isotopic compositions of four iron groups are shown in Figure 7. The IAB iron group was discussed previously under "Achondrites," since the silicates in IAB irons belong to the same oxygen class as the winonaites (Clayton et al 1983b). Dayton, a IIID meteorite, and Kendall County, classified as "Anomalous," also appear to belong to this group. This appears to be a 1:1 association, implying either that the IAB irons and

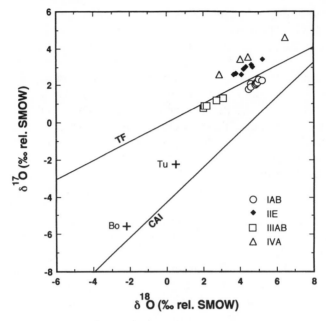

Figure 7 Oxygen isotopic composition of silicates, oxides, and phosphates from iron meteorites. Terrestrial fractionation (TF) and refractory inclusion (CAI) lines shown for reference. Group IVA overlaps L and LL chondrites; Group IIE overlaps H chondrites; Group IIIAB coincides with the HED achondrites; Group IAB coincides with winonaites; Tucson (Tu) is near the CR chondrites and Bocaiuva (Bo) is near the CV chondrites and the Eagle Station pallasites.

winonaites are from a single parent body or that the IAB breccias are the result of a single collision.

Oxygen isotopic compositions of chromite in IIIAB irons fall on the same fractionation line as the main-group pallasites and HED achondrites, and coincide with the value for chromite from the Brenham pallasite (Clayton et al 1986b). Most of the compositions are consistent with isotopic equilibration between chromite and olivine at about 1300°C, although one sample from Cape York is fractionated toward "heavier" isotopic compositions. The internal consistency of $\Delta^{17}O$ values for IIIAB irons, pallasites, and HED achondrites argues for origin in the same parent body.

Isotopic compositions of silicates in IIE irons coincide with those from H chondrites. Since there are no known achondrites in the same isotope class as the H chondrites, it may be unlikely that the IIE irons are differentiated in situ from H chondrites, and thus a two-body collision should be considered. Particularly significant is the observation that Kodaikanal and Watson—IIE irons with H-chondrite-like inclusions (Olsen et al 1991)—have cosmic-ray exposure ages of ~ 8 Ma, characteristic of H chondrites, rather than > 100 Ma, characteristic of most irons (Schultz & Weber 1992). The implication is that these irons were part of the H-chondrite parent at the time of its disruption.

Oxygen isotopic abundances in silicates from three IVA iron meteorites match those in L or LL chondrites (Clayton et al 1983b). The nature of this association is not yet established.

The Tucson iron meteorite, classified as an anomalous nickel-rich ataxite, contains abundant globular silicate inclusions of 10–100 μm in diameter. The oxygen isotopic composition of the silicates ($\delta^{18}O = +0.5‰$, $\delta^{17}O = -2.2‰$) falls among the data points for CR anhydrous silicates (Figure 7) and near the composition of the primitive carbonaceous chondrite LEW-85332. The Tucson silicates are, therefore, the only connection between iron meteorites and primitive stony meteorites.

Bocaiuva is another unique iron meteorite, possibly related to the IIF group (Malvin et al 1985). It contains mm- to cm-sized clasts of chondritic silicates with oxygen isotopic compositions in the range of CO-CV chondrites and near the Eagle Station pallasites. This meteorite may be an impact-produced stony-iron breccia.

ISOTOPIC RESERVOIRS

One striking feature of the oxygen isotope anomalies in meteoritic materials is the presence of the effect, with similar magnitude, on all scales studied: from micrometer distances in individual mineral grains to astronomical unit distances in different planets. Mechanisms must have existed to maintain heterogeneities even in the presence of efficient mixing processes in the solar nebula. It is difficult to imagine any alternative to physical separation of different phases (e.g. gas and dust) as the principal underlying mechanism. The number of such reservoirs is not constrained by any independent theory, but the dominance of a single linear mixing array for common chondrules and CAI suggests that only two major reservoirs were present. Identification of two reservoirs does not require that each had a permanently fixed isotopic composition. All of the observed variations in isotopic composition attest to the *interaction* of these reservoirs, and hence an evolution of their isotopic compositions with time. Furthermore, since the magnitude of the primordial (nucleosynthetic) isotopic variations was not greatly different from the magnitudes of mass-dependent isotope fractionation effects, interactions between reservoirs must result in a combination of mixing and fractionation processes, so that solar system samples do not all fall along a single mixing line. Given these provisos, it seems possible to account for all the observed variability in oxygen isotopic compositions in terms of a two-phase nebula, with an ^{16}O-enriched dust component and an ^{16}O-depleted gas component.

The mean isotopic composition of the dust was similar to that now seen in the most ^{16}O-rich minerals in CAI and other refractory minerals ($\delta^{18}O = -40‰$, $\delta^{17}O = -42‰$). The isotopic composition of the gas cannot be directly measured, but must be assessed from the compositions of the most ^{16}O-poor solids with which the gas interacted. The highest observed values of $\Delta^{17}O$ in meteoritic matter are about $+2.9‰$ in the Carlisle Lakes chondrites (Weisberg et al 1991). A value of $+1.9‰$ was found in a cristobalite inclusion in an L6 chondrite (Olsen et al 1981). The largest value found in carbonaceous chondrites is 1.8‰, for magnetite in Orgueil, Ivuna, and Essebi (Rowe et al 1989). The Carlisle Lakes data imply a gas composition near: $\delta^{18}O = +8‰$; $\delta^{17}O = +7‰$. This approach gives an estimate of the gas composition only after its exchange with the ^{16}O-rich solid reservoir. In order to estimate the pre-exchange composition, it is necessary to use a specific model of the exchange inter-action, specifying the relative sizes of the gas and dust reservoirs. This approach was taken by Clayton & Mayeda (1984), who assumed that the gas/dust ratio was that given by unfractionated solar system abundances. They calculated an initial gas composition of: $\delta^{18}O = +30.0$, $\delta^{17}O = +24.2$, $\Delta^{17}O = +8.6‰$, based on data from minerals in Murchison (CM2). The large difference between these values and those derived from the Carlisle Lakes meteorites is due to the assumption that all mete-oritic solids had the same ^{16}O-enrichment as the spinel in Murchison. It is important to continue to search for high-$\Delta^{17}O$ materials which can improve our estimates of the initial composition of the gas reservoir. At issue is the relative size of the exchanged dust reservoir, in particular whether *all* the observed solids in stony meteorites and terrestrial planets went through a major gas/solid exchange process, either during infall onto the solar accretion disk or during chondrule formation (Cassen & Boss 1988).

An additional consideration is the chemical state of oxygen in the solar nebula. The relevant chemical equilibrium is governed by the reaction:

$$CO + 3H_2 \leftrightarrow CH_4 + H_2O. \tag{3}$$

High temperatures favor the assemblage on the left, with the right side favored below about 700 K. Given a solar ratio of C/O of 0.42 (Anders & Grevesse 1989), the high-temperature gas should be an almost equimolar mixture of CO and H_2O, whereas CO should be in very low abundance at low temperature. However, observations of low-temperature molecular clouds show that the dominant carbon-bearing gas is CO (Irvine et al

1985). In warm regions of the Orion nebula ($T = 100–200$ K), CO and H_2O are of comparable abundance, i.e. the gas at low temperature and pressure is not at chemical equilibrium. Lewis & Prinn (1980) have shown that the slowness of reaction (3) implies that CO remained abundant throughout the nebula. Thus the gases in the nebula were not in chemical equilibrium, and therefore the oxygen fugacity was not well defined. Thus manipulation of gas/dust ratios in order to satisfy the oxidation conditions of meteoritic silicates (Kring 1985, Wood 1985) is not justified unless the kinetics of gas-dust interactions are considered.

FUTURE RESEARCH

Almost all of the oxygen isotopic studies discussed in this review have been based on conventional analytical procedures, which require sample sizes of a milligram or greater. It is likely that substantial new information will be available when microprobe techniques are developed which will allow in situ analysis on single crystals or parts of crystals in the < 10 μm size range. Such techniques would permit a more detailed investigation of interactions of chondrules and CAI with their nebular environment, and, in particular, would allow study of the fine-grained matrix of unequilibrated chondrites of all types. A major goal remains the identification and characterization of presolar grains which may have been carriers of isotopic anomalies in oxygen and lithophile elements.

ACKNOWLEDGMENTS

This work has been supported by National Science Foundation grant EAR 8920584. The predominance of University of Chicago results in the literature of the subject under review reflects the extraordinary skill and diligence of Toshiko K. Mayeda.

Literature Cited

Akai, J. 1988. Incompletely transformed serpentine-type phyllosilicates in the matrix of Antarctic CM chondrites. *Geochim. Cosmochim. Acta* 52: 1593–99

Anders, E., Grevesse, N. 1989. Abundances of the elements: meteoritic and solar. *Geochim. Cosmochim. Acta* 53: 197–214

Anders, E., Virag, A., Zinner, E., Lewis, R. S. 1991. ^{26}Al and ^{16}O in the early solar system: clues from meteoritic Al_2O_3. *Astrophys. J.* 373: L77–80

Ando, K., Oishi, Y. 1974. Self-diffusion coefficients of oxygen ion in single crystals of $MgO \cdot nAl_2O_3$ spinels. *J. Chem. Phys.* 61: 625–29

Audouze, J., Vauclair, S. 1980. *An Introduction to Nuclear Astrophysics.* Dordrecht: Reidel

Baedecker, P. A., Wasson, J. T. 1975. Elemental fractionations among enstatite chondrites. *Geochim. Cosmochim. Acta* 29: 735–65

Basov, N. G., Belenov, E. M., Gavrilina, L. K., Isakov, V. A., Markin, E. P., et al. 1974. Isotope separation in chemical reactions occurring under thermodynamic non-equilibrium condition. *JETP Lett.* 19: 190–91

Begemann, F. 1980. Isotopic anomalies in meteorites. *Rep. Prog. Phys.* 43: 1309–56

Bergman, R. C., Homicz, G. F., Rich, J. W., Wolk, C. L. 1983. ^{13}C and ^{18}O isotopic enrichment by vibrational energy exchange pumping of CO. *J. Chem. Phys.* 78: 1281–92

Bhattacharya, S. K., Thiemens, M. H. 1989. New evidence for symmetry dependent isotope effects: $O+CO$ reaction. *Z. Naturforsch.* 44a: 435–44

Bischoff, A., Palme, H., Ash, R. D., Clayton, R. N., Schultz, L., et al. 1992. Renazzotype carbonaceous chondrites from the Sahara Desert. *Geochim. Cosmochim. Acta* Submitted

Brearley, A. J. 1992. Phyllosilicates in the matrix of the unusual carbonaceous chondrite, LEW 85332 and possible affinities to CI chondrites. *Lunar Planet. Sci. XXIII*, pp. 155–56

Buseck, P. R. 1977. Pallasite meteorites— mineralogy, petrology, and geochemistry. *Geochim. Cosmochim. Acta* 41: 711–40

Cassen, P., Boss, A. P. 1988. Protostellar collapse, dust grains and solar-system formation. See Kerridge & Matthews 1988, pp. 304–28

Chiba, H., Chacko, T., Clayton, R. N., Goldsmith, J. R. 1989. Oxygen isotope fractionations involving diopside, forsterite, magnetite, and calcite: application to geothermometry. *Geochim. Cosmochim. Acta* 53: 2985–95

Clayton, D. D. 1988. Stellar nucleosynthesis and chemical evolution of the solar neighborhood. See Kerridge & Matthews 1988, pp. 1021–62

Clayton, R. N., Grossman, L., Mayeda, T. K. 1973. A component of primitive nuclear composition in carbonaceous meteorites. *Science* 182: 485–88

Clayton, R. N., Hinton, R. W., Davis, A. M. 1988. Isotopic variations in the rockforming elements in meteorites. *Philos. Trans. R. Soc. London Ser. A* 325: 483–501

Clayton, R. N., Kieffer, S. W. 1991. Oxygen isotopic thermometer calibrations. In *Stable Isotope Geochemistry*, ed. H. P. Taylor Jr., J. R. O'Neil, I. R. Kaplan, pp. 3–10. Geochem. Soc. Spec. Publ. No. 3

Clayton, R. N., Mayeda, T. K. 1977. Correlated oxygen and magnesium isotopic anomalies in Allende inclusions, I: Oxygen. *Geophys. Res. Lett.* 4: 295–98

Clayton, R. N., Mayeda, T. K. 1978a. Multiple parent bodies of polymict brecciated meteorites. *Geochim. Cosmochim. Acta* 42: 325–27

Clayton, R. N., Mayeda, T. K. 1978b. Genetic relations between iron and stony meteorites. *Earth Planet. Sci. Lett.* 40: 168–74

Clayton, R. N., Mayeda, T. K. 1983. Oxygen isotopes in eucrites, shergottites, nakhlites, and chassignites. *Earth Planet. Sci. Lett.* 62: 1–6

Clayton, R. N., Mayeda, T. K. 1984. The oxygen isotope record in Murchison and other carbonaceous chondrites. *Earth Planet. Sci. Lett.* 67: 151–66

Clayton, R. N., Mayeda, T. K. 1985. Oxygen isotopes in chondrules from enstatite chondrites: possible identification of a major nebular reservoir. *Lunar Planet. Sci. XVI*, pp. 142–43

Clayton, R. N., Mayeda, T. K. 1988. Formation of ureilites by nebular processes. *Geochim. Cosmochim. Acta* 52: 1313–18

Clayton, R. N., Mayeda, T. K. 1989. Oxygen isotope classification of carbonaceous chondrites. *Lunar Planet. Sci. XX*, pp. 169–70

Clayton, R. N., Mayeda, T. K., Goswami, J. N., Olsen, E. J. 1986a. Oxygen and silicon isotopes in chondrules from ordinary chondrites. *Terra Cognita* 6: 130

Clayton, R. N., Mayeda, T. K., Goswami, J. N., Olsen, E. J. 1991. Oxygen isotope studies of ordinary chondrites. *Geochim. Cosmochim. Acta* 55: 2317–37

Clayton, R. N., Mayeda, T. K., Nagahara, H. 1992. Oxygen isotope relations among primitive achondrites. *Lunar Planet. Sci. XXIII*, pp. 231–32

Clayton, R. N., Mayeda, T. K., Olsen, E. J., Prinz, M. 1983b. Oxygen isotope relationships in iron meteorites. *Earth Planet. Sci. Lett.* 65: 229–32

Clayton, R. N., Mayeda, T. K., Prinz, M., Nehru, C. E., Delaney, J. S. 1986b. Oxygen isotope confirmation of a genetic association between achondrites and IIIAB iron meteorites. *Lunar Planet. Sci. XVII*, pp. 141–42

Clayton, R. N., Mayeda, T. K., Rubin, A. E. 1984a. Oxygen isotopic compositions of enstatite chondrites and aubrites. *J. Geophys. Res.* 89: C245–49

Clayton, R. N., Mayeda, T. K., Yanai, K. 1984b. Oxygen isotopic compositions of some Yamato meteorites. *Natl. Inst. Polar Res. Spec. Issue No. 35*, pp. 267–71. Tokyo: N.I.P.R.

Clayton, R. N., Onuma, N., Grossman, L., Mayeda, T. K. 1977. Distribution of the presolar component in Allende and other carbonaceous chondrites. *Earth Planet. Sci. Lett.* 34: 209–24

Clayton, R. N., Onuma, N., Ikeda, Y., Mayeda, T. K., Hutcheon, I. D., et al. 1983a. Oxygen isotopic composition of chondrules in Allende and ordinary chondrites. In *Chondrules and Their Origins*, ed. E. A. King, pp. 37–43. Houston: Lunar Planet. Inst.

Clayton, R. N., Onuma, N., Mayeda, T. K.

1976. A classification of meteorites based on oxygen isotopes. *Earth Planet. Sci. Lett.* 30: 10–18

Davis, A. M., Grossman, L., Ganapathy, R. 1977. Yes, Kakangari is a unique chondrite. *Nature* 265: 230–32

Davis, A. M., Hashimoto, A., Clayton, R. N., Mayeda, T. K. 1990. Isotope mass fractionation during evaporation of Mg_2SiO_4. *Nature* 347: 655–58

Dodd, R. T. 1981. *Meteorites.* Cambridge: Cambridge Univ. Press. 368 pp.

Gooding, J. L., Mayeda, T. K., Clayton, R. N., Fukuoka, T. 1983. Oxygen isotopic heterogeneities, their petrological correlations, and implications for melt origins of chondrules in unequilibrated ordinary chondrites. *Earth Planet. Sci. Lett.* 65: 209–24

Graham, A. L., Hutchison, R. 1974. Is Kakangari a unique chondrite? *Nature* 251: 128–29

Grimm, R. E., McSween, H. Y. 1989. Water and the thermal evolution of carbonaceous chondrite parent bodies. *Icarus* 82: 244–80

Grossman, L., Fahey, A. J., Zinner, E. 1988. Carbon and oxygen isotopic compositions of individual spinel crystals from the Murchison meteorite. *Lunar Planet. Sci. XIX,* pp. 435–36

Halbout, J., Robert, F., Javoy, M. 1990. Hydrogen and oxygen isotope compositions in kerogen from the Orgueil meteorite: clues to a solar origin. *Geochim. Cosmochim. Acta* 54: 1453–62

Hashimoto, A. 1990. Evaporation kinetics of forsterite and implications for the early solar nebula. *Nature* 347: 53–55

Hassanzadeh, J., Rubin, A. E., Wasson, J. T. 1990. Compositions of large metal nodules in mesosiderites: links to iron meteorite group IIIAB and the origin of mesosiderite subgroups. *Geochim. Cosmochim. Acta* 54: 3197–3208

Hayashi, T., Muehlenbachs, K. 1986. Rapid oxygen diffusion in melilite and its relevance to meteorites. *Geochim. Cosmochim. Acta* 50: 585–91

Heidenreich, J. E. III, Thiemens, M. H. 1982. A non-mass-dependent isotope effect in the production of ozone from molecular oxygen. *J. Chem. Phys.* 78: 892–95

Hewins, R. H., Newsom, H. E. 1988. Igneous activity in the early solar system. See Kerridge & Matthews 1988, pp. 73–101

Hutcheon, I. D. 1982. Ion probe magnesium isotopic measurements of Allende inclusions. In *Nuclear and Chemical Dating Techniques: Interpreting the Environmental Record,* Am. Chem. Soc. Symp. Ser. No. 176, pp. 95–128

Ireland, T., Compston, W., East, T. M. 1986.

Magnesium isotopic compositions of olivine, spinel, and hibonite from the Murchison carbonaceous chondrite. *Geochim. Cosmochim. Acta* 50: 1412–21

Irvine, W. M., Schloerb, F. P., Hjalmarson, A., Herbst, E. 1985. The chemical state of dense interstellar clouds: an overview. In *Protostars and Planets II,* ed. D. C. Black, M. S. Matthews, pp. 579–620. Tucson: Univ. Ariz. Press

Ishii, T., Miyamoto, M., Takeda, H. 1976. Pyroxene geothermometry and crystallization—subsolidus equilibration temperature of lunar and achondritic pyroxenes. *Lunar Sci. Conf. VII,* 408–10

Johnson, C. A., Prinz, M., Weisberg, M. K., Clayton, R. N., Mayeda, T. K. 1990. Dark inclusions in Allende, Leoville, and Vigarano: evidence for nebular oxidation of CV3 constituents. *Geochim. Cosmochim. Acta* 54: 819–30

Kallemeyn, G. W. 1987. Compositional comparisons of metamorphosed carbonaceous chondrites. N.I.P.R. Spec. Issue No. 46, pp. 151–61, Tokyo

Karlsson, H. R., Clayton, R. N., Gibson, E. K. Jr., Mayeda, T. K. 1992. Water in SNC meteorites: evidence for a Martian hydrosphere. *Science* 255: 1409–11

Kerridge, J. F. 1985. Carbon, hydrogen, and nitrogen in carbonaceous chondrites: abundances and isotopic compositions in bulk samples. *Geochim. Cosmochim. Acta* 49: 1707–14

Kerridge, J. F., Matthews, M. S., eds. 1988. *Meteorites and the Early Solar System.* Tucson: Univ. Ariz. Press

Kring, D. A. 1985. Heterogeneity of O/H in the solar nebula, as indicated by mafic mineral compositions in chondrules. *Lunar Planet. Sci. XVI,* pp. 469–70

Kung, C. C., Clayton, R. N. 1978. Nitrogen abundances and isotopic compositions in stony meteorites. *Earth Planet. Sci. Lett.* 38: 421–35

Kyser, T. K., ed. 1987. *Stable Isotope Geochemistry of Low Temperature Fluids.* Toronto: Mineral. Assoc. Canada. 452 pp.

Lattimer, J. M., Schramm, D. N., Grossman, L. 1978. Condensation in supernova ejecta and isotopic anomalies in meteorites. *Astrophys. J.* 219: 230–49

Lee, T. 1978. A local proton irradiation model for isotopic anomalies in the solar system. *Astrophys. J.* 224: 217–26

Lewis, J. S., Prinn, R. G. 1980. Kinetic inhibition of CO and N_2 reduction in the solar nebula. *Astrophys. J.* 238: 357–64

Lewis, R. S., Anders, E., Wright, I. P., Norris, S. J., Pillinger, C. T. 1983. Isotopically anomalous nitrogen in primitive meteorites. *Nature* 305: 767–71

Lewis, R. S., Tang, M., Wacker, J. F., Anders, E., Steel, E. 1987. Interstellar diamonds in meteorites. *Nature* 326: 160–62

Malvin, D. J., Wasson, J. T., Clayton, R. N., Mayeda, T. K., da Silva Curvello, W. 1985. Bocaiuva—a silicate-inclusion-bearing iron meteorite related to the Eagle Station pallasites. *Meteoritics* 20: 259–73

Mayeda, T. K., Clayton, R. N. 1980. Oxygen isotopic compositions of aubrites and some unique meteorites. *Proc. Lunar Planet Sci. Conf. 11*, pp. 1145–51. Oxford: Pergamon

Mayeda, T. K., Clayton, R. N., Molini-Velsko, C. A. 1983. Oxygen and silicon isotopes in ALHA 81005. *Geophys. Res. Lett.* 10: 799–800

Mayeda, T. K., Clayton, R. N., Rubin, A. E. 1984. Oxygen isotopic compositions of enstatite chondrites and aubrites. *J. Geophys. Res.* 89: C245–49

Mayeda, T. K., Clayton, R. N., Sodonis, A. 1989. Internal oxygen isotope variations in two unequilibrated chondrites. *Meteoritics* 24: 301

Mayeda, T. K., Clayton, R. N., Yanai, K. 1987. Oxygen isotopic compositions of several Antarctic meteorites. *Natl. Inst. Polar Res. Spec. Issue No. 46*, pp. 144–50. Tokyo: N.I.P.R.

McCoy, T. J., Keil, K., Mayeda, T. K., Clayton, R. N. 1992. Monument Draw and the formation of the acapulcoites. *Lunar Planet. Sci. XXIII*, pp. 871–72

McCulloch, M. T., Wasserburg, G. J. 1978a. Barium and neodymium isotopic anomalies in the Allende meteorite. *Astrophys. J.* 220: L15–19

McCulloch, M. T., Wasserburg, G. J. 1978b. More anomalies from the Allende meteorite: Samarium. *Geophys. Res. Lett.* 5: 599–602

McSween, H. Y. Jr. 1985a. Constraints on chondrule origin from petrology of isotopically characterized chondrules in the Allende meteorite. *Meteoritics* 20: 523–40

McSween, H. Y. Jr. 1985b. SNC meteorites: clues to Martian petrologic evolution? *Rev. Geophys.* 23: 391–416

Mittlefehldt, D. W. 1980. The composition of mesosiderite olivine clasts and implications for the origin of pallasites. *Earth Planet. Sci. Lett.* 51: 29–40

Mittlefehldt, D. W., Chou, C.-L., Wasson, J. T. 1979. Mesosiderites and howardites: igneous formation and possible genetic relationships. *Geochim. Cosmochim. Acta* 43: 673–88

Navon, O., Wasserburg, G. J. 1985. Self-shielding in O_2—a possible explanation for oxygen isotopic anomalies in meteorites? *Earth Planet. Sci. Lett.* 73: 1–16

Niederer, F., Papanastassiou, D. A. 1984. Ca

isotopes in refractory inclusions. *Geochim. Cosmochim. Acta* 48: 1279–93

Olsen, E. J., Bunch, T. E. 1984. Equilibration temperatures of ordinary chondrites: a new evaluation. *Geochim. Cosmochim. Acta* 48: 1363–65

Olsen, E. J., Mayeda, T. K., Clayton, R. N. 1981. Cristobalite-pyroxene in an L6 chondrite: implications for metamorphism. *Earth Planet. Sci. Lett.* 56: 82–88

Olsen, E. J., Schwade, J., Davis, A. M., Clayton, R. N., Mayeda, T. K., et al. 1991. Watson: a new link in the IIE iron chain. *Lunar Planet. Sci. XXII*, pp. 999–1000

Onuma, N., Clayton, R. N., Mayeda, T. K. 1972a. Oxygen isotope studies of ordinary chondrites. *Geochim. Cosmochim. Acta* 36: 157–68

Onuma, N., Clayton, R. N., Mayeda, T. K. 1972b. Oxygen isotope cosmothermometer. *Geochim. Cosmochim. Acta* 36: 169–88

Palme, H., Larimer, J. W., Lipschutz, M. E. 1988. Moderately volatile elements. Kerridge & Matthews 1988, pp. 436–61

Prinz, M., Weisberg, M. K., Nehru, C. E., MacPherson, G. J., Clayton, R. N., Mayeda, T. K. 1989. Petrologic and stable isotope study of the Kakangari (K-group) chondrite: chondrules, matrix, CAI's. *Lunar Planet. Sci. XX*, pp. 870–71

Rowe, M. W., Clayton, R. N., Mayeda, T. K. 1989. Oxygen isotopes in separated components of CI and CM chondrites. *Meteoritics* 24: 321

Rubin, A. E., Jerde, E. A. 1987. Diverse eucritic pebbles in the Vaca Muerta mesosiderite. *Earth Planet. Sci. Lett.* 84: 1–14

Rubin, A. E., Kallemeyn, G. W. 1990. Lewis Cliff 85332: a unique carbonaceous chondrite. *Meteoritics* 25: 215–25

Rubin, A. E., Wasson, J. T., Clayton, R. N., Mayeda, T. K. 1990. Oxygen isotopes in chondrules and coarse-grained chondrule rims from the Allende meteorite. *Earth Planet. Sci. Lett.* 96: 247–55

Sander, R. K., Loree, T. R., Rockwood, S. D., Freund, S. M. 1977. ArF laser enrichment of oxygen isotopes. *Appl. Phys. Lett.* 30: 150–52

Schultz, L., Weber, H. W. 1992. Noble gases in metal and silicates of the IIE iron meteorite Watson. *Lunar Planet. Sci. XXIII*, pp. 1229–30

Scott, E. R. D. 1977. Pallasites—metal composition, classification, and relationships with iron meteorites. *Geochim. Cosmochim. Acta* 41: 349–60

Steele, I. M., Smith, J. V. 1977. Mineralogy of the Ibitira eucrite and comparison with other eucrites and lunar samples. *Earth Planet. Sci. Lett.* 33: 67–78

Stolper, E. 1977. Experimental petrology of

eucritic meteorites. *Geochim. Cosmochim. Acta* 41: 587–612

Stolper, E. 1982. Crystallization sequences of Ca-, Al-rich inclusions from Allende: an experimental study. *Geochim. Cosmochim. Acta* 46: 2159–80

Takeda, H. 1987. Mineralogy of Antarctic ureilites and a working hypothesis for their origin and evolution. *Earth Planet. Sci. Lett.* 81: 358–70

Tang, M., Anders, E., Hoppe, P., Zinner, E. 1989. Meteoritic silicon carbide and its stellar sources; implications for galactic chemical evolution. *Nature* 339: 351–54

Taylor, G. J., Keil, K., Berkley, J. L., Lange, D. E. 1979. The Shaw meteorite: history of a chondrite consisting of impact-melted and metamorphic lithologies. *Geochim. Cosmochim. Acta* 43: 323–37

Taylor, H. P. Jr., Duke, M. B., Silver, L. T., Epstein, S. 1965. Oxygen isotope studies of minerals in stony meteorites. *Geochim. Cosmochim. Acta* 29: 489–512

Thiemens, M. H., Heidenreich, J. E. III. 1983. The mass-independent fractionation of oxygen: a novel isotope effect and its possible cosmochemical implications. *Science* 219: 1073–75

Tomeoka, K., Buseck, P. R. 1990. Phyllosilicates in the Mokoia CV carbonaceous chondrite: evidence for aqueous alteration in an oxidizing environment. *Geochim. Cosmochim. Acta* 54: 1745–54

Tomeoka, K., Kojima, H., Yanai, K. 1989a. Yamato-82162: a new kind of CI carbonaceous chondrite found in Antarctica. *Proc. Natl. Inst. Polar Res. Symp. Antarct. Meteorites* 2: 36–54

Tomeoka, K., Kojima, H., Yanai, K. 1989b. Yamato-86720: a CM carbonaceous chondrite having experienced extensive aqueous alteration and thermal metamorphism. *Proc. Natl. Inst. Polar Res. Symp. Antarct. Meteorites* 2: 55–74

Valley, J., O'Neil, J. R., Taylor, H. P. Jr., eds. 1986. *Stable Isotopes in High Temperature Geological Processes.* Reviews in Mineralogy, No. 16. Washington, DC: Mineral. Soc. Am. 570 pp.

Wang, J., Davis, A. M., Hashimoto, A., Clayton, R. N. 1991. The role of diffusion in the isotopic fractionation of magnesium during evaporation of forsterite. *Lunar Planet. Sci. XXII*, pp. 1461–62

Wasserburg, G. J., Lee, T., Papanastassiou, D. A. 1977. Correlated O and Mg isotopic anomalies in Allende inclusions: II. Magnesium. *Geophys. Res. Lett.* 7: 299–302

Watters, T. R., Prinz, M. 1979. Aubrites: their origin and relationship to enstatite chondrites. *Proc. Lunar Planet. Sci. Conf. 10th*, pp. 1073–93. Oxford: Pergamon

Weinbruch, S., Zinner, E., El Goresy, A., Palme, H. 1989. Oxygen isotopic compositions of individual forsterite grains, fayalitic rims, and matrix olivines from the Allende meteorite. *Lunar Planet. Sci. XX*, pp. 1187–88

Weisberg, M. K., Prinz, M., Clayton, R. N., Mayeda, T. K. 1992. The CR2 (Renazzo-type) carbonaceous chondrite group and its implications. *Geochim. Cosmochim. Acta* Submitted

Weisberg, M. K., Prinz, M., Kojima, H., Yanai, K., Clayton, R. N., Mayeda, T. K. 1991. The Carlisle Lakes-type chondrites: a new grouplet with high $\Delta^{17}O$ and evidence for nebular oxidation. *Geochim. Cosmochim. Acta* 55: 2657–69

Wilson, L., Keil, K. 1991. Consequences of explosive eruptions on small Solar System bodies: the case of the missing basalts on the aubrite parent body. *Earth Planet. Sci. Lett.* 104: 505–12

Wood, J. A. 1985. Meteoritic constraints on processes in the solar nebula. In *Protostars and Planets II*, ed. D. C. Black, M. S. Matthews, pp. 687–702. Tucson: Univ. Ariz. Press

Yurimoto, H., Morioka, M., Nagasawa, H. 1989. Diffusion in single crystals of melilite: I. Oxygen. *Geochim. Cosmochim. Acta* 53: 2387–94

Zolensky, M. E., Score, R., Schutt, J. W., Clayton, R. N., Mayeda, T. K. 1989. Lea County 001, an H5 chondrite, and Lea County 002, an ungrouped type 3 chondrite. *Meteoritics* 24: 227–32

Annu. Rev. Earth Planet. Sci. 1993. 21:151–74

ACID RAIN[1]

Owen P. Bricker

U.S. Geological Survey, MS 432, Reston, Virginia 22092

Karen C. Rice

U.S. Geological Survey, 1936 Arlington Blvd., Rm. 118, Charlottesville, Virginia 22903

KEY WORDS: groundwater acidification, streamwater acidification, anthropogenic emissions

INTRODUCTION

Acid deposition, or acid rain as it is more commonly referred to, has become a widely publicized environmental issue in the U.S. over the past decade. The term usually conjures up images of fish kills, dying forests, "dead" lakes, and damage to monuments and other historic artifacts. The primary cause of acid deposition is emission of SO_2 and NO_X to the atmosphere during the combustion of fossil fuels. Oxidation of these compounds in the atmosphere forms strong acids—H_2SO_4 and HNO_3—which are returned to the Earth in rain, snow, fog, cloud water, and as dry deposition.

Although acid deposition has only recently been recognized as an environmental problem in the U.S., it is not a new phenomenon (Cogbill & Likens 1974). As early as the middle of the 17th century in England, the deleterious effects of industrial emissions on plants, animals, and humans, and the atmospheric transport of pollutants between England and France had become issues of concern (Evelyn 1661, Graunt 1662). It is interesting that well over three hundred years ago in England, recommendations were made to move industry outside of towns and build higher chimneys to

---

[1] The US Government has the right to retain a nonexclusive, royalty-free license in and to any copyright covering this paper.

spread the pollution into "distant parts." Increasing the height of smoke-stacks has helped alleviate local problems, but has exacerbated others. In the U.S. the height of the tallest smokestack has more than doubled, and the average height of smokestacks has tripled since the 1950s (Patrick et al 1981). This trend occurred in most industrialized nations during the 20th century and has had the effect of transforming acid rain from a local urban problem into a problem of global scale.

The first extensive scientific studies of acid deposition and its effects were made in the mid-19th century by the English chemist Robert Angus Smith. In 1852, Smith, a fellow of the Royal Society and Inspector-General of the Alkali Works, published a paper detailing the chemistry of rain in the city of Manchester and the surrounding countryside (Smith 1852). He observed that there are "three kinds of air, that with carbonate of ammonia in the fields at a distance, that with sulfate of ammonia in the suburbs, and that with sulfuric acid, or acid sulfate, in the town." In 1872, his research on rain chemistry in England, Scotland, and Germany culminated in a book entitled *Air and Rain, the Beginnings of a Chemical Climatology*, in which he summarized the previous 20 years of investigations and first used the term "acid rain" (Smith 1872). In this remarkable work he documented many effects of acid deposition that have been "rediscovered" in the past several decades. Smith observed that the acidity of rain "is caused almost entirely by sulfuric acid—and in country places to a small extent by nitric acid and by acids from the combustion of wood, peat, turf, etc." With respect to effects on materials he observed that "the presence of free sulfuric acid in the air sufficiently explains the fading of colors in prints and dyed goods, the rusting of metals, and the rotting of blinds." In the biological area, he noted that "mosses may be seen growing in the acid rain of towns when trees, shrubs and grasses disappear" and suggested that the discoloration on bark and leaves might result from fungal attack on trees weakened by acid rain. He also speculated that acid rain might interfere with the microbes that cause decay. As a result of his studies in Scotland he postulated that the high rate of human mortality in Glasgow at that time was related to poor air quality as reflected in Glasgow's rain. Smith's work soon fell into obscurity, but acid deposition continued unabated and grew from a problem localized around industrialized urban areas then, to one of global scale today.

From Smith's time until the mid-20th century there was no focused systematic program of research on acid deposition. The connection between acid deposition and damage to ecological systems was once again demonstrated a century after Smith's pioneering work by an American limnologist, Eville Gorham, and his colleagues in a series of papers published between 1955 and 1965 (e.g. Gorham 1961, also see Gorham 1976).

Their work, like that of Smith, did not receive much attention at that time. In the late 1960s a Swedish soil scientist, Svante Oden, established a network to measure surface-water chemistry in Scandinavia. When he combined data from this network with those from the existing European Air Chemistry network, he observed some disturbing correlations. He found that long range atmospheric transport of sulfur and nitrogen compounds was occurring over much of Europe and Scandinavia, that acid deposition was a large scale, regional phenomenon with well-defined source and sink areas, and that both precipitation and surface waters were becoming more acidic. Oden's work received wide publicity and generated much concern and controversy in the scientific community and in the public at large.

About the same time that Oden was documenting the acid rain problem in Scandinavia and Europe, researchers were discovering similar problems in the eastern U.S., particularly in the northeastern States (Gambell & Fisher 1966, Fisher et al 1968, Pearson & Fisher 1971, Likens et al 1972). Cogbill & Likens (1974) focused attention on the severity of the acid deposition problem in the northeastern U.S. In 1976, a National Acid Deposition Program (NADP) was initiated by Galloway & Cowling (1978), in cooperation with a number of universities and state and federal agencies, to coordinate long-term deposition monitoring in the U.S. The NADP network concentrated on sites east of the Mississippi River, particularly in the northeastern portion of the country where acid deposition was thought to be most severe. In 1980, Congress mandated a more comprehensive ten year federal program (P.L.96-294), the National Acid Precipitation Assessment Program (NAPAP). This program incorporated the original NADP network into a nationwide network called the National Trends Network (NTN) and initiated extensive research on the effects of acid rain on surface waters, aquatic organisms, forests, agricultural crops, and building materials. In 1990, at the conclusion of NAPAP, a comprehensive report was issued on the findings, and Congress authorized indefinite extension of the program including the NTN network (NAPAP 1990). Additional accounts of the history and politics of the acid rain phenomenon can be found in Cowling (1982) and Gould (1985). This paper reviews the current state of knowledge of acid deposition and its effects on aquatic and terrestrial resources.

ATMOSPHERIC DEPOSITION

Atmospheric deposition is a term that refers to solid, liquid, and gaseous materials deposited by atmospheric processes. Rain and snow are most commonly thought of in this context, but fog, cloud water, aerosols, particulates, and gases are also important with respect to transport of materials through, and deposition from the atmosphere. Rain and snow comprise wet deposition, whereas settleable particulates, aerosols, and gases comprise dry deposition.

The chemical composition of atmospheric deposition is influenced by natural processes (Placet et al 1990). Sulfur compounds (e.g. SO_2, H_2S, CH_3SCH_3, COS, CS_2) and nitrogen compounds (e.g. NO, NO_2) are released to the atmosphere by natural processes, including volcanic emissions, lightning, forest fires, and from biogenic sources (Adams et al 1981, Bates et al 1992, Cadle 1980, Albritton et al 1984). These compounds are called acid precursors because they can be oxidized by atmospheric processes and combine with water to form acids (Placet & Streets 1987). Rainwater in equilibrium with the unpolluted atmosphere (containing only natural acid precursors, CO_2, and organic acids) is slightly acidic, with a pH of about 5 to 5.7. In regions where alkaline dust is prevalent in the atmosphere, rain pH can exceed 7.

Although natural processes affect the chemical composition of atmospheric deposition, anthropogenic pollutants in some parts of the world have become overwhelmingly important. Table 1 shows the composition of rain from some minimally impacted areas (analyses a–f) and from heavily impacted areas (analyses g–i). In proximity to and downwind of heavily industrialized areas, rain is substantially more acidic than would be expected from natural processes (Patrinos 1985, Sisterson & Shannon 1990). The major anthropogenic process that releases SO_2 and NO_X to the atmosphere is the combustion of fossil fuels. Burning of oil and low-grade, high-sulfur coal and peat releases substantial amounts of these compounds, and smelting of nonferrous metals is an additional source of sulfur emissions (USEPA 1990). Exhaust emissions from cars, trucks, planes, and other forms of transportation are the largest source of nitrogen compounds to the atmosphere (USEPA 1990). Application of agricultural fertilizer to fields and the decomposition of animal wastes also contribute nitrogen compounds to the atmosphere.

Today, 95–98% of the sulfur emissions in North America result from human activities (Venkatram et al 1990). In 1985, man-made emissions amounted to nearly 25 million megagrams of SO_2. Approximately 70% of the total U.S. emissions are contributed by coal-burning power plants in the generation of electricity. Sulfur emissions in the U.S. are relatively uniform throughout the year with less than 5% seasonal variation, but the spatial distribution of sources is not uniform across the country.

Like sulfur, emissions of NO_X in North America are dominated by human activities (Venkatram et al 1990). Approximately 90% of nitrogen emissions are anthropogenic. In 1985, nearly 21 million megagrams (as

Table 1 Precipitation-weighted mean annual concentrations 1989 (mg/L) (NADP/NTN 1990)

	a American Samoa	b Olympic Nat. Park WA	c Loch Vale CO	d South Pass City CO	e Everglades Nat. Park FL	f Manua Loa HI	g Leading Ridge PA	h Bennett Bridge NY	i Catoctin Mountain MD
Ca^{2+}	0.08	0.02	0.17	0.28	0.13	0.02	0.11	0.15	0.07
Mg^{2+}	0.199	0.037	0.023	0.033	0.062	0.005	0.020	0.027	0.013
K^+	0.063	0.014	0.014	0.016	0.078	0.003	0.028	0.019	0.053
Na^+	1.677	0.320	0.087	0.23	0.487	0.038	0.068	0.077	0.037
NH_4^+	0.02	0.02	0.14	0.13	0.16	0.03	0.35	0.36	0.30*
NO_3^-	0.04	0.08	0.70	0.61	0.65	0.09	2.14	2.33	1.24
Cl^-	3.05	0.58	0.08	0.13	0.85	0.06	0.17	0.18	0.17
SO_4^{2-}	0.57	0.21	0.60	0.91	0.77	0.95	3.15	2.71	2.11
pH	5.52	5.45	5.31	5.42	5.17	4.82	4.16	4.23	4.35

* Estimated from charge balance.
a–h from NADP/NTN 1990.
i unpublished data.

NO_2) were emitted, of which about 93% was emitted as NO. NO_X emissions are relatively uniform throughout the year. Nearly half of the NO_X emissions are from the transportation sector, so NO_X is somewhat more evenly distributed over the continent than SO_2.

These sulfur and nitrogen compounds may be transported tens to hundreds of kilometers from their source by air masses before they finally return to the earth as acid deposition (Rodhe 1972, Logan 1983). During transit, the sulfur and nitrogen compounds are oxidized and combine with water vapor to form sulfuric and nitric acids, sulfate aerosols, and particulate sulfate (Rodhe et al 1981). Much of the NO_X is emitted closer to ground level than SO_2 and is generally oxidized more rapidly than SO_2. Consequently, NO_X is deposited closer to its source.

Our understanding of the atmospheric chemistry of sulfur and nitrogen compounds is far from complete, but knowledge of some of the oxidation mechanisms is beginning to emerge. The oxidation of sulfur and nitrogen compounds in the atmosphere is accomplished by a sequence of gas-phase and aqueous-phase processes. Most of the oxidants are produced photochemically in the gas phase (Durham et al 1982, McMullen et al 1986). The most important oxidant that causes gas-phase oxidation of sulfur and nitrogen compounds is the hydroxyl radical. Ozone and hydrogen peroxide are the major oxidants responsible for the aqueous-phase oxidation of sulfur dioxide to sulfuric acid (Hicks et al 1990). Many of the photochemical reactions that create oxidants involve volatile organic compounds (Atkinson 1986). Consequently, natural and man-made volatile organic compounds increase photochemical production of oxidants.

There is a question of whether the rate of production and availability of oxidants might be limiting in the conversion of gaseous emissions to acidic compounds (National Research Council 1983). In a region where oxidants are sparse, a reduction in emissions of acid precursors would not be directly reflected in a concomitant reduction in acidic products in air and rain (Hidy 1984). This has been called the nonlinearity problem (Hidy et al 1984, Venkatram et al 1990). A large fraction of the sulfate in rain is formed by aqueous-phase oxidation of SO_2 dissolved in cloud water. Availability of hydrogen peroxide, the major aqueous-phase oxidant, limits the amount of SO_2 that can be oxidized. Near some SO_2 source areas ambient SO_2 concentrations frequently exceed H_2O_2 concentrations. Under these conditions, a change in SO_2 concentration does not result in an equivalent change in the wet deposition rate of sulfur. The spatial scale of wet deposition of sulfur, therefore, is larger than that of dry deposition of sulfur because oxidant availability limits the amount of sulfur that can be deposited in precipitation near source areas (Smith 1985). Although there is nonlinearity between emissions and wet-deposition of sulfur near source areas, on a larger regional scale, however, reductions in sulfur emissions appear to be reflected by linear reductions in total sulfur deposition (Venkatram & Pleim 1985).

Reliable methods for the collection and analysis of wet deposition have been used over the past 15 years (Peden 1986). There is a comprehensive data base for wet-deposition chemistry from 1978 to the present (1992) in the region east of the Mississippi River (NADP network), and for the entire U.S. from 1985 to the present (NADP/NTN network).

Prior to 1978, the chemistry of wet deposition in the U.S. was not well documented. Cogbill & Likens (1974) attempted to calculate pH for the period 1955–1956 from the imbalance between anions and cations using the precipitation chemistry data of Junge (1958) and Junge & Werby (1958), because pH or acidity was not directly measured in these studies. The calculations of Cogbill & Likens (1974) suggest that the pH of rain in the northeastern U.S. decreased in the decade between the mid-1950s and mid-1960s. Measured pH values for rain in this region in 1978 were lower than the calculated pH values for the mid-1960s, but no change has been observed in measured pH values from 1978 to 1990 (Sisterson et al 1990).

The region currently receiving the most acidic precipitation in the U.S. is centered around western Pennsylvania (Figure 1a). Although the annual average pH in this region is <4.2, individual storms with pH values of <3.5 are not uncommon (NADP/NTN 1990). This region also receives the highest precipitation-weighted mean concentrations and loadings of H^+, SO_4^{2-}, and NO_3^- in the Nation (Figures 1a–f). West of the Mississippi River, precipitation pH generally exceeds 4.8 and precipitation-weighted mean concentrations and annual loadings of H^+, SO_4^{2-}, and NO_3^- are much lower than those to the east (Figures 1a–f).

Cloud and fog waters contribute substantially to the water budget in high-elevation systems and in some coastal regions. Under some conditions, cloud water deposition may exceed the amount of rainfall (Schemenauer et al 1988). Typically, cloud water and fog contain 5 to 10 times the ionic concentrations normally found in rain (Weathers et al 1988, Vong 1990, Vong et al 1991). Some of the lowest pH values observed in natural waters were measured in cloud water at Whiteface Mountain, New York (Mohnen & Kadlecek 1989) and Mt. Mitchell, North Carolina (Saxena et al 1989). At both localities, the pH of cloud waters was as low as 2.

WATERSHED PROCESSES

Atmospheric deposition in the form of rain or snow is the ultimate source of virtually all surface waters and groundwaters. The chemical composition

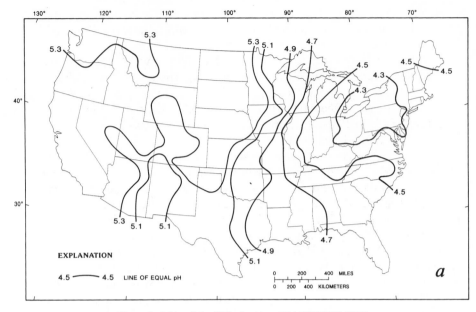

Figure 1   Map of the U.S. showing: (NADP/NTN 1990)
(*a*)   1990 annual precipitation-weighted mean hydrogen ion concentrations as pH.

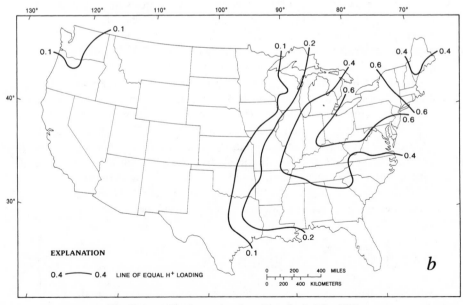

(*b*)   Estimated hydrogen ion deposition (kg/ha) for 1990.

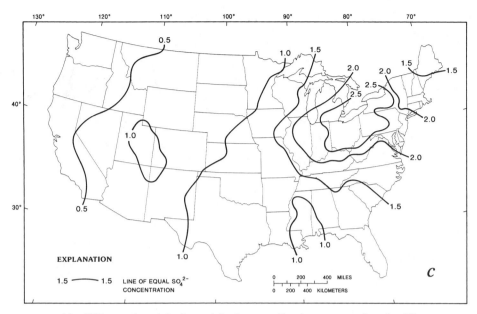

(*c*)   1990 annual precipitation-weighted mean sulfate ion concentrations (mg/L).

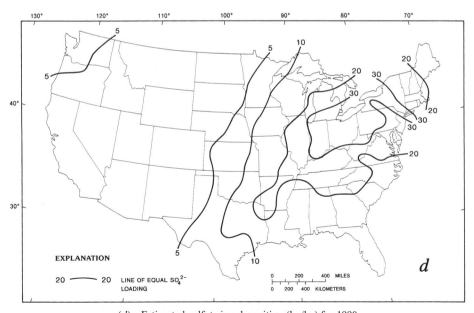

(*d*)   Estimated sulfate ion deposition (kg/ha) for 1990.

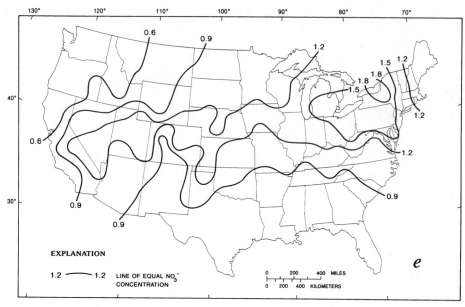

(e) 1990 annual precipitation-weighted mean nitrate ion concentrations (mg/L).

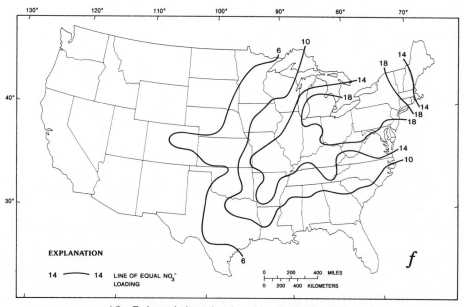

(f) Estimated nitrate ion deposition (kg/ha) for 1990.

of these waters is a function of the chemistry of wet and dry deposition input modified by watershed processes (Turner et al 1990). In a region receiving equal atmospheric inputs, it is common to find surface waters that are quite different in pH and chemical composition (Bricker & Rice 1989). These differences arise from interactions among atmospheric deposition and watershed components, such as vegetation, soils, and bedrock.

A raindrop falling on a typical watershed in the northeastern U.S. first contacts the vegetation canopy where it dissolves dry deposition and leaf exudates that have accumulated since the last rain event (Lindberg et al 1986). The chemistry of the rain that was in contact with the vegetation canopy (throughfall) may be substantially changed depending upon the period of time since the last rain, the type of vegetation, and the acidity of the rain itself. Commonly, the acidity of rain is decreased by passage through deciduous canopies and increased by passage through coniferous canopies (Puckett 1990). Table 2 shows the chemical changes in throughfall under different canopy types at a watershed in northern Virginia. This

Table 2 Chemical changes in throughfall under different canopy types in a watershed in northern Virginia during 1983

		Percent of ion contributed by:	
Ion	Canopy type[a]	Rain	Vegetation canopy
Ca^{2+}	C	24	76
	D	26	74
Mg^{2+}	C	13	87
	D	11	89
Na^+	C	44	56
	D	72	28
K^+	C	21	79
	D	8	92
H^+	C	64	36
	D	141	−41
Cl^-	C	65	35
	D	79	21
NO_3^-	C	53	47
	D	80	20
SO_4^{2-}	C	50	50
	D	66	33

[a] C = coniferous; D = deciduous (modified from Puckett 1990).

throughfall then drips onto the forest floor and passes through the organic litter horizon where the precipitation acquires additional dissolved organic material in the form of dissolved organic carbon (DOC) and, commonly, K^+ and SO_4^{2-}. As the water droplet continues its downward path through the soil horizon, some of the H^+ may be exchanged for base cations from soil exchange sites or react with mineral particles, thereby partially neutralizing the water. Deeper in the soil profile, microbial oxidation of organic matter decreases the DOC content of the water and adds CO_2 to the soil atmosphere. The CO_2 dissolves in soil water, increasing the concentration of H_2CO_3, thereby enhancing weathering processes. If the water and the soil are strongly acidic, Al^{3+} may be displaced from exchange sites or be acquired by dissolution of an aluminum oxyhydroxide phase. Deeper in the soil profile, the drop may react with less weathered bedrock minerals, further increasing the concentration of base cations and decreasing acidity. The chemistry of the water that reaches streams and lakes or recharges the groundwater system, therefore, is governed more by the watershed materials that the water reacts with and the length of time the water is in contact with those materials, than by the original composition of the atmospheric deposition (Bricker 1987). This supports the observation that the chemistry of surface waters and groundwaters in watershed systems reflects, to a large extent, the composition and reactivity of the materials comprising the watershed (Garrels & Mackenzie 1971, Bricker 1986). In regions receiving acidic atmospheric deposition, water in watersheds that develop on bedrock types or surficial materials that contain few reactive minerals that have the capacity to neutralize H^+ will tend to become acidified, whereas water in watersheds containing reactive minerals that have the capacity to neutralize the acid inputs will not become acidified. Waters in these latter systems, however, will reflect the increased atmospheric H^+ inputs by an increase in the dissolved products of weathering and elevated concentrations of SO_4^{2-}.

The mineralogy of the watershed substrate is a good predictor of the sensitivity of waters in the system to chronic acidification by acid deposition (Puckett & Bricker 1992, Bricker & Rice 1989, Norton 1980). Waters in watersheds developed on carbonate rocks (e.g. limestones, dolomites, marbles) or reactive silicate rocks (e.g. dunites, serpentinites, gabbros, amphibolites) are not likely to become acidified because the minerals comprising these rocks rapidly react with H^+. Waters in watersheds developed on silicious clastic rocks (e.g. sandstones, quartzites, siltstones, shales) or sialic crystalline rocks (e.g. granites, gneisses, schists, phyllites) are at risk of acidification because the minerals comprising these rocks do not react, or only react slowly, with H^+. Stream baseflow chemistry strongly reflects the mineralogy of the watershed and is a good predictor of the susceptibility of a system to chronic acidification by acid deposition (Lynch & Dise 1985, Bricker & Rice 1989).

During storms the chemistry of streamflow may change drastically from its baseflow composition (Wigington et al 1990, Kennedy et al 1986). Streams that are well buffered under baseflow conditions may experience a precipitous drop in pH and alkalinity at the peak of storm discharge or snowmelt runoff. Acute depressions of pH and alkalinity associated with runoff events may last for hours to weeks and adversely affect the stream biota (Baker et al 1990).

Storm-induced changes in stream chemistry appear to be related to differences in the paths along which water flows through the system (Wheater et al 1990, Vogt et al 1990). Flow paths determine which materials the water contacts in its passage to the stream and the length of time available for the water to react with these materials. Under normal baseflow conditions, groundwater that feeds the stream is supplied through discharges from the mineral soil and bedrock. During storms, the water table rises and an increasing amount of stream flow is contributed by groundwater from the shallow mineral soil and organic horizon. Water traveling along shallow flow paths through organic-rich material and strongly weathered mineral soil in the near-surface acquires an elevated concentration of dissolved organic carbon and remains acid because it has little opportunity to react with unweathered minerals at depth. In general, the baseflow chemistry of streams reflects the composition of the geological substrate of the watershed, whereas stormflow chemistry reflects processes occurring in the more highly weathered and organic-rich surficial materials.

The chemistry of lakes is also strongly related to the watershed processes discussed above; however, additional in-lake processes (e.g. sulfate reduction) contribute to lake-water chemistry (Schindler 1986). The chemistry of lakes with large watershed area relative to lake surface area is generally dominated by watershed processes, whereas in-lake processes may significantly affect the chemistry of lakes with small watershed area to lake surface area. Lakes integrate the inputs from atmospheric deposition, groundwater, and streams if the lake has a surface inlet. Each lake, depending on its size and rate of inflow and outflow, has a finite water residence time. The longer the residence time, the greater the potential for in-lake processes to modify the lake-water chemistry. In contrast, the water in streams has a short residence time and is less affected by in-stream processes. Therefore, streams will exhibit more rapid chemical changes than lakes in response to changes in atmospheric deposition chemistry, and are better suited than lakes for monitoring short-term effects of acid rain.

ACID RAIN EFFECTS

Surface Waters

Acidification of surface waters occurs when the acidic input to the system exceeds its buffering capability. A surface-water body is considered acidic if the acid-neutralizing capacity (ANC) is less than or equal to zero. Surface waters may become acidic by two processes: (*a*) chronic acidification is caused by acid deposition over a relatively long period of time, e.g. years; and (*b*) episodic acidification is caused by a temporary reduction or loss of ANC of the lake or stream in response to acidic wet deposition (rainfall or snowmelt) that occurs over a relatively short period of time.

The major effect of the acidification of surface waters is the change that occurs in biological communities (Baker et al 1990). Acidification of surface waters affects all major groups of aquatic organisms, which contain zooplankton, algae, bacteria, mollusks, insects, amphibians, and finfish, including important sport fish species. As the pH levels decrease, the species diversity of the biological community is impoverished but populations of acid-tolerant species increase so that commonly little net change in total community productivity occurs. More difficult to detect, however, is the effect on larger organisms of the loss of the lower food chain before it is replaced by acid-tolerant species. As insect, amphibian, and fish populations die out, mammals and birds that rely on these organisms as a food source also may die out or be seriously reduced in numbers.

A decline in the numbers of fish and other aquatic organisms can occur through chronic and episodic acidification of the surface-water body. The organisms are negatively impacted by the high concentrations of H^+ ions and toxic metals (aluminum, cadmium, copper, lead, manganese, and mercury) that may be present in acidic waters. The elevated concentrations of these ions in the water cause osmoregulatory failure in aquatic organisms. In finfish, this leads to the irreversible loss of NaCl from body fluids, subsequently causing death (Baker et al 1990). In addition, the early life stages of fish, the eggs and larvae, are especially sensitive to changes in water chemistry so that an episodic acidification event at spawning time may be particularly damaging. The effects of acid deposition on trout populations in sensitive waters in the southern Appalachians have been described by Camuto (1991).

Groundwaters

Acidification of surface waters by acid deposition has been well documented in many industrialized countries; however, less is known about the effects of acid deposition on groundwater systems. In many parts of the United States and the world groundwater is used for domestic, municipal, and industrial purposes. It is a major source of water in rural areas without municipal distribution systems. Acidification of groundwaters may have deleterious effects on water quality through solubilization of trace metals from the soil and bedrock and from metal pipes and solder joints in water distribution systems. Metals mobilized by acidic waters can pose a threat to human health (Graham et al 1990). Acidic groundwaters used for domestic purposes cause rapid deterioration of metal plumbing if neutralizers are not installed in supply systems, thereby substantially increasing the costs of potable water.

Investigations suggesting the possibility of acidification of groundwater by acid deposition are beginning to emerge from various countries. Bachman & Katz (1986) investigated water quality in the Columbia aquifer, an unconsolidated sand and gravel aquifer system located on the coastal plain deposits of eastern Maryland in the U.S. They observed acidic groundwaters with pH values as low as 4.3, a value just slightly higher than the pH of precipitation falling in this region. Waters from the Columbia aquifer also commonly contain elevated levels of dissolved aluminum. No other trace metals were investigated in the study.

Holmberg (1986) reviewed the potential effects of acid deposition on groundwater and discussed approaches for modeling groundwater acidification. Hendriksen & Kirkhusmo (1986) examined data from the Norwegian Ground Water Observation Network and found low pH values with high aluminum concentrations in groundwaters from areas receiving acid deposition. Jonasson et al (1985, 1987) reported large decreases in alkalinity of well waters in Sweden between 1949 and 1984—an indication of acidification. The acidification trends in Swedish groundwaters from 1950–1985 have been reviewed by von Brömssen (1989). Two common factors at all of the locations where acid groundwaters have been observed are: (a) high loadings of acid from acidic deposition, and (b) watersheds that developed on materials that exhibit little reactivity (neutralizing capacity) with respect to hydrogen H^+ ions.

Forests

In parts of North America, Europe, and Scandinavia, forest damage, purportedly caused directly or indirectly by acid rain, has been observed in various species of trees (Barnard et al 1990). Direct damage by air pollution to tree leaves and needles may be caused by a reduction in foliar nutrition from an increase in leaching of the canopy and, under extreme conditions, by tissue necrosis (Shriner et al 1990). These effects have been demonstrated under experimental conditions but generally require higher levels of acidity than are currently found in the environment (Jacobson et al 1990, Peterson et al 1989). Indirect damage to trees occurs when toxic metals (especially aluminum) interfere with root function and inhibit the uptake of essential elements such as magnesium, calcium, and phosphorus. Ulrich et al (1980) observed a direct correlation between soluble aluminum in soils and the death of feeder roots in birch, fir, and spruce in the Black Forest of Germany. This research attributed the increased concentration of aluminum in the soil and widespread decline in the growth of these forests to increases in the acidity of rain. Another indirect effect is related to the acidification of soil and leaching of soil nutrients by acid rain to a depth below the root zone (Paces 1986, Moldan & Schnoor 1992). Direct and indirect damage to trees by air pollution affects the reproductive processes of trees. Although a direct causal link between forest damage and acid deposition has not been made, enough damage by acid deposition may occur to weaken the tree and enhance its susceptibility to factors such as cold climate shocks, insect attacks, and other air pollutants such as ozone and SO_2 (Shriner et al 1990).

Although the NAPAP investigations (Barnard et al 1990) found that the majority of forests in the U.S. and Canada are not currently affected by acid deposition, high-elevation red spruce forests in the northeastern U.S. have been declining for more than 30 years (McLaughlin et al 1987, Johnson & McLaughlin 1986). Red spruce in the southern Appalachians are also showing signs of decline (Saxena et al 1989). The decline has been attributed to extensive exposure to cloud waters which, in areas of atmospheric pollution, typically have pH values as low as 2 and ionic concentrations that are 5–10 times those found in rainwater (Vong et al 1991, Mohnen & Kedlecek 1989).

Perhaps the most convincing example of the effects of acid deposition on forests is in Czechoslovakia where more than 100,000 ha of Norway spruce forest in the Erzgebirge region has died since the early 1960s (Moldan & Schnoor 1992). Moldan (1990) estimated that 54% of the forests in the Czech Republic have suffered irreversible damage. This region has been subjected to severe air pollution and acid deposition for more than 50 years. The relative effects of wet and dry deposition and the above-ground and soil-acidification and nutrient-leaching processes have not yet been quantified. It is clear, however, that the combination of wet and dry deposition has had a devastating effect on the forests of Czechoslovakia as well as those in former East Germany and Poland.

Although agricultural crops typically are exposed to the same air pollution as forests, acidic precipitation at current ambient levels has not been implicated in any reduction of crop growth or yield (Shriner et al 1990), probably because of the addition of soil amendments to agricultural croplands that nullify the effects of acidic precipitation. However, it has been shown that for some sensitive crop species, reductions in growth and yield can result from exposure to ozone. Although it is believed that pollutant mixtures (e.g. $SO_2 + O_3$, or $SO_2 + NO_2$) have the potential for additive as well as interactive effects on crops, there is a well-established need for additional research (Shriner et al 1990).

Materials

Natural weathering processes cause degradation of virtually all materials that are exposed in the outdoor environment. Research on the effects of acidic deposition suggests that the degradation of materials is accelerated if the wet and dry deposition is acidic (Baedecker et al 1990). Research by NAPAP has focused on quantifying the effects of H^+, SO_2, and NO_X, deposited as wet and dry deposition, on mass loss and other forms of alteration of materials. The NAPAP research concentrated on four categories of damage: (a) damage to metals, (b) damage to carbonate stone, (c) damage to painted wood, and (d) damage to painted metals.

Metals were found to be sensitive to pollution-induced corrosion, especially that caused by SO_2 (Kucera & Mattsson 1987, Mikhailovskii & Sokolov 1985). The metals most susceptible to corrosion include iron, copper, zinc, and aluminum-based metals and alloys. Metal corrosion occurs as a result of reactions with sulfur dioxide, wet acid deposition, and natural acidity from the presence of dissolved CO_2.

Carbonate stone was found to be affected by chemical dissolution and physical grain loss, and carbonate stone erosion is accelerated by acidic deposition (Amoroso & Fassina 1983, Reddy 1988). Field experiments involving marble and limestone slabs which were exposed at a 30° angle from the horizontal revealed the processes responsible for the dissolution of the slabs were: 10% from wet deposition of H^+, 5 to 20% from dry deposition of SO_2 between rain events, 2 to 6% from dry deposition of HNO_3, and the remaining approximately 70% of dissolution was from water in equilibrium with atmospheric CO_2. On areas of carbonate stone sheltered from rainfall, dry deposition of SO_2 was found to cause the formation of gypsum crusts that incorporated carbonaceous material. These crusts, which are ubiquitous on buildings and monuments, are the most visible signs of the effects of acidic pollutants on carbonate stone.

For painted wood, it was found that alkaline pigments ($CaCO_3$ and ZnO) dissolve when exposed to acid, increasing the rate of paint erosion and possibly shortening the service life of the paint (Edney 1989). Reactions involving high concentrations of SO_2 at the paint/wood interface can degrade the wood and affect paint adhesion; however, it is not known whether this reaction affects long-term paint performance. The photochemical degradation of wood is increased by the contact of sulfuric, nitric, and sulfurous acids at low pH (<3.5). It was also found that H_2S and SO_2 can cause color change in pigments containing lead. However, this color change is apparently not an important factor in the degradation of paint coatings.

It is expected that the degradation of painted metals will be accelerated by acidic pollutants if the metallic substrate is exposed by degradation of the paint coating (Munger 1984). Degradation of alkyd-painted steel is accelerated when it is exposed to higher than ambient levels of SO_2 (1 ppm) and 95% relative humidity. Examples of degradation that result are loss of tensile adhesion strength, blistering and rusting (especially at defects), decreases in electrochemical resistance, sulfur adsorption, and discoloration of the paint.

REMEDIATION

Techniques used to mitigate the effects of acid rain fall into two categories: (a) source, and (b) receptor. Remedying the problem at the source involves control of emissions. Dealing with the problem at the receptor end involves the application of an acid buffering material to the affected water body or watershed.

Methods of dealing with the problem at the source involve the reduction of emissions of NO_X and SO_2 released from the fossil fuel combusting source. In the case of coal-fired power plants, technologies for reducing emissions include coal cleaning and upgrading, SO_2 and NO_X control, fluidized-bed combustion, and integrated gasification/combined cycle. To help reduce emissions from cars and trucks, states have required that catalytic converter systems be installed in exhaust systems. Once all emission controls are in place, it may take years or decades before results are observed. Thornton et al (1990) suggested that a considerable amount of time may pass between emission reductions and the improvement of aquatic conditions to the point of restoration of fish populations. Although the results of controlling emissions at the source would not be immediately observable, the results achieved would continue as long as the emissions were controlled.

Several techniques have been implemented at the receptor end to mitigate the effects of acid rain (Olem 1991). Unlike mitigation at the source, the results of the application of a buffering material to the receptor are immediately observable. Although the results are immediate, they are not permanent as long as the acidic input continues. Liming is the most commonly used technique for mitigating acidic surface waters. Liming refers to the application of any base material (e.g. ground limestone, soda ash) to surface waters, sediments, or soils with the intention of neutralizing the acidity.

Neuralization of acidic lakes is accomplished by applying liming material to the water surface by boat or barge, truck or tractor (on ice), by helicopter, or by fixed-wing aircraft (Olem 1990). These same methods are also used when the watershed surrounding an acidic lake is limed.

Liming strategies for streams are slightly different than those for lakes and watersheds. When flowing water is acidic, base materials are added directly to the stream by means of a doser, diversion well, limestone barrier, or rotary drum (Olem 1990).

SUMMARY

Acid deposition and its effects were recognized at the local scale as early as the middle of the 17th century. The problem was well defined and documented in Great Britain and parts of Europe nearly two centuries later by Robert Angus Smith during the period between 1852 and 1872. It then rapidly sank from visibility in both the scientific and public sectors, but surfaced again a century later having grown in the intervening time period from a local-scale problem to one of global proportions.

In the U.S., concern about the environmental effects of acid deposition first led to the initiation of an acid deposition network, the NADP in 1978. This was followed in 1980 by a major research program (NAPAP) designed to identify the causes and extent of the problem and recommend options for managing "acid rain." The original NADP network covered the northeastern U.S. but was expanded under NAPAP to cover the entire U.S.; it was then designated the NTN. Information gained during the 10 year duration of the NAPAP, augmented by results from acid deposition programs in other countries (e.g. Norway SNSF, Great Britain SWAP, Canada CAPMON) and from research at universities and in the private sector, has provided a broad understanding of the acid deposition phenomenon. Many of the details of various atmospheric, terrestrial, aquatic, and biologic processes have yet to be elucidated. However, a general picture has emerged.

In most industrialized countries of the world, anthropogenic emissions of sulfur and nitrogen have overwhelmed those from natural sources. Reduced forms of S and N are oxidized and transformed by atmospheric processes into acidic compounds that return to the Earth's surface as wet and dry deposition. On a local scale near emission sources, there may not be a 1:1 relationship between emissions and acidity in deposition because of a paucity of atmospheric oxidants. On a regional scale, however, the relationship does appear to be linear. The acidity of precipitation appears to have increased in the period between the 1950s and 1970s on the basis of pH values calculated from rain chemistry for the early period and direct measurements of pH for the latter period. Since 1978 when the NADP/NTN was first put into operation, no systematic trends in the acidity of precipitation in the U.S. have been observed.

The effect of acid deposition on surface and ground waters is not uniform. Within a region receiving a given amount of acid deposition, some waters will be acidified, whereas others will not. The degree of acidification strongly depends on the geologic characteristics of the watershed materials and the hydrologic pathways that incoming precipitation follows to the stream, lake, or groundwater system. The mineralogy and composition of the watershed rocks and soils determine the reactivity (capacity for neutralization), and the hydrologic pathways (deep as opposed to shallow) determine which minerals the water contacts and the time available for reaction. Waters draining watersheds that developed on rocks resistant to weathering (e.g. silicic crystalline rocks or noncalcareous clastic rocks) are highly susceptible to acidification by acid deposition, whereas those waters draining reactive rocks (e.g. calcareous rocks, gabbros and ultramafic rocks) are not susceptible to acidification.

The effects of acid deposition on aquatic ecosystems have been well documented and are understood at the large scale. Although many details at the fine scale have yet to be elucidated, there is no longer any doubt that acid deposition can have devastating effects on aquatic communities in poorly-buffered waters that are susceptible to both episodic and chronic acidification.

Forest decline in a number of countries in Europe and North America has been attributed to acid deposition, although the evidence is not as strong as it is with respect to aquatic ecosystems. In Cechoslovakia, much of the forest in the Erzgebirge region has died and the remainder shows severe damage. In Germany, the Black Forest and forests in parts of Bavaria have undergone extensive dieback. In North America, forest damage appears to be confined largely to high-elevation red spruce forests in the eastern and northeastern parts of the continent. The details of the extent of forest damage directly caused by acid deposition relative to gaseous air pollutants (e.g. O_3, SO_2) is not clear; however all of these pollutants largely originate directly or indirectly from common sources.

Most materials are adversely affected by acid deposition and associated gaseous air pollutants. Building stone, particularly limestone and marble, rapidly deteriorates in areas where acid rain and air pollution are prevalent. A good example is the city of Prague where virtually all of the historic buildings and monuments constructed from marble have experienced severe damage. The useful life of other materials, paints and coatings, metals, wood, and cement is reduced in regions experiencing acid deposition, which may substantially increase maintenance costs.

Measures to mitigate acid deposition effects have, to date, focused more on the symptoms than on the causes. In the U.S., Canada, and Scandinavia liming has been extensively used to protect sensitive water bodies. This procedure generally has been effective in the short term, but is a temporary measure at best. Efforts being made to develop more acid-tolerant fish and trees, to manufacture acid-resistant paints and coatings, and to mitigate acid deposition at the receptor end address only the symptoms and not the causes. The long term solution is to control emissions of acid precursors at the source. Although some advances have been made in removing acidifying combustion products from emissions at the source and developing cleaner combustion technology, acid deposition remains a serious environmental problem.

Literature Cited

Adams, D. F., Farwell, S. O., Robinson, E., Pack, M. R., Bamesberger, W. L. 1981. Biogenic sulfur source strengths. *Environ. Sci. Technol.* 15: 1493–98

Albritton, D. L., Liu, S. C., Kley, D. 1984. Global nitrate deposition from lightning. In *Proc. Conf. Environ. Impact of Natural Emissions, 7–9 March, Research Triangle Park, N.C.* Pittsburgh, PA: Air Pollution Control Assoc.

Amoroso, G., Fassina, V. 1983. *Stone Decay and Conservation.* Materials Sci. Monogr. 11. New York: Elsevier. 453 pp.

Atkinson, R. 1986. Kinetics and mechanisms of the gas-phase reactions of the hydroxyl radical with organic compounds under atmospheric conditions. *Chem. Rev.* 86: 69–201

Bachman, L. J., Katz, B. G. 1986. Relationship between precipitation quality, shallow ground-water geochemistry, and dissolved aluminum in Eastern Maryland. Maryland Power Plant Siting Prog. *PPSP-AD-14.* 37 pp.

Baedecker, P. A., Edney, E. O., Simpson, T. C., Moran, P. J., Williams, R. S. 1990. Effects of acidic deposition on materials. *NAPAP State of Science and Technology,* Rep. 19. Washington, DC: Natl. Acid Precip. Assess. Prog. 280 pp.

Baker, J. P., Bernard, D. P., Christensen, S. W., Sale, M. J. 1990. Biological effects of changes in surface water acid-base chemistry. *NAPAP State of Science and Technology,* Rep. 13. Washington, DC: Natl. Acid Precip. Assess. Prog. 381 pp.

Barnard, J. E., Lucier, A. A., Brooks, R. T.,

Dunn, P. H., Johnson, A. H., Karnosky, D. F. 1990. Changes in forest health and productivity in the United States and Canada. *NAPAP State of Science and Technology,* Rep. 16. Washington, DC: Natl. Acid Precip. Assess. Prog. 186 pp.

Bates, T. S., Lamb, B. K., Guenther, A. B. 1992. Sulfur emissions to the atmosphere from natural sources. *J. Atmos. Chem.* 14: 315–37

Bricker, O. P. 1986. Geochemical investigations of selected Eastern United States watersheds affected by acid deposition. *Geol. Soc. London.* 143: 621–26

Bricker, O. P. 1987. Influence of hydrologic pathways on watershed geochemistry. In *Proc. Int. Symp. on Acidification and Water Pathways.* Norwegian Natl. Comm. Hydrol. Bolkesjo, Norway. May 4–8, 1987

Bricker, O. P., Rice, K. C. 1989. Acidic deposition to streams—a geology-based method predicts their sensitivity. *Environ. Sci. Technol.* 23(4): 379–85

Cadle, R. D. 1980. A comparison of volcanic with other fluxes of atmospheric trace gases. *Rev. Geophys. Space Phys.* 18: 746–52

Camuto, C. 1991. Dropping acid in the southern Appalachians. *Trout* 32: 16–39

Cogbill, C. V., Likens, G. E. 1974. Acid precipitation in the northeastern United States. *Water Resour. Res.* 10(6): 1133–37

Cowling, E. B. 1982. Acid precipitation in historical perspective. *Environ. Sci. Technol.* 16(2): 110A–23A

Durham, J. L., Demerjian, K. L., Barnes,

H. M., Wilson, W. E. 1982. Physical and chemical properties of sulfur oxides and particulate matter, Chapter 2. In *Air Quality Criteria for Particulate Matter and Sulfur Oxides. EPA 600/8-82-0296b.* Washington, DC: Criticality Assess. Off. USEPA

Edney, E. O. 1989. Paint coatings: Controlled field and chamber experiments. *U.S. EPA Rep. EPA600/53-89/032.* Atmos. Res. Exposure Assess. Lab. Res. Triangle Park, N.C. 11 pp.

Evelyn, J. 1661. *Fumifugium.* London: Bedel and Collins

Fisher, D. W., Gambell, A. W., Likens, G. E., Bormann, F. H. 1968. Atmospheric contributions to water quality of streams in the Hubbard Brook Experimental Forest, New Hampshire. *Water Resour. Res.* 4(5): 1115–26

Galloway, J. N., Cowling, E. B. 1978. The effects of precipitation on aquatic and terrestrial ecosystems—a proposed precipitation network. *J. Air Pollut. Control Assoc.* 28: 229–35

Gambell, A. W., Fisher, D. W. 1966. Chemical composition of rainfall in eastern North Carolina and southeastern Virginia. *USGS WSP 1535-K.* 41 pp.

Garrels, R. M., Mackenzie, F. T. 1971. *Evolution of Sedimentary Rocks.* New York: Norton. 397 pp.

Gorham, E. 1961. Factors influencing the supply of major ions to inland waters, with special reference to the atmosphere. *Geol. Soc. Am. Bull.* 72: 795–840

Gorham, E. 1976. Acid precipitation and its influence upon aquatic ecosystems—an overview. *Water, Air, Soil Pollut.* 6: 457–81

Gould, R. 1985. *Going Sour: Science and Politics of Acid Rain.* Boston: Birkhäuser. 153 pp.

Graham, J. A., Grant, L. D., McKee, D. J., Schlesinger, R. B., Lounsbury, S. W., et al. 1990. Direct health effects of air pollutants associated with acidic precursor emissions. *NAPAP State of Science and Technology*, Rep. 22. Washington, DC: Natl. Acid Precip. Assess. Prog.

Graunt, J. 1662. *Natural and Political Observations Mentioned in a Following Index, and Made Upon the Bills of Mortality.* London: Martin, Allestry, and Dicas

Hendriksen, A., Kirkhusmo, L. A. 1986. Water chemistry of acidified aquifers in southern Norway. *Water Qual. Bull. II.* 43 pp.

Hicks, B. B., Draxler, R. R., Albritton, D. L., Fehsenfeld, F. C., Dodge, M., et al. 1990. Atmospheric processes research and process model development. *NAPAP State of Science and Technology*, Rep.

2. Washington, DC: Natl. Acid Precip. Assess. Prog.

Hidy, G. 1984. Source-receptor relationships for acid deposition: pure and simple? A critical review. *J. Air Pollut. Control Assoc.* 34: 518–31

Hidy, G., Hanson, D., Henry, R., Ganesan, K., Collins, J. 1984. Trends in historical acid precursor emissions and their products. *J. Air Pollut. Control Assoc.* 34: 333–53

Holmberg, M. 1986. The impact on acid deposition of groundwater: A review. *WP-86-31.* Int. Inst. Appl. Sys. Anal. A-2361. Laxenburg, Austria

Jacobson, J. S., Heller, L. I., Yamada, K. E., Osmeloski, J. F., Bethard, T., Lassoie, J. P. 1990. Foliar injury and growth response of red spruce to sulfate and nitrate acidic mist. *Can. J. For. Res.* 20: 58–65

Johnson, A. H., McLaughlin, S. B. 1986. The nature and timing of the deterioration of red spruce populations in Appalachian forests. In *Monitoring and Assessing Trends in Acidic Deposition,* Committee on Acid Deposition, Long-Term Trends, J. H. Gibson, Chairman, pp. 200–30. Washington, DC: Natl. Acad. Sci.

Jonasson, S., Lang, L.-O., Swedberg, S. 1985. Factors affecting pH and alkalinity—an analysis of acid well waters in south-west Sweden. *Natl. Swedish Environ. Protect. Board Rep. 3021*

Jonasson, S., Lang, L.-O., Swedberg, S. 1987. Analysis of well water data from Värmland. Dug wells 1949–84. Situation report. *Chalmers Univ. Technol., Dep. Geol. Publ. B308,* Göteborg, Sweden

Junge, C. E. 1958. The distribution of ammonia and nitrate in rainwater over the United States. *Eos Trans. Am. Geophys. Union.* 39(2): 241–48

Junge, C. E., Werby, R. T. 1958. The concentration of chloride, sodium, potassium, calcium, and sulfate in rainwater over the United States. *J. Meteorol.* 15(5): 417–25

Kennedy, V. C., Kendall, C., Zellweger, G. W., Wyerman, T. A., Avanzino, R. J. 1986. Determination of the components of stormflow using water chemistry and environmental isotopes, Mattole River Basin, California. *J. Hydrol.* 84: 107–40

Kucera, V., Mattsson, E. 1987. Atmospheric corrosion. In *Corrosion Mechanisms,* ed. F. Mansfield, pp. 211–84. New York: Dekker

Likens, G. E., Bormann, F. H., Johnson, N. M. 1972. Acid rain. *Environment* 14(2): 33–40

Lindberg, S. E., Lovett, G. M., Richter, D. D., Johnson, D. W. 1986. Atmospheric deposition and canopy interactions of major ions in a forest. *Science* 231: 141–45

Logan, J. A. 1983. Nitrogen oxides in the troposphere: global and regional budgets. *J. Geophys. Res.* 88: 10,785–10,807

Lynch, D. D., Dise, N. B. 1985. Sensitivity of stream basins in Shenandoah National Park to acid deposition. *USGS WRIR 85-4115.* 61 pp.

McLaughlin, S. B., Downing, B. J., Blasing, T. J., Cook, E. R., Adams, H. S. 1987. An analysis of climate and competition as contributors to decline of red spruce in high elevation Appalachian forests of the eastern United States. *Oecologia* 72: 487–501

McMullen, T. B., Robinson, E., Tilton, B. E., Westberg, H., Winer, A. M., et al. 1986. Properties, chemistry, and transport of ozone and other photochemical oxidants and their precursors. In *Air Quality Criteria for Ozone and Other Photochemical Oxidants,* Chap 3. *EPA/600/8-84/020bF.* Washington, DC: Environ. Criticality Assess. Off. USEPA.

Mikhilovskii, Y. N., Sokolov, N. A. 1985. New ideas on the mechanism by which sulfur dioxide stimulates the atmospheric corrosion of metals. *Prot. Met.* 21: 214–20

Mohnen, V. A., Kadlecek, J. A. 1989. Cloud chemistry research at Whiteface Mountain. *Tellus* 41B: 79–91

Moldan, B. 1990. *Environment of the Czech Republic,* Parts I and II. Prague: Academic

Moldan, B. 1991. *Atmospheric Deposition: A Biogeochemical Process.* Prague: Academic

Moldan, B., Schnoor, J. L. 1992. Czechoslovakia: Examining a critically ill environment. *Environ. Sci. Technol.* 26: 14–21

Munger, C. G. 1984. *Corrosion Prevention by Protective Coatings.* Houston, TX: Natl. Assoc. Corrosion Eng. 512 pp.

NADP/NTN 1990. *Annual Data Summary.* NADP/NTN Coordination Off. Natural Resour. Ecol. Lab. Colo. State Univ., Fort Collins, Colo.

NAPAP 1990. *Acidic Deposition: State of Science and Technology. Vol. I–IV.* Washington, DC: Gov. Print. Off.

National Research Council 1983. *Acid Deposition: Atmospheric Processes in Eastern North America,* ed. J. Calvert, Washington, DC: Natl. Acad. Press

Norton, S. A. 1980. Geologic factors controlling the sensitivity of aquatic ecosystems to acid precipitation. In *Atmospheric sulfur deposition. Environmental impact and health effects,* ed. D. S. Shriner, C. R. Richmond, S. E. Lindberg, pp. 521–31. Ann Arbor, MI: Ann Arbor Sci. Publ.

Olem, H. 1990. Liming acidic surface waters.

In *NAPAP State of Science and Technology,* Rep. 15. Washington, DC: Natl. Acid Precip. Assess. Prog.

Olem, H. 1991. *Liming Acidic Surface Waters.* Chelsea, Michigan: Lewis. 331 pp.

Paces, T. 1986. Weathering rates of gneiss and depletion of exchangeable cations in soils under environmental acidification. *J. Geol. Soc. London.* 143: 673–77

Patrick, R., Binetti, V. P., Halterman, S. G. 1981. Acid lakes from natural and anthropogenic causes. *Science* 211: 446–48

Patrinos, A. 1985. The impact of urban and industrial emissions on mesoscale precipitation chemistry. *J. Air. Pollut. Control Assoc.* 35: 719–27

Pearson, F. J. Jr., Fisher, D. W. 1971. Chemical composition of atmospheric precipitation in the northeastern United States. *USGS WSP 1535-P.* 23 pp.

Peden, M. E. 1986. Methods of collection and analysis of wet deposition. *Ill. State Water Survey. Champaign, Ill. Rep. 381*

Peterson, C. E., Mattson, K. G., Mickler, R. A. 1989. Seedling response to sulfur, nitrogen, and associated pollutants. *EPA Rep. no. EPA/600/3-89/081.* Corvallis, OR: U.S. Environ. Protect. Agency. 104 pp.

Placet, M., Battye, R. E., Fehsenfeld, F. C., Bassett, G. W. 1990. Emissions involved in acid deposition processes. *NAPAP State of Science and Technology,* Rep. 1. Washington, DC: Natl. Acid Precip. Assess. Prog.

Placet, M., Streets, D. G. 1987. Emissions of acidic deposition precursors. In *Vol. II, NAPAP Interim Assessment.* Washington, DC: Natl Acid Precip. Assess. Prog.

Puckett, L. J. 1990. Estimates of ion sources in deciduous and coniferous throughfall. *Atmos. Environ.* 24A(3): 545–55

Puckett, L. J., Bricker, O. P. 1992. Factors controlling the major ion chemistry of streams in the Blue Ridge and Valley and Ridge physiographic provinces of Virginia and Maryland. *Hydrol. Process.* 6: 79–98

Reddy, M. M. 1988. Acid rain damage to carbonate stone: a quantitative assessment based on the aqueous geochemistry of rainfall runoff from stone. *Earth Surf. Process. Landf.* 13: 335–54

Rodhe, H. 1972. A study of the sulfur budget for the atmosphere over Northern Europe. *Tellus* 14: 128–38

Rodhe, H., Crutzen, P., Vanderpol, A. 1981. Formation of sulfuric and nitric acid in the atmosphere during long range transport. *Tellus* 33: 132–41

Saxena, V. K., Stogner, R. E., Hendler, A. H., DeFelice, T. P., Yeh, R. J.-Y., Lin, N.-H. 1989. Monitoring of the chemical climate of the Mt. Mitchell State Park for evaluation of its impact on forest decline. *Tellus* 41B: 92–109

Schemenauer, R. S., Schuepp, P. H., Kermasha, S., Cereceda, P. 1988. Measurements of the properties of high elevation fog in Quebec, Canada. In *Acid Deposition at High Elevation Sites*, ed. M. H. Unsworth, D. Fowler. NATO ASI Ser. C. Vol. 252, pp. 359–74. Boston: Kluwer

Schindler, D. W. 1986. The significance of in-lake production of alkalinity. *Water, Air, Soil Pollut.* 30: 931–44

Shriner, D. S., Heck, W. W., McLaughlin, S. B., Johnson, D. W., Irving, P. M., et al. 1990. Response of vegetation to atmospheric deposition and air pollution. In *NAPAP State of Science and Technology*, Rep. 18. Washington, DC: Natl. Acid Precip. Assess. Prog.

Sisterson, D. L., Bowersox, V., Meyers, T., Olsen, T., Vong, R. J. 1990. Deposition monitoring: methods and results. *NAPAP State of Science and Technology*, Rep. 6. Washington, DC: Natl. Acid Precip. Assess. Prog. 338 pp.

Sisterson, D., Shannon, J. 1990. A comparison of urban and suburban precipitation chemistry. *Atmos. Environ.* 24B: 389–94

Smith, F. B. 1985. The response of long-term depositions to non-linear processes inherent in the wet removal of airborne acidifying pollutants. *Tech. Note, Meteorol. Off.*, Bracknell, UK

Smith, R. A. 1852. On the air and rain of Manchester. *Mem. Manchester Lit. Philos. Soc.* Ser. 1(10): 207–17

Smith, R. A. 1872. *Air and Rain—The Beginnings of a Chemical Climatology*. London: Longmans & Green. 600 pp.

Thornton, K. W., Marmorek, D., Ryan, P. 1990. Methods for projecting future changes in surface water acid-base chemistry. *NAPAP State of Science and Technology*, Rep. 14. Washington, DC: Natl. Acid Precip. Assess. Prog.

Turner, R. S., Cook, R. B., Van Miegroet, H., Johnson, D. W., Elwood, J. W., et al. 1990. Watershed and lake processes affecting surface water acid-base chemistry. *NAPAP State of Science and Technology*, Rep. 10. Washington, DC: Natl. Acid Precip. Assess. Prog. 167 pp.

Ulrich, B. 1980. Production and consumption of hydrogen ions in the ecosphere. In *Effects of Acid Precipitation on Terrestrial Ecosystems*, ed. T. C. Hutchinson, M. Havas, pp. 255–82. New York: Plenum

USEPA 1990. National air pollutant emission estimates, 1940–1988. *Report EPA-450/4-90-001.* U.S. Environ. Protect. Agency. Res. Triangle Park, N.C.

Venkatram, A., McNaughton, D., Karamchandani, P. K., Shannon, J., Fernau, M., Sisterson, D. L. 1990. Relationships between atmospheric emissions and deposition/air quality. *NAPAP State of Science and Technology*, Rep. 8. Washington, DC: Natl. Acid Precip. Assess. Prog. 110 pp.

Venkatram, A., Pleim, J. 1985. Analysis of observations relevant to long-range transport and deposition of pollutants. *Atmos. Environ.* 19: 659–67

Vogt, R. G., Andersen, D. O., Andersen, S., Christophersen, N., Mulder, J. 1990. Streamwater, soil-water chemistry, and water-flow paths at Birkenes during a dry-wet hydrological cycle. In *The Surface Waters Acidification Programme*, ed. B. J. Mason, pp. 149–54. Cambridge: Cambridge Univ. Press

von Brömssen, U. 1989. Acidification trends in Swedish groundwaters. *Natl. Swedish Environ. Protect. Board Rep. 3547.* 67 pp.

Vong, R. J. 1990. Mid-latitude Northern Hemisphere background sulfate concentration in rainwater. *Atmos. Environ.* 21: 1353–62

Vong, R. J., Sigmon, J. T., Mueller, S. F. 1991. Cloud water deposition to Appalachian forests. *Environ. Sci. Technol.* 25: 1014–21

Weathers, K. C., Likens, G. E., Bormann, F. H., Bicknell, S. H., Bormann, B. T., et al. 1988. Cloudwater chemistry from ten sites in North America. *Environ. Sci. Technol.* 22: 1018–26

Wheater, H. S., Langan, S. J., Miller, J. D., Ferrier, R. C., Jenkins, A., et al. 1990. Hydrological processes on the plot and hillslope scale. In *The Surface Waters Acidification Programme*, ed. B. J. Mason, pp. 121–35. Cambridge: Cambridge Univ. Press

Wigington, P. J. Jr., Davies, T. D., Tranter, M., Eshleman, K. N. 1990. Episodic acidification of surface waters due to acidic deposition. In *NAPAP State of Science and Technology*, Rep. 12. Washington, DC: Natl. Acid Precip. Assess. Prog. 200 pp.

Annu. Rev. Earth Planet. Sci. 1993. 21:175–204

MANTLE AND SLAB CONTRIBUTIONS IN ARC MAGMAS

C. J. Hawkesworth, K. Gallagher, J. M. Hergt, and F. McDermott

Department of Earth Sciences, The Open University, Walton Hall, Milton Keynes, MK7 6AA, United Kingdom

KEY WORDS: radiogenic isotopes, trace elements, slab dehydration, fluid percolation, element fluxes

INTRODUCTION

Destructive plate margins are major sites of terrestrial magmatism that have long had a key role in models for the generation of continental crust and the development of chemical heterogeneities in the upper mantle. In these models it is important to evaluate the nature and size of contributions from preexisting crust in the generation of magmas above subduction zones, and to constrain the size of the fluxes of different elements from the subducted crust and the overlying mantle wedge in arc magmas (Kay 1980, Hawkesworth & Ellam 1989, Hawkesworth et al 1991a, McCulloch Gamble 1991).

The more mobile components of the subducted oceanic crust are the products of element fractionation processes in the upper continental crust, the ocean basins, and hydrothermal systems on the ocean floor. They therefore have distinctive isotope and even trace element ratios that, in principle, should be readily distinguished from those in peridotite in the mantle wedge prior to subduction. In practice, however, element ratios are almost certainly fractionated during shallow level dehydration of subducted material, and at greater depths where material from the slab appears to contribute in the generation of arc magmas. Moreover, the isotope signal of slab-derived material may be heavily diluted by exchange with

175

rocks in the mantle wedge. The result is that there are few unambiguous tracers for slab components in arc magmas, and that most estimates of the relative contributions of the mantle wedge and preexisting subducted crust are linked to preferred models for the movement of fluids and melts through the mantle wedge, and the causes of melt generation.

Along mid-ocean ridges and in intra-plate settings magmas are generated by decompression in response to lithosphere extension, and/or by the development of anomalously high temperatures at shallow levels, as in a mantle plume (McKenzie & Bickle 1988). Arc magmas differ in that they are generated along convergent plate boundaries in areas where the upper mantle has been chilled by the introduction of subducted oceanic crust. The current consensus is that most arc rocks crystallized from parental magmas generated in the mantle wedge, and that melting took place in the presence of water so that the temperature at the solidus was less than that beneath mid-ocean ridges and oceanic islands (Plank & Langmuir 1988, Davies & Stevenson 1992). The inference is that the water was released from the subducted oceanic crust, and that other more mobile elements were transported in the hydrous fluids, but the size and nature of the contribution from the subducted crust, and even how it may be recognized, remain highly contentious. In this review we consider some of the ways in which contributions from the subducted slab and the mantle wedge may be recognized, before looking again at some of the estimated fluxes from subducted material. The emphasis is on the isotope and trace element geochemistry of the rocks themselves, because that must be the basis for assessing the size and nature of the mantle and slab-derived components.

GENERAL GEOCHEMISTRY

A number of authors have emphasized the similarities in the major element compositions of primitive island arc basalts (IAB) and those erupted along mid-ocean ridges (MORB) (Perfit et al 1980, Plank & Langmuir 1988), and concluded that they were derived from similar source regions in the upper mantle. The development of tholeiitic and calc alkaline trends is beyond the scope of this review, but it is increasingly argued that both are derivatives of basaltic parental magmas and that the different trends reflect factors such as differences in water content and the pressure of differentiation (Grove & Kinzler 1986, Meen 1990), and/or the degree of melting of the parental magmas (Miller et al 1992). However, island arc rocks have trace element characteristics that are markedly different from those of MORB and ocean island basalts (OIB)—most notably lower high field strength element (HFSE) abundances relative to the rare earth elements (REE) and the large ion lithophile elements (LILE) (Figure 1). This is the basis for many geochemical schemes to distinguish subduction related magmas from those generated in other tectonic settings (e.g. Pearce & Cann 1973), and it clearly has some link with the subduction process.

The most distinctive feature on mantle-normalized diagrams, such as Figure 1, is the negative anomaly at Nb and Ta which has been interpreted differently by different authors. Some have noted that, apart from their relatively low HFSE abundances, IAB have similar trace element patterns to ocean island basalts. Thus, some island arc basalts may have been derived from slightly enriched source regions in the mantle wedge, but in the presence of an HFSE bearing residual phase (e.g. Morris & Hart 1983, Foley & Wheeler 1990). In contrast, others have viewed the IAB trace element pattern in terms of the selective *addition* of LILE and REE in what is generally termed the subduction component (e.g. Pearce 1983, McCulloch & Gamble 1991, Saunders et al 1991). In these models the HFSE are left behind in the subducted slab, either because it contains a distinctive HFSE bearing phase, or because the fluids released from the subducted slab do not mobilize the HFSE significantly. Such views illustrate recent debates, and in particular highlight some of the underlying questions on the nature of the mantle wedge prior to subduction, the

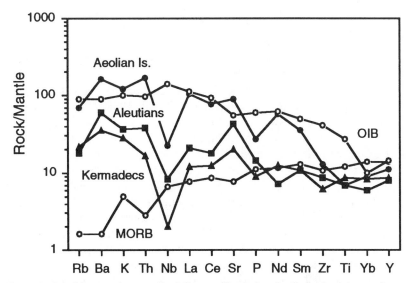

Figure 1   Primitive mantle normalized diagram illustrating the distinctive minor and trace element contents of selected island arc basalts, compared with those of average MORB and OIB (from Sun & McDonough 1989).

nature of the subduction zone component, and ultimately how continental crust is generated.

1. The nature of the mantle wedge prior to the effects of subduction. In many models of the Earth's mantle, chemically and isotopically depleted MORB-type mantle is widespread in the uppermost mantle (Allègre 1982, Zindler & Hart 1986); it should therefore be the dominant material in the mantle wedge above subduction zones. Along continental margins the picture is complicated by the potential presence of trace element enriched material, typically inferred to be in the continental mantle lithosphere. There may also be areas in the oceans where trace element enriched mantle persists at sufficiently shallow levels to be present in the mantle wedge beneath some oceanic arc systems.

2. The nature of the subduction zone component. The subduction zone component is characterized by higher LILE/HFSE and LREE/HFSE ratios (where LREE are light rare earth elements) than those typically observed in mantle-derived melts in other tectonic settings. Thus it is necessary to invoke either different residual phases during partial melting of the slab or the mantle wedge, or the production of fluids of different composition to silicate melts. In one group of models, slab-derived melts, which are presumably fairly siliceous, interact with mantle peridotite (Kelemen et al 1990); in another, the subduction signature in arc basalts is attributed to the introduction of hydrous fluids into the mantle wedge source of the IAB (Tatsumi et al 1986). One cause of confusion has been that because fluids and melts rising from the subducted slab may also scavenge elements from the mantle wedge, some portion of individual element abundances in the so-called subduction zone component may in practice have been derived from the mantle wedge (Le Bel et al 1985, Arculus & Powell 1986, Hawkesworth & Ellam 1989, Hawkesworth et al 1991a). This becomes important in the interpretation of the observed variations in radiogenic isotopes.

3. The generation of continental crust. Most estimates of the bulk composition of the continental crust indicate that it is andesitic in composition, and that it has elevated Rb/Sr and Th/U ratios relative to those of the Earth's mantle (Taylor & McLennan 1985). Yet the dominant flux from mantle to crust along recent subduction zones is basaltic, and higher silica contents and Rb/Sr ratios result from differentiation processes *within the crust* (Perfit et al 1980, Gill 1981, Arculus & Powell 1986, Kay & Kay 1986, Kushiro 1987, Ellam & Hawkesworth 1988a, Ellam et al 1990). It follows either that much of the continental crust was generated by different processes earlier in Earth history, consistent with its average model Nd age of ~2 Ga (Reymer & Schubert 1984),

or that the delamination of gabbroic cumulates from the lower crust back into the upper mantle is sufficiently common for the resultant new crust to be andesitic in composition.

The rare earth elements are widely used in petrogenetic models for the generation of igneous rocks, both because they are a geochemically coherent group and because they include the radioactive decay scheme of ^{147}Sm to ^{143}Nd. The REE data for arc rocks fall broadly into two groups: one with low Ce/Yb in which Ce and Yb vary together, and a second with much higher and more variable Ce contents but similar Yb abundances (Figure 2, Hawkesworth et al 1991b). This is in many ways a crude distinction in that it masks important differences between arc suites within each group, and some arcs, most noticeably the Lesser Antilles and the Sunda Arc, include rocks from both the high and low Ce/Yb groups. Moreover, it has been argued that in some arcs the high Ce/Yb rocks were generated in slightly unusual local tectonic settings (DeLong et al 1975). Nonetheless, the presence of two general trends in Figure 2 requires major differences in either the REE profiles of the source rocks for the two groups, and/or in the bulk distribution coefficients and the degree of melting.

Several authors have subdivided basalts and andesites on the basis of their K_2O abundances (see Gill 1981). Overall there is a broad positive correlation between K_2O and Ce/Yb. The boundary between the high and low Ce/Yb groups in Figure 2 is at Ce/Yb ~ 15. This corresponds to ~ 0.8% K_2O in rocks with < 53% SiO_2. In the scheme of Gill (1981) low-K rocks have < 0.54% K_2O, and the medium-K rocks have 0.54–1.58% K_2O, at 53% SiO_2.

RADIOGENIC ISOTOPES

Island arc and continental margin rocks may contain contributions from fresh and hydrothermally-altered oceanic crust, subducted sediments, and variably enriched or depleted material in the mantle wedge. The altered oceanic crust and sediments include the more mobile material in the downgoing slab, and as they contain a component from the upper continental crust, either directly as continental detritus, or via seawater, they tend to have distinctive radiogenic isotope signatures (White et al 1985, Ben Othman et al 1989, Hart & Staudigel 1989). Trace element calculations suggest that most of the Sr and Pb contents, of IAB in particular, are in the subduction zone component (Figure 1, and Pearce 1983), and that radiogenic isotope ratios are highly sensitive to contributions from different source components. Nonetheless, the general observation is that the Sr,

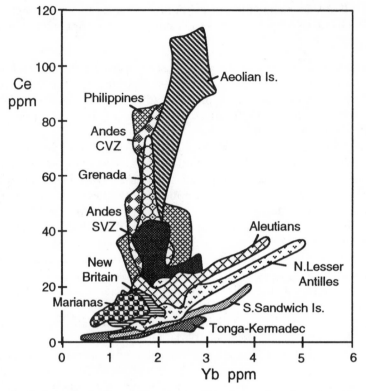

Figure 2   Variations in the Ce and Yb contents of island arc and continental margin rocks with ⩽53% SiO_2. See Appendix for data sources.

Nd, Pb, and Th isotope ratios of subduction related rocks are remarkably similar to those generated in other tectonic settings.

Nd, Sr, and Pb Isotopes

Figures 3–6 summarize published Nd, Sr, Pb, and Th isotope ratios on island arc and continental margin rocks for which major and trace element data are also available. The effects of crustal contamination remain contentious. Thus we have simply selected samples for which it has been argued that such effects were insignificant, excluding on that basis andesites from, for example, Martinique (Davidson 1986), and including those from Cerro Galan and San Pedro in the Central Andes (Rogers & Hawkesworth 1989, Francis et al 1989). Some have disagreed with the suggestion that basaltic andesites with initial Sr isotope ratios of ~0.706 in N. Chile were derived from old segments of trace element enriched mantle (Hildreth & Moorbath 1988, Davidson et al 1990) but, irrespective of such debate over individual sample suites, the key points are that the difference between the high and low Ce/Yb trends in Figure 2 is not obviously due to contamination processes and the two groups tend to have different isotope ratios.

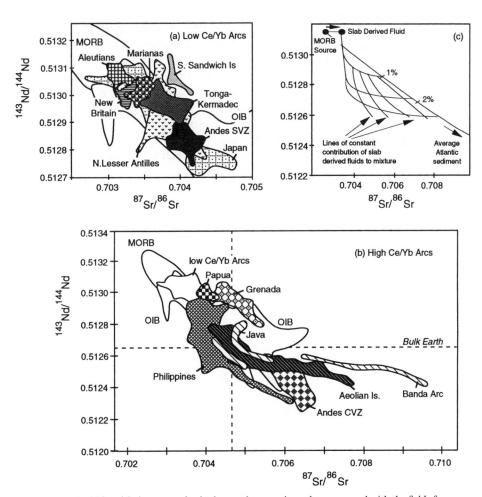

Figure 3 Nd and Sr isotope ratios in destructive margin rocks, compared with the fields for MORB and OIB, after Zindler & Hart (1986). (*a*) Low Ce/Yb arc rocks; (*b*) high Ce/Yb arc rocks; and (*c*) a three-component mixing model between MORB type mantle in the wedge, average Atlantic sediment, and a slab derived fluid (Ellam & Hawkesworth 1988b). Data sources are summarized in the Appendix.

The low Ce/Yb suites have remarkably restricted $^{87}Sr/^{86}Sr$ and $^{143}Nd/^{144}Nd$ and, with the exception of some of the N. Lesser Antilles samples which may contain an upper crustal component, relatively homogeneous Pb isotope ratios. The average Nd, Sr, and Pb isotope ratios of these low Ce/Yb arc rocks are also unexpectedly similar to those of the PREMA mantle component identified by Zindler & Hart (1986) (Hawkesworth et al 1991a). In contrast, the high Ce/Yb rocks have a much greater range in radiogenic isotopes (Figures 3–5), indicating that they were derived from source regions characterized by different trace element ratios. In general these rocks have higher potassium contents, and are more alkalic in composition, at least consistent with their derivation from source regions with higher incompatible element contents. The isotope ratios are enriched relative to depleted mantle, and in the simplest interpretation the high Ce/Yb rocks therefore contain a contribution from material which in many cases was both old enough to have developed different isotope ratios, and was LREE enriched. Such features have been attributed both to subducted sediments and to trace element enriched source regions in the mantle wedge.

Hydrothermally-altered MORB is characterized by high $^{87}Sr/^{86}Sr$ and high $^{143}Nd/^{144}Nd$ ratios, and oceanic sediments have high $^{87}Sr/^{86}Sr$ and low $^{143}Nd/^{144}Nd$ (O'Nions et al 1978, White et al 1985, Ben Othman et al 1989, Hart & Staudigel 1989). Primitive IAB tend to have high Sr/Nd ratios, and mixing between the mantle wedge and high Sr/Nd material from altered MORB and/or sediment in the subducted slab should result in strongly convex upwards mixing lines on Nd-Sr isotope diagrams (Figure 3c). Such mixing processes have therefore been widely invoked to explain the range in Sr isotopes at restricted $^{143}Nd/^{144}Nd$ in low Ce/Yb arc rocks such as in the Aleutians, Marianas, South Sandwich Islands, and Tonga (Figure 3a). However, they do not account for the isotope ratios of arc rocks with lower $^{143}Nd/^{144}Nd$, nor for the general observation that the Nd isotope ratios of arc rocks are lower than those in depleted MORB.

Oceanic sediments have relatively high $^{207}Pb/^{204}Pb$ and $^{208}Pb/^{204}Pb$ ratios compared with those in MORB (Figure 4). Some island arc suites exhibit steeper arrays on Pb–Pb isotope diagrams than those commonly observed in MORB and OIB, strongly suggesting that the lead isotope ratios of such rocks include a significant contribution from subducted sediment (Kay et al 1978, Woodhead & Fraser 1985, White & Dupré 1986). Ellam & Hawkesworth (1988b) observed a weak overall correlation between the size of the displacement to higher $^{207}Pb/^{204}Pb$ in some arc rocks, and their $^{143}Nd/^{144}Nd$ ratios. This implies that the low Nd isotope ratios of arc rocks compared with depleted MORB are due at least in part to a sedimentary component from the subducted slab, and it encouraged

Ellam & Hawkesworth (1988b) to formulate the three-component model illustrated in Figure 3c. Such a model does not offer a unique solution, not least because the mantle wedge and the subducted material may have variable isotope ratios. However, it does demonstrate that the isotope and trace element features of arc rocks can be explained by contributions from a high Sr/Nd component, sediment from the subducted slab, and depleted material in the mantle wedge.

The relatively restricted range of Pb isotope ratios in the low Ce/Yb rocks contrasts with that in MORB and OIB in which $^{206}Pb/^{204}Pb$ varies from 17.3 to 21.1 (Zindler & Hart 1986). It appears that (a) the Pb isotope signature, particularly of the more primitive arc rocks, is dominated by a contribution from a distinctive source component that is available along destructive plate margins in different parts of the world, and/or (b) there is something about the processes associated with arc magmatism that homogenizes Pb isotope ratios very effectively. The effect is presumably the same as that which governed the development of the lead-ore growth curve (Stacey & Kramers 1975), since that curve implies very restricted Pb isotope compositions at any one time, and it appears to have been associated with island arc magmatism. The reason for the restricted Pb isotope ratios in low Ce/Yb arcs is still a matter for conjecture, although in most models they are attributed to mixtures of oceanic sediment and MORB-type mantle, and deep ocean sediments and sulfides have both high Pb contents and relatively restricted Pb isotopes (Kay 1980, Ben Othman et al 1989).

Th Isotopes

The radiogenic isotopes of Nd, Sr, and Pb are the products of decay schemes with half lives of $\sim 10^9-10^{11}$ years. Thus, differences in their isotope ratios require 100s of millions of years to develop, and those observed in IAB are unlikely to reflect element fractionation processes that took place either in the oceanic crust or during subduction. Rather the relatively high ^{87}Sr and ^{207}Pb abundances, for example, in hydrothermally-altered MORB and deep-sea sediments are primarily derived from the preexisting continental crust. However, there are geologically useful decay schemes with much shorter half-lives, and Th isotopes have been the subject of a number of studies to contrast changes in Th/U on time scales of tens of thousands of years with those of hundreds of millions of years inferred from Pb isotopes.

On a $(^{230}Th/^{232}Th)-(^{238}U/^{232}Th)$ diagram, samples in secular equilibrium plot on the equiline (Figure 5a). Many young MORB and OIB are displaced to the left of the equiline, indicating that in the simplest model U/Th in the liquid is less than that in the source during partial melting in the upper mantle (Condomines et al 1988). Some island arc rocks are unusual in that they plot to the right of the equiline, and this has been regarded as a distinctive feature of subduction related magmatism (Allègre & Condomines 1982). In more detail it is clear that the relatively high $(^{238}U/^{232}Th)$ values occur in rocks with relatively low incompatible element contents, and low Ce/Yb (McDermott & Hawkesworth 1991). Rocks with

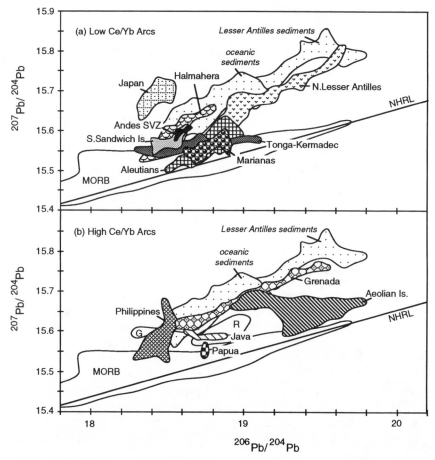

Figure 4   Pb isotope ratios in destructive plate margin rocks, compared with selected fields for MORB and OIB (after Zindler & Hart 1986; G—Gough; R—Reunion), and sediments (White et al 1985, Ben Othman et al 1989, McDermott & Hawkesworth 1991). NHRL is the Northern Hemisphere Reference Line for oceanic basalts from Hart (1984). (*a*) and (*c*) are low Ce/Yb suites, and (*b*) and (*d*) are high Ce/Yb. See Appendix for data sources.

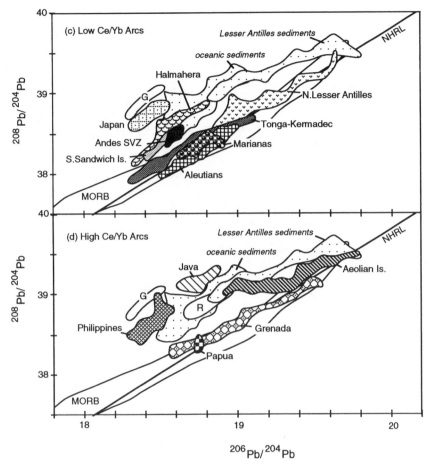

Figure 4 continued

higher incompatible element contents, and in particular those with high Ce/Yb, have lower U/Th, similar to that of the bulk continental crust, and most of them are close to secular equilibrium. The range in Th isotopes in arc rocks is similar to that in MORB and OIB (Gill & Williams 1990, McDermott & Hawkesworth 1991). This contrasts with the very high values of $(^{230}Th/^{232}Th) \geqslant 10$ inferred for altered MORB and marine carbonates from their measured U/Th ratios (Hart & Staudigel 1989).

Pb isotopes may be expressed as $^{208}Pb^*/^{206}Pb^*$, which reflects variations in Th/U on the time scale of 100s of Ma (Allègre et al 1986), and then

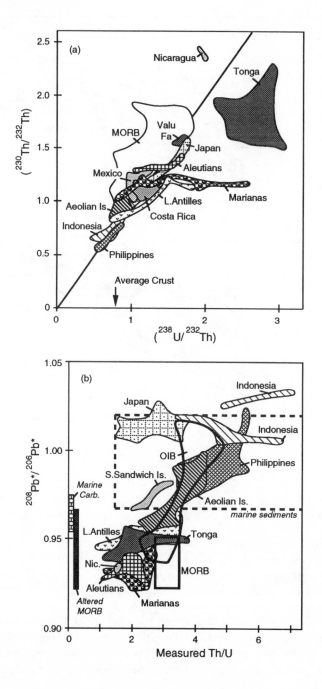

compared with both measured Th/U in arc rocks, and time-integrated Th/U (k_{Th}) estimated from Th isotopes. As in MORB and OIB there is a broad correlation between $^{208}Pb^*/^{206}Pb^*$ and k_{Th} (Allègre et al 1986, Gill & Williams 1990, McDermott & Hawkesworth 1991), as demonstrated by a plot of $^{208}Pb^*/^{206}Pb^*$ against measured Th/U, for which there are many more data (Figure 5b). Thus much of the major (six-fold) variation in measured Th/U between rocks above different subduction zones (for example, between Tonga and the Philippines) does not primarily reflect young Th/U fractionation processes that might be associated with subduction. Rather, it must have been established several 100 Ma before the magmas were generated in order to develop the observed range in $^{208}Pb^*/^{206}Pb^*$. In detail, however, the slope of the array between $^{208}Pb^*/^{206}Pb^*$ and measured Th/U appears to be shallower for arcs than that for MORB and OIB (Figure 5b), due either to differences in the partial melting processes and/or to contributions from different source materials.

Many MORB and IAB have similar Th isotope ratios: Thus much of the difference between measured Th/U in MORB and low Ce/Yb arcs with similar Pb isotope signatures is due to the degree of U-Th disequilibrium (Figure 5a). The interpretation of such disequilibrium depends on the nature of the partial melting processes: For simple melting models Th/U in MORB and OIB is greater than that in their source regions, whereas in some dynamic melting models Th/U in the liquid is similar to that in the source, and the degree of U-Th disequilibrium reflects ingrowth of ^{230}Th during partial melting and melt extraction (Williams & Gill 1989). Carbonate and hydrothermally-altered MORB in the subducted crust have very low Th/U ratios (Figure 5b), and U may be preferentially mobilized from the slab into the mantle wedge. Thus the low Th/U ratios of the

---

Figure 5   (a) A ($^{230}Th/^{232}Th$)—($^{238}U/^{232}Th$) diagram summarizing the available results for subduction related magmas. ^{230}Th is generated within the natural decay chain from ^{238}U to ^{206}Pb, and it has a half life of 75,200 years. In secular equilibrium (^{230}Th) = (^{238}U) (where the parentheses denote activities, which are a function of the decay constant λ, and the number of atoms N, such that $A = \lambda N$), and equilibrium is restored within ~300,000 yr of any change in Th/U. Note that the majority of arc rocks have ($^{238}U/^{232}Th$) values that are significantly higher than most estimates of the bulk continental crust (Taylor & McLennan 1985). (b) $^{208}Pb^*/^{206}Pb^*$ vs measured Th/U ratios in selected IAB rocks, marine carbonates, fresh and altered MORB (Hart & Staudigel 1989), and detrital sediments (Ben Othman et al 1989). Carbonate sediments are assumed to have the present-day Pb isotope composition of the Pb-ore growth curve. Nic.—Nicaragua. Data sources: Oversby & Gast 1968; Sun 1980; Condomines et al 1981; Allègre & Condomines 1982; Capaldi et al 1983; Newman et al 1983, 1984; Krishnaswami et al 1984; Allègre et al 1986; Hemond 1986; Condomines et al 1988; Rubin et al 1989; Gill & Williams 1990; Sigmarsson et al 1990; McDermott & Hawkesworth 1991; and in the Appendix.

low Ce/Yb arcs are likely also to reflect a contribution from subducted material.

Resolving the contributions of such processes to the different Th/U ratios in MORB and the low Ce/Yb IAB is still a matter for debate. In the simplest model it is inferred that, because MORB and the low Ce/Yb IAB have similar Th and Pb isotope signatures, their source regions plotted in similar positions on the U-Th equiline prior to the onset of partial melting, or the introduction of a subduction component. Partial melting beneath MORB then resulted in displacement to high Th/U (low $^{238}U/^{232}Th$, Figure 5a), whereas the introduction of a subduction component beneath island arcs resulted in the displacement to low Th/U observed in some arc suites (i.e. to high $^{238}U/^{232}Th$ in Figure 5). This displacement to low Th/U is best developed in IAB with the lowest incompatible element contents, consistent with an approximately constant U flux from the subducted slab which has a variable effect on U/Th in the mantle wedge depending on the preexisting U and Th contents of the wedge (McDermott & Hawkesworth 1991). In contrast to MORB and OIB, many IAB are close to U-Th equilibrium. This observation may be due to slower rates of melt extraction above subduction zones, or to the length of time required for slab-derived fluids to traverse the wedge prior to melting. Finally, marine sediments have a wide range of measured Th/U (~ 1.5 to > 7), but a paradoxically narrow range in $^{208}Pb^*/^{206}Pb^*$ of 0.965–1.015 (Ben Othman et al 1989, White et al 1985), and they may be responsible for the relatively high Th/U ratio seen in some of the high Ce/Yb suites (Figure 5b, and McDermott et al 1992).

^{10}Be and Boron

^{10}Be is a radioactive nuclide (half-life 1.5 Ma) formed by spallation of oxygen and nitrogen in the atmosphere, and it rapidly becomes concentrated in clay-rich sediments. Thus, the uppermost oceanic sediments have high ^{10}Be concentrations with an average $^{10}Be/Be$ atom ratio of $\sim 5000 \times 10^{-11}$, which is several orders of magnitude higher than that in mantle-derived magmas from mid-ocean ridges, ocean islands, and active continental rifts ($< 5 \times 10^{-11}$) (Tera et al 1986, Morris et al 1990, Morris 1991). A distinctive feature of subduction related magmas is that many of them have markedly higher $^{10}Be/Be$ ratios than those in MORB and OIB (Figure 6), and this requires the recent incorporation of a contribution from young sediments. However, while high $^{10}Be/Be$ ratios are a good indication of a contribution from subducted sediments, they are tracers only for young sediments, and magmas with low $^{10}Be/Be$ may also contain a significant contribution from subducted sediments which are simply older than ~ 6 Ma.

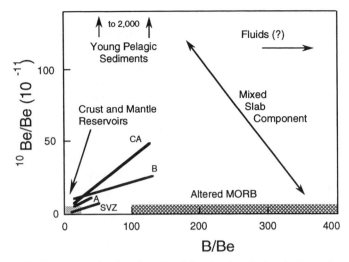

Figure 6 ^{10}Be/Be atom ratio plotted against B/Be showing the trends from selected arcs, fields for crustal and mantle reservoirs (MORB, OIB and the continental crust), altered MORB, and pelagic sediments (from Morris et al 1990). CA—Central America; B—Bismarck; A—Aleutians; SVZ—Andes Southern Volcanic Zone. The ^{10}Be/Be ratios for pelagic sediments are corrected for radioactive decay of ^{10}Be over 2.5 Ma.

Boron is in some respects analogous to ^{10}Be, in that it occurs in high concentrations in oceanic sediments and altered MORB, and it has been used in combination with Be to investigate the subducted component in arc lavas (Morris et al 1990). Figure 6 illustrates the variation in ^{10}Be/Be and B/Be in selected arc suites, and compares them with typical values for oceanic sediments, altered MORB, and crust and mantle reservoirs. The striking observations are: first, that surprisingly good correlations are observed in selected arc suites, and second, that the B/Be ratios of at least some arc rocks are higher than those in subducted sediments. The simplest interpretation is that these straight lines reflect mixing between a high B/Be component from the subducted slab, and a low B/Be component, presumably from the mantle wedge. The high B/Be component would appear to have been relatively homogeneous in each of the areas studied, and because of the elevated ^{10}Be/Be, it must contain a contribution from young subducted sediments. However, the B/Be ratios are greater than those of bulk sediment, and so this component has been interpreted as either a mixture of altered MORB and pelagic sediment, and/or a fluid derived from such material which has a relatively high B/Be ratio (Morris et al 1990). The significance of the ^{10}Be and B data is that while they are highly sensitive tracers for some contribution from the hydrous material in the subducted slab, their signal is not sensitive to the trace element composition of the mantle wedge. Thus, they may offer a way of evaluating whether the size of the hydrous flux is different in the sites of magma generation beneath different arcs. The combination of ^{10}Be and U-Th disequilibria data constrain the length of time between the fractionation of U/Th—presumably during the release of hydrous fluids from the subducted slab—and eruption in an arc magma. Sigmarsson et al (1990) observed that rocks from the southern Andes with high ^{10}Be/Be also had high (^{238}U/^{230}Th), and concluded that the time scale for dehydration, melting, and eruption in this arc was probably less than $\sim 20{,}000$ yr.

We summarize the isotopic evidence as follows:

1. Low Ce/Yb arc suites tend to occur in relatively primitive arc systems and to exhibit a restricted range in Sr, Nd, and Pb isotope ratios. High Ce/Yb arc suites tend to be more alkalic in composition and to possess a much greater range in radiogenic isotopes.
2. Given the diversity of isotope ratios in sediment and altered MORB in the subducted slab, island arc rocks have isotope ratios that are surprisingly similar to those in MORB and OIB. The distinctive features of IAB are that their Nd isotope ratios tend to be lower than those in depleted MORB, and that some arc suites exhibit steeper Pb-Pb arrays than those commonly seen in oceanic basalts. Both features are probably the result of contributions from subducted oceanic sediment.
3. The majority of island arc and continental margin rocks analyzed for Th isotopes are close to secular equilibrium; the remainder tend to be displaced to relatively high U/Th. The latter is attributed to a high U/Th component from subducted material, and the time constraints from the U-Th decay scheme indicate that this high U/Th ratio developed within the last 100,000–200,000 yr.
4. ^{10}Be is at present the one, apparently unambiguous piece of evidence that some material from the subducted slab finds its way into island arc magmas. However, the Be contents of island arc basalts, relative to for example the LREEs, are not noticeably higher than those in MORB and OIB. In an arc rock with an elevated ^{10}Be/Be $= 50 \times 10^{-11}$ (see Figure 6), less than 1% of the Be may have been derived from subducted sediment.

DISCUSSION

In order to evaluate the contributions from the subducted slab and the mantle wedge it is necessary to combine minor and trace element data

(Figure 7) with the isotope results. The simplest approach is first to infer that variations in trace element ratios that are similar to those in MORB and selected OIB, reflect variations in the mantle source regions and/or in partial melting processes that were not affected significantly by subduction. Second, we assume that subduction-related processes were responsible for the distinctive minor and trace element features of arc rocks, such as the high LILE/HFSE ratios.

The element ratios plotted in Figure 7 were selected to illustrate variations in elements with different chemical characteristics. We note the following details: (a) the variation in Na/Yb and Ti/Yb is not significantly different in the high and low Ce/Yb arc suites. Most IAB have Na/Ti in the range 2.5–4, similar to that in average MORB (Na/Ti ~ 3, Table 1 in McKenzie & Bickle 1988), and significantly higher than those in many OIB which have Na/Ti ~ 1. (b) Sm/Ti ratios in many of the low Ce/Yb arc rocks are similar to those in average MORB and OIB, but the high Ce/Yb rocks are displaced to higher Sm/Ti. Such high Sm/Ti ratios might be attributed to a contribution from subducted sediment, but the presence of elevated Sm/Ti in some suites with high ^{143}Nd/^{144}Nd, such as Grenada, suggests that they can also develop in other ways. (c) Ba/Nb and K/Ce ratios are higher in arc rocks than in MORB and OIB. The higher ratios are best developed in samples with lower incompatible element contents and, for example, lower Nb/Y and Ce/Sm ratios. (d) Low Ce/Yb rocks exhibit a much steeper trend on a diagram of Nb/Zr vs Nb than the high Ce/Yb rocks. The latter is consistent with relative bulk partition coefficients for Nb and Zr similar to those inferred from variations in MORB and OIB. However, the steep trend for the low Ce/Yb rocks requires either significantly different relative partition coefficients or, more likely, variable Nb abundances in the source of the low Ce/Yb arc rocks due to, for example, previous depletion events. (e) Samples with high LILE/REE (e.g low Nd/Sr, Figure 7) and high LILE/HFSE ratios tend to have more restricted, and more depleted Nd, Sr, and Pb isotope ratios (see also Hawkesworth et al 1991b). Thus, even though altered MORB and sediment have distinctive, and in many cases relatively enriched isotope ratios, and they may be expected to be readily mobilized from the subducted slab, it is found that those IAB which are inferred to contain the greatest relative contribution from the subduction component (in that they have the highest LILE/LREE and LILE/HFSE ratios) are characterized by the more depleted radiogenic isotope ratios.

This apparent conflict between the size of the subduction component inferred from minor and trace elements, and from radiogenic isotopes, is at the heart of the current debate over the nature and size of element fluxes from subducted material in arc magmas. In his pioneering work, Kay

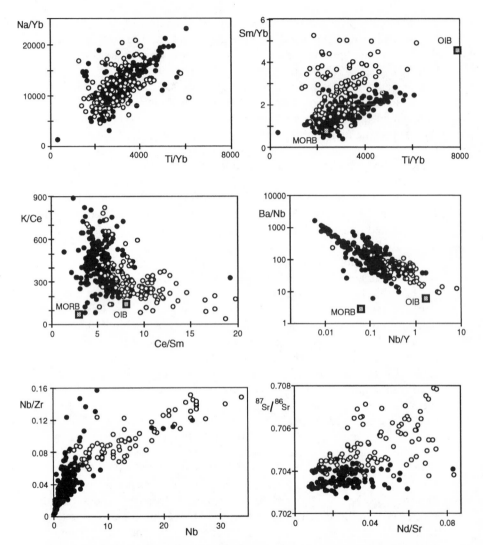

Figure 7   Variations in selected element abundances, element ratios, and $^{87}Sr/^{86}Sr$ in high
(○) and low (●) Ce/Yb arc rocks, compared with average values for depleted, normal (N)-
MORB and OIB from Sun & McDonough (1989).

(1980) argued that the excess potassium responsible for the high K/REE ratios in arc magmas was derived from subducted material, with as much as 90% probably of continental origin. Pearce (1983) extended Kay's approach and proposed that the excess large ion lithophile elements responsible for the high LILE/Nb ratios in arc rocks were from the slab (see also McCulloch & Gamble 1991). He estimated that 70–90% of the Sr, Th, Ba, Rb, and K in a typical Chilean basalt could have been derived from subducted material. Several authors have been concerned that these high values are difficult to reconcile with isotope ratios of many arc rocks, and so inferred that much of the Sr, for example, in the so-called sub-duction zone component, was scavenged from the mantle wedge, and was not necessarily derived from subducted material (Le Bel et al 1985, Arculus & Powell 1986, Hawkesworth & Ellam 1989, Hawkesworth et al 1991a). One consequence of such models is that the size of the slab contribution will be difficult to establish because it may be masked by the contribution from the mantle wedge. Hawkesworth & Ellam (1989) suggested that the slab-derived flux was best estimated in areas where the mantle wedge had been highly depleted in incompatible elements (e. g. Tonga). In such areas all the measured LILE contents of the arc magmas could be inferred to have been derived from the slab, and the relative contribution of such a flux in an estimate of the LILE abundances in average new crust was sufficiently small ($\leqslant 25\%$) that it could be reconciled with the observed isotope ratios.

The relatively low Nb, Ta, and other HFSE abundances in IAB might be due to the presence of a distinctive residual phase which preferentially retained these elements during melt generation. However, the within-suite variations suggest that such elements are incompatible during partial melt-ing (McCulloch & Gamble 1991), and the available experimental data indicate that very high TiO_2 contents are required for rutile saturation, for example, in basaltic melts (Ryerson & Watson 1987). Thus, most authors have attributed the distinctive trace element features of IAB to different relative D values, compared with those inferred from MORB and OIB, which in turn are typically ascribed to the presence of hydrous fluids released from the subducted slab. McCulloch & Gamble (1991) further suggested that the order of elements on a mantle-normalized diagram, such as Figure 1, should be changed for IAB to reflect the different inferred order of incompatibility. The causes of such differences in the presence of a fluid phase are likely to be many and complex. Some elements, such as boron and uranium, may be preferentially mobilized by the development of complexes (e.g. borates), or a change in oxidation state, but there is also a reasonable correlation between the relative enrichment of an element in

IAB and ionic radius, at least in the range 69–167 × 10^{-12}m (Pearce 1983, Hawkesworth et al 1992).

Models that provide a basis for assessing the relative contributions from the subducted slab and the mantle wedge in the generation of IAB vary widely, but a number which serve to illustrate the approaches are summarized in Figures 8–10. One follows the trace element approach of Kay (1980) and Pearce (1983) in inferring that the relative element abundances greater than those of, for example Nb, on a MORB-normalized diagram are in the subduction zone component (Figure 8). This implies that greater than 80% of the large ion lithophile elements were derived from subducted material but, as indicated above, such large proportions may be difficult to reconcile with the available isotope data. For example, if the subducted slab typically has ~1% sediment and ~20% altered MORB, these may contain ~80% of the water in the downgoing slab. Since the high LILE/HFSE ratios of arc magmas appear to be linked to the release of hydrous fluids (e.g. Tatsumi et al 1986), it follows that the slab contribution should be dominated by material from altered MORB and sediment. Moreover, altered MORB and sediment tend to have distinctive isotope ratios, such as elevated $^{87}Sr/^{86}Sr$, and these are difficult to reconcile with the high estimates of the proportion of LILE in the subduction component and the measured isotope ratios in arc rocks.

To overcome this problem, a second approach assumes that the average arc basalt consists of a mixture of depleted MORB-type partial melt from the mantle wedge, contributions from sediment and altered MORB in the subducted slab, and a component scavenged by hydrous fluids from the mantle wedge (e.g. Hawkesworth et al 1991a). Some isotope ratios are more sensitive to the size of the slab contribution than others, because these depend on the differences between the isotope ratios and the element abundances in the slab component, the mantle wedge, and the arc magma. Thus the slab contributions calculated from Nd and even Pb isotopes are less well constrained than those for Sr and Th isotopes, but the latter are ~20% and ~10% respectively (Figure 8).

A third approach is to consider the effects of element fractionation and isotope exchange as fluids released by slab dehydration migrate through the mantle wedge. Recent models for the generation of arc magmas suggest that hydrous fluids traverse the wedge horizontally by a combination of vertical movement as a fluid phase and oblique movement fixed in amphiboles carried by induced mantle flow (Figure 9, and Davies & Stevenson 1992). Only in mantle warmer than amphibole stability may melts escape upwards. During the movement first of hydrous fluids, and subsequently partial melts, trace elements may be fractionated because they

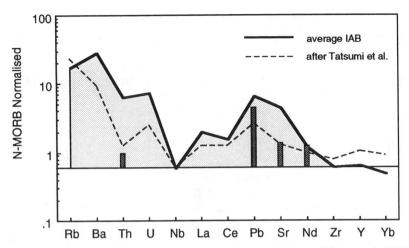

Figure 8 A MORB-normalized trace element diagram (after Sun & McDonough 1989) illustrating the composition of an average low Nb, low Ce/Yb arc basalt, and the maximum contribution of Th, Pb, Sr, and Nd from altered MORB and sediment in the subducted slab (*vertical bars*), calculated from representative radiogenic isotope ratios, and assuming no isotope exchange with the wedge peridotite (Hawkesworth et al 1991a). The shaded area is the subduction contribution calculated by the method of Pearce (1983). The dashed line is the estimated slab contribution, from altered MORB and subducted sediment, which is consistent with both the isotope data and the relative mobilities of different trace elements during dehydration (Tatsumi et al 1986).

have different effective velocities relative to the solid matrix (McKenzie 1984, Navon & Stolper 1987). Several authors have argued that there is little evidence that IAB magmas have been in equilibrium with mantle with a different mineralogy than, for example, MORB (Perfit et al 1980, Plank & Langmuir 1988), and thus the distinctive minor and trace element features of IAB are more likely to reflect the migration of hydrous fluids. In this model, the fluid in the wedge migrates at some direction between its initial vertical vector, and the vector of the mantle matrix sub-parallel to the downgoing slab. The concentration fronts for individual elements travel with vectors intermediate between those of the fluid and the downgoing wedge peridotite, because of their generally lower velocities. For an element to move across from the slab to the zone of partial melting as illustrated in Figure 9, and hence avoid being dragged down with the wedge, it must travel with a velocity greater than the product of the dip of the subducting slab and the velocity of the wedge. The effective velocities for individual elements are inversely dependent on their bulk rock/fluid partition coefficients (*D*), such that those with the lower partition coefficients have the higher effective velocities and are more likely to migrate laterally across the mantle wedge.

The distribution coefficients for peridotite/fluid, and hence wedge/fluid are not well known, and models have been developed for both low and high inferred D values. If the minor and trace element signature of arc rocks is entirely in the fluid, either local equilibrium between the wedge and the fluid is not achieved, or $D_{wedge/fluid}$ must be small (< 1, and probably $\ll 1$). Moreover, models in which some of the trace element subduction component (Figure 8) is scavenged by fluids from the mantle wedge (Le Bel et al 1985, Arculus Powell 1986, Hawkesworth & Ellam 1989, Hawkesworth et al 1991a) further require that $D_{wedge/fluid} \lesssim 0.1 \times D_{slab/fluid}$. Experimental data on the distribution of selected elements between olivine and fluids of different compositions emphasize that partition coefficients are very sensitive to the composition of the fluids (Brenan & Watson 1991), and they indicate that $D_{peridotite/fluid}$ values may be high (~ 0.1–1000). If the $D_{wedge/fluid}$ values are high the trace element content of the fluid are low, and so: (*a*) the trace element characteristics of the inferred IAB source cannot be simply due to bulk addition of fluid (because at equilibrium the trace element contents of the in situ fluid are too low), but rather they reflect modification of the matrix by interaction with the migrating fluid;

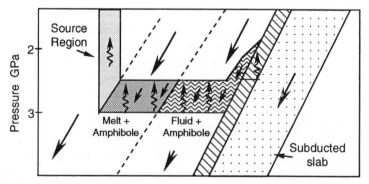

Figure 9   Schematic cross section of the mantle wedge illustrating the release of hydrous fluids from the subducted slab, and the lateral migration to the zone of melt generation (after Davies & Bickle 1991). The fluids traverse laterally by a combination of vertical movement as a fluid, and downward movement fixed in amphibole carried down by induced mantle flow. Element fractionation may occur during the percolation of hydrous fluids, and subsequently partial melts, through the peridotite matrix, and then during partial melting in the magma source regions. Models are discussed in the text for a mantle wedge porosity of 0.5%, and fluid to solid velocity ratios of ~4.

(*b*) modification of a peridotite matrix by a fluid that has relatively low trace element contents requires large amounts of fluid, i.e. high fluid fluxes; and (*c*) high fluid velocities are required to ensure that even the more incompatible elements are not dragged down with the mantle wedge. Hawkesworth et al (1992) used the inferred $D_{peridotite/fluid}$ values to calculate the fluid velocities required to develop the distinctive minor and trace element features of IAB in wedge peridotite. However, the calculated velocities of at least 10^3 km my^{-1} are unrealistically high, suggesting either that the $D_{peridotite/fluid}$ values calculated from the olivine-fluid experimental data of Brenan & Watson (1991) are 2–3 orders of magnitude too high, or that the assumption of local equilibrium was not valid for element partitioning.

A striking feature of arc rocks, and in particular those in the low Ce/Yb group, is that the high LILE/HFSE ratios are best developed in rocks with low HFSE contents, and that variations in Ba, for example, are less than those in Nb (Figure 7). One interpretation is therefore that the more incompatible large ion lithophile elements approached equilibrium peridotite/fluid values in the IAB source prior to melting, and that these may be used crudely to estimate $D_{wedge/fluid}$. Davies & Bickle (1991) and Davies & Stevenson (1992) estimated fluid fluxes from the subducted slab of the order of 0.01–1 kg m^{-2} yr^{-1}, approximately equivalent to porespace velocities of 2–200 km my^{-1}, for a porosity of 0.5%. The velocity of subducting slabs is ~ 30–70 km my^{-1} (Davies & Bickle 1991, Davies & Stevenson 1992), and thus for a typical subduction velocity of 50 km my^{-1}, the fluid to solid velocity ratio in the mantle wedge is likely to be <4, at least for a porosity of 0.5%. The maximum values of $D_{wedge/fluid}$ permissible for an element to maintain a velocity vector directed above the horizontal, and hence to make it across the wedge to the zone of IAB magma generation in the scheme illustrated in Figure 9, is a function of the fluid to solid velocity ratio and the porosity (Figure 10). Thus, for a porosity of 0.5%, and the relatively high fluid to solid velocity ratio of 4, $D_{wedge/fluid}$ is $\sim 10^{-2}$ (Figure 10). For the large ion lithophile elements, this is two orders of magnitude less than that estimated from the olivine/fluid experimental data of Brenan & Watson (1991) by Hawkesworth et al (1992). Additionally, if some LILE are scavenged from the mantle wedge in order to dilute the isotope signature of the slab, a $D_{wedge/fluid}$ value of $\sim 10^{-2}$ implies that $D_{slab/fluid}$ for these elements is >0.1 (see also McCulloch & Gamble 1991).

In summary, most models for assessing the relative contributions of the mantle wedge and the subducted slab in arc magmas concur that the role of fluids is important. However, the degree of interaction between the mantle wedge and slab-derived fluids is not well resolved. The models of Kay (1980), Pearce (1983), Hawkesworth & Ellam (1989), Hawkesworth

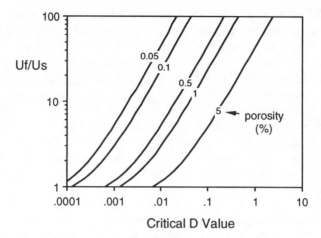

Figure 10 The variation in the critical value of $D_{\text{wedge/fluid}}$ permissible for an element to maintain a velocity vector directed above the horizontal, as a function of the fluid to solid velocity ratio U_f/U_s, and the porosity in the mantle wedge. If $D_{\text{wedge/fluid}}$ is greater than the calculated critical value, the element will be dragged down in the mantle wedge, and so not make it across the wedge to the zone of IAB magma generation, as illustrated in Figure 9.

et al (1991a), and McCulloch & Gamble (1991) are based on the assumption that the fluids do not equilibrate with the wedge, and that the relative contributions can be determined from initial mantle wedge and slab fluid concentrations by mass balance. In contrast, the fluid percolation models of Navon & Stolper (1987) and Hawkesworth et al (1992) assume that complete local equilibrium is achieved between the fluid and the mantle wedge, and that their respective initial concentrations are modified as the two components interact. In reality, it is likely that these are end-member models and that individual elements, and indeed isotope ratios, will achieve different levels of equilibration, depending on the relative velocities and volumes of the fluid and the matrix, as well as additional factors such as crystal chemistry, grain size, and temperature.

CONCLUSIONS

This review has sought to concentrate on those isotopes and minor and trace elements that are more widely used in the study of arc and continental margin rocks, and which have been the basis for many Earth evolution models: It is not a comprehensive review in terms of all possible tracers for contributions from subducted material. We note:

1. Island arc and continental margin rocks have distinctive minor and trace element compositions which may be related to the subduction process, and it appears that the relative contributions from the subducted slab and the mantle wedge are very different for different elements.

2. Magmas erupted along destructive plate margins may be usefully subdivided into high and low Ce/Yb groups. Low Ce/Yb arc suites tend to occur in relatively primitive arc systems, and they exhibit a strikingly restricted range in Sr, Nd, and Pb isotope ratios, often over considerable distances. High Ce/Yb arc rocks have higher K_2O contents, and they exhibit a much greater range in radiogenic isotope ratios, locally even within an individual island.

3. In contrast to their restricted radiogenic isotope ratios, low Ce/Yb rocks have variable Nb and Nb/Y. The latter appear to reflect trace element variations that were too short lived to affect the radiogenic isotope ratios, and in some areas they have been attributed to magma extraction in the associated back arc system.

4. Rocks that contain a larger relative contribution from the subduction component (i. e. have higher LILE/HFSE ratios) tend to have depleted Nd, Sr, and Pb isotope ratios. Moreover, this occurs even though hydrothermally-altered MORB and sediment have distinctive isotope compositions, such as elevated $^{87}Sr/^{86}Sr$, and they are expected to be readily mobilized from the subducted slab.

5. The Nd isotope ratios of island arc rocks are consistently lower than those of depleted MORB, and this is generally interpreted in terms of a small contribution (2–4%) from subducted sediment in arc magmas. However, the isotope and trace element compositions of IAB cannot be modeled by two-component mixing of sediment and N-MORB from the mantle wedge, and they require an additional high LILE/HFSE component which is typically attributed to the release of hydrous fluids from the subducted slab.

6. A number of models have been developed to investigate the effects of fluid percolation in the mantle wedge. $D_{wedge/fluid}$ values are not well constrained but, if local equilibrium occurs, the available data on arc rocks are consistent with $D_{wedge/fluid} \sim 10^{-2}$ for the LILE. Several authors have proposed that some of the trace elements in the subduction component were scavenged from the mantle wedge, in order to reconcile the different contributions from subducted materials inferred from radiogenic isotope and trace element data. In the models outlined here this suggests that $D_{slab/fluid}$ for the LILE is >0.1.

7. Island arcs may evolve with time from relatively primitive low K_2O,

low Ce/Yb suites with restricted isotope ratios, to more evolved and isotopically complex systems. In some areas the range in Nd isotopes, for example, is too great for it to have developed by in situ decay in response to subduction-related changes in Sm/Nd, and so it would appear to require an increased contribution from subducted sediment in the magma source regions.

ACKNOWLEDGMENTS

Our thanks go to many friends and colleagues who over the years have shared a wealth of knowledge and insight into the petrogenesis of arc magmas, and in particular to Rob Ellam, Tony Ewart, Jim Gill, Julie Morris, Julian Pearce, and the late Richard Thorpe, who is sadly missed. This manuscript was greatly improved by the detailed comments of Rob Ellam, Peter Hooper, David Peate, Nick Rogers, Steve Eggins, Simon Turner and Peter van Calsteren, and it was typed by Janet Dryden.

APPENDIX: Principal data sources

Aeolian Is: Ellam 1986; Ellam et al 1988; Ellam et al 1989; Luais 1988
Aleutians: Kay 1980; McCulloch & Perfit 1981; Miller et al 1992; Kay & Kay 1992 and references therein
Andes: Hickey et al 1986; Francis et al 1989; Rogers & Hawkesworth 1989
Fiji: Gill 1984; Gill & Whelan 1989
Indonesia: Morris et al 1982; Wheller et al 1987; Vukadinovic & Nicholls 1989
— Muriah: Edwards et al 1991
— Halmahera: Morris et al 1982
Japan: Tatsumoto & Knight 1969; Nohda & Wasserburg 1981; Ishizaka & Carlson 1983; Tatsumi et al 1988
Lesser Antilles: Davidson 1986; Hawkesworth & Powell 1980; Hawkesworth et al 1979; Thirlwall & Graham 1984; White & Dupré 1986
Marianas: Woodhead 1989; Woodhead & Fraser 1985
New Britain: DePaolo & Johnson 1979
New Hebrides: Dupuy et al 1982; Gorton 1977
Papua New Guinea—Manam Island: Johnson et al 1985
                 —Lihir Island: Kennedy et al 1990
Philippines: Defant et al 1989; Defant et al 1990; McDermott et al 1992
South Sandwich Is.: Hawkesworth et al 1977; Cohen & O'Nions 1982; Barreiro 1983
Tonga–Kermadecs–New Zealand: Ewart et al 1977; Ewart & Hawkesworth 1987

Literature Cited

Allègre, C. J. 1982. Chemical geodynamics. *Tectonophysics* 81: 109–32

Allègre, C. J., Condomines, M. 1982. Basalt genesis and mantle structure studied through Th isotope geochemistry. *Nature* 299: 21–24

Allègre, C. J., Dupré, B., Lewin, E. 1986. Thorium/Uranium ratio of the Earth. *Chem. Geol.* 56: 219–27

Arculus, R. J., Powell, R. 1986. Source component mixing in the regions of arc magma generation. *J. Geophys. Res.* 91(B6): 5913–26

Barreiro, B. 1983. Lead isotopic compositions of South Sandwich Island volcanic rocks and their bearing on magma genesis in intra-oceanic island arcs. *Geochim. Cosmochim. Acta.* 47: 817–22

Ben Othman, D., White, W. M., Patchett, J. 1989. The geochemistry of marine sediments, island arc magma genesis, and crust-mantle recycling. *Earth Planet. Sci. Lett.* 94: 1–21

Brenan, J. M., Watson, E. B. 1991. Partitioning of trace elements between olivine and aqueous fluids at high P-T conditions: implications for the effect of fluid composition on trace element transport. *Earth Planet. Sci. Lett.* 107: 672–88

Capaldi, G., Cortini, M., Pece, M. 1983. U and Th decay-series disequilibria in historical lavas from the Eolian Islands, Tyrrhenian Sea. *Isotope Geosci.* 1: 39–55

Cohen, R. S., O'Nions, R. K. 1982. Identification of recycled continental material in the mantle from Sr, Nd and Pb isotope investigations. *Earth Planet. Sci. Lett.* 61: 73–84

Condomines, M., Morand, P., Allègre, C. J. 1981. ^{230}Th–^{238}U radioactive disequilibria in tholeiites from the FAMOUS zone (M.A.R. 36°N 50'N): Th and Sr isotope geochemistry. *Earth Planet. Sci. Lett.* 55: 247–56

Condomines, M., Hemond, Ch., Allègre, C. J. 1988. U-Th-Ra radioactive disequilibrium and magmatic processes. *Earth Planet. Sci. Lett.* 90: 243–62

Davidson, J. P. 1986. Isotopic and trace element constraints on the petrogenesis of subduction–related lavas from Martinique, Lesser Antilles. *J. Geophys. Res.* 91(B6): 5943–62

Davidson, J. P., McMillan, N. J., Moorbath, S., Worner, G., Harmon, R. S., Lopez-Escobar, L. 1990. The Nevados de Payachata volcanic region (18° S/69° W, N. Chile) II. Evidence for widespread crustal involvement in Andean magmatism. *Contrib. Mineral. Petrol.* 105: 412–32

Davies, J. H., Bickle, M. J. 1991. A physical model for the volume and composition of melt produced by hydrous fluxing above subduction zones. *Spec. Publ. R. Soc. London Ser. A* 335: 355–64

Davies, J. H., Stevenson, D. J. 1992. Physical model of source region of subduction zone magmatism. *J. Geophys. Res.* 97(B2): 2037–70

Defant, M. J., Jacques, D., Maury, R. C., De Boer, J., Joron, J. L. 1989. Geochemistry and tectonic setting of the Luzon Arc, Philippines. *Geol. Soc. Am. Bull.* 101: 663–72

Defant, M. J., Maury, R. C., Joron, J. L., Feigenson, M. D., Leterrier, J., et al 1990. The geochemistry and tectonic setting of the northern section of the Luzon Arc (the Philippines and Taiwan). *Tectonophysics* 183: 187–205

DeLong, S. E., Hodges, F. N., Arculus, R. J. 1975. Ultramafic and mafic inclusions, Kanaga Island, Alaska, and the occurrence of alkaline rocks in island arcs. *J. Geol.* 83: 721–36

DePaolo, D. J., Johnson, R. W. 1979. Magma genesis in the New Britain island arc: constraints from Nd and Sr isotopes and trace element patterns. *Contrib. Mineral. Petrol.* 70: 367–79

Dupuy, C., Dostal, J., Marcelot, G., Bougault, H., Joron, J. L., Treuil, M. 1982. Geochemistry of basalts from central and southern New Hebrides arc: implications for their source rock composition. *Earth Planet. Sci. Lett.* 60: 207–25

Edwards, C., Menzies, M., Thirlwall, M. 1991. Evidence from Muriah, Indonesia, for the interplay of supra-subduction zone and intraplate processes in the genesis of potassic alkaline lavas. *J. Petrol.* 32: 3555–92

Ellam, R. M. 1986. *Comparative study of Roman and Aeolian Island volcanic centres.* PhD thesis. Open Univ. Milton Keynes, UK

Ellam, R. M., Hawkesworth, C. J. 1988a. Is average continental crust generated at subduction zones? *Geology* 16: 314–17

Ellam, R. M., Hawkesworth, C. J. 1988b. Elemental and isotope variations in subduction related basalts: evidence for a three component model. *Contrib. Mineral. Petrol.* 98: 72–80

Ellam, R. M., Menzies, M. A., Hawkesworth, C. J., Leeman, W. P., Rosi, M., Serri, G. 1988. The transition from calc-alkaline to potassic orogenic magmatism in the Aeolian Islands, Southern Italy. *Bull. Volcanol.* 50: 386–98

Ellam, R. M., Hawkesworth, C. J., Menzies, M. A., Rogers, N. W. 1989. The volcanism of Southern Italy: the role of subduction and the relationship between potassic and sodic alkaline magmatism. *J. Geophys. Res.* 94(B4): 4589–4601

Ellam, R. M., Hawkesworth, C. J., McDermott, F. 1990. Pb isotope data from the Proterozoic subduction-related rocks: implications for crust-mantle evolution. *Chem. Geol.* 83: 165–81

Ewart, A. W., Hawkesworth, C. J. 1987. The Pleistocene to Recent Tonga-Kermadec arc lavas: interpretation of new isotope and rare earth element data in terms of a depleted mantle source model. *J. Petrol.* 28: 495–530

Ewart, A. W., Brothers, R. N., Mateen A. 1977. An outline of the geology and geochemistry, and the possible petrogenetic evolution of the volcanic rocks of the Tonga-Kermadec-New Zealand island arc. *J. Volcanol. Geotherm. Res.* 2: 205–50

Foley, S. F., Wheller, G. E. 1990. Parallels in the origin of the geochemical signatures of island arc volcanics and continental potassic rocks: the role of residual titanites. *Chem. Geol.* 85: 1–18

Francis, P. W., Sparks, R. S. J., Hawkesworth, C. J., Thorpe, R. S., Pyle, D. M., et al 1989. Petrology and geochemistry of volcanic rocks of the Cerro Galan caldera, northwest Argentina. *Geol. Mag.* 126(5): 515–47

Gill, J. B. 1981. *Orogenic Andesites and Plate Tectonics.* Berlin: Springer-Verlag. 390 pp

Gill, J. B. 1984. Sr-Pb-Nd isotopic evidence that both MORB and OIB sources contribute to oceanic island arc magmas in Fiji. *Earth Planet. Sci. Lett.* 68: 443–58

Gill, J. B., Whelan, P. 1989. Early rifting of an oceanic island arc (Fiji) produced shoshonitic to tholeiitic basalts. *J. Geophys. Res.* 94: 4561–78

Gill J. B., Williams, R. W. 1990. Th isotopes and U-series studies of subduction-related volcanic rocks. *Geochim. Cosmochim. Acta* 54: 1427–42

Gorton, M. P. 1977. The geochemistry and origin of Quaternary volcanism in the New Hebrides. *Geochim. Cosmochim. Acta* 41: 1257–70

Grove, T. L., Kinzler, R. J. 1986. Petrogenesis of Andesites. *Annu. Rev. Earth Planet. Sci.* 14: 417–54

Hart, S. R. 1984. A large-scale isotope anomaly in the Southern Hemisphere mantle. *Nature* 309: 753–57

Hart, S. R., Staudigel, H. 1989. Isotope characterisation and identification of recycled components. In *Crust/Mantle Recycling at Convergence Zones*, ed. S. R. Hart, L. Gülen, pp. 15–28. Nato Workshop Vol. Dordrecht: Kluwer

Hawkesworth, C. J., Ellam, R. M. 1989. Chemical fluxes and wedge replenishment rates along Recent destructive plate margins. *Geology* 17: 46–49

Hawkesworth, C. J., Powell, M. 1980. Magma genesis in the Lesser Antilles island arc. *Earth Planet. Sci. Lett.* 51: 297–308

Hawkesworth, C. J., O'Nions, R. K., Pankhurst, P. J., Hamilton, P. J., Evensen, N. M. 1977. A geochemical study of island-arc and back-arc tholeiites from the Scotia Sea. *Earth Planet. Sci. Lett.* 36: 253–62

Hawkesworth, C. J., O'Nions, R. K., Arculus, R. J. 1979. Nd and Sr isotope geochemistry of island arc volcanics, Grenada, Lesser Antilles. *Earth Planet. Sci. Lett.* 45: 237–48

Hawkesworth, C. J., Hergt, J. M., Ellam, R. M., McDermott, F. 1991a. Element fluxes associated with subduction related magmatism. *Spec. Publ. R. Soc. London Ser. A* 335: 393–405

Hawkesworth, C. J., Hergt, J. M., McDermott, F., Ellam, R. M. 1991b. Destructive margin magmatism and the contributions from the mantle wedge and subducted crust. *Aus. J. Earth Sci.* 38: 577–94

Hawkesworth, C. J., Gallagher, K., Hergt, J. M., McDermott, F. 1992. Trace element fractionation processes in the generation of island arc basalts. *Philos. Trans. R. Soc. London.* In press

Hemond, Ch. 1986. *Géochemie isotopique du thorium et du strontium dans la série tholéiitique d'Islande et dans des series calco-alcalines diverses.* Thèse 3eme Cycle, Univ. Paris VII. 151 pp.

Hickey, R. L., Frey, F. A., Gerlach, D. C. 1986. Multiple sources for basaltic arc rocks from the southern volcanic zone of the Andes (34°–41°S): trace element and isotopic evidence for contributions from subducted oceanic crust, mantle, and continental crust. *J. Geophys. Res.* 91: 33–45

Hildreth, W., Moorbath, S. 1988. Crustal contributions to arc magmatism in the Andes of Central Chile. *Contrib. Mineral. Petrol.* 98: 455–89

Ishizaka, K., Carlson, R. W. 1983. Nd-Sr systematics of the Setouchi volcanic rocks, southwest Japan: a clue to the origin of orogenic andesite. *Earth Planet. Sci. Lett.* 64: 327–40

Johnson, R. W., Jaques, A. L., Hickey, R. L., McKee, C. O., Chappell, B. W. 1985. Manam Island, Papua New Guinea: petrology and geochemistry of a low-TiO_2 basaltic island-arc volcano. *J. Petrol.* 26: 283–323

Kay, R. W. 1980. Volcanic arc magmas: implications of a melting-mixing model for element recycling in the crust–upper mantle system. *J. Geol.* 88: 497–522

Kay, R. W., Kay, S. M. 1986. Petrology and geochemistry of the lower continental crust: an overview. In *The nature of the lower continental crust*, ed. J. B. Dawson, D. A. Carswell, J. Hall, K. H. Wedepohl. *Geol. Soc. London Spec. Publ.* 24: 147–59

Kay, S. M., Kay, R. W. 1992. Aleutian magmas in space and time. *Decade of North American Geology.* In press

Kay, R. W., Sun, S. S., Lee-Hu, C. N. 1978. Pb and Sr isotopes in volcanic rocks from Aleutian Islands and Pribilof Islands, Alaska. *Geochim. Cosmochim. Acta* 42: 263–73

Kelemen, P. B., Johnson, K. T. M., Kinzler, R. J., Irving, A. J. 1990. High field strength element depletions in arc basalts due to mantle-magma interaction. *Nature* 345: 521–24

Kennedy, A. W., Hart, S. R., Frey, F. A. 1990. Composition and isotopic constraints on the petrogenesis of alkaline arc lavas: Lihir Island, Papua New Guinea. *J. Geophys. Res.* 95: 6929–42

Krishnaswami, S., Turekian, K. K., Bennett, J. T. 1984. The behaviour of ^{230}Th and the ^{238}U decay chain nuclides during magma formation and volcanism. *Geochim. Cosmochim. Acta* 48: 505–11

Kushiro, I. 1987. A petrological model of the mantle wedge and lower crust in the Japanese island arcs. *Geochem. Soc. Spec. Publ.* 1: 165–81

Le Bel, L., Cocherie, A., Baubron, J. C., Fouillac, A. M., Hawkesworth, C. J. 1985. A high-K mantle derived plutonic suite from "Linga," near Arequipa (Peru). *J. Petrol.* 26: 124–48

Luais, B. 1988. Mantle mixing and crustal contamination as the origin of the high-Sr radiogenic magmatism of Stromboli (Aeolian Arc). *Earth Planet. Sci. Lett.* 88: 93–106

McCulloch, M. T., Perfit, M. R. 1981. ^{143}Nd/^{144}Nd, ^{87}Sr/^{86}Sr and trace element constraints on the petrogenesis of Aleutian island arc magmas. *Earth Planet. Sci. Lett.* 56: 167–79

McCulloch, M. T., Gamble, J. A. 1991. Geochemical and geodynamical constraints on subduction zone magmatism. *Earth Planet. Sci. Lett.* 102: 358–74

McDermott F., Hawkesworth C. J. 1991. Th, Pb and Sr isotope variations in young arc volcanics and oceanic sediments. *Earth Planet. Sci. Lett.* 104: 1–15

McDermott, F., Defant, M. J., Hawkesworth, C. J., Maury, R. C., Joron, J. L. 1992. Isotope and trace element evidence for three-component mixing in the genesis of the N. Luzon lavas (Philippines). *Contrib. Mineral. Petrol.* In press

McKenzie, D. P. 1984. The generation and compaction of partially molten rock. *J. Petrol.* 25: 713–65

McKenzie, D. P., Bickle, M. J. 1988. The volume and composition of melt generated by extension of the lithosphere. *J. Petrol.* 29: 625–79

Meen, J. 1990. Elevation of potassium content of basaltic magma by fractional crystallization: the effect of pressure. *Contrib. Mineral. Petrol.* 104: 309–31

Miller, D. M., Langmuir, C. H., Goldstein, S. L., Franks, A. L. 1992. The importance of parental magma composition to calc-alkaline and tholeiitic evolution: evidence from Umnak Island in the Aleutians. *J. Geophys. Res.* 97(B1): 321–43

Morris, J. D. 1991. Applications of cosmogenic ^{10}Be to problems in the Earth Sciences. *Annu. Rev. Earth. Planet. Sci.* 19: 313–50

Morris, J. D., Hart, S. R. 1983. Isotopic and incompatible trace element constraints on the genesis of island arc volcanics from Cold Bay and Amak Island, Aleutians, and implications for mantle structure. *Geochim. Cosmochim. Acta* 47: 2015–30

Morris, J. D., Jezek, P. A., Hart, S. R., Gill, J. B. 1982. The Halmahera Island Arc, Molucca Sea Collision Zone, Indonesia: a geochemical survey. In *The Tectonic and Geologic Evolution of Southeast Asian Seas and Islands. Geophys. Monogr. Ser.*, Part 2, 27: 373–87

Morris, J. D., Leeman, W. P., Tera, F. 1990. The subducted component in island arc lavas: constraints from Be isotopes and B-Be systematics. *Nature* 344: 31–36

Navon, O., Stolper, E. 1987. Geochemical consequences of melt percolation: the upper mantle as a chromatographic column. *J. Geol.* 95: 285–307

Newman, S. M Finkel, R. C., Macdougall, J. D. 1983. ^{230}Th-^{238}U disequilibrium systematics in oceanic tholeiites from 21°N of the East Pacific Rise. *Earth Planet. Sci. Lett.* 65: 17–33

Newman, S., Finkel, R. C., MacDougall, J. D. 1984. Comparison of ^{230}Th/^{238}U disequilibrium systematics in lavas from three hotspot regions: Hawaii, Prince Edward and Samoa. *Geochim. Cosmochim. Acta* 48: 315–24

Nohda, S., Wasserburg, G. J. 1981. Nd and Sr isotopic study of volcanic rocks from Japan. *Earth Planet. Sci. Lett.* 52: 264–76

O'Nions, R. K., Carter, S. R., Cohen, R. S., Evensen, N. M., Hamilton, P. J. 1978. Pb, Nd and Sr isotopes in oceanic ferromanganese deposits and ocean floor deposits. *Nature* 273: 435–38

Oversby, V. M., Gast, P. W. 1968. Lead isotope composition and uranium decay series disequilibrium in recent volcanic rocks. *Earth Planet. Sci. Lett.* 5: 199–206

Pearce, J. A. 1983. Role of the sub-continental lithosphere in magma genesis at active continental margins. In *Continental Basalts and Mantle Xenoliths*, ed. C. J. Hawkesworth, M. J. Norry, pp. 230–49. Nantwich, England: Shiva

Pearce J. A, Cann, J. 1973. Tectonic setting of basic volcanic rocks determined using trace element analyses. *Earth Planet. Sci. Lett.* 19: 290–300

Perfit, M. R., Gust, D. A., Bence, A. R., Arculus, R. J., Taylor, S. R. 1980. Chemical characteristics of island arc basalts: implications for mantle sources. *Chem. Geol.* 30: 227–56

Plank, T., Langmuir, C. H. 1988. An evaluation of the global variations in the major element chemistry of arc basalts. *Earth Planet. Sci. Lett.* 90: 349–70

Reymer, A., Schubert, G. 1984. Phanerozoic addition rates to the continental crust and crustal growth. *Tectonics* 3(1): 63–77

Rogers, G. R., Hawkesworth, C. J. 1989. A geochemical traverse across the north Chilean Andes: evidence for crust generation from the mantle wedge. *Earth Planet. Sci. Lett.* 91: 271–85

Rubin, K. H., Wheeler, G. E., Tanzer, M. O., Macdougall, J. D., Varne, R., Finkel, R. 1989. ^{238}U decay series systematics of young lavas from Batur Volcano, Sunda Arc. *J. Volcanol. Geotherm. Res.* 39: 215–26

Ryerson, F. J., Watson, E. B. 1987. Rutile saturation in magmas: implications for Ti-Nb-Ta depletion in island-arc basalts. *Earth Planet. Sci. Lett.* 86: 225–39

Saunders, A. D., Norry, M. J., Tarney, J. 1991. Fluid influence on the trace element compositions of subduction zone magmas. *Philos. Trans. R. Soc. London Ser. A* 335: 377–92

Sigmarsson, O., Condomines, M., Morris, J. D., Harmon, R. S. 1990. Uranium and ^{10}Be enrichments by fluids in Andean arc magmas. *Nature* 346: 163–65

Stacey, J. S., Kramers, J. D. 1975. Approximation of terrestrial lead isotope evolution by a two-stage mode. *Earth Planet. Sci. Lett.* 26: 207–21

Sun, S. S. 1980. Lead isotopic study of young volcanic rocks from mid-ocean ridges, ocean islands and island arcs. *Philos. Trans. R. Soc. London Ser. A* 297: 409–45

Sun, S. S., McDonough, W. F. 1989. Chemical and isotopic systematics of oceanic basalts: implications for mantle composition and processes. In *Magmatism in Ocean Basins*, ed. A. D. Saunders, M. J. Norry, 42: 313–45. London: Geol. Soc. Spec. Publ.

Tatsumi, Y., Hamilton, D. L., Nesbitt, R.

W. 1986. Chemical characteristics of fluid phase released from a subducted lithosphere and origin of arc magmas: evidence from high-pressure experiments and natural rocks. *J. Volcanol. Geotherm. Res.* 29: 293–309

Tatsumi, Y., Nohda, S., Ishazaka, K. 1988. Secular variation of magma source compositions beneath the northeast Japan Arc. *Chem. Geol.* 68: 309–16

Tatsumoto, M., Knight, R. J. 1969. Isotopic composition of lead in volcanic rocks from central Honshu—with regard to basalt genesis. *Geochem. J.* 3: 53–86

Taylor, S. R., McLennan, S. M. 1985. *The Continental Crust: Its Composition and Evolution*. Oxford: Blackwell. 312 pp

Tera, F., Brown, L, Morris, J., Sacks, S., Klein, J., Middleton, S. 1986. Sediment incorporation in island-arc magmas: inferences from ^{10}Be. *Geochim. Cosmochim. Acta* 50: 535–50

Thirlwall, M. F., Graham, A. M. 1984. Evolution of high-Ca, high-Sr C-series basalts from Grenada, Lesser Antilles: the effects of intra-crustal contamination. *J. Geol. Soc. London* 141: 427–45

Vukadinovic, D. , Nicholls, I. A. 1989. The petrogenesis of island arc basalts from Gunung Slamet volcano, Indonesia: trace element and $^{87}Sr/^{86}Sr$ constraints. *Geochim. Cosmochim. Acta* 53: 2349–64

Wheeler, G. E., Varne, R., Foden, J. D., Abbott, M. J. 1987. Geochemistry of Quaternary volcanics in the Sunda-Banda arc, Indonesia, and three component genesis of island-arc basaltic magmas. *J. Volcanol. Geotherm. Res.* 32: 137–60

White, W. M., Dupré, B. 1986. Sediments subduction and magma genesis in the Lesser Antilles: isotopic and trace element constraints. *J. Geophys. Res.* 91(B6): 5927–41

White, W. M., Dupré, B., Vidal, Ph. 1985. Isotope and trace element geochemistry of sediments from the Barbados Ridge–Demerara Plain region, Atlantic Ocean. *Geochim. Cosmochim. Acta* 49: 1875–86

Williams, R. W., Gill, J. B. 1989. Effects of partial melting on the uranium decay series. *Geochim. Cosmochim. Acta* 53: 1607–19

Woodhead, J. D. 1989. Geochemistry of the Mariana arc (western Pacific): source composition and processes. *Chem. Geol.* 76: 1–24

Woodhead, J. D., Fraser, D. G. 1985. Pb, Sr and ^{10}Be isotopic studies of volcanic rocks from the northern Mariana Island. Implications for magma genesis and crustal recycling in the Western Pacific. *Geochim. Cosmochim. Acta* 49: 1925–30

Zindler, A., Hart, S. 1986. Chemical geodynamics. *Annu. Rev. Earth Planet. Sci.* 14: 493–571

Annu. Rev. Earth Planet. Sci. 1993. 21:205–25

TRENDS AND PATTERNS OF PHANEROZOIC ICHNOFABRICS

Mary L. Droser

Department of Earth Sciences, University of California, Riverside, California 92521

David J. Bottjer

Department of Geological Sciences, University of Southern California, Los Angeles, California 90089–0740

KEY WORDS: trace fossils, bioturbation, infaunal ecospace

INTRODUCTION TO THE ICHNOFABRIC CONCEPT

The record of biogenic structures in sedimentary rocks typically consists of a variety of identifiable trace fossils associated with a background of unidentifiable bioturbation structures, which in previous studies have commonly been termed *burrow mottling.* The object of most ichnological studies to date has been to document, identify, and interpret these identifiable trace fossils in a sedimentary unit (elite trace fossils; Bromley 1990). Assemblages of trace fossils have then commonly been synthesized for paleoenvironmental analysis in the context of ichnofacies (e.g. Seilacher 1967, Frey & Pemberton 1984).

This total record of sedimentary rock fabric resulting from bioturbation has recently been termed *ichnofabric.* Ichnofabric was first defined as "that aspect of a sediment's texture and internal structure that arises from bioturbation and bioerosion at all scales related to the level of biological input" (Ekdale & Bromley 1983). It includes discrete identifiable trace fossils, along with mottled bedding, all produced through the activities of organisms forming surface tracks and trails and infaunal burrows (and borings). While discrete identifiable trace fossils provide important sedimentological and paleoenvironmental information, a great deal of data are

205

lost by recording only this type of information. A variety of different factors can contribute to the ichnofabric that is observed in a sedimentary rock. Biological factors include: the rate of bioturbation, the style of the ichnofabric (i.e. is it controlled by vertical or horizontal elements)—which to some extent is a function of ichnofacies, and the evolution of bioturbating organisms. Physical factors include: sedimentation rate, episodicity of sedimentation (and thickness of resulting episodic beds), composition of the substrate, water depth, and variations in oxygen content and salinity.

Studies of ichnofabrics have concentrated on the record of bioturbation as viewed in vertical cross-section (e.g. Figure 1). Thus, the contribution to ichnofabric of burrows that have a vertical component has been emphasized, a bias that will be reflected in this article. Similarly, just as it is difficult to evaluate discrete identifiable trace fossils in certain lithofacies such as

---

Figure 1   Characteristic recurrent ichnofabrics, as viewed in vertical cross section. (*A*) Interval from Seegrasschiefer beds of Jurassic Posidonienschiefer (southern Germany), which were deposited in oxygen-deficient paleoenvironments. Small burrows in lower part of photograph are *Chondrites*, larger burrows in uppper part are *Thalassinoides*. Composite ichnofabric shows smaller *Chondrites* cross-cutting larger *Chondrites*, and both sizes of *Chondrites* cross-cutting *Thalassinoides*. Analysis of tiering relationships from composite ichnofabric indicates that *Chondrites* existed in lower tiers, and that *Thalassinoides* was emplaced in an upper tier. Ichnofabric index is 4 at bottom of photograph, and 5 at top. (Modified from Savrda & Bottjer 1989a.) (*B*) Interval from chalk of Upper Cretaceous Annona Formation (southwestern Arkansas), deposited in outer shelf paleoenvironments. Trace fossils with preservation indicating that they were emplaced in the transition layer include those expressed as larger ovals (*Planolites*), straight to meandering "stripes" (*Zoophycos*), and faint "dots" and burrow segments (*Chondrites*). Analysis of composite ichnofabric reveals *Chondrites* cross-cutting *Zoophycos* and *Planolites*, and *Zoophycos* crosscutting *Planolites*, indicating that *Chondrites* occupied the deepest tier, *Zoophycos* occupied an intermediate tier, and *Planolites* occupied the shallowest tier of the transition layer. Ichnofabric index is 5. (*C*) Interval from Lower Cambrian Zabriskie Quartzite, southern Nopah Range, southeastern California, deposited in a high-energy nearshore siliciclastic paleoenvironment, showing the trace fossil *Skolithos* and piperock ichnofabric. Ichnofabric index is 4. (*D*) Interval from Eocene Torrey Sandstone, Solana Beach, California, deposited in a high-energy nearshore siliciclastic paleoenvironment, showing the trace fossil *Ophiomorpha* and *Ophiomorpha* ichnofabric. Ichnofabric index is 2. (*E*) Interval from cored Triassic strata recovered on ODP Leg 122 showing thin event beds of parallel- to cross-laminated siltstones to fine-grained sandstone deposited in a prodelta setting, showing the trace fossil *Anconichnus* (dark fecal core with pale burrow fill) and *Anconichnus* ichnofabric. Ichnofabric index is 5 at base of lower bed and 4 at top; ii is 2 in uppper bed. (*F*) Interval from Poleta Formation of the White-Inyo Mountains, California, deposited in an inner shelf carbonate paleoenvironment. Bioturbation (light-colored patches) shown by indistinct horizontal *Planolites* and *Thalassinoides* that has been diagenetically enhanced. Ichnofabric index is 3. The lengths of the scale bars are equivalent to 2 cm (*A*), 3 cm (*B*), 10 cm (*C,D,F*), and 4 cm (*E*).

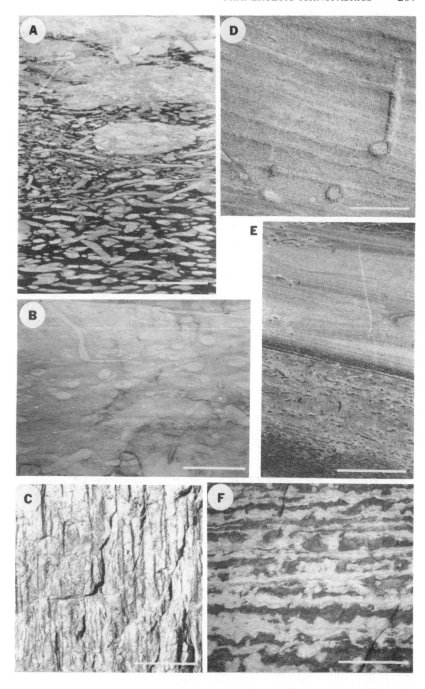

massive carbonates, it is equally difficult to evaluate ichnofabrics in other facies such as shales (particularly from outcrops); this limitation is also reflected in our existing knowledge of ichnofabric.

Recently, workers have recognized that ichnofabric studies provide a unique approach for investigating research problems in paleobiology, stratigraphy, and sedimentology. For example, ichnofabric studies have been effectively utilized to examine the nature of the history of the metazoan radiation into the infaunal habitat and have also begun to be used as a corroborative data base in basin analysis studies. Ultimately, ichnofabric is the source of information on the complete history of the activity of bioturbators recorded in sedimentary rocks. This article discusses the insights that can be gained by utilizing the variety of new methods and approaches which have been developed for examining ichnofabric, those ichnofabrics which occur commonly, and several temporal trends in ichnofabric which have been documented from the stratigraphic record.

METHODS OF ICHNOFABRIC ANALYSIS

Ichnofabric and Tiering

Paleoecologists have determined that marine organisms living in ancient environments were tiered (Ausich & Bottjer 1982, Bottjer & Ausich 1986). Tiering has been defined as the distribution of organisms in niches at different levels above and below the substrate (Ausich & Bottjer 1982, Bottjer & Ausich 1986). This phenomenon is well-known as stratification among ecologists who study modern ecosystems; because in geological studies stratification has other broad meanings, tiering is the preferred term for ancient settings (Ausich & Bottjer 1982). Research on trace fossils has for some time indicated that information on tiering of the ancient infauna can be reconstructed from patterns of bioturbation (e.g. Seilacher 1962). However, such studies have only recently become common (Savrda & Bottjer 1986, 1987, 1989a,b; Ekdale & Bromley 1983, 1991; Bromley & Ekdale 1986; Wetzel & Aigner 1986; Wetzel 1984)

Vertical profiles of modern marine sediments exhibit differences in sediment water content, tiering, and preservation of trace fossils. The surface layer, termed the "mixed layer," is a few centimeters to a decimeter thick, and because it has a relatively high water content, does not preserve biogenic sedimentary structures well (Ekdale et al 1984). The underlying transition layer, typically ranging from one to several decimeters thick, has less water content and is characterized by tiers of organisms that live or feed at greater depths in the sediment column; biogenic sedimentary structures emplaced in this layer have sharper boundaries than those in the mixed layer, and are thus typically better preserved (Figure 2A) (Ekdale et al 1984). With sediment accretion and upward migration of these two actively bioturbated zones, these burrows become part of the historical layer—a zone in which no new burrowing takes place (Ekdale et al 1984).

Because infauna are strongly tiered, this upward migration of the sediment column creates what has been termed a "composite ichnofabric" (Bromley & Ekdale 1986), where burrows of organisms in the lower tiers cross-cut burrows of organisms in the shallower tiers, as mixed and transition layers migrate upward with steady sediment accretion (Figure 2*B*). Due to differences in preservation, such as sharpness of burrow walls, burrows emplaced in the mixed layer can also typically be distinguished from those emplaced in the transition layer.

Documentation of original tiering relationships from the composite ichnofabric, through analysis of cross-cutting relationships, has revo-

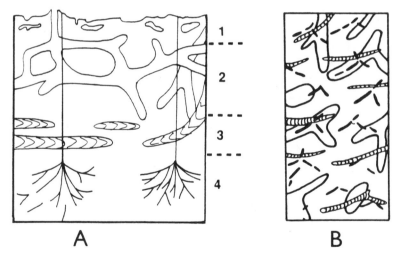

Figure 2 Tiering of infauna as recognized from study of trace fossils. (*A*) Schematic vertical profile of modern marine sediments, showing four tiers, labeled on right. Tier 1 contains poorly preserved burrows of the mixed layer, tier 2 *Thalassinoides*, tier 3 *Zoophycos*, tier 4 *Chondrites*; tiers 2–4 are in the transition layer. Thickness of sediment profile showing mixed and transition layers can be quite variable, but in this example it is 15 cm. (*B*) Schematic showing composite ichnofabric (transition layer burrows only) that is produced with sediment accretion and upward migration of tiers in (*A*). *Chondrites* (tier 4) cross-cuts *Zoophycos* (tier 3) and *Thalassinoides* (tier 2), *Zoophycos* (tier 3) cross-cuts *Thalassinoides* (tier 2), and *Thalassinoides* (tier 2) cross-cuts no transition layer burrows. Modified from Savrda & Bottjer (1986).

lutionized the study of biogenic sedimentary structures in ancient sedimentary rocks. These studies have led to the realization that different sedimentary settings have physical and biological conditions which have led to different tiering relationships of infauna, and hence different but characteristic ichnofabrics for each setting. Several characteristic ichnofabrics have been documented, and a number of examples are discussed in a later section.

Ichnofabric Indices: Evaluation of the Extent of Bioturbation

Bioturbation disrupts the original physical sedimentary fabric and structures. In many previous studies observations on the extent of bioturbation (or ichnofabric) recorded in sedimentary rocks were only subjectively reported, using terms such as "poorly bioturbated," "moderately bioturbated," and "well bioturbated." While such data on ichnofabric may be internally consistent within any one study, it may not be comparable to similar data from other studies. Utilization of this type of ichnofabric data thus has not evolved to the level of synthesis employed with identifiable trace fossils and ichnofacies. Classification schemes of ichnofabric based on the amount of disturbed original sedimentary fabric have been proposed a number of times (e.g. Moore & Scrutton 1957, Reineck 1967, Howard & Frey 1975, Frey & Pemberton 1984), but none of these approaches have included a standardized field and laboratory methodology for semi-quantitatively measuring the extent of bioturbation as reflected in ichnofabric.

To render ichnofabric data on the extent of bioturbation more comparable between different facies within and between basins, as well as between facies of different ages, the ichnofabric index (ii) methodology was developed. This methodology allows the semi-quantitative ranking of the extent of bioturbation as reflected by ichnofabric (Droser & Bottjer 1986) (Figure 3). Through examination of numerous outcrops of different sedimentary facies, six natural categories of ichnofabric were defined (Droser & Bottjer 1986; Figure 3). These categories reflect the extent of bioturbation, based on the degree to which the original physical sedimentary structures were disrupted by biogenic reworking, from no bioturbation (ii1) to complete homogenization (ii6). The indices can be represented by schematic diagrams such as those shown in Figure 3, which have been developed for a number of sedimentary facices (Droser & Bottjer 1986, 1989b, 1991). By using a standardized scheme for recording ichnofabric data on the preserved extent of bioturbation, this information becomes amenable for comparative studies and syntheses of the extent of bioturbation recorded by ichnofabrics between facies within one basin, several basins of the same age, or basins of different ages.

Measurement of Ichnofabric Indices

In outcrops or split cores, ichnofabric indices can be recorded from vertical sequences in a manner similar to the continuous logging of other proper-

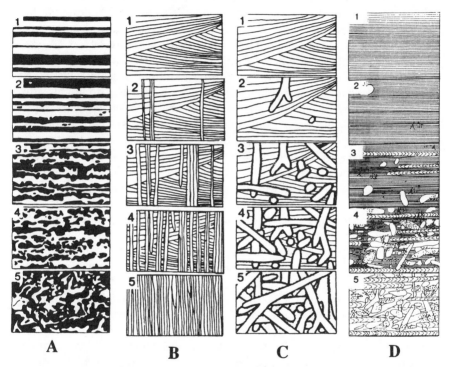

Figure 3   Schematic diagrams of ichnofabric indices 1 through 5 for strata deposited in: (*A*) shelf environments; (*B*) high-energy nearshore sandy environments dominated by *Skolithos*; (*C*) high-energy nearshore sandy environments dominated by *Ophiomorpha*; (*D*) deep-sea deposits. Ichnofabric indices are defined as follows: 1. No bioturbation recorded—all original sedimentary structures. 2. Discrete, isolated trace fossils—up to 10% of original bedding disturbed. 3. Approximately 10–40% of original bedding disturbed. Burrows are generally isolated, but locally overlap. 4. Last vestiges of bedding discernible; approximately 40–60% disturbed. Burrows overlap and are not always well defined. 5. Bedding is completely disturbed, but burrows are still discrete in places and the fabric is not mixed. Ichnofabric index 6, for which there is no schematic diagram, is nearly totally homogenized sediment. Modified from Droser & Bottjer (1991).

ties, such as rock color or lithology. On outcrops, ichnofabric is measured either during or as an adjunct to measurement and description of stratigraphic sections. Because ichnofabric varies laterally, a standard horizontal field-of-view for recording ichnofabric indices should be maintained. In outcrop studies, the standard to date has been to visually record ichnofabric indices from a 50-cm wide vertical strip (Droser & Bottjer 1986, 1988, 1989a,b). When ichnofabric index data is reported, the width of the field of view should always be indicated. This recording of ichnofabric index data is done by visually scoring continuous vertical intervals of the field-of-view with respect to its ichnofabric index, and the thickness of each scored vertical interval is also recorded (one centimeter is typically the thinnest vertical interval recorded) (Figure 4A).

Because ichnofabric indices directly reflect the amount of biogenic disruption of the original sedimentary fabric, additional schematic diagrams of ichnofabric indices (e.g. Figure 3) for facies where they do not already exist should be produced before work begins on a novel facies type. Use of these idealized diagrams (Figure 3) is not necessary for recording ichnofabric indices in the field, but they are provided as an aid to the eye towards increasing the standarization of data collection between different observers. Ichnofabric indices are most useful in strata that does not show complete bioturbation. In cases with complete bioturbation, ichnofabric index 5 represents complete disruption of physical sedimentary structures (Figure 3). Although a series of additional ichnofabric index flashcards has been developed for a variety of ichnofabrics that show complete bioturbation, these have no biological meaning (other than indicating complete bioturbation), and are useful strictly for descriptive purposes (Droser & Bottjer 1991).

Ichnofabric index data can be tallied over a given thickness and then normalized in terms of percent of total measured thickness. This normalized ichnofabric index data is best presented in the form of histograms, termed "ichnograms," which provide a useful summary of ichnofabric measurements when computed from data collected from strata deposited in a single genetically-defined sedimentary environment (Figure 4B) (Bottjer & Droser 1991, Droser & O'Connell 1992). Ichnograms allow ichnofabric index data to be synthesized for comparative facies studies, in much the same way that trace fossil assemblages are used in the ichnofacies concept (e.g. Frey & Pemberton 1984).

Ichnofabric indices are defined according to a natural gradation that is readily recognizable in sedimentary rocks. As this is not an ordinal scale, a statistical average of measured ichnofabric indices cannot be computed in the usual sense. Instead, an average ii can be determined for a particular sedimentary facies using the following formula:

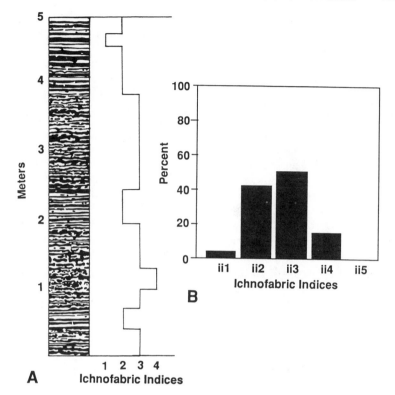

Figure 4 Measurement of ichnofabric indices. (*A*) Schematic diagrams of hypothetical stratigraphic section logged for ichnofabric indices. (*B*) Ichnogram computed from this hypothetical data. Logging is based on a 50-cm-wide field of view. Of 5 m measured, ii1 was recorded from 15 cm, ii2 from 200 cm, ii3 from 255 cm, and ii4 from 30 cm. Average ichnofabric index is 2.6. Modified from Bottjer & Droser (1991).

Average Ichnofabric Index = %strata represented by ii1 [1]+ %strata represented by ii2 [2] + · · · + %strata represented by ii6 [6].

Average ichnofabric index can be used together with ichnograms as a simple way to characterize the extent of bioturbation in strata deposited in a single genetically-defined sedimentary environment (e.g. Figure 4*B*).

RECURRING ICHNOFABRICS

In the same way that sedimentary facies, characterized by a typical association of sedimentary structures, occur in strata of different ages and from different geographic regions, ichnofabrics, which are types of biogenic sedimentary fabrics, also recur. Because ichnofabrics are produced by biological processes, the temporal or stratigraphic distribution of a given ichnofabric is controlled to some extent by evolution. Recurrent ichnofabrics are also affected by very specific locally constrained biological and physical parameters, and thus are not as broadly applicable as Seilacher's ichnofacies (Seilacher 1967). However, because of these constraints, they can be extremely useful in sedimentological and paleobiological studies.

Ichnofabrics of Strata Deposited in Oxygen-Deficient Settings

Stagnant oceanographic conditions commonly develop in settings where fine-grained sediments are being deposited. In conditions of reduced availability of bottom-water dissolved oxygen, the size and depth of penetration of burrows below the seafloor decreases as bottom-water oxygenation decreases (Savrda & Bottjer 1986). In such environments, mixed/transition/historical layers (as outlined above) also occur below the seafloor, but as bottom-water oxygenation decreases, either along a lateral seafloor redox gradient or through time, the thickness of the surface mixed layer decreases as the redox boundary rises to the seafloor (Savrda & Bottjer 1986). Interaction of these factors leads to an infaunal tiering structure in which the occupants of the shallowest tier produce the largest burrows, and the occupants of the deepest tier produce the smallest burrows (Figure 1B) (Savrda & Bottjer 1986, 1987, 1989a,b). Depending upon the level of oxygenation, such settings can typically have from one to three infaunal tiers. In oxygen-deficient settings with minimal oxygen concentrations, only one tier of small-diameter burrows exists, typically *Chondrites* (Bromley & Ekdale 1984; Savrda & Bottjer 1986, 1987, 1989a,b). As bottom-water oxygen concentrations increase a second tier of larger burrows, such as *Zoophycos*, develops, which then overlies the tier of smaller burrows. Further increases in oxygenation commonly generate the development of yet another tier of even larger burrows above the two tiers already discussed; burrows in this tier can include *Thalassinoides* and *Planolites*. Thus, ichnofabrics in strata deposited in settings with the lowest levels of bottom-water dissolved oxygen will consist typically of *Chondrites*, which were emplaced in a shallow tier just below the seafloor. Ichnofabrics of strata deposited in such settings with several burrow types of different sizes indicate greater levels of dissolved oxygen concentrations, and the original tiering structure is easily deciphered through analysis of cross-cutting relationships (Figure 1B). Ichnofabric indices from oxygen-deficient strata typically can range from ii1, where only laminations are present, to ii2 or ii3 in *Chondrites*-dominated ichnofabric, to ii4 or ii5 where ichnofabrics are the product of cross-cutting by several tiers of trace fossils (Figure 1B) (Bottjer and Droser 1991).

Chalk Ichnofabrics

The ichnofabric concept was largely developed through the study of biogenic structures in chalks (Ekdale & Bromley 1983, 1991; Ekdale et al 1984; Bromley & Ekdale 1986; Bromley 1990). Chalk is initially deposited as a fine-grained pelagic carbonate sediment in deeper-water environments, although in the Cretaceous much of it was deposited on shelves. The ichnofabrics of chalks are the best-studied of all sedimentary rocks, and since chalk is deposited in settings ranging from well-oxygenated to oxygen-deficient (e.g. ROCC Group 1986), a variety of chalk ichnofabrics exist.

In oxygen-deficient settings chalk ichnofabric occurs as described above (e.g. Savrda & Bottjer 1989b). Well-oxygenated settings differ in that the redox boundary occurs at much greater depths in the sediment than is found in oxygen-deficient settings. Thus, there are a greater number of tiers of trace fossils, the mixed and transitional layers are relatively thick, and there is less likely to be a consistent size-grading of traces from shallow to deeper tiers as described for oxygen-deficient environments (e.g. Bromley 1990, Ekdale & Bromley 1991). Composite ichnofabrics of chalks that were deposited in well-oxygenated environments can thus record cross-cuting relationships of several (e.g. Figure 1C) to many tiers of trace fossils. Indeed, one example of ichnofabric from Upper Cretaceous chalk in Denmark has been described with cross-cutting relationships indicating that nine tiers of trace fossils were emplaced during sedimentation (Ekdale & Bromley 1991). Typical trace fossils found in chalk ichnofabrics include *Thalassinoides*, *Taenidium*, *Zoophycos*, and *Chondrites* (Bromley 1990). Chalk, like many deep-sea deposits, is usually intensely bioturbated, showing an ii of 5 or 6 (Droser & Bottjer 1991). Recognition of ichnofabric indices less than 5 in chalk sections indicates a sudden change in conditions of sedimentation, such as the onset of oxygen-deficiency or deposition of a turbidite. Upper Cretaceous chalks also can show the development of complex hardgrounds, which because they involve burrowing, seafloor cementation, and mineralization, as well as boring and physical erosion, may have some of the most complex ichnofabrics in the sedimentary record (e.g. Kennedy & Garrison 1975).

Ichnofabrics in Nearshore Sandstones

The nearshore terrigenous clastic setting, or *Skolithos* ichnofacies, is characterized by shifting substrates, moderate to high-energy settings, and episodic erosion or deposition in shallow to intertidal marine environments. Body fossils are commonly poorly preserved and the record of bioturbation thus provides critical evidence for the paleontological history of this setting. In the Paleozoic, well-developed ichnofabric first appears with the advent of *Skolithos*. Dense assemblages of *Skolithos* produce a characteristic ichnofabric, commonly called "piperock," in this facies (Figure 1). Once this behavior is established in the Early Cambrian, it sets the stage for the rest of the Paleozoic. In the post-Paleozoic, *Ophiomorpha* dominates the *Skolithos* ichnofacies (Figure 1) and *Skolithos* piperock is rare (e.g. Curran 1984). *Ophiomorpha* consists of three-dimensional boxwork networks (Frey et al 1978) as opposed to the two-dimensional vertical *Skolithos* burrows. The different geometries of these burrows result in the production of different types of ichnofabrics (Figure 5) which characterize Paleozoic and post-Paleozoic nearshore sandstones.

While *Skolithos* can be found in the Vendian, piperock first appears in the Lower Cambrian. Although piperock is most commonly formed by *Skolithos* several other vertical burrows may occur densely packed and form piperock, such as *Monocraterion* and the U-shaped burrows *Arenicolites* and *Diplocraterion* (Droser 1991). Piperock ichnofabric is generally confined to terrigenous clastic sandstones with only rare occurrences in carbonates (A. Curran 1991 and R. Pickerill 1991, personal communications). Piperock is generally considered to be indicative of shoreface/littoral conditions. However, it also occurs in a variety of facies within the broad framework of the shallow marine-nearshore setting (Droser 1991). Piperock has even been reported from shelf deposits occuring in amalgamated storm beds and with hummocky cross-stratification. This ichnofabric also occurs within estuarine and lagoonal deposits, tidal channels, and intertidal sediments, as well as rarely in nonmarine facies.

The high-energy nature of the nearshore environment sets the stage for unique stratification. While bioturbation by itself cannot be viewed as an independent event, the overriding influence of physical events in this setting creates conditions for the production of "event bioturbation." This is particularly true for the Paleozoic. In Paleozoic strata, although there are cases for which ii4 and ii5 are recorded from nearly the entire unit—and rarely are units composed of ii2 and ii3, commonly, ichnofabric index 1 consititutes approximately 50% of logged strata (Droser & Bottjer 1989b) (Figure 5). In these cases ii4 and ii5 are recorded from the remainder of the strata (Figure 5). This results in discrete beds on the order of centimeters to a meter in thickness with either all physical structures preserved (ii1) or piperock (ii3, ii4, and ii5) (Figure 5). Thus, where complete or nearly complete disturbance of the original physical sedimentary structures by bioturbation occurred, bioturbation acted to produce a type of stratification. Such features have been termed "biostratification structures" (Frey 1978) and also include biogenic graded bedding (e.g. Rhoads & Stanley 1965) and biogenic stratification (Meldahl 1987). Piperock as a

Figure 5 Schematic of characteristic distribution of ichnofabric indices within: (*A*) sandstone strata with *Skolithos* as the dominant trace fossil, typical of Paleozoic nearshore settings; and (*B*) sandstone strata with *Ophiomorpha* as the dominant trace fossil, typical of post-Paleozoic nearshore settings. Modified from Droser & Bottjer (1989).

biostratification feature appears to be limited to Paleozoic rocks, while only unusual examples of densely packed *Skolithos* exist in the post-Paleozoic.

In post-Paleozoic nearshore sandstones, similar to the Paleozoic, ichnofabric index 1 is commonly recorded from approximately 50% of a given unit (Figure 5). However, intervals with completely disturbed bedding are relatively rare suggesting that the original physical event stratification of post-Paleozoic nearshore sandstones is rarely completely destroyed, particularly if the dominant trace fossil is *Ophiomorpha*.

Anconichnus *Ichnofabric*

While many ichnofabrics consist of several burrow types (e.g. most oxygen-deficient and chalk ichnofabrics), others are characterized by a single trace fossil and commonly do not have associated traces (such as many examples of the *Ophiomorpha* and *Skolithos* ichnofabrics described above). Another example of an ichnofabric that typically consists of just one trace type is the *Anconichnus* ichnofabric, which is characteristic of post-Paleozoic siliciclastic strata deposited between the upper offshore and lower shore-face in shelf depositional environments (Goldring et al 1991). The characteristic trace of this ichnofabric is *Anconichnus horizontalis*, which is endogenic and generally emplaced in a shallow tier, and appears as a narrow, discontinuous, twisting muddy fecal string within a poorly defined burrow fill (Figure 1*E*) (Goldring et al 1991). Normally this trace fossil occurs in association with a variety of other trace fossils. However, in event beds deposited in shelf environments, it occurs alone to form the *Anconichnus* ichnofabric (Goldring et al 1991). Although this ichnofabric has only recently been recognized and described (Goldring et al 1991), it appears to be common and has subsequently been described from other examples (Figure 1*E*) (Droser & O'Connell 1992).

ICHNOFABRIC TRENDS THROUGH TIME

The Metazoan Radiation into the Infaunal Habitat

Discrete trace fossils have proven their worth in understanding the early history of metazoans, particularly during the interval of time from the late Precambrian through the early Paleozoic. The increase in trace fossil diversity and complexity across the Precambrian-Cambrian boundary is well documented (e.g Crimes 1992a,b) and, indeed, the boundary is now formally based on the first appearance of key ichnogenera (see Narbonne et al 1987). Discrete trace fossils provide important information about complexity and diversity of behaviors as well as modes of locomotion. However, it is through evaluation of ichnofabric that it is possible to evaluate changes in the extent of bioturbation and hence the degree to which the infaunal habitat was being utilized.

In order to determine the nature of the metazoan radiation into the infaunal habitat, ichnofabric index data were recorded from Cambrian and Ordovician carbonate strata of the Great Basin (western United States) (Droser & Bottjer 1988, 1989a). Studied sections were categorized as having been deposited in inner, middle, or outer shelf environments. Inner shelf environments are defined as being below fair-weather wave base and above mean storm wave base; middle shelf environments are below mean storm wave base and above maximum storm wave base; and outer shelf environments are below maximum storm wave base, but landward of the shelf edge. Assignments of strata to each of these depositional environments were made based on the analysis of associated bedding features, physical sedimentary structures, and stratigraphic context within facies sequences (Droser & Bottjer 1988, 1989a).

In this study the greatest amount of ii data collected was from carbonate strata deposited in inner shelf environments, including 834 m of ii data from Cambrian stratigraphic sections and 819 m of ii data from Ordovician stratigraphic sections (Droser & Bottjer 1989a). The average ii computed from this data are shown in Figure 6. Note the large increase in bioturbation within the Lower Cambrian, with a second major increase between the Middle and Upper Ordovician. This increase in the extent of bioturbation through the early Paleozoic could simply be the result of a constant rate of bioturbation coupled with a systematic decrease in sediment accumulation rate. However, by examining lower Paleozoic carbonate strata of similar composition which were deposited under equivalent physical conditions, observed variations in sedimentation rate should have been minimized. In addition, these trends in the amount of bioturbation were accompanied by changes in other features of bioturbation—particularly type and size of discrete trace fossils—confirming that biological changes in bioturbation were occurring throughout the early Paleozoic. Droser & Bottjer (1989a) thus interpreted this trend in average ii as indicating, for inner shelf environments, increased utilization of infaunal ecospace during the early Paleozoic.

For the studied time intervals there was much less ii data attainable for middle and outer carbonate shelf enviroments than was collected for the inner shelf (i.e. no data was attainable for middle and outer shelf strata of Early Cambrian age or from outer shelf carbonate strata of Early and Middle Ordovician age). However, average ii was calculated for each of the studied time intervals for all three carbonate shelf environments, when data were available (Droser 1987; Droser & Bottjer 1988, 1989a). These average ii results were rounded to the nearest integer and are schematically displayed in Figure 7, with the ii "flash card" (Figure 3A) of the average ii (rounded to the nearest integer) displayed for each time and environment.

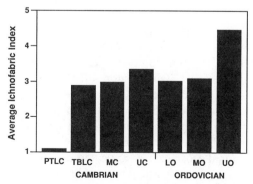

Figure 6 Average ichnofabric index for carbonate inner shelf paleoenvironments for Cambrian and Ordovician strata of the Great Basin. Modified from Droser & Bottjer (1989a).

ENVIRONMENT

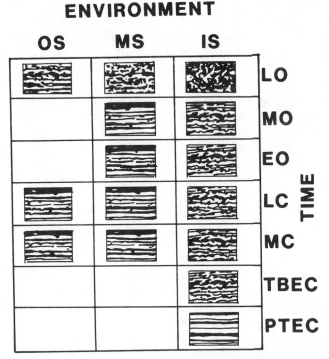

Figure 7 Time-environment diagram showing average ii for Great Basin Cambrian and Ordovician carbonate shelf paleoenvironments. Average ii results are schematically displayed with the "flash card" of the average ii (rounded to the nearest integer) for each time and environment where data were available. PTEC is pre-trilobite Early Cambrian, TBEC is trilobite-bearing Early Cambrian, MC is Middle Cambrian, LC is Late Cambrian, EO is Early Ordovician, MO is Middle Ordovician, LO is Late Ordovician. Modified from Droser (1987).

Figure 7, a time-environment diagram, shows average ii for the inner shelf, as in Figure 6, with average ii for the middle and outer shelf less than that for the inner shelf at any time interval. Average ii increases through time for all three environments. This onshore-offshore pattern indicates that for carbonate shelf environments, not only did bioturbation increase through the early Paleozoic, but that increases in bioturbation moved from onshore to offshore through time (Droser 1987, Droser & Bottjer 1989a). These patterns most likely indicate that innovations which led to increased utilization of infaunal ecospace first appeared onshore and then migrated offshore; such patterns are consistent with other early Paleozoic trace fossil onshore-offshore patterns reported by other authors (Bottjer et al 1988, Crimes & Anderson 1985).

Nearshore Terrigenous Clastic Setting

As discussed above, *Ophiomorpha* is dominant in post-Paleozoic nearshore sandstones and *Skolithos*, along with other vertical trace fossils, dominate such sandstones from the Paleozoic. Although piperock occurs throughout the Paleozoic, there is a distinct temporal trend associated with this ichnofabric. While workers have previously recognized that piperock is typical of Cambrian high-energy nearshore terrigenous clastic deposits (Hallam & Swett 1966, Hantzchel 1975, Miller & Byers 1984, Crimes & Anderson 1985, Droser 1987), a recent compilation demonstrates that piperock is indeed most common in the Cambrian (Figure 8) and decreases through the remainder of the Paleozoic (Droser 1991). This distribution may be the result of large-scale physical or biological factors. For example, preserved nearshore sandstones may decrease through this time interval. In addition,

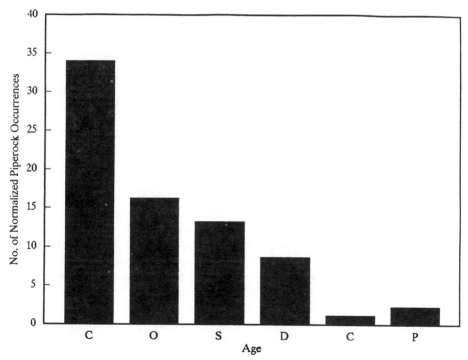

Figure 8 Distribution of piperock through the Paleozoic. Histogram represents the number of piperock occurrences as a function of age normalized to correct for differences in map area (following Blatt & Jones 1975, Raup 1977) and duration (Harland 1989, Cowie & Basset 1989) such that the area and duration of the Cambrian is set equal to 1. Data source listed in Droser (1991).

the decline of piperock corresponds with the Ordovician faunal diversification when bioturbation increased in other environments and, thus, may reflect the appearance of a new and better adapted fauna in the nearshore terrigenous clastic setting. An initial compilation of the distribution of *Ophiomorpha* does not show a similar distribution (Droser & Bottjer, in preparation).

BASIN-WIDE PATTERNS IN INCHNOFABRICS

Perhaps the best known geologic application of trace fossils is their use as paleoenvironmental indicators. In particular, trace fossils have traditionally been employed in basin-wide studies from the perspective of Seilacher's (1967) ichnofacies model. While, as discussed above, workers have recorded the extent of bioturbation in a qualitative fashion, these data were not in a form that facilitated comparison between or within basins. Ichnofabrics have only recently been utilized on a basin-wide scale, both through the analysis of the distribution of ichnofabric indices as well as through the process of mapping out characteristic ichnofabric types which can be related to lithofacies, energy of deposition, and reservoir distribution.

Ichnograms: *The Distribution of Ichnofabric Indices Within a Basin*

Ichnological studies that combine the examination of identifiable trace fossils, ichnofacies, and composite ichnofabrics with the measurement of ichnofabric indices and ichnograms yield a more complete summary of biogenic sedimentary structures in a sedimentary unit than has been previously possible. As discussed above, a variety of different physical and biological factors lead to the final development of the ichnofabric that is observed in a sedimentary rock. Proportionately different contributions of each of these factors in the overall process that leads to production of the preserved record of bioturbation in a sedimentary rock will result in different ichnograms for rocks deposited in different environments. In each sedimentary environment some factors are greater contributors to the resulting ichnofabric than others. Knowledge of those factors allows interpretation of variations in ichnograms measured from different strata but deposited in similar environments (Bottjer & Droser 1991).

An Example from The Jurassic of the Norwegian North Sea

Basin analyses utilizing ichnofabric data will ultimately combine studies of measurements on ichnofabric indices (as discussed above) with those that determine the distribution of recurring ichnofabric types within a basinal setting. This has been done for a Norwegian offshore Jurassic sequence in the Sognefjord Formation of the Troll Field (Bockelie 1991). For this study five ichnofabrics (*Helminthoida, Anconichnus, Palaeophycus, Ophiomorpha,* and *Skolithos* ichnofabrics) were identified and their distribution was mapped within reservoir zones in the Troll Field. These maps of ichnofabric were utilized, in conjunction with other data, to produce quantitative base maps for computerized reservoir models, as well as for facies predictions (Bockelie 1991). This study is a good indicator of the potentially powerful new view that integration of ichnofabric studies into basin analysis can give.

CONCLUSIONS

The recognition of ichnofabric as a significant component of sedimentary rocks which had been neglected in previous studies (e.g. Ekdale & Bromley 1983, Ekdale et al 1984) subsequently led to the independent development of a variety of approaches for studying ichnofabric. An understanding of composite ichnofabrics through use of the tiering concept and analysis of cross-cutting relationships of trace fossils has fostered research that has revealed much useful paleoenvironmental information (e.g. Bromley & Ekdale 1986, Savrda & Bottjer 1986, Wetzel 1986). The ichnofabric index approach for measuring the extent of bioturbation in a sedimentary rock was initially developed to study changes in the recorded extent of bioturbation in late Precambrian through Ordovician sedimentary rocks (Droser & Bottjer 1986, 1988, 1989a). Measurements of ichnofabric indices are now being used to supply paleoenvironmental information as well as assist in basin analysis (Droser & Bottjer 1989b, 1991; Bottjer & Droser 1991; Droser & O'Connell 1992). Mapping of recurrent ichnofabrics within stratigraphic sections and sedimentary basins has also proved to be a very useful approach (Bockelie 1991). With the development of these various approaches, the next phase in ichnofabric analysis will be one in which several approaches are integrated into single studies. Ultimately, it will become routine for data on ichnofabric to be combined with information gleaned from the various methods for studying discrete identifiable trace fossils, to reveal the full extent of ichnologic information available in sedimentary rocks.

ACKNOWLEDGMENTS

MLD acknowledges support from the National Geographic Society, the Petroleum Research Foundation of the American Chemical Society, and the White Mountain Research Station. DJB acknowldeges the support of the National Geographic Society. This paper benefited from our discussion and interaction with participants from the First International Ichnofabric Workshop (Norway, September 1991), particularly organizers A. A. Ekdale and J. F. Bockelie.

Literature Cited

Ausich, W. I., Bottjer, D. J. 1982. Tiering in suspension-feeding communities on soft substrata throughout the Phanerozoic. *Science* 216: 173–74

Blatt, H., Jones, R. L. 1975. Proportions of exposed igneous, metamorphic, and sedimentary rocks. *Geol. Soc. Am. Bull.* 86: 1085–88

Bockelie, J. F. 1991. Ichnofabric mapping and interpretation of Jurassic reservoir rocks of the Norwegian North Sea. *Palaios* 6: 206–15

Bottjer, D. J., Ausich, W. I. 1986. Phanerozoic development of tiering in soft substrata suspension-feeding communities. *Paleobiol.* 12: 400–20

Bottjer, D. J., Droser, M. L., Jablonski, D. 1988. Paleoenvironmental trends in the history of trace fossils. *Nature* 333: 252–55

Bottjer, D. J., Droser, M. L. 1991. Ichnofabric and basin analysis. *Palaios* 6: 199–95

Bromley, R. G. 1990. *Trace Fossils.* London: Unwin Hyman. 280 pp.

Bromley, R. G., Ekdale, A. A. 1984. *Chondrites*: a trace fossil indicator of anoxia in sediments. *Science* 224: 872–74

Bromley, R. G., Ekdale, A. A. 1986. Composite ichnofabrics and tiering of burrows. *Geol. Mag.* 123: 59–65

Cowie, J. W., Basset, M. G. 1989. International Union of Geological Sciences 1989 Global Stratigraphic Chart. *Episodes* 12

Crimes, T. P. 1992a. Changes in the trace fossil biota across the Proterozoic-Phanerozoic boundary. *Geol. Soc. London Q. J.* 149: 637–46

Crimes, T. P. 1992b. The record of trace fossils across the Proterozoic-Cambrian boundary. In *Origins and Early Evolution of the Metazoa*, ed. J. H. Lipps, P. W. Signor, pp. 177–202. New York: Plenum

Crimes, T. P., Anderson, M. M. 1985. Trace fossils from Late Precambrian-Early Cambrian strata of southeastern Newfoundland (Canada): Temporal and environmental implications. *J. Sediment. Petrol.* 59: 310–43

Curran, H. A. 1984. Ichnology of Pleistocene carbonates on San Salvador. *J. Paleontol.*

58: 312–21

Droser, M. L. 1987. *Trends in depth and extent of bioturbation in Great Basin Precambrian-Ordovician strata, California, Nevada and Utah.* PhD dissertation, Univ. Southern Calif., Los Angeles, CA, 365 pp.

Droser, M. L. 1991. Ichnofabric of the Paleozoic Skolithos ichnofacies and the nature and distribution of *Skolithos* piperock. *Palaios* 6: 316–25

Droser, M. L., Bottjer, D. J. 1986. A semiquantitative classification of ichnofabric. *J. Sediment. Petrol.* 56: 558–59

Droser, M. L. Bottjer, D. J. 1988. Trends in extent and depth of bioturbation in Cambrian carbonate marine environments, western United States. *Geology* 16: 233–36

Droser, M. L., Bottjer, D. J. 1989a. Ordovician increase in extent and depth of bioturbation: implications for understanding early Paleozoic ecospace utilization. *Geology* 17: 850–52

Droser, M. L., Bottjer, D. J. 1989b. Ichnofabric of sandstones deposited in high-energy nearshore environments: measurement and utilization. *Palaios* 4: 598–94

Droser, M. L., Bottjer, D. J. 1991. Trace fossils and ichnofabric in Leg 119 cores In *Proc. Ocean Drilling Prog., Sci. Results*, ed. J. Barron, J. Anderson, J. G. Baldouf, B. Larsen, 119: 635–41. College Station, TX: Ocean Drilling Prog.

Droser, M. L., O'Connell, S. O. 1992. Trace fossils and ichnofabric in Triassic sediments from cores recovered on Leg 122 In *Proc. Ocean Drilling Prog., Sci. Results*, ed. U. von Rad, B. U. Haq, R. B. Kidd, S. O'Connell, 122: 475–85. College Station, TX: Ocean Drilling Prog.

Ekdale, A. A., Bromley, R. G. 1983. Trace fossils and ichnofabric in the Kjolby Gaard Marl, Upper Cretaceous, Denmark. *Bull. Geol. Soc. Denmark* 31: 107–19

Ekdale, A. A., Bromley, R. G., Pemberton, S. G. 1984. *Ichnology: The Use of Trace Fossils in Sedimentology and Stratigraphy.* Soc. Econ. Paleontol. Mineral., Short Course 15. 317 pp.

Ekdale, A. A., Bromley, R. G. 1991. Analysis of composite ichnofabrics: an example in uppermost Cretaceous chalk of Denmark. *Palaios* 6: 232–49

Frey, R. W. 1978. Behavioral and ecological implications of trace fossils. In *Trace Fossil Concepts*, ed. Basan, P. B., pp. 43–66. Soc. Econ. Paleontol. Mineral. Short Course 6

Frey, R. W., Howard, J. D., Pryor, W. A. 1978. *Ophiomorpha*: its morphologic, taxonomic, and environmental significance. *Palaeogeogr. Palaeoclimatol. Palaeoecol.* 23: 199–229

Frey, R. W., Pemberton, S. G. 1984. Trace fossil facies models. In *Facies Models*, ed. R. G. Walker. Geosci. Canada Reprint Ser. 1: 189–207

Goldring, R., Pollard, J. E., Taylor, A. M. 1991. *Anconichnus horizontalis*: a pervasive ichnofabric-forming trace fossil in post-Paleoozic offshore siliclastic facies. *Palaios* 6: 250–63

Hallam, A., Swett, K. 1966. Trace fossils from the Lower Cambrian Pipe Rock of the north-west highlands. *Scott. J. Geol.* 2: 101–6

Hantzchel, W. 1975. Trace Fossils and problematica. In *Treatise on Invertebrate Paleontology*, ed. R. C. Moore, Part W. Lawrence, KS: Univ. Kansas Press. 269 pp.

Harland, W. B., Armstrong, R. L., Cox, A. V., Craig, L. E., Smith, A. G., Smith, D. G. 1989. *A Geologic Time Scale*. New York: Cambridge Univ. Press. 263 pp.

Howard, J. D., Frey, R. W. 1975. Estuaries of the Georgia coast, U.S.A.: Sedimentology and biology, II. Regional animal-sediment characteristics of Georgia estuaries. *Senckenb. Maritima* 7: 33–103

Kennedy, W. J., Garrison, R. E. 1975. Morphology and genesis of nodular chalks and hardgrounds in the Upper Cretaceous of southern England. *Sedimentology* 22: 311–86

Meldahl, K. H. 1987. Sedimentologic and taphonomic implications of biogenic stratification. *Palaios* 2: 350–58

Miller, M. F., Byers, C. W. 1984. Abundant and diverse early Paleozoic infauna indicated by the stratigraphic record. *Geology* 25: 271–94

Moore, D. G., Scrutton, P. C. 1957. Minor internal structures of some unconsolidated sediments. *Am. Assoc. Petrol. Geol.* 41: 2723–51

Narbonne, G. M., Myrow, P., Landing, E., Anderson, M. M. 1987. A candidate stratotype for the Precambrian-Cambrian boundary, Fortune Head, Burin Penin-

sula, southeastern Newfoundland. *Can. J. Earth Sci.* 24: 1277–93

Raup, D. M. 1977. Species diversity in the Phanerozoic: an interpretation. *Paleobiol.* 2: 289–97

Reineck, H.-E. 1967. Parameter von Schichtung und bioturbation. *Geol. Rundsch.* 56: 420–38

Rhoads, D. C., Stanley, D. J. 1965. Biogenic graded bedding. *J. Sediment. Petrol.* 35: 956–63

ROCC (Research on Cretaceous Cycles) Group (M. A. Arthur, D. J. Bottjer, W. E. Dean, A. G. Fischer, D. E. Hattin, E. G. Kauffman, L. M. Pratt, P. A. Scholle) 1986. Rhythmic bedding in Upper Cretaceous pelagic carbonate sequences: varying sedimentary response to climatic forcing. *Geology* 14: 153–56

Savrda, C. E., Bottjer, D. J. 1986. Trace-fossil model for reconstruction of paleo-oxygenation in bottom waters. *Geology* 14: 3–6

Savrda, C. E., Bottjer, D. J. 1987. Trace fossils as indicators of bottom-water redox conditions in ancient marine environments. In *New Concepts in the use of Biogenic Sedimentary Structures for Paleoenvironmental Interpretation*, ed. D. J. Bottjer, pp. 3–26. Los Angeles: SEPM Pac. Soc. 65 pp.

Savrda, C. E., Bottjer, D. J. 1989a. Anatomy and implications of bioturbated beds in "black shale" sequences: examples from the Jurassic Posidonienschiefer (southern Germany). *Palaios* 4: 330–42

Savrda, C. E., Bottjer, D. J. 1989b. Trace-fossil model for reconstructing oxygenation histories of ancient marine bottom waters: application to Upper Cretaceous Niobrara Formation, Colorado. *Palaeogeogr. Palaeoclimatol. Palaeoecol.* 74: 49–74

Seilacher, A. 1962. Paleontological studies on turbidite sedimentation and erosion. *J. Geol.* 70: 227–34

Seilacher, A. 1967. Bathymetry of trace fossils. *Mar. Geol.* 5: 413–28

Wetzel, A. 1984. Bioturbation in deep-sea fine-grained sediments: influence of sediment texture, turbidite frequency and rates of environmental change. In *Fine-Grained Sediments: Deep-Water Processes and Facies*, ed. D. A. V. Stow, D. J. W. Piper, Geol. Soc. London Spec. Pub. 15: 595–608

Wetzel, A., Aigner, T. 1986. Stratigraphic completeness: tiered trace fossils provide a measuring stick. *Geology* 14: 234–37

Annu. Rev. Earth Planet. Sci. 1993. 21:227–54

THE ROLE OF POLAR DEEP WATER FORMATION IN GLOBAL CLIMATE CHANGE

William W. Hay

GEOMAR, Christian-Albrechts-Universität, D-2300 Kiel 14, Germany, and Department of Geological Sciences, CIRES and Museum, University of Colorado, Boulder, Colorado 80309

KEY WORDS: Arctic, Antarctic, palaeoceanography

INTRODUCTION

The present ocean is thermally stratified. The waters become colder with depth, and the great mass of ocean deep water is near 2°C—almost the same as that of surface waters in the polar regions. The idea that the waters have a polar origin goes back to Benjamin Thompson (1800) (Count Rumford). He reasoned that the cold waters of the interior of the ocean must form by sinking in the polar regions and must drive poleward flow of surface waters. The idea was refined by Alexander von Humboldt (1814), who noted that the density of sinking cold polar waters must exceed the density of more saline waters in lower latitudes.

The first person to have seriously questioned the assumption that the deep waters of the ocean have always been cold and hence formed in the polar regions was T. C. Chamberlin (1906), who suggested that salinity rather than thermal differences may have driven the ocean.

The first direct evidence that ocean deep waters might have been significantly different in the past was provided by Emiliani & Edwards (1953) and Emiliani (1954). They noted that Tertiary deep water temperatures recorded by oxygen isotopes in the shells of benthic foraminifers from the Pacific were significantly warmer (10.5°C) than that of the deep waters at present (1.75°C). They suggested that this indicated warmer polar seas in the past.

227

Henry Stommel (1961) quantified the argument that both temperature and salinity differences might be driving mechanisms for ocean convection. He concluded that the thermohaline convection of the ocean has two stable regimes of flow, one in which the temperature differences dominate the density differences, and the other where salinity dominates the density differences.

Claes Rooth (1982, p. 132) discussed the climatic consequences of a change in polar deep water formation. He noted that "the most cursory examination of any oceanographic atlas or standard textbook quickly establishes an awareness of temperature as the predominant factor controlling oceanic density stratification. That such should be the case is intuitively appealing in view of the large range of climatic conditions on the surface of the earth, but on closer examination, the inevitability of this state of affairs is not at all self-evident." Rooth introduced several new ideas. He noted that thermal anomalies of the sea surface cause changes in both the latent and sensible transfer of heat between the ocean and atmosphere. Because the water content of saturated air doubles with every $10°C$ temperature increase, the evaporation is important in modifying the surface water masses in the tropics, but its effects are dominated by those of the meridional temperature gradient. The effect of the excess of rainfall and runoff over evaporation at high latitudes resulting from the import of water from humid lower latitudes, is to locally freshen the surface of the ocean. In the polar regions, where the temperature of the water is everywhere about the same, salinity effects dominate. "A dominance of salt effects would in fact almost certainly be hypothesized by a land-bound naturalist with access only to observations from estuaries and from warm hypersaline tropical lagoons . . ." (p. 133). Rooth noted that although the driving force for the main ocean gyres is wind stress, the circulation of the ocean is not global. Rather, it is broken up into basin-wide gyres which lie between the extremes of zonal wind stress and the equator. In the present thermally stratified ocean, the effect of the wind stress is concentrated in the surface mixed layer. In the absence of other forces, this would mean that poleward heat transport in the oceans would not extend beyond 45°N and S latitude. It is the thermohaline circulation that allows the ocean heat transport to extend to higher latitudes. On the basis of data presented by Lazier (1973 and later personal communications), Rooth discussed the sensitivity of the deep water production in the Labrador Sea to changing salinities in the North Atlantic. He concluded that the deep water production in the North Atlantic is delicately balanced, and can be shut down by relatively small changes in the fresh-water balance. Most significantly, he observes that if the thermohaline circulation were to slow down significantly, changes in the hydrology at high latitudes would have a positive feedback effect, producing a sub-polar halocline which would make it more difficult to restart the thermohaline circulation.

The atmosphere and ocean transport heat from the equatorial to the polar regions. The ocean presently accounts for about 2/3 of poleward flux of energy in the tropics. In the subtropics, the ocean accounts for about 1/2 of the poleward energy flux. Between the subtropical and polar convergences, oceanic energy transport declines, and beyond the polar convergences (about 60°N and S) poleward energy transport by the ocean is negligible. The role of the ocean in poleward energy transport is limited by the presence of oceanic convergences along which waters sink, interrupting the poleward flow. The rate at which polar waters sink is a major factor in determining the rate at which tropical-subtropical waters are advected to high latitudes.

Hsiung (1985) has shown that the meridional heat transport in the different ocean basins is remarkably different. As shown in Figure 1, heat is transported northward throughout the Atlantic and southward throughout the Indian Ocean. In the Pacific Ocean, heat is transported northward north of 5°S. These differences are to a large extent determined by the presence of sites of formation of ocean deep waters in the Antarctic and North Atlantic.

FORMATION OF OCEAN DEEP WATER

Stommel (1962) noted that although the thermohaline vertical circulation of the ocean is one of its major features, the sites of sinking of ocean deep waters are quite small. The thermohaline circulation is driven by formation of dense water in small regions where isolation and transformation of the water masses can occur. These small areas are shown in Figures 2 and 3.

Peterson (1979) modeled the sinking of dense plumes and concluded that it is the buoyancy flux, defined as the product of the excess density times the volume flux, that determines whether a given source will produce ocean bottom water. The plume with the highest buoyancy flux fills the ocean from below; plumes with a lesser buoyancy flux produce intermediate waters that are recycled more rapidly to the surface.

Killworth (1983) has summarized the processes by which waters sink to become part of the deep convection system of the ocean. He distinguished two types of formation of deep water: that which takes place near ocean boundaries and that which occurs in the open ocean.

Deep Water Formation Near an Ocean Boundary

For deep water formation near an ocean boundary, Killworth (1983) cited five requirements: a reservoir, a source of dense water within the reservoir,

Meridional   Heat Transport

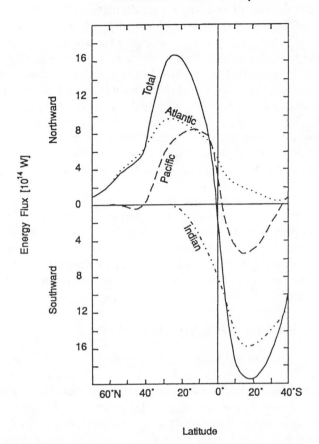

Figure 1   Meridional heat transport in the oceans (after Hsiung 1985).

a reason for the dense water to leave the reservoir, that more than one water mass be involved, and lastly, that no obstructions prevent the dense water from reaching the ocean bottom.

A reservoir is a restricted site within which the water can be maintained and modified. The continental shelf is such a site, and the wider the shelf the better it is as a site for transformation of ocean waters. Shallow shelf seas may also serve as appropriate sites. The shelf or shallow sea must

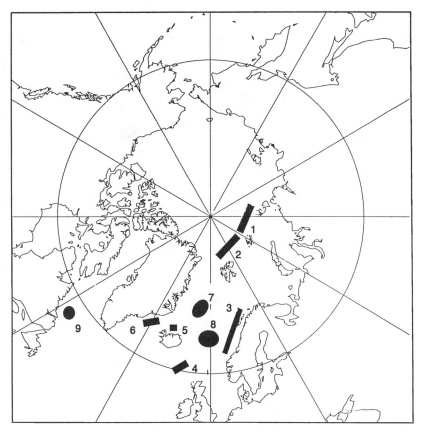

Figure 2 Sites of deep water formation in the northern hemisphere (after Killworth 1983). Rectangles are sites near the ocean boundary: (1) Kara Sea shelf, (2) Barents Sea shelf, (3) Norwegian Sea shelf, (4) Iceland-Scotland Ridge, (5) Icelandic shelf, (6) Greenland shelf. Ellipses are open ocean sites: (7) Greenland Sea, (8) Norwegian Sea, (9) Labrador Sea. The Iceland Shelf and Labrador Sea sites are the main contributors to North Atlantic Deep Water formation.

either have a negative or negligible fresh water balance (evaporation > precipitation + runoff). Most of the Arctic shelves have a positive fresh water balance and are well supplied with runoff so that they cannot become sites of deep water formation. The shelves surrounding the ice-covered Antarctic, with a negligible fresh water balance and receiving no runoff, have a much greater potential to be sites of deep water formation.

Within the reservoir dense water forms mostly through air-sea inter-

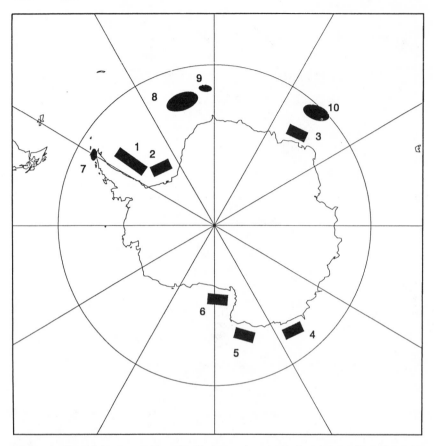

Figure 3   Sites of deep water formation in the southern hemisphere (after Killworth 1983). Rectangles are sites near the ocean boundary: (1) Weddell Sea shelf along the Antarctic Peninsula, (2) Weddell Sea shelf in front of Ronne and Filchner Ice Shelves, (3) off Enderby Land, (4) off Wilkes Land, (5) off Adélie Land, (6) Ross Sea shelf. Ellipses are open ocean sites: (7) Bransfield Strait, (8) Weddell Polynya above Maude Rise, (9) site of Weddell Sea chimneys reported by Gordon (1978), (10) off Enderby Land. The Weddell Sea sites are the main contributors to Antarctic Bottom Water formation.

action: cooling by conduction, cooling by evaporation, increase of salinity by evaporation, and increase of salinity by formation of sea ice. As suggested by Brennecke (1921) and Mosby (1934), the main mechanism for increasing the density of sea water in the polar regions today, is freezing of sea ice. Sea ice has a salinity < 6‰ compared with the minimum salinity of polar waters of 27.4‰ (Gow & Tucker 1990).

The reason for dense water to leave the reservoir and descend to the ocean floor is not always obvious. Mosby (1934) assumed that as the waters in the reservoir became more dense than those of the surrounding ocean, gravity would pull them down the slope to the deep ocean floor. However, as the dense waters begin to move they are acted upon by the Coriolis force and soon a geostrophic balance is attained between the downslope buoyancy force and an upslope Coriolis force so that the water would tend to move along depth contours rather than descend into the basin (Killworth 1977). Killworth (1983) suggested that a necessary prerequisite for dense water to leave a reservoir and descend to the ocean floor is preexisting circulation that can drive it off the shelf. Another possibility is that turbulent friction of flow along the bottom may act as a third force to destroy the geostrophic balance. Another possible factor is the greater compressibility of cold water: the density of descending cold water can increase faster than that of the surrounding warmer water, thus destroying the geostrophic balance.

In the present day oceans formation of deep water always involves more than one water mass. The process of *caballing*—the formation of denser water by mixing of waters of equal density but different temperature and salinity characteristics—may be important. Because bottom flows are inherently turbulent and must involve entrainment of the surrounding waters, their density must always decrease after leaving the site where they were modified by air-sea interaction.

The last factor discussed by Killworth is the lack of obstructions, such as sills and other topographic features, which might trap the dense waters and prevent them from reaching the deep seafloor.

Killworth (1983) lists the following as sites where deep water formation takes place near an ocean boundary: 1. the western Weddell Sea, 2. the Ross Sea, 3. the Adélie Coast and Enderby Land, 4. Wilkes Land, 5. the Arctic, 6. the Norwegian, Greenland, and Iceland shelves, 7. the Adriatic, 8. the shelf of the Gulf of Lions, 9. the Gulf of Suez, and 10. the Persian Gulf. The first six of these are polar sites of cold water formation; the last four are subtropical.

THE ANTARCTIC SITES   The Weddell Sea was identified by Mosby (1934) as the most important site of deep water formation today. It produces the bulk of the ± 38 Sv (1 Sv $= 10^6$ m^3 s^{-1}) of Antarctic Bottom Water (AABW) that enters the deep Atlantic, Indian, and Pacific Oceans. The Weddell Sea has a narrow (< 90 km) eastern shelf, a broad (> 400 km) shelf off the Filchner and Ronne Ice Shelves, and a 200 km wide shelf off the Antarctic (Palmer) Peninsula. Although these shelves occupy 25% of the 2.3×10^6 km^2 area of the Weddell Sea, the water over them is only 4% of the 7.6×10^6 km^3 volume of the basin. The surface water layer develops a thickness of 50 m in the summer and has a variety of temperatures ($-2°$ to $1°$) and salinities ($< 34.4‰$). It is underlain by Winter Water (WW); a halocline separates the less saline surface water from the saltier WW ($-1.8°C$, $34.4‰$). The WW overlies Warm Deep Water (WDW) which here typically has temperatures of $0°$ to $0.8°$ and salinities of 34.6–$34.7‰$. The two water masses are separated by an intermediate layer of Modified Warm Deep Water (MWDW). The base of the WDW lies at a depth of 1000–2000 m. The bulk of the basin (60%) is filled with Antarctic Bottom Water (AABW: $0°$ to $-0.8°C$, 34.6–$34.7‰$). The coldest and densest water in the basin is Weddell Sea Bottom Water (WSBW) ($-1.4°$ to $-0.8°$, $34.65‰$) which occurs on the slopes and southern and western edges of the basin (Gill 1973, Carmack & Foster 1977).

The formation of AABW has been discussed by Gill (1973), Foster & Carmack (1977), Killworth (1977, 1983), and Carmack (1990). Mosby (1934) assumed that the driving mechanism was formation of sea ice over the shelf in the winter. Fofonoff (1956) and Foster (1972) suggested that caballing at the shelf edge played a major role. Seabrooke et al (1971) proposed that freezing of seawater onto the bottom of the ice shelves is significant in increasing the salinity of the shelf waters. Gill (1973) noted that the Antarctic Coastal Current flows as a ribbon of fresher water along the edge of the shelf, separating the saline shelf water from the Weddell Sea surface water along a V-shaped front (Figure 4). At the base of the front, the cold, saline shelf waters mix with warmer, fresher MWDW and flow down the slope into the deep basin at a rate of about 1 Sv. He also emphasized that the pack ice is continually being blown offshore, presenting fresh open water surface for the formation of new sea ice. The annual brine release equals the freezing of 1 m or more of ice. Foster & Carmack (1976) described the interleaving of shelf waters with MWDW in the southern and southwestern shelves, and noted the presence of a thin layer of cold WSBW on the adjacent continental slopes. They argue that it is the mixing of the shelf and MWDW waters that produces WSBW, which then flows clockwise around the basin entraining WDW to form AABW. Killworth noted that the onshore Ekman transport driven by the easterly winds produces fluxes of 1–1.5 Sv and is an important factor in setting conditions for sinking.

Warren (1981) suggested that the water entrained by outflow from the Ross Sea shelf is warmer and more saline than that entrained by Weddell Sea shelf water at the shelf break. Because of these differences the Ross Sea may be a minor contributor to AABW.

Around much of the Antarctic margin there is a depression between the coast and the shelf break. The weight of the Antarctic ice sheet has depressed the continent and inner shelf so that there is an upward slope

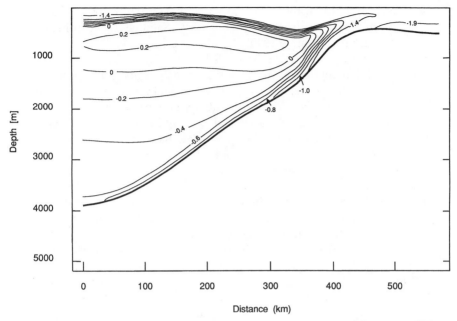

Figure 4 Potential temperature section across the shelf break in the Weddell Sea near 51°W showing a tongue of Modified Warm Deep Water interleaving with cold waters in front of the Ronne Ice Shelf (after Foster & Carmack 1977).

to the shelf break at about 400 m. This situation is well developed off the Adélie Coast (Gordon & Tchernia 1972) and Enderby Land (Jacobs & Georgi 1977). In these places the shelf depressions act as reservoirs for saline shelf water, but these are small compared with the Weddell and Ross Sea shelves. Their contribution to AABW is probably small.

Off Wilkes Land, just west of the Ross Sea, a plume of cold water descends the slope, but does not reach the bottom and spreads horizontally at 2000 m (Carmack & Killworth 1978). The plume lacks the salinity required to be dense enough to reach the bottom. Killworth (1983) suggested that similar processes may go on around much of the Antarctic margin.

THE ARCTIC SITES   Significantly, the Earth's only truly polar ocean, the Arctic Ocean, is not a site of formation of deep water that enters the world's oceans (Coachman & Aagaard 1974). In contrast to the ocean surrounding Antarctica, the Arctic Ocean has an extensive perennial sea ice cover. The surface layer is termed Arctic Water, and can be divided into surface, subsurface, and lower layers. Arctic Surface Water is up to

50 m thick. Because of the perennial sea ice cover, the temperatures are always about freezing—from $-1.5°$ to $-1.8°C$. Salinities range from 28‰ to 33.5‰ (Pickard & Emery 1990). The higher salinities lie off the Barents shelf, and decrease toward the Bering Strait. The Arctic Subsurface layer extends to 150 m and has a strong halocline, with salinities increasing to 34‰. The Arctic Lower layer extends to 200 m and is the transition to the underlying Atlantic Water. The low salinity surface layer of the Arctic Ocean results from melting sea ice and the large volume of river runoff during the summer. The runoff into the Arctic Basin, estimated by Alyushinskaya & Ivanov (1978) to be 5220 km^3 yr^{-1}, is equivalent to a 35 cm layer. It effectively prevents any large-scale deep water formation.

Beneath the Arctic Water lies Atlantic Water, which enters from the Norwegian Sea as the West Spitsbergen Current. It has a temperature of $1°$ to $0.4°C$ and salinites between 34.8‰ and 35.0‰. It flows along the Eurasian Margin and spreads across the Lomonosov Ridge into the Canadian Basin. On the Eurasian Margin it is introduced via submarine canyons onto the shelf, where it can be modified (Aagaard et al 1985).

Below 800 m the Arctic Ocean is filled with Arctic Bottom Water (ABW), with potential temperatures of $0.4°$ to $-0.8°C$ and salinities between 34.90 and 34.99‰ (Pickard & Emery 1990). Aagaard et al (1985) suggest that ABW is a mixture of waters from four sources: Greenland Sea Deep Water (GSDW), Norwegian Sea Deep Water (NSDW), Eurasian Basin Deep Water (EBDW), and Canadian Basin Deep Water (CBDW). As discussed below, GSDW and NSDW are formed in the Greenland-Iceland-Norwegian (GIN) Seas. They are introduced into the Arctic Basin through Fram Strait. EBDW and CBDW which form at sites on the Arctic shelves away from the influence of major rivers may also be from sites of generation of saline shelf waters through brine expulsion resulting from the freezing of sea ice (Aagaard 1981, Melling & Lewis 1982). The saline shelf waters do not descend into the deep but spread laterally beneath the cold fresher surface water, contributing to the development of the Arctic Ocean halocline at 250 m. The rate of formation of saline shelf waters is estimated by these authors to be about 2.5 Sv.

It is likely that dense cold saline waters also form on the shelves surrounding the GIN Sea.

Deep Water Formation in the Open Ocean

For deep waters to form in the open ocean Killworth (1983) also cites five requirements: 1. background cyclonic circulation, 2. preconditioned gyre waters, 3. intense surface forcing, 4. the involvement of more than one water mass, and 5. breakup of the convective structure, with sinking and spreading of the transformed water mass. He cites the following areas as examples of regions where open ocean convection occurs: the Labrador Sea, the Bransfield Strait, the Weddell Sea, the Greenland Sea, and the Mediterranean.

Cyclonic circulation means that the waters in the center of the gyre are moving upward. At first this might seem to be contradictory to sinking, but the dynamics of the upward motion induced in the gyre center bring denser waters closer to the surface, reducing the intensity of wind forcing required for upwelling to occur. The upward doming of the isopycnals also reduces the vertical stability or stratification in the centers of the gyres.

Preconditioning is a modification of the surface waters over a period of weeks to a few months that further reduces the vertical stability of the water. The preconditioning is usually a result of atmospheric interactions. Cyclonic eddies superposed on the cyclonic circulation can further weaken the stability, as has been described by Clarke & Gascard (1983) in the Labrador Sea. Wind-driven upwelling might serve to precondition the water. It is also possible that the submarine topography may precondition the water by affecting the gyral circulation. This has been proposed by Hogg (1973) to explain the reduction in stratification in the Gulf of Lions above the Rhone Fan. Development of a *polynya*, an open water area that forms in the midst of sea ice, also preconditions the water, but the cause of polynyas remains uncertain. Gordon (1982) has cited submarine topography (the Weddell Polynya seems to form consistently above Maud Rise), reduced precipitation, increase in ice formation, an increase in the flux of deep saline water, and a shift in the position of the Weddell Sea gyre as possible causes.

Deep water formation in the open ocean always involves more than one water mass. The different water masses initially provide stability, but thermal stabilization can be overcome by cooling and salinity stabilization can be overcome by cooling or salinization of the fresher water through sea ice formation. Killworth (1979) noted that the greater compressibility of cold water may allow surface forcing to mix down waters that are neutrally buoyant at the surface but become more dense than the surrounding waters at depth. The initial effect would be a gradual increase in the thickness of the mixed layer until the lower layer is reached. Then the depth of mixing might increase rapidly to depths of 2000 m or more.

Intense surface forcing occurs over a period of days to weeks. Very cold winds blowing over warmer waters chill the water both by sensible and latent heat loss. Brine ejection from freezing sea ice can be another form of intense surface forcing.

The surface forcing results in an episode of violent vertical mixing. This occurs within a *vertical chimney*—an unstable convective feature a few tens of kilometers across in which large convective fluxes can occur. The chimney is associated with cyclonic circulation set up in the surface waters and presumed to be also present at depth. Within the chimney waters can flow to depth and be replaced by deep waters without being stopped by a pycnocline.

Convection stops with breakup of the chimney structure and the development of vertical stability. The transformed water masses that sank through the chimney spread in the deep sea, ending the episode of open ocean convection. Relics of the chimney may persist for months.

THE ANTARCTIC SITES   Two kinds of open ocean convection have been described from the Weddell Sea. The general circulation of the Weddell Sea is cyclonic. Gordon (1978) described a cyclonic eddy containing a central chimney about 30 km in diameter (Figure 5). Although when observed it was vertically stable, it was clearly a relic from the winter convection. Gordon estimated that up to 30 such chimneys might develop in the Weddell gyre each year, and that they might be responsible for deep water production at a rate of 0.6 Sv. The Weddell Polynya, which forms over Maud Rise on the east side of the gyre, has been present during many winters. Martinson et al (1981) have suggested that the freezing of sea ice adjacent to the polynya results in a downward flux of saline water at a rate of 1 Sv causing overturning and bringing $0°$ water to the surface. Gordon (1982) has suggested average rates of flux of saline water from freezing of sea ice at 3.8–7.7 Sv during the wintertime.

The Bransfield Strait lies west of the tip of the Antarctic Peninsula, and is floored by three troughs ranging from 1100 to 2800 m in depth. It does not have an internal cyclonic circulation but does experience surface to bottom convection in the winter (Gordon & Nowlin 1978). It contains a nearly homogeneous water mass of lower temperature and salinity and higher density than the surrounding waters. The water is trapped in the basins by relatively shallow sills.

THE ARCTIC SITES   The Greenland Sea has a cyclonic circulation resulting in formation of a pycnocline dome located at $0°E$ and $75°N$. The dome contains water colder than $0°$ and having a salinity of 34.9‰. The bottom water of the Greenland Sea is colder ($< -1°C$). According to Peterson & Rooth (1976) tritium measurements indicate that deep mixing must take place in the middle of the gyre. They estimated the time scale of deep convective mixing in the Greenland Sea to be about 30 years, implying that 100 m of surface waters are mixed down each year. Attempts to locate surface waters with the same temperature and salinity characteristics as the Greenland Sea bottom waters have failed, and hence the origin of the Greenland Sea bottom water remains uncertain. However, several interesting new ideas have recently been proposed. Reid (1979) has shown

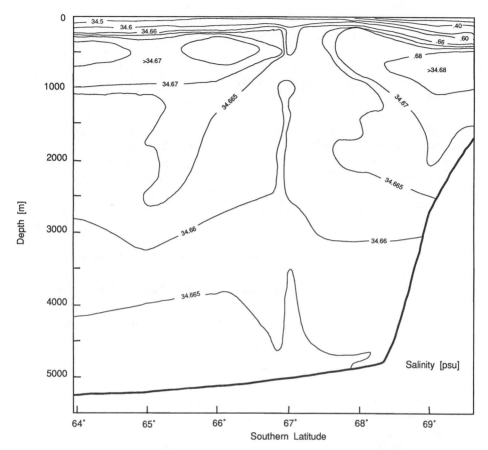

Figure 5   Salinity section northward from the Antarctic margin in the Weddell Gyre showing a "chimney" (after Gordon 1978).

that warm saline Mediterranean water that enters the Nordic Seas between the Faroes and Scotland may be advected westward to become a significant contributor to the water that sinks in the Greenland Sea. Hakkinen (1987) has noted that upwelling occurs in the ice marginal zone because of the larger drag coefficient of the ice, suggesting that the cooling of this upwelled water may produce deep water. Rudels (1989) has proposed that brine expulsion during the early stages of sea ice formation may create an unstable surface boundary layer which then sinks as a turbulent plume.

The deep waters of the Norwegian Sea are warmer, more saline, but slightly less dense than those of the Greenland Sea. Worthington (1970) suggested that the warm saline waters entering the Norwegian Sea are cooled without appreciable freshening. Reid (1979) showed that the higher salinity comes both from the North Atlantic Drift at the surface and modified Mediterranean outflow at depth. Peterson & Rooth (1976) noted that the time scale of mixing of the Norwegian Sea is considerably longer (100 years) than that of the Greenland Sea.

The circulation in the Labrador Sea is also cyclonic. Clarke & Gascard (1982) described a region of homogeneous water 50 km wide and 2000 m deep, with internal eddies on a variety of scales down to 5 km. It was suggested that the region of homogeneous water was an eddy generated by intensive heat loss [480 W m^{-2} according to a personal communication from J. C. Gascard cited by Killworth (1983)] which was caused by extremely cold, dry winds blowing off the ice cover of the northern Labrador Sea.

FORMATION OF NORTH ATLANTIC DEEP WATER   North Atlantic Deep Water (NADW) is a mixture of water from several sources (Figure 6). Most of it originates in flows over the Greenland-Scotland Ridge (Meincke 1983). The densest water is contributed from the Greenland Sea via overflow of the Greenland-Iceland Ridge through the Denmark Strait as Denmark Strait Overflow Water (DSOW) with a potential temperature 1–2°C and a salinity of 34.9‰. Swift et al (1980) noted that the densest water above the sill is Norwegian Sea Deep Water ($<0°$, 34.9‰), but that this contributes only 10% of the outflow. The major outflow is Arctic Intermediate Water formed north of Iceland during the winter. The Denmark Strait outflow is estimated at 4 Sv by Warren (1981). DSOW flows down the slope and along the eastern margin of south Greenland, entraining another 1 Sv of Atlantic water to form North-West Atlantic Bottom Water. Contributions from the Norwegian Sea flow through passages between Iceland and the Faroes and between the Faroes and Shetlands. This outflow is less dense than that from the Greenland Sea, and follows a tortuous route being held against the right-hand slopes by the Coriolis force. About 1 Sv of water passes through the Faroes Bank Channel, with a sill depth of about 850 m (Warren 1981). Some of it flows down and south along the eastern margin of Rockall Bank, entraining another 4 Sv to become North-East Atlantic Deep Water (NEADW). Turning north along the west side of Hatton Bank, it reaches the south side of Iceland-Faroes Ridge. There it is joined by 1 Sv of water flowing over the shallower (400 m) ridge between Iceland and the Faroes (Warren 1981). These waters entrain another 3 Sv as they flow south again along Rekjanes Ridge. At 52°N they turn west to flow through the Charlie Gibbs fracture zone into the western basin. There they override and mix with the water from the Denmark

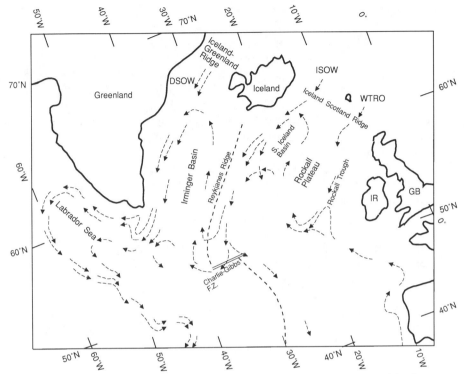

Figure 6 Deep waters in the North Atlantic, showing the paths of overflows of the Green-land-Iceland-Scotland Ridge. Water flowing northward past Iberia is AABW.

Strait, forming NADW. The combined flow around Greenland and into the Labrador Sea is estimated to be about 10 Sv (Warren 1981).

Deep water formed by ocean convection in the Labrador Sea flows into the Western North Atlantic Basin, overriding the Greenland-Scotia ridge overflow waters as middle NADW. Outflow of saline water from the Mediterranean spreads in the Central and North Atlantic at depths of 1000 to 1500 m, and is termed Mediterranean Water or Upper NADW (Kellogg 1987). Mixtures of all of these waters form North Atlantic Deep Water.

KNOWN PERTURBATIONS OF DEEP WATER PRODUCTION

The Great Salinity Anomaly

A modern example of the delicate balance of the thermohaline circulation occurred in the late 1960s, 1970s, and early 1980s during an event that has come to be known as the Great Salinity Anomaly, described by Dickson et al (1988).

In 1962 the Denmark Strait Overflow Water (DSOW) had the "normal" temperatures and salinities cited above. However, from 1965 to 1971 extremely heavy ice conditions developed in the sea around Iceland. These severe ice years were characterized by abnormally low surface salinities believed to be advected into the area from the northwest. As the salinity decreased below 34.7‰, winter overturning of the surface water column ceased. With the lack of convective overturn, advection of fresher waters further freshened the surface. The low salinity surface waters became incorporated into the southward flowing East Greenland Current. In the Labrador Sea, the salinity decrease reached a threshold in 1968, when yearround salinity stratification developed, preventing the production of Labrador Sea Deep Water. The near surface salinity fell to 34.4‰ and convection was limited to the upper 200 m (Lazier 1980). The minimum salinities were reached in the Labrador Sea in 1972, whereupon the salinity began to increase again.

Entering the system of eastward flow across the North Atlantic, the salinity anomaly spread across the Subarctic Convergence, the northern boundary of the North Atlantic Current. Dickson et al (1988) attributed this to interactions between the Gulf Stream System and the slope water south and east of Flemish Cap. In 1972, the number of icebergs drifting south of 48°N was the largest ever recorded, and the iceberg season was the longest on record (Wolford 1982).

By 1975 the salinity anomaly had reached the eastern side of the North Atlantic, off Ireland, whereupon it split, part going south with the Canaries Current and passing Iberia in 1977, and part flowing north into the Norwegian Sea. The anomaly crossed the Iceland Faroes Ridge in 1976, reached the Barents Sea in 1978, and passed back into the Arctic Basin in 1979 (Dickson et al 1988).

The rate of deep water formation in the Greenland Sea was reduced in the 1980s, starting sometime between 1978 and 1982 (Schlosser et al 1991). Although the cause is uncertain, it may be related to the return of the freshened waters of the Great Salinity Anomaly to this area.

At its maximum in the Labrador Sea the salinity anomaly represented a salt deficit of 72×10^9 tons; on its return across the Iceland Faroes Ridge, it represented a deficit of 47×10^9 tons. Pollard & Pu (1985) advocated in situ freshening as an explanation of the salinity anomaly, but calculated that by 1977 the freshening was equivalent to a layer of fresh water 1.4 m thick having been added to the surface of the Eastern Atlantic. Equivalent to an addition of 0.16 m of fresh water per year for each of the nine years from 1967–1976, this would require a 20–30% increase in precipitation or reduction in evaporation, which was not observed. They noted that the salinity anomaly freshened a large volume of ocean water and wondered where the ocean had become saltier. Dickson et al (1988) suggested an alternative: the advection hypothesis. They proposed that the freshening occurred as a result of enhanced sea ice formation in the high Arctic. The sea ice formation transferred the salt that had been brought into the region by northward flowing currents entering the Norwegian Sea into the intermediate and deep waters, freshening the surface waters and providing an excess of low salinity polar surface water for export into the Atlantic. Alternatively, they suggest that the salinity anomaly might have been caused by increased southward transport of polar water under strong winds during the 1960s.

The climatic effects of short-term (decadal to centennial) changes in sea surface temperature associated with the El Niño and the Southern Oscillation (ENSO) (a tropical phenomenon which influences the climate into the subtropics and perhaps beyond) have been demonstrated (Rasmussen 1985, Folland et al 1986, Flohn 1987, Philander 1989, Peixoto & Oort 1992). There is less information about the effects of short-term changes in sea surface temperature at high latitude associated with salinity anomalies and deep water production, and the climatic effect of the passages of the Great Salinity Anomaly have not been directly documented. Newell & Hsiung (1987) have discussed the relation between changes of sea surface temperatures (SSTs), free air temperatures, and precipitation for 1949–1979. They noted that the main variability of SST and air temperatures is associated with El Niño. However, they divided the SST eigenvector analysis into two periods, 1949–1964 and 1965–1979, which have been characterized as "wet" and "dry" or warmer and cooler respectively, with the latter period being associated with the Sahel drought. The latter of these periods corresponds to the life of the Great Salinity Anomaly. It is not known whether the change in climate caused the Great Salinity Anomaly or vice versa. Zonal westerly winds over the North Atlantic were stronger during the first period. This should have resulted in Ekman transport of cooler high latitude waters to the south and greater evaporation over the ocean surface, but these were the warm years for SSTs in the North Atlantic. It may be that the cooling of the North Atlantic and onset of dryer conditions might be related to the salinity anomaly and advection of fresher cooler waters into the North Atlantic.

The Younger Dryas Cooling Event

Much of the discussion about the climatic effects of changes in polar deep water production has centered on seeking explanation for the Younger Dryas cooling episode that affected the North Atlantic (Ruddimann &

McIntyre 1981) and surrounding land areas (Mercer 1969, Mangerud et al 1974, Mangerud 1987) from 11,000 to 10,000 ^{14}C years B.P. [= 13,000–11,700 calendar years B.P. according to Fairbanks (1990)]. Prior to the Younger Dryas the climate of northern Europe had become much like it is today. The cooling of summer temperatures experienced in northern Europe during the Younger Dryas was in the order of 5–6°C. In western Norway the mean annual temperature declined by 13°C. This climate change occurred within one or two centuries (Flohn 1986).

Kennett (1990) has reviewed the state of knowledge of conditions during the Younger Dryas and discussed some possible causes. The immediate cause is generally agreed to be a relocation of the North Atlantic Current to its glacial position. During the last glacial period it flowed east at about 45°N (McIntyre et al 1976). Its nothern edge marked the boundary between the open Central Atlantic Ocean and ice-covered North Atlantic. Perennial ice cover would inhibit deep water formation by insulating the ocean from the influence of the atmosphere. Prior to the Younger Dryas the North Atlantic Current flowed northward into the Norwegian Sea although it was not as strong as it is today (Jansen 1987). The climatic changes that occurred in northern Europe during the Younger Dryas are well explained by assuming that the North Atlantic Current returned to its glacial position.

Why did this change occur? Fresh water outflow through the St. Lawrence increased as the glacial drainage systems became reorganized (Broecker et al 1988). Teller (1990) analyzed the changes in runoff from the Laurentide ice sheet before, during, and after the Younger Dryas. The runoff from the southern side of the ice sheet was concentrated into the Mississippi drainage prior to the Younger Dryas. At about 11,000 B.P. the ice retreat exposed a passage between Lake Aggasiz on the southwest side of the ice sheet, and the Great Lakes (Broecker et al 1989). This allowed the major part of the southern drainage to be diverted around the southern margin of the ice and to the northeast, into the St. Lawrence valley. Teller estimates the earlier St. Lawrence outflow to have been about 900 km^3 yr^{-1}. The additional meltwater increased this flow to 1730 km^3 yr^{-1} during the Younger Dryas. The connection from Lake Agassiz into the Great Lakes region was subsequently severed by the Marquette re-advance of the ice, which returned flow to the Mississippi. Broecker & Denton (1989) have assumed that this would freshen the surface of the North Atlantic sufficiently to turn off the production of deep water—a conclusion supported by ocean model studies (Maier-Reimer & Miko-lajewicz 1989). The increment during the Younger Dryas amounts to a thickness of 0.16 m yr^{-1} over the North Atlantic between 45° and 60°N. This is the amount that Pollard & Pu (1985) reckoned to be required to produce the Great Salinity Anomaly as observed in the eastern North Atlantic. The higher inflow during the Younger Dryas lasted a thousand years whereas that associated with the Great Salinity Anomaly lasted less than a decade.

Sources other than the increased meltwater outflow through the St. Lawrence have been suggested as potentially significant supplies of fresh water. Miller & Kaufman (1990) suggest that the entry of warmer waters into the Labrador Sea resulted in large-scale calving of icebergs. Another mechanism by which the cessation of North Atlantic Deep Water formation during the Younger Dryas may have affected climate is by altering the CO_2 content of the atmosphere, as will be discussed below. Stauffer et al (1984) described a decrease in the CO_2 content of the air trapped in the Dye 3 ice core from Greenland during the Younger Dryas, although they noted that melting may have affected the record.

The Glacial-Interglacial Record

During the last glacial maximum, the polar oceans had more extensive sea ice cover than they do at present. Most of the present areas of deep and bottom water formation were covered by perennial sea ice. Kellogg (1987) has discussed the possible sites and modes of deep water formation during the last glacial maximum. The lowered sea level (~ 130 m) greatly reduced the area of continental shelves available as sites for deep water formation. It is clear that the different climate of the glacial maximum would profoundly affect deep water formation, but it is not clear whether the changes in deep water formation would significantly affect the ice-age climate.

A decade ago it was discovered that the atmospheric CO_2 content during the last glacial maximum was about 80 ppm less than its modern pre-industrial level (Berner et al 1979, Delmas et al 1980, Neftel et al 1982). Then it was discovered that rapid variations (a few hundred years) of the atmospheric CO_2 level were recorded in ice cores (Stauffer et al 1984). The hypotheses that had been suggested to explain the long-term glacial-interglacial changes in atmospheric CO_2, such as alternating sequestering and resupply of nutrients to the ocean from the continental shelves with sea level change (Broecker 1981), the coral reef buildup hypothesis of Berger (1982), the change in the Redfield ratio originally suggested by Weyl (1968), the denitrification and model of Berger & Kier (1984), and the change in rain rate of organic and inorganic carbon of Berger & Kier (1984) all have long time constants and cannot explain short-term fluctuations of the magnitude observed. A consensus has arisen that the answer must lie in changes in ocean circulation.

Oeschger et al (1984) noted that phosphate and presumably other nutrients are not depleted in waters where active upwelling and overturning takes place. They cited equatorial Pacific waters as having an available ("preformed") phosphate content of 20% of the deep sea value and Antarctic waters as having a preformed phosphate content 70% that of deep water. They argue that biological productivity in these areas is not limited by the availability of the nutrients, but is controlled by the rate of convection of the water. They estimate that complete depletion of the phosphate in the surface waters of the ocean would reduce atmospheric CO_2 from 300 ppm to 200 ppm. They suggest that if the ocean changed from a relatively stagnant state to one with vigorous themohaline circulation, this would result in a situation where the nutrients brought to the surface could no longer be fully utilized, resulting in higher ΣCO_2 in the surface waters and feeding CO_2 into the atmosphere. A shutdown in deep water formation implies a lowered level of atmospheric CO_2 and intensive thermohaline circulation implies higher levels of atmospheric CO_2.

Broecker & Takahashi (1984) explored the behavior of ocean box models, each with an atmosphere, a warm surface ocean reservoir, and a deep ocean reservoir in communication with the atmosphere and laterally with the warm ocean reservoir at high latitudes. They considered the differences between two different types of model: (a) a Thermodynamic Ocean, in which the CO_2 exchange is dependent only on the temperature differences between the reservoirs and (b) a Redfield Ocean, where the atmospheric transport of CO_2 is assumed to be ineffective, and the differences in ΣCO_2 are assumed to be strictly driven by lateral exchange of water between the warm and cold oceans and by the organic particle fluxes from the warm surface ocean to the deep sea. They conclude that changing the transfer rate between the warm and cold water reservoirs causes the introduction of nutrients and "biological pumping" of carbon from the warm reservoir into the deep sea. They calculated that a significant increase in transfer of water between the warm and cold reservoirs would result in a lowering of atmospheric CO_2 to glacial maximum levels.

Sarmiento & Toggweiler (1984) provided a quantitative model with boxes to represent the atmosphere, low-latitude and high-latitude surface oceans, and the deep ocean. They emphasize the importance of complete use of nutrients upwelled into the low-latitude ocean contrasted with only partial use of the nutrients upwelled into the high-latitude ocean. Their model demonstrates the importance of high-latitude ocean convection in controlling atmospheric CO_2 levels. They note that at present the upwelling subpolar cyclonic gyres (within which deep water formation must take place) are centered on 60° latitude, at the edge of polar night. They suggest that during the glacial maximums these were displaced equatorward, into regions where they received more light, enhancing the growth of phytoplankton.

Siegenthaler & Wenk (1984) have designed a different four box model with atmosphere, warm and cold surface oceans, and a deep ocean. They note that the ΣCO_2 of surface seawater is 10% to 20% lower than that of deep water because of the continuous removal of carbon by organisms and the subsequent organic particulate flux to the deep sea. Broecker (1981) had suggested that during times of lowered sea level (glacials) the ocean receives an extra supply of nutrients from the erosion of the exposed continental shelves. This increased nutrient supply raised the overall ocean nutrient content and this allowed the contrast between ΣCO_2 in the surface and deep waters to increase, effectively depleting surface water CO_2 and lowering atmospheric CO_2. They note that this process would require thousands of years to be effective, and although this might explain the long-term trends between glacials and interglacials it cannot explain the short-term variations observed in ice cores. For their standard case, assumed to mimic the present condition, they found a global upwelling flux of 15 Sv and an equal Antarctic deep water formation rate. They assume an Antarctic vertical convection rate of 44 Sv. This produces an atmospheric CO_2 concentration of 282 ppm. If the rate of vertical convection in the Antarctic is reduced by 50% to 22 Sv, while assuming the organic particulate rain rate to remain constant, the surface ΣCO_2 and atmospheric CO_2 are reduced, the latter by 52 ppm. They also experimented with a change in the rate of upwelling/deep water formation, doubling it to 30 Sv. This results in a decrease of atmospheric CO_2 by 25 ppm. The enhanced low-latitude upwelling results in a greater particulate flux to the deep sea. In this case increasing the rate of global thermohaline circulation results in a decrease of atmospheric CO_2. It must be borne in mind that all of these models are gross simplifications of the circulation of the real ocean. They include only deep water formation and do not include formation of intermediate waters that are directly involved in low-latitude upwelling.

THE CONVEYOR BELT

Broecker et al (1985) have suggested that the global thermohaline circulation system can be thought of as a conveyor belt (Figure 7). The large-scale circulation of water in the deep sea is reflected in the distribution of PO_4, NO_3, Ba, dissolved SiO_2, ΣCO_2, alkalinity, and O_2 in the deep waters (Broecker & Peng 1982). All of these, except O_2, have their lowest concentrations in the North Atlantic and their highest concentrations in the North Pacific, with intermediate values in the South Atlantic, Indian, and South Pacific Oceans. The concentration of O_2 is the reverse. Oxygen is consumed as organic particulate material is decomposed in the deep

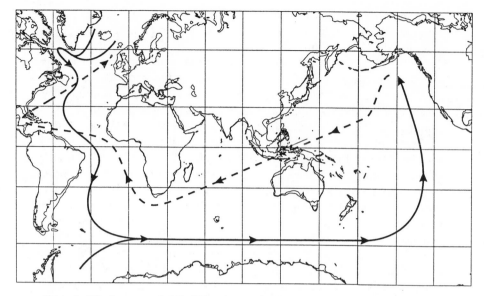

Figure 7 The "conveyor belt" of Broecker et al (1985). Deep transport shown by solid arrows and the very generalized surface transport by dashed arrows.

sea, releasing the other constituents. The concentration gradients are a reflection of the age of the water since it left the surface. It is suggested that the water that sinks in the GIN Sea flows southward through the deep Atlantic to the Southern Ocean, where it mixes with water sinking in the Weddell Sea. It then flows across the Indian Ocean, passes south of Australia, enters the South Pacific and finally reaches the North Pacific. Along the route water upwells to the surface and follows a return path from the Pacific through the Southeast Asian Seas, across the Indian Ocean, around the tip of Africa, and north through the Atlantic to the GIN Sea. Gordon (1986) has discussed this global circulation scheme showing that the return flow takes place both through surface currents and in the thermocline. The transfer of warm water from the Indian Ocean to the South Atlantic via the Agulhas Current makes a significant contribution to the northward meridional heat flux of the Atlantic, and thus a direct impact on the regional climate. They argue that the hypotheses requiring reduced nutrients in high-latitude waters to achieve the lower glacial levels of CO_2 in the atmosphere are not supported by geochemical evidence which indicates that the surface waters during the glacial were not nutrient depleted. They believe that the Cd/Ca nutrient proxy data of Boyle & Keigwin (1982) provide compelling evidence that the production of deep water in the North Atlantic was greatly reduced. Subsequent studies offer additional support (Boyle 1988, 1992).

Broecker et al (1985) report the results of an atmospheric General Circulation Model (GCM) with present-day ocean temperatures everywhere except in the North Atlantic, where they used glacial (18 ka) temperatures of CLIMAP (1981). They believe the results to be consistent with the climatic data for the Younger Dryas in the North Atlantic region. They speculate on the possibility that if one aspect of the reduction of North Atlantic deep water formation is a lowered salinity of the surface water, the opposite might be true for the North Pacific. If this were the case, then the global thermohaline circulation system might reverse, with deep water formation taking place in the North Pacific and the deep North Atlantic becoming the site of concentration of nutrients.

The St. Lawrence meltwater diversion hypothesis has been called into question by Fairbanks (1989) who noted that sea level rose more rapidly both before and after than during the Younger Dryas. Taking this into account, Broecker et al (1990) have suggested that the conveyor belt may have an internal oscillation built in. They suggest that when the conveyor is turned on (NADW is being produced) the introduction of warm waters to high latitudes causes the ice sheets surrounding the North Atlantic to retreat. The running conveyor belt continuously exports salt from the North Atlantic. The export of salt, along with the inflow of meltwater, causes the salinity of the North Atlantic to decrease. When the system becomes delicately balanced, a meltwater diversion can cause the conveyor to stop (NADW production ceases). This reduces the meridional heat flux to the North Atlantic and the salt export. The ice sheet retreat slows or stops, reducing the meltwater input. As vapor transport to the Pacific continues, the salinity again rises until the conveyor starts again. These ideas have been further developed and explored by Broecker (1989, 1990), Broecker et al (1990), Birchfield & Broecker (1990), and Birchfield et al (1990).

Ocean modelers have shown that multiple states of the thermohaline circulation system are possible. Marotzke & Willebrand (1991) used a general circulation model with two idealized oceans of equal size extending north from a southern circumglobal seaway and separated by two continents of equal size. They found that the easiest to achieve condition was one with sinking in the north in both oceans. They initially perturbed the system by increasing the evaporation over the "Atlantic" by 0.18 m yr^{-1} and increasing the precipitation over the Pacific by an equal amount, thereby reproducing the conveyor belt. By reversing these conditions they could reverse the conveyor belt. They also found that by initially decreasing the salinities in the north by 0.5‰ the model could be forced to a condition where sinking took place only in the south. Stocker & Wright (1991) have used a model with two zonally averaged basins representing the Atlantic and Pacific. They found that with an Atlantic to Pacific fresh water flux of 0.3 Sv, the conveyor belt was maintained, and that it was insensitive to changes until the interocean vapor transfer was reduced to 0.03 Sv, whereupon it stopped. In this state deep water formation took place only in the south.

WHEN AND WHERE DID POLAR DEEP WATER FORMATION BEGIN?

Hay (1988) presented a general review of paleoceanography. A variety of evidence suggests that polar oceans were warm during the Mesozoic, and that the cooling began in the early Cenozoic. A chilling of deep water occurred in the South Atlantic at the end of the Paleocene, and the ocean basins began to fill with cooler and cold waters at the end of the Eocene. The Weddell Sea has long had the appropriate shape and position to be a source of cold deep waters, but very little is known about the sources that must have existed in other parts of the world. The GIN Sea, which plays such an important role in NADW production today, had shallow, narrow connections with the North Atlantic until 10 Ma (Thiede & Eldholm 1983). Much more will have to be known about the paleogeography of the continental shelves and the connections between marginal seas and the ocean basins before ancient sites of polar and subtropical deep water formation can be determined.

ACKNOWLEDGMENTS

The author is indebted to Tom Rossby and Michael Schulz who reviewed drafts of the manuscript. This review was written while the author was a Senior Research Scientist and Gastprofessor at GEOMAR, Kiel, FRG, under the auspices of the Alexander Von Humboldt Foundation.

Literature Cited

Aagaard, K. 1981. On the deep circulation in the Arctic Ocean. *Deep-Sea Res.* 29: 251–68

Aagaard, K., Swift, J. H., Carmack, E. C. 1985. Thermohaline circulation in the Arctic Mediterranean Seas. *J. Geophys. Res.* 90: 4833–46

Alyushinskaya, N. M., Ivanov, V. V. 1978.

Water inflow from land. In *World Water Balance and Water Resources of the Earth*, ed. V. I. Korzun, pp. 568–74. UNESCO

Berger, W. H. 1982. Increase of carbon dioxide in the atmosphere during deglaciation: the coral reef hypothesis. *Naturwissenschaften* 69: 87–88

Berger, W. H., Kier, R. S. 1984. Glacial-

Holocene changes in atmospheric CO_2 and the deep sea record. See Hansen & Takahashi 1984, pp. 337–51

Berner, W., Stauffer, B., Oeschger, H. 1979. Past atmospheric composition and climate, gas parameters measured on ice cores. *Nature* 275: 53–55

Birchfield, G. E., Broecker, W. S. 1990. A salt oscillator in the glacial Atlantic? A "scale analysis" model. *Paleoceanography* 6: 835–44

Birchfield, G. E., Wang, H., Wyant, M. 1990. A bimodal climate response controlled by water vapor transport in a coupled ocean-atmosphere box model. *Paleoceanography* 5: 383–95

Boyle, E. A. 1988. The role of vertical chemical fractionation in controlling late Quaternary atmospheric carbon dioxide. *J. Geophys. Res.* 93: 15,701–14

Boyle, E. A. 1992. Cadmium and $\delta^{13}C$ paleochemical ocean distributions during the stage 2 glacial maximum. *Annu. Rev. Earth Planet. Sci.* 20: 245–87

Boyle, E. A., Keigwin, L. D. 1982. Deep circulation of the North Atlantic over the last 200,000 years: geological evidence. *Science* 218: 784–87

Brennecke, W. 1921. Die ozeanographischen Arbeiten der deutschen Antarktischen Expedition 1911–12. *Arch. Dtsch. Seewarte* 39: 1–214

Broecker, W. S. 1981. Glacial to interglacial changes in ocean and atmosphere chemistry. In *Climatic Variations and Variability: Fact and Theories*, ed. A. Berger, pp. 111–21. Dordrecht: Reidel

Broecker, W. S. 1989. The salinity contrast between the Atlantic and Pacific Oceans during glacial time. *Paleoceanography* 4: 207–12

Broecker, W. S. 1990. Salinity history of the northern Atlantic during the last deglaciation. *Paleoceanography* 5: 459–68

Broecker, W. S., Denton, G. H. 1989. The role of ocean-atmosphere reorganizations in glacial cycles. *Geochim. Cosmochim. Acta* 53: 2465–2601

Broecker, W. S., Peng, T.-H. 1982. *Tracers in the Sea.* Palisades, NY: Eldigio. 690 pp.

Broecker, W. S., Takahashi, T. 1984. Is there a tie between atmospheric CO_2 content and ocean circulation. See Hansen & Takahashi 1984, pp. 314–36

Broecker, W. S., Peteet, D. M., Rind, D. 1985. Does the ocean-atmosphere system have more than one stable mode of operation? *Nature* 315: 21–26

Broecker, W. S., Andree, M., Wolfli, W., Oeschger, H., Bonani, G., et al. 1988. The chronology of the last deglaciation: Implications to the cause of the Younger Dryas Event. *Paleoceanography* 3: 1–19

Broecker, W. S., Kennett, J. P., Flower, B. P., Teller, J., Trumbore, S., et al. 1989. The routing of Laurentide ice-sheet meltwater during the Younger Dryas cold event. *Nature* 341: 318–21

Broecker, W. S., Bond, G., Klas, M. 1990. A salt oscillator in the glacial Atlantic? 1, The concept. *Paleoceanography* 5: 469–78

Carmack, E. C. 1990. Large-scale physical oceanography of the polar regions. In *Polar Oceanography: Part A: Physical Science*, ed. W. O. Smith, Jr., pp. 171–222. San Diego: Academic

Carmack, E. C., Foster, T. D. 1977. Water masses and circulation in the Weddell Sea. In *Polar Oceans*, ed. M. J. Dunbar, pp. 151–65. Calgary, Alta., Canada: Arctic Inst. North Am.

Carmack, E. C., Killworth, P. D. 1978. Formation and interleaving of water masses off Wilkes Land, Antarctica. *Deep-Sea Res.* 25: 357–69

Chamberlin, T. C. 1906. On a possible reversal of deep-sea circulation and its influence on geologic climates. *J. Geol.* 14: 363–73

Clarke, R. A., Gascard, J.-C. 1982. The formation of Labrador Sea Water. Part I: large scale processes. *J. Phys. Oceanogr.* 13: 1764–78

CLIMAP Project Members 1981. Seasonal reconstruction of the Earth's surface at the last glacial maximum. In *Geological Society of America Map and Chart, Series 36.* Boulder, Colo.: Geol. Soc. Am.

Coachman, L. K., Aagaard, K. 1974. Physical oceanography of Arctic and Subarctic seas. In *Marine Geology and Oceanography of the Arctic Seas*, ed. Y. Herman, pp. 1–72. Berlin: Springer-Verlag

Delmas, R. J., Ascencio, J. M., Legrand, M. 1980. Polar ice evidence that atmospheric CO_2 20,000 BP was 50% of present. *Nature* 284: 155–57

Dickson, R. R., Meincke, J., Malmberg, S.-A., Lee, A. J. 1988. The "Great Salinity Anomaly" in the northern North Atlantic 1968–1982. *Progr. Oceanogr.* 20: 103–51

Emiliani, C. 1954. Temperatures of Pacific bottom waters and polar superficial waters during the Tertiary. *Science* 119: 853–55

Emiliani, C., Edwards, G. 1953. Tertiary ocean bottom temperatures. *Nature* 171: 887–89

Fairbanks, R. G. 1989. A 17,000-year long glacio-eustatic sea level record: influence of glacial melting rates on the Younger Dryas event and deep-ocean circulation. *Nature* 342: 637–42

Fairbanks, R. G. 1990. The age and origin of the "Younger Dryas Climate Event" in Greenland ice cores. *Paleoceanography* 5: 937–48

Flohn, H. 1986. Singular events and catastrophes now and in climatic history. *Naturwissenschaften* 73: 136–49

Flohn, H. 1987. Air-sea interaction processes as models for abrupt climatic changes. In *Abrupt Climatic Change*, ed. W. H. Berger, L. D. Labeyrie, pp. 23–30. Dordrecht: Riedel

Fofonoff, N. P. 1956. Some properties of seawater influencing the formation of Antarctic Bottom Water. *Deep-Sea Res.* 4: 32–35

Folland, C. K., Palmer, T. N., Parker, D. E. 1986. Sahel rainfall and worldwide sea temperatures 1901–85. *Nature* 320: 602–6

Foster, T. D. 1972. An analysis of the cabeling instability in sea water. *J. Phys. Oceanogr.* 2: 294–301

Foster, T. D., Carmack, E. C. 1976. Frontal zone mixing and Antarctic Bottom Water formation in the southern Weddell Sea. *Deep-Sea Res.* 23: 301–17

Foster, T. D., Carmack, E. C. 1977. Antarctic Bottom Water formation in the Weddell Sea. In *Polar Oceans*, ed. M. J. Dunbar, pp. 167–77. Calgary, Alberta, Canada: Arctic Inst. North Am.

Gill, A. E. 1973. Circulation and bottom water production in the Weddell Sea. *Deep-Sea Res.* 28: 111–40

Gordon, A. L. 1978. Deep Antarctic convection west of Maude Rise. *J. Phys. Oceanogr.* 8: 600–12

Gordon, A. L. 1982. Weddell Sea deep water variability. *J. Mar. Res.* 40: 199–217 (Suppl.)

Gordon, A. L. 1986. Interocean exchange of thermocline water. *J. Geophys. Res.* 91: 5037–46

Gordon, A. L., Nowlin, W. D. 1978. The basin waters of the Bransfield Strait. *J. Phys. Oceanogr.* 8: 258–64

Gordon, A. L., Tchernia, P. 1972. Waters of the continental margin off Adélie Coast, Antarctica. In *Antarctic Res. Ser.*, *19, Antarctic Oceanology II: The Australian-New Zealand Sector*, ed. D. E. Hayes, pp. 59–69. Washington, DC: Am. Geophys. Union

Gow, A. J., Tucker, W. B., III. 1990. Sea ice in the polar regions. In *Polar Oceanography: Part A: Physical Science*, ed. W. O. Smith, Jr., pp. 47–122. San Diego: Academic

Hakkinen, S. 1987. A coupled dynamic-thermodynamic model of an ice-ocean system in the marginal ice zone. *J. Geophys. Res.* 92: 9469–78

Hansen, J. E., Takahashi, T., eds. 1984. *Geophysical Monograph 29, Maurice Ewing Volume 5, Climate Processes and Climate Sensitivity*. Washington, DC: Am. Geophys. Union

Hay, W. W. 1988. Paleoceanography: a review for the GSA Centennial. *Geol. Soc. Am. Bull.* 100: 1934–56

Hogg, N. G. 1973. The preconditioning phase of MEDOC 1969, II, Topographic effects. *Deep-Sea Res.* 20: 449–59

Hsiung, J. 1985. Estimates of global oceanic meridional heat transport. *J. Phys. Oceanogr.* 15: 1405–13

Jacobs, S. S., Georgi, D. T. 1977. Observations of the southwest Indian/Antarctic Ocean. In *A Voyage of Discovery*, ed. M. Angel, pp. 43–84. New York: Pergamon

Jansen, E. 1987. Rapid changes in the inflow of Atlantic water into the Norwegian Sea at the end of the last glaciation. Ir *Abrupt Climatic Change*, ed. W. H. Berger, L. D. Labeyrie, pp. 299–310. Dordrecht: Riedel

Kellogg, T. B. 1987. Glacial-interglacial changes in global deepwater circulation. *Paleoceanography* 2: 259–71

Kennett, J. P. 1990. The Younger Dryas Cooling Event: an introduction. *Paleoceanography* 5: 891–95

Killworth, P. D. 1977. Some models of bottom water formation. In *Polar Oceans*, ed. M. J. Dunbar, pp. 179–89. Calgary, Alta., Canada: Arctic Inst. North Am.

Killworth, P. D. 1979. On "chimney" formations in the ocean. *J. Phys. Oceanogr.* 9: 531–54

Killworth, P. D. 1983. Deep convection in the world ocean. *Rev. Geophys. Space Phys.* 21: 1–26

Lazier, J. R. N. 1973. The renewal of Labrador-Sea Water. *Deep-Sea Res.* 20: 341–53

Lazier, J. R. N. 1980. Oceanographic conditions at Ocean Weather Ship *Bravo* 1964–74. *Atmosphere-Ocean* 18: 227–38

Maier-Reimer, E., Mikolajewicz, V. 1989. Experiments with an OGCM on the cause of the Younger Dryas. In *Oceanography, 1988*, ed. A. Ayala-Castañares, W. Wooster, A. Yanez-Arancibia, pp. 87–100. Mexico City: UNAM

Mangerud, J. 1987. The Alleröd/Younger Dryas boundary. In *Abrupt Climatic Change*, ed. W. H. Berger, L. D. Labeyrie, pp. 163–71. Dordrecht: Riedel

Mangerud, J., Anderson, S. T., Berglund, B. E., Donner, J. J. 1974. Quaternary stratigraphy of Norden, a proposal for terminology and classification. *Boreas* 3: 109–28

Marotzke, J., Willebrand, J. 1991. Multiple equilibria of the global thermohaline circulation. *J. Phys. Oceanogr.* 21: 1372–85

Martinson, D. G., Killworth, P. D., Gordon, A. L. 1981. A convective model for the Weddell polynya. *J. Phys. Oceanogr.* 11: 466–88

McIntyre, A., Kipp, N. G., Bé, A. W. H.,

Crowley, T., Kellogg, T. B., et al. 1976. Glacial North Atlantic 18,000 years ago: a CLIMAP reconstruction. *Geol. Soc. Am. Mem.* 145: 43–76

Meincke, J. 1983. The modern current regime across the Greenland-Scotland Ridge. In *Structure and Development of the Greenland Scotland Ridge: New Methods and Concepts*, ed. M. H. P. Bott, S. Saxov, M. Talwani, J. Thiede, pp. 637–50. New York: Plenum

Melling, H., Lewis, E. L. 1982. Shelf drainage flows in the Beaufort Sea and their effect on the Arctic Ocean pycnocline. *Deep-Sea Res.* 29: 967–86

Mercer, J. H. 1969. The Alleröd oscillation: a European climatic anomaly. *Arc. Alp. Res.* 1: 227–34

Miller, G. H., Kaufman, D. S. 1990. Rapid fluctuations of the Laurentide ice sheet at the mouth of Hudson Strait: new evidence for ocean/ice-sheet interactions as a control on the Younger Dryas. *Paleoceanography* 5: 907–19

Mosby, H. 1934. The waters of the Atlantic Antarctic Ocean. *Sci. Results Nor. Antarct. Exped. 1927–1928* 1: 1–131

Neftel, A., Oeschger, H., Staffelbach, T., Stauffer, B. 1982. CO₂ record in the Byrd ice core, 50,000–5,000 years BP. *Nature* 331: 609–11

Newell, R. E., Hsiung, J. 1987. Factors controlling the free air and ocean temperature of the last 30 years and extrapolation to the past. In *Abrupt Climatic Change*, ed. W. H. Berger, L. D. Labeyrie, pp. 67–87. Dordrecht: Riedel

Oeschger, H., Beer, J., Siegenthaler, U., Stauffer, B., Dansgaard, W., Langway, C. C. 1984. Late glacial climate history from ice cores. See Hansen & Takahashi 1984, pp. 299–306

Peixoto, J. P., Oort, A. H. 1992. *Physics of Climate*. New York: Am. Inst. Physics. 520 pp.

Peterson, W. H. 1979. *A steady state thermohaline convection model*. PhD thesis. Rosenstiel School Mar. Atmos. Sci., Univ. Miami. 160 pp.

Peterson, W. H., Rooth, C. 1976. Formation and exchange of deep water in the Greenland and Norwegian seas. *Deep-Sea Res.* 23: 273–83

Philander, S. G. H. 1989. *El Niño, La Niña, and the Southern Oscillation*. New York: Academic. 293 pp.

Pickard, G. L., Emery, W. J. 1990. *Descriptive Physical Oceanography*. Oxford: Pergamon. 320 pp.

Pollard, R. T., Pu, S. 1985. Structure and circulation of the upper Atlantic Ocean northeast of the Azores. *Progr. Oceanogr.* 14: 443–62

Rasmussen, E. M. 1985. El Niño and variations in climate. *Am. Sci.* 73: 168–77

Reid, J. L. 1979. On the contribution of the Mediterranean Sea outflow to the Norwegian-Greenland Sea. *Deep-Sea Res.* 26: 1199–1223

Rooth, C. 1982. Hydrology and ocean circulation. *Prog. Oceanogr.* 11: 131–49

Ruddimann, W. F., McIntyre, A. 1981. The North Atlantic Ocean during the last deglaciation. *Palaeogeogr. Palaeoclim. Palaeoecol.* 35: 145–214

Rudels, B. 1989. Haline convection in the Greenland Sea. *Deep-Sea Res.* 37: 1491–1511

Sarmiento, J. L., Toggweiler, R. 1984. A new model for the role of the oceans in determining atmospheric pCO₂. *Nature* 308: 621–24

Schlosser, P., Bönisch, G., Rhein, M., Bayer, R. 1991. Reduction of deepwater formation in the Greenland Sea during the 1980s: evidence from tracer data. *Science* 251: 1054–56

Seabrooke, J. M., Hufford, G. L., Elder, R. B. 1971. Formation of Antarctic bottom water in the Weddell Sea. *J. Geophys. Res.* 76: 2164–78

Siegenthaler, U., Wenk, T. 1984. Rapid atmospheric CO₂ variations and ocean circulation. *Nature* 308: 624–26

Stauffer, B., Hofer, H., Oeschger, H., Schwander, J. 1984. Atmospheric CO₂ concentration during the last deglaciation. *Ann. Glaciology* 5: 160–64

Stocker, T. F., Wright, D. G. 1991. Rapid transitions of the ocean's deep circulation induced by changes in surface water fluxes. *Nature* 351: 729–32

Stommel, H. 1961. Thermohaline convection with two stable regimes of flow. *Tellus* 13: 224–38

Stommel, H. 1962. On the smallness of sinking regions in the ocean. *Proc. Natl. Acad. Sci. USA* 48: 766–72

Swift, J. H., Aagaard, K., Malmberg, S.-A. 1980. The contribution of the Denmark Strait overflow to the deep North Atlantic. *Deep-Sea Res.* 27: 29–42

Teller, J. T. 1990. Volume and routing of late-glacial runoff from the southern Laurentide Ice Sheet. *Quat. Res.* 34: 12–23

Thiede, J., Eldholm, O. 1983. Speculations about the paleodepth of the Greenland-Scotland Ridge during late Mesozoic and Cenozoic times. In *Structure and Development of the Greenland-Scotland Ridge*, ed. M. H. P. Bott, S. Saxov, M. Talwani, J. Thiede, pp. 445–56. New York: Plenum

Thompson, Benjamin (Count Rumford). 1800. *Essays, Political, Economical, and Philosophical*. London

von Humboldt, A. 1814. *Voyages aux*

Regions Equinoctales du Noveau Continent, fait en 1799–1804, 3 vols. Paris

Warren, B. A. 1981. Deep circulation of the world ocean. In *Evolution of Physical Oceanography*, ed. B. A. Warren, C. Wunsch, pp. 6–41. Cambridge, MA: MIT Press

Weyl, P. K. 1968. The role of the ocean in climatic change: a theory of the ice ages. *Meterol. Monogr.* 8: 37–62

Wolford, T. C. 1982. Sea ice and iceberg conditions, 1970–79. *Northwest Atl. Fish. Organ. Sci. Counc. Stud.* 5: 39–42

Worthington, L. V. 1970. The Norwegian Sea as a mediterranean basin. *Deep-Sea Res.* 17: 77–84

Annu. Rev. Earth Planet. Sci. 1993. 21:255–305

MATRICES OF CARBONACEOUS CHONDRITE METEORITES

Peter R. Buseck and Xin Hua[1]

Departments of Geology and Chemistry, Arizona State University, Tempe, Arizona 85287

KEY WORDS: transmission electron microscopy, matrix mineralogy, interplanetary dust, phyllosilicates

1.   INTRODUCTION

Carbonaceous chondrite meteorites (CCs) are the most chemically primitive macroscopic solids in the solar system. They are nebular leftovers and are thus invaluable recorders of some of the oldest and best kept secrets regarding the formation and early history of the solar system. With a U/Pb age of 4.56×10^9 years, they are also the oldest known materials, 0.6 billion years older than the most ancient terrestrial rocks (Chen & Wasserburg 1981, Bowring et al 1990). They contain mineral grains that occurred in and even preceded the solar nebula. As relics of material that formed the sun and planets, they are a totally unique resource and have thus received intensive study. Yet, in spite of revealing many intimate secrets of solar system formation, they have, paradoxically, also proven remarkably intractable. Matrix is perhaps the least understood major component of CC meteorites.

Matrix is the background "sea" in which the chondrules, inclusions, and larger mineral grains occur. It is finely comminuted material that mainly consists of the anhydrous phases of meteorites, such as olivine, pyroxene, and Fe/Ni metal, together with lesser amounts of Fe oxides and sulfides, carbonates, sulfates, and a variety of other minerals that are generally present in only minor amounts. Some meteorites contain significant amounts of phyllosilicates, and these have been the focus of much

---

[1] Permanent address: China University of Geosciences (Wuhan), PR China

0084–6597/93/0515–0255$02.00

study and attention. The interplanetary dust particles (IDPs) are in some ways similar to the CC matrix in both mineralogy and composition, but the IDPs are even finer grained, which suggests that they are a related but distinct class of materials. The focus of this review is the matrix of the CCs.

There is much interest in matrix because (*a*) it contains a wealth of mineralogical and chemical information about the early history and subsequent development of the CCs (some of the minerals are intact from the time of the solar nebula, which is the earliest stage of the solar system that we can ever hope to understand in reasonable detail); (*b*) it forms a significant fraction of the highly important CCs; (*c*) it has long been difficult to study—a result of its extremely fine-grained, opaque character; and (*d*) it provides a model for studying alteration processes in the early solar system.

Much matrix appears to have developed from the primary anhydrous minerals olivine, pyroxene, and metal. It appears that accretion brought these minerals to a variety of meteorite parent bodies. It is probable that cometary debris and similar hydrated materials were included among the samples collected by the parent bodies and subsequently buried as other rocks rained onto the parent body surfaces. A variety of events, not all of which are known, then combined to produce alteration reactions. Regolith gardening, produced by the infall of other meteorites, resulted in collisional heating, fragmentation, and mixing of surficial rocks. These sources of energy, possibly combined with gentle heating from radioactive decay of unstable nuclides, provided the heat necessary for alteration.

Three features of CC matrices are of particular interest:

1. Tiny diamonds and crystals of SiC have recently been found in these meteorites (Lewis et al 1987, Tang & Anders 1988b, Bernatowicz et al 1987); their isotopic compositions indicate that they had their origin in circumstellar or interstellar space, outside of the solar system (Tang et al 1989, Zinner et al 1989, Lewis et al 1989, Huss & Lewis 1990). These are the first minerals about which it can be said with confidence that they are older than the materials of the solar system.

2. It has long been known that the bulk chemical composition of the CI meteorites closely matches that of the solar photosphere (Holweger 1977, Anders & Ebihara 1982), and yet recent work has shown that aqueous alteration is pervasive in at least some of these meteorites (Tomeoka & Buseck 1988, Zolensky & McSween 1988). It is apparent that the alteration must have occurred in a largely closed system so that it did not change the bulk composition significantly. Yet there are questions whether some of the alteration occurred prior to incorporation into a parent body. Paradoxically, the meteorite group that is

most primitive chemically (CI) is not the same as those that are physically most pristine (CO and CV). (The symbols for the meteorite types are defined in Section 3.)

3. The minerals of the matrix, because of their fine-grained, reactive character, provide a sensitive indicator of secondary processes such as heating, shock, and aqueous alteration. Unraveling these effects is required before a comprehensive understanding of the early solar system can be achieved.

Because much matrix appears to have developed as a result of alteration, or was profoundly affected by it, considerable emphasis in this review is placed on alteration minerals, processes, and mechanisms. Determining the temperatures, pressures, oxidation conditions, and durations of these alteration reactions is one of the challenging aspects of research on CC matrices. The resulting information provides insights into activities in the early solar system.

On a personal note, the senior author has long been both intrigued and frustrated regarding the character of CC matrix. Prior to the mid–1970s, the available mineralogical techniques were inadequate to provide much insight into the character of CC matrix. However, with the advent of the transmission electron microscope (TEM) as a viable mineralogical tool, especially through the use of high-resolution transmission electron microscopy (HRTEM), it became possible to image and study minerals as fine grained as those in the CC matrices (Buseck & Iijima 1974).

In order to achieve the goal of understanding the HRTEM images of the minerals in CC matrices, we embarked on a systematic series of studies to obtain the requisite knowledge and experience. Our investigations of pyroxene (Iijima & Buseck 1975, Buseck & Iijima 1975), micas (Iijima & Buseck 1978), serpentines (Veblen & Buseck 1979), and pyrrhotite (Pierce & Buseck 1974) were all part of that effort. Because of the problems of radiation damage, efforts to image clays (Mckee & Buseck 1978) of the types thought to occur in CCs had only limited success. However, in the late 1970s and early 1980s, several TEM studies of CC matrices appeared (Mackinnon & Buseck 1979; Barber 1981; Mackinnon 1982; Tomeoka & Buseck 1982a,b, 1983c). These and subsequent studies established the details of the secondary minerals that occur in the matrices.

It is now widely accepted that the TEM is a necessary tool for studying the details of matrix mineralogy, but only relatively few such studies have been completed (although the number is increasing steadily). As a result, the following discussion and generalizations are based disproportionately on those CC meteorites whose matrices have been studied using TEM techniques.

The main differences in the alteration products in matrix among different CC samples are (a) their amounts, (b) the phyllosilicates (generally either serpentine or saponite, a clay), and (c) the Fe-bearing phases (troilite, magnetite, maghemite, ferrihydrite, or a variety of poorly crystallized oxyhydroxides of uncertain character). Other alteration minerals that show differences in occurrence, abundance, and distribution include carbonates and sulfates.

Recent reviews that describe aspects of the matrices of the carbonaceous chondrites are by Scott et al (1988) on the matrix material itself, Barber (1985) on layer silicates within stony meteorites, Zolensky & McSween (1988) on the aqueous alteration, Grimm & McSween (1989) on the thermal evolution of carbonaceous chondrite parent bodies, and Tomeoka et al (1989) on alteration mineralogy of the CM meteorites.

2. MORPHOLOGY AND GENERAL CHARACTERISTICS

When viewed in hand specimen or at moderate magnifications using a conventional petrographic microscope, matrix is opaque, black, and relatively featureless—an appearance that visually distinguishes the CCs from other meteorites. The grain sizes of its constituent minerals range from ~ 5 μm down to the limit of optical resolution. Using a TEM, it is evident that grains as small as a few nm in diameter can occur. Grain sizes vary with CC type (Scott et al 1988); they are smallest in the CRs (Zolensky 1991) and largest in the CVs. The minerals of matrix include typical CC anhydrous silicates such as olivine and pyroxene, as well as lesser oxides, carbonates, and sulfates. Also prominent are alteration products such as serpentines, other layer silicates, and hybrid phases such as tochilinite.

Definitions of matrix vary widely (see Scott et al 1988 for a review). For present purposes we include polyphasic material with grains smaller than 5 μm in diameter, in agreement with many others (Huss et al 1981, Nagahara 1984, Scott et al 1988, Brearley 1989); in many cases such material is interstitial to coarser crystals or crystal fragments, composite inclusions, or chondrules. By polyphasic we include regions where several grains are clustered and exclude small monophasic inclusions within larger crystals. The 5-μm size is arbitrary, and in specific cases the grains could be larger (e.g. Kallemeyn et al 1991). McSween & Richardson (1977) on the basis of their study of matrix compositions, redefined matrix as "all those low-temperature components which are genetically related and which, taken together, properly represent the composition of the primitive matter from which they formed." While intriguing, this is a difficult definition to use because it presupposes we know which parts of the meteorite are primordial and also which parts are "genetically related." Subsequently, McSween (1979a) defined matrix "operationally" as everything that is not recognizable petrographically as something else (i.e. chondrules, inclusions, lithic and mineral fragments, or opaque minerals).

Scott et al (1984) distinguished no fewer than seven types of matrix occurrences, ranging from lumps and clasts to rims of chondrules for the ordinary chondrites and the Vigarano (CV) meteorite. Clasts are common and can be of other types of meteorites than their hosts, giving rise to the interpretation that some CC meteorites were part of a regolith breccia (McSween & Richardson 1977, Bunch & Chang 1980), much like the surface of the moon. The clasts have typically been altered in whole or in part to phyllosilicate minerals, thereby complicating their recognition, although their outlines are retained. Bunch & Chang (1980) point out the common juxtapositions of altered and unaltered clasts, lending support to the mixing of fresh and altered material such as occurs on a parent body that is repeatedly impacted by falling bodies.

3. CLASSIFICATION

During the past three decades several versions of CC nomenclature have been used. For example, depending on when the paper was written, the Murchison meteorite is grouped with the type II, C2, or CM meteorites. Since these are all more or less equivalent, the terminology can be confusing. Therefore, in reviewing the literature of CCs it seems prudent to provide a brief summary of the terminology that is used so commonly for the CCs.

Based largely on the amounts of H_2O, S, and C, Wiik (1956) subdivided the CCs into types I, II, and III. In their "chemical-petrologic" classification of stony meteorites, Van Schmus & Wood (1967) retained Wiik's groupings but changed their names to C1, C2, and C3, respectively, to be consistent with the other meteorite groups. The numerical designations reflect petrographic and mineralogical changes and extend to type 6. Type–3 meteorites are physically the most pristine, whereas types 2 and 1 contain increasing amounts of minerals typical of aqueous alteration. Types 4 to 6 show features that are compatible with progressively increasing intensities of thermal metamorphism. The C3 group was subsequently subdivided into C3(V) and C3(O) groups by Van Schmus (1969). That nomenclature was revised by Wasson (1974) to CI (for C1), CM (for C2), and CV and CO (for C3), where the second letters designate representative meteorites (Ivuna, Mighei, Vigarano, and Ornans, respectively). It is these last terms that are in current use. The several groups have distinctive mineralogical, chemical, and isotopic characteristics, although there is

Table 1 The classification of chondrites[a] (a listing of some key taxonomic parameters)

	Al/Si	Mg/Si	Ni/Si	Zn/Si	FeO$_x$ FeO$_x$+MgO (mol%)	δ^{17}O (‰)	δ^{18}O (‰)
		(normalized ratios)					
GROUP							
CV	1.34	1.00	0.85	0.25	35	-3	1
CO	1.07	0.97	0.87	0.21	33	-4	0
CM	1.10	0.97	0.92	0.48	43[b]	1	7
CI	$\equiv 1.00$	$\equiv 1.00$	$\equiv 1.00$	$\equiv 1.00$	45[b]	9	17

[a] From *Philos. Trans. R. Soc. London Ser. A* 325: 535–44 (1988).
[b] Estimated FeO$_x$/(FeO$_x$ + MgO) for equilibrium assemblage.

overlap. Table 1 gives some salient chemical and isotopic characteristics for the major groups.

Recently, the CR (for Renazzo) and CK (for Karoonda) groups have been established for "anomalous" meteorites that do not comfortably fit into the above grouping. Their characteristics are given in Section 5.5. Both contain fewer meteorites than do the other CC groups, and so they have not received the extensive attention that the other major groups have had.

4. CHEMISTRY

4.1 *Inorganic Chemistry*

The chemical composition of the CCs, and the surprisingly similar abundances of their nonvolatile elements to those of the solar photosphere, is a major reason why the CCs are of such great interest and led to the conclusion that these meteorites are chemically primitive (Suess 1949, Suess & Urey 1956). Since the sun contains 99 + % of the mass of the solar system, its composition is considered representative of the solar system. The CCs contain the highest abundances of volatile elements of any meteorite group, an observation that is also supportive of their primitive origin since these elements tend to be depleted during geological processing. As a major component of CCs, matrix plays an important role in the search for pristine nebular materials and in understanding the early stages of the solar system. Other meteorites have experienced chemical fractionation such that their compositions deviate widely from that of the sun, as is also the case for all terrestrial rocks.

Although much has been written about the inorganic chemistry of the CCs, it is almost exclusively based on "whole-rock" analyses, and so it is difficult to generalize about matrix composition using such data. Moreover, many CCs are brecciated and have a polymict character so that their compositions, while relatively homogeneous within a given clast, vary considerably from one clast to another. As a result, it is difficult to obtain reliable and representative matrix abundance data. However, in situ chemical analyses of CC matrices have been obtained by using broad-beam electron microprobe analysis (EMPA) (Fuchs et al 1973, Kerridge 1976, McSween & Richardson 1977, Rubin et al 1988, Scott et al 1988, Rubin & Kallemeyn 1990). Separated matrix has been analyzed by wet chemistry (Clarke et al 1970) and by instrumental neutron activation analysis (INAA) (Rubin & Wasson 1987, 1988).

The most thorough study of matrix compositions is by McSween & Richardson (1977), who analyzed 32 CCs using a defocussed beam with their electron microprobe. Although physical separation of matrix is not required with the microprobe, trace element abundances are not easily obtained. Rubin & Wasson (1987, 1988) determined the bulk matrix composition of CO3s and CV3s by INAA. However, their data differ significantly from the EMPA data of McSween & Richardson (1977). Rubin & Wasson indicate that these apparent discrepancies may arise from differences in methods of sampling and analysis, which points out a significant problem in the determination of average matrix compositions.

Larimer & Anders (1970) suggested that the CIs consist entirely of matrix, so their bulk analyses should provide direct matrix compositions. Compilations of CI elemental abundances are provided by Cameron (1982), Anders & Ebihara (1982), Wasson (1985), Anders & Grevesse (1989), and Wasson & Kallemeyn (1988); an evaluation of much prior work was made by Burnett et al (1989), who confirmed the previous abundance curves.

Figure 1 summarizes chemical data for the CC matrices. Since the water and organic contents within and among the several CC groups are variable, direct comparisons of matrix analyses are difficult. To minimize this problem, we adopt the method of McSween & Richardson (1977) by using Si-normalized weight ratios of the elements. The Na/Si ratios are depleted relative to the whole-rock analyses, with CI matrix the most depleted and CM the least. Na/Si ratios in CMs show much variation. Mg/Si ratios are unusual in being rather constant, with an average value of 0.822 ± 0.053, similar to the value of ~ 0.910 from bulk analyses of CCs (McSween & Richardson 1977). Al/Si ratios show large variations, especially in the CM and CVs, and differ from the average Al/Si ratio of 0.083 in bulk analyses of CCs. K/Si ratios vary such that CI > CM > CO, CV. The K/Na ratios in the matrices are almost five times higher than those in the whole rock (Wiik 1956, Bunch & Chang 1980). Ti/Si, Cr/Si, and Mn/Si ratios are

262

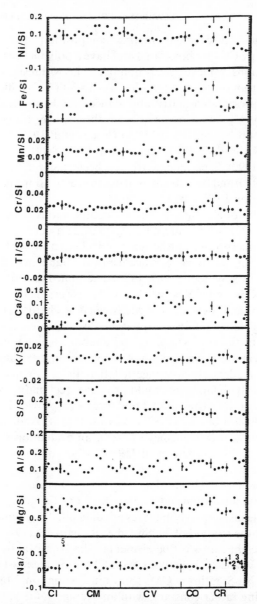

Figure 1 Compilation of chemical analyses of CC matrices, all nomalized to Si. The average value for each subgroup is indicated with a bar. Meteorites that either do not belong to the indicated CC groups or that are of uncertain type are numbered: 1. Karoonda (CK); 2. Warrenton (CO?); 3. Ningqiang (CK?); 4. Lewis Cliff 85332; 5. Kakangari.

Sources of data are: 1. McSween & Richardson (1977, EMPA); 2. Kerridge (1976, EMPA); 3. Fuchs et al (1973, EMPA); 4. Clarke et al (1970, WCA); 5. Scott et al (1988, EMPA); 6. Rubin & Wasson (1988, INAA); 7. Rubin et al (1988, EMPA); and 8. Rubin & Kallemeyn (1990, EMPA). EMPA: electron microprobe analysis; WCA: wet chemistry analysis; INAA: instrumental neutron activation analysis.

nearly constant from group to group, but the low bulk concentrations of Ti and Mn make the Ti/Si and Mn/Si values uncertain. Ca/Si ratios vary in the sequence CI < CM < CO, CV. This sequence is explained in part by exclusion of minerals in veins from the CI matrix analyses. Fe/Si in CIs (1.04 ± 0.06) is distinctly lower than the 1.857 ± 0.297 of the CMs, CVs, and COs; Ni/Si is much higher in CMs than in CIs, COs, and CVs; S/Si varies in CMs. Both the Ni/Si and S/Si ratios in CMs suggest the existence of a Ni-, S-, Fe-bearing phase.

McSween & Richardson (1977) compared published bulk compositions of the CCs to their EMPA analyses of matrix and found a difference exists for the CIs. Moreover, the CI bulk compositions agree better with the solar abundances than does the matrix, which is strongly depleted in Na, Ca, S, and Fe relative to solar abundances and to the other CCs. This depletion presumably reflects the CI compositional heterogeneity that occurs on the scale of microscopic clasts plus the fact that McSween & Richardson assumed that magnetite, calcium sulfate, and carbonate vein-fillings are not matrix components and so avoided obtaining EMPA analyses in regions where they occur. The CM matrix compositions are, in fact, closer to solar values than those of the CIs. This difference results from a combination of the extensive aqueous alteration and consequent chemical fractionation of the CIs plus the selective EMPA analysis mentioned above. Clearly, bulk rather than matrix CI compositions must be considered when comparisons to solar abundances are made.

The major element abundances of matrix group into clusters that correspond to the CC types (McSween & Richardson 1977). The CIs show considerable compositional variations among the various clasts that comprise what are probably polymict regolith breccias. CV and CO matrices contain lower amounts of Mg, Al, Ca, Ti, and S than do CIs and CMs, presumably because these elements have not been as extensively mobilized into matrix from chondrules, inclusions, and sulfides. Even though McSween & Richardson found that the differences in matrix chemistry among the various CC types is significant, they concluded that, in general, the chemical variations are relatively small.

4.2 *Organic Chemistry*

The CC matrix is host to a large variety of organic matter—partly soluble material but mainly insoluble macromolecular carbon. A fascinating aspect of such compounds is that some may be precursors of life, thereby giving their study extra significance. Murchison is the most thoroughly studied meteorite for its organic chemistry, but Orgueil, Ivuna, Murray, Renazzo, Cold Bokkeveld, and Allende have also received attention.

Although discrete organic compounds in CCs occur in amounts less than 1 wt%, they are widespread and sufficiently abundant to be analyzed in detail by current techniques. They are most concentrated in those meteorites and in those portions of the meteorites that experienced the most intense aqueous alteration (Cronin et al 1988, Bunch & Chang 1980). These are the CM and CI matrices, in which 70 wt% of the carbon occurs in extremely fine-grained ($\leqslant 50$ nm) acid-insoluble macromolecular matter containing H, N, O, S, and perhaps halogens (Chang & Bunch 1986), and 30% is in organic compounds that are soluble in nonpolar organic solvents like CCl_4 or polar solvents like CH_3OH and water (Hayes 1967). However, the polar solvents used for the extractions do not clearly separate organic compounds from inorganic salts, leading to an overestimate of the soluble fraction; thus, the actual insoluble to soluble ratio is probably greater than 7:3 (J. Cronin 1992, personal communication).

Significantly smaller amounts of carbonaceous matter occur in COs and CVs; much of the carbon in the Allende CV3 occurs in its elemental form as poorly ordered, turbostratic graphite (Lumpkin 1981, Smith & Buseck 1981). Smith & Buseck (1981) showed a TEM image (Figure 2) of carbon

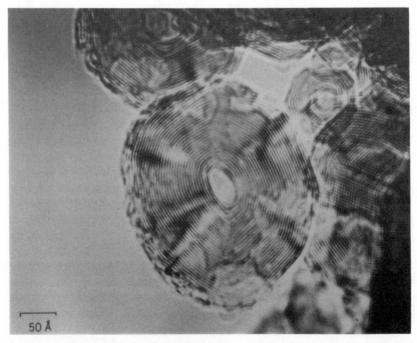

50 Å

Figure 2   HRTEM image of round layered carbon, found in an acid-residue from Allende (Smith & Buseck 1981).

that resembles a hypothesized cross-section of the layered fullerenes that were recently proposed by Kroto et al (1985).

The macromolecular matter consists of both amorphous and poorly crystalline components including condensed aromatic, heteroaromatic, and hydroaromatic ring systems, cross-linked by short methylene chains, ethers, sulfides, and biphenyl groups (Hayatsu et al 1977, 1980; Cronin & Pizzarello 1986; Zenobi et al 1989; de Vries et al 1991). The soluble compounds include amino acids (Cronin et al 1988); aliphatic hydrocarbons (Cronin & Pizzarello 1990); aromatic hydrocarbons (Basile et al 1984); various carboxylic acids (Lawless et al 1974, Peltzer & Bada 1978, Yuen et al 1984); nitrogen heterocycles, amines, and amides (Hayatsu et al 1975, Jungclaus et al 1979); and alcohols and carbonyl compounds (Jungclaus et al 1976, Cronin et al 1988). Most data regarding the organic fraction of CCs have been obtained by analyses of bulk samples and are well reviewed by Cronin et al (1988) and Cronin & Chang (1992); a brief summary is given by Chang & Bunch (1986).

It would be desirable to know the exact location of the various organic compounds within different parts of the matrix. However, in situ studies of carbon and organic compounds are almost nonexistent—the result of the extreme challenge of analyzing the light elements in small sample regions. Two methods are currently available for such in situ analyses at high spatial resolution: (a) electron microscopy and (b) laser desorption combined with an analytical method such as mass spectrometry.

There is evidence from transmission electron microscopy that graphitic layers coat some mineral grains in Allende (Green et al 1971). However, currently the best hope for understanding in situ occurrences of organic materials in CCs are studies done by two-step, laser-desorption, mass spectrometry (L^2MS) (Hahn et al 1988, Zenobi et al 1989). This technique uses a CO_2 laser to volatilize neutral molecules intact from surfaces. After a 20-μs delay, a second laser (Nd:YAG) using the principle of resonance-enhanced multiphoton ionization (REMPI) selectively ionizes polycyclic aromatic hydrocarbons (PAHs) from among the vaporized desorbed species. These ions are passed into a time-of-flight mass spectrometer, which records species with masses up to ~ 1500 amu. The desorbing laser can be focused to a spot ~ 40 μm in diameter, and the position of this spot can be monitored with an optical microscope with attached video camera. There is thus the possibility of selectively analyzing PAHs from specific places on the surface of a meteorite and correlating the results with mineralogical and textural observations, in effect giving rise to a microprobe for organic molecules. By using lasers having different wavelengths (or tunable lasers), it should be possible to analyze different types of organic species, especially if they are relatively volatile and ionizable.

L^{2}MS studies indicate that a wide range of PAHs with extensively alkylated rings are heterogeneously distributed within the matrix of Allende. Major species include naphthalene, phenanthrene/anthracene, and their alkyl-substituted derivative, as well as a large range of heavier PAHs (Zenobi et al 1989). Specific correlations of particular organic species and inorganic minerals or structural features within the meteorites remain for the future.

Interstellar clouds, the solar nebula, and parent bodies within the solar system have each been proposed as locations where meteoritic organic matter formed (Miller 1955, Studier et al 1968, Khare & Sagan 1973, Hayatsu & Anders 1981, Greenberg 1984, Peltzer et al 1984). Based on the intriguing excess of deuterium in insoluble acid residues of meteorites, a relationship was suggested between meteoritic and interstellar organic matter (Kolodny et al 1980, Becker & Epstein 1982, Robert & Epstein 1982, Yang & Epstein 1983, Yuen et al 1984, Epstein et al 1987, Kerridge et al 1987). Some noble gases that have anomalous isotopic compositions are enriched in the acid-insoluble residues—an observation that also supports an interstellar origin (Anders 1981, Lewis & Anders 1981, Tang et al 1988). Shock & Schulte (1990) provide evidence that extraterrestrial amino acids and phyllosilicates formed in the same aqueous alteration events, presumably on the parent bodies of meteorites, as discussed in Sections 5 and 6. It seems clear that the formation of organic materials in the meteorites involved a combination of both interstellar and solar system abiogenic processes and environments.

δD and δ^{13}C values correlate with molecular species (Krishnamurthy et al 1991), whereas the macromolecular carbon is lighter and lies off the correlation line. The enrichment in deuterium implies an interstellar synthesis and suggests that at the radial distance of the asteroids (2–4 AU), in the approximate region where the primitive meteorites occur, the environment was not too harsh to preclude the survival of preexisting C-C and C-H bonded material. This material, which presumably predated the solar system, was then modified by the aqueous alteration that is so prominent in the CM and CI chondrites, consistent with the correlation of organic matter with extent of aqueous alteration (Cronin & Chang 1992).

4.3   Isotope Chemistry

During solar system formation and subsequent evolution, several processes took place that were recorded by the isotopic compositions of meteoritic material. As summarized by Clayton (1986), processes occurring at $T < 50$ K in a pre-solar molecular cloud caused isotopic variations of H, C, and N; evaporation caused by transient heating in the nebula and subsequent recondensation led to large isotopic fractionation effects in Mg and Si and smaller effects in Ca and Ti; oxygen underwent large-scale isotopic exchange between solid and gaseous reservoirs. Continuous isotopic exchange of oxygen occurred within planetary bodies during melting and igneous differentiation, thermal metamorphism, and aqueous alteration. We focus on the isotopes of O, H, C, and N in the following section because they seem most likely to have recorded changes that occurred in the matrix. Isotopic data for organic species were discussed in Section 4.2.

Many isotopic measurements of CCs exist. Some are of particular fractions (e.g. magnetic, nonmagnetic, carbonates, acid-insoluble residues) while others are of unseparated pieces, producing "whole-rock" values. Certain measurements occur over a range of temperatures—the result of step-heating experiments. This potpourri of data is difficult to compare and summarize because the values are from different fractions or they were obtained under varied conditions. (For reviews of the implications of isotopic measurements in meteorites see Begemann 1980, Anders 1987, Clayton 1986, Grady et al 1988). Our Table 2 and Figure 3 indicate the ranges in $\delta^{18}O$, $\delta^{13}C$, $\delta^{15}N$, and δD values by CC type. For details, we refer the reader to the references listed in the table. It is evident that the isotopic results are widely scattered and, not surprisingly, for a given CC type they are dispersed in proportion to the number of meteorites studied. It is difficult to be confident about trends from such a mixed set of measurements, but certain patterns can be discerned. In general, ^{17}O, ^{18}O, ^{15}N, and ^{13}C show the greatest enrichments among the meteorites that have most clearly been exposed to hydrous, low-temperature environments.

The whole-rock $\delta^{18}O$ measurements show the only clear trend. They indicate a progressive increase through the series CO, CV, CR, CM, and CI. The $\delta^{18}O$ trend for the carbonates, on the other hand, is weaker and somewhat different, going from CV, CO/CI, CM, and at the high end, CR. For the $\delta^{13}C$ whole-rock measurements, the corresponding trend is CO/CV, CI/CM, and CR. For the $\delta^{15}N$ whole-rock measurements, the trend is similar to that for $\delta^{13}C$: CO/CV, CI, CR, and CM; the last two types are interchanged. Finally, for the δD whole-rock measurements, the trend is CI/CV, CM, and CR. CM and CI values are interchanged for the acid-insoluble residues.

The isotope compositions can be used to provide information regarding the sources and subsequent events involved in the formation of CC matrices. Although the basis of their interpretation has been challenged by Thiemens (1988), Clayton & Mayeda (1984) proposed a widely accepted model of two oxygen reservoirs and two episodes of fluid-solid interaction for the origin of CMs. The reservoirs initially had different oxygen-isotope abundances, with the solid-phase reservoir enriched in ^{16}O, while the gas-

Table 2 Summary of $\delta^{18}O$, $\delta 13C$, $\delta^{15}N$, and δD measurements of carbonaceous chondrite matrices

Group	Meteorite	Fraction	$\delta^{18}O_{SMOW}$ (‰)	Refs	$\delta^{13}C_{PDB}$ (‰)	Refs	$\delta^{15}N_{AIR}$ (‰)	Refs	δD_{SMOW} (‰)	Refs
CI	Alais	whole rock	*15.85/16.84 [3]	1	-6.6/-15.6 [1]	2, 3	31/52 [3]	4	44/223 [3]	4, 5
	Ivuna	carbonate	11.9/30.55 [3]	6	23.3/70.2 [3]	6, 7				
	Orgueil	non-magnetic	17.82 [1]	1						
		magnetic	17.35/19.26 [1]	1						
		HF-HCl residues	-23.6/3.7 [1]	15	-19.4 [1]	3	27 [1]	3	950/1150 [1]	3
CM	26 meteorites	bulk matrix	10.90/12.61 [3]	1	-0.4/-22.0 [15]	4	13/47† [11]	4	-176/990 [15]	4
	analyzed	carbonate	19.2/35.3 [26]	6	1.9/80.7 [26]	6				
		HF-HCl residues			-13.0/-14.6 [3]	3	18/29 [3]	3	650/830 [3]	3
CR	Al-Rais	whole rock	1.5/6.8 [4]	13	-9.9/-11.8 [2]	4, 8	140/190 [2]	4, 3	520/1014 [2]	5, 9
	Renazzo	carbonate	23.1/33.8 [3]	6	54.3/65.4 [3]	6				
	Y 790112	HF-HCl residues			-21.0 [7]	6				
CV	7 meteorites	bulk matrix	2.10/3.72 [1]	10, 11	-12.8/-21.6 [7]	3	150/175.6 [3]	3, 14	2500 [1]	3
	analyzed	carbonate	12.3/18.9 [1]	6	-6.3/-7.4 [1]	4	-43/24 [7]	4, 12	-77/440 [7]	4
CO	5 meteorites	whole rock	-2.3/0 [4]	13	-14.2/-19.3 [5]	4				
	analyzed	carbonate	17.5/25.0 [3]	6	-16.9/5.2 [3]	6	-30/13 [5]	4	-147/2150 [5]	4

* The slash signifies a range; e.g. 15.85/16.84 indicates the smallest and largest values.
† The 335 value for $\delta^{15}N$ in Bells is far higher than any other CM meteorite and has not been included in the table.
[#] indicates the number of measured meteorites to which this range applies; in several cases multiple measurements were made.
References:
1. Clayton & Mayeda (1984); 2. Boato (1954); 3. Robert & Epstein (1982); 4. Kerridge (1985); 5. Yang & Epstein (1983); 6. Grady et al (1988); 7. Smith & Kaplan (1970); 8. Grady et al (1983); 9. McNaughton et al (1982); 10. Clayton et al (1976); 11. Clayton et al (1983); 12. Kung & Clayton (1978); 13. Clayton & Mayeda (1989); 14. Grady et al (1991); 15. Halbout et al (1986).

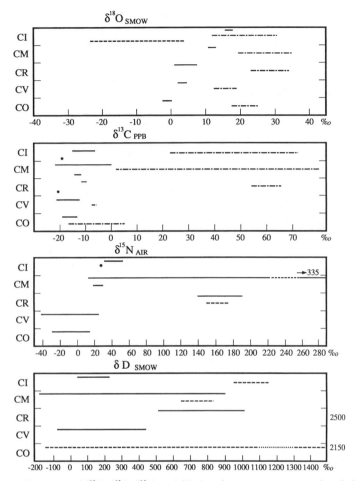

Figure 3   Summary of $\delta^{18}O$, $\delta^{13}C$, $\delta^{15}N$, and δD data for measurements on the whole rock or matrix, carbonates, and acid-insoluble residues of CCs. Solid line: data for whole rock or matrix; dashed line or the single dots: data for HF-HCl residues; dashed and dotted line: data for carbonates. The values are from Table 2.

phase reservoir, consisting principally of H_2O and CO, was depleted in ^{16}O. The two episodes were: 1. an incomplete exchange at high temperatures between solid (or melt) and gas to yield the oxygen isotope compositions of anhydrous silicates, and 2. alteration, much of which occurred at 20°C or less and that required at least 44% by volume of water to make the phyllosilicates and carbonates. In contrast, the CIs are more strongly depleted in ^{16}O than CMs, which suggests the alteration occurred in a warmer, wetter environment, with temperatures ranging from 100 to 150°C. The CO and CV chondrites are relatively unaltered and therefore required only the high-temperature step.

Many assignments of Antarctic meteorites to specific groups based on oxygen isotope compositions differ from similar assignments based on mineralogy. These differences suggest an ambiguity in the distinction between CI and CM chondrites and a possible continuum of chemical, mineralogical, and isotopic features (Mayeda et al 1987). For example, Belgica–7904 is classified as a CM meteorite based on its mineralogy, but it is likely to be a CI meteorite according to its oxygen isotope composition; Yamato–82042 is just the reverse.

Unusually high D/H and $^{15}N/^{14}N$ ratios were found by Epstein et al (1987) in the amino acids and monocarboxylic acids extracted from the matrix of Murchison ($\delta D = 1370‰$, $\delta^{15}N = 90‰$). These values provide evidence for a relationship between the organic matter in CC matrix and interstellar clouds. As shown by Robert & Epstein (1982), D is extremely high (up to 3500‰) in acid residues. The D/H ratios of gases released on step heating vary with temperature above 400°C, which indicates that different hydrogen hosts in the acid residues have different isotopic compositions.

The meteoritic carbonates in CI, CM, and CR matrices are enriched in ^{13}C, with $\delta^{13}C$ ranging from 40 to 50‰ for CMs, 50 to 80‰ for CIs, and most values below 60‰ for CRs (Grady et al 1988). Carbonates are rare or absent in CV and CO meteorites. The carbonates in matrix are more ^{13}C-enriched than those derived from alteration of high-temperature inclusions, presumably because the ^{13}C-rich fluid could more easily percolate through the porous and fine-grained matrix than through the more indurated CAIs (Grady et al 1988).

Stepped combustion gives unusual isotopic results for C in the high-temperature fraction (above 650°C) of acid residues. These refractory carbonaceous materials have a $\delta^{13}C$ value of at least $+1500‰$ (Swart et al 1983, Yang & Epstein 1984), and in situ ion probe measurements show values as high as $+7000‰$ (Niederer et al 1985, Zinner et al 1986). These high values are presumably derived from refractory interstellar materials such as SiC and perhaps other carbides. The tiny diamond (2–10 nm) and larger SiC (0.1–1.0 μm) grains were found in acid residues from primitive meteorites (CCs, types 1 to 3 ordinary chondrites, and type 4 enstatite chondrites) (Huss et al 1981; Bernatowicz et al 1987; Tielens et al 1987; Tang & Anders 1988a,b; Huss 1990). Significantly, diamond and SiC as well as some amorphous carbon also show associated Ne and Xe anomalies and enrichment of heavy N, all of which point to an interstellar origin (Tang et al 1988, Anders et al 1988).

5.   CC TYPE VS MATRIX MINERALOGY

5.1   *Matrix in CIs*

The CIs have the largest fraction of matrix among all known meteorites; they consist almost entirely of matrix, containing neither chondrules nor aggregates of anhydrous minerals. The rarity of CI meteorites may arise from the abundance of extremely fine-grained minerals, which leads to fragility and susceptibility to alteration. As a result, there have been relatively few detailed studies of their matrix mineralogy, although thorough studies exist for Orgueil, Ivuna, Alais, and Yamato–82162 (see Table 3 for references); the most carefully studied CI meteorite is Orgueil, which fell in 1864 in France.

Minerals occurring in CIs are summarized in Table 3. Phyllosilicates are major phases, but most are poorly crystalline and occur as extremely small crystals (see Table 4). Thus their identities are controversial. Using different techniques, a variety of conclusions have been reached, some of which conflict. For example, major phyllosilicates reportedly are serpentine and montmorillonite (Bass 1971—x-ray powder diffraction, Mackinnon 1985—analytical electron microscopy), smectite (Caillère & Rautureau 1974—electron diffraction, Fanale & Cannon 1974—adsorption measurements), Fe-rich chlorite or serpentine (Boström & Fredriksson 1966, Kerridge 1976—microprobe analyses), and chamosite (Zaikowski 1979—infrared spectroscopy).

A review of the papers describing matrix and alteration features in Orgueil is given by Tomeoka & Buseck (1988), who also did a HRTEM study of the matrix phases in an attempt to resolve questions regarding the matrix phyllosilicates and Fe-bearing phases. They found the major minerals to be Fe-bearing serpentine, saponite (a variety of trioctahedral smectite) (Figures 4 and 5), and ferrihydrite—a poorly crystallized S- and Ni-bearing ferric hydroxide of somewhat variable composition, together with lesser amounts of magnetite and sulfides (Figure 6). The serpentine and saponite occur in roughly equal molar proportions, are coherently intergrown, and occur in intimate association with the ferrihydrite, suggesting a related origin.

Ferrihydrite is dispersed in particles <8 nm in diameter; it contains $\sim 60\%$ of the Fe in matrix, which causes the superparamagnetic character of CIs (Wdowiak & Agresti 1984, Madsen et al 1986). Most remaining Fe is in the saponite, which accounts for 40% of the matrix Fe in CIs (Tomeoka & Buseck 1988).

The ferrihydrite is extremely fine grained, and it is probable that the high surface area acted as a "getter" for volatiles and minor elements, thereby contributing to its somewhat variable compositional and structural

Table 3 Matrix mineralogy of CCs

CC type	Major phases	Minor phases	Phyllosilicates	Meteorites represented	References
CV	olivine pyroxene	magnetite, Fe-Ni sulfides, enstatite, diopside, fassaite, andradite, chromite, spinel, anorthite, sodalite	montmorillonite*, saponite, Na-phlogopite*, margarite*, clintonite*	Allende, Kaba, Mokoia, Vigarano.	1 to 8
CO	olivine	Ca-rich & Ca-poor clinopyroxene, kamacite, taenite, magnetite, maghemite, ferrihydrite, ferroxhyte	serpentine	Lancé, Ornans, ALHA77307, Kainsaz, Warrenton, Isna; inclusions in Murchison, Isna.	5, 10, 11
CM	phyllosilicate tochilinite	olivine, pyroxene, pentlandite, pyrrhotite, magnetite, calcite, aragonite, dolomite, brucite, iron-hydroxide, alkali-halides (NaCl, KCl), carbonaceous matter	cronstedtite, greenalite, antigorite, chrysotile	Murchison, Cold Bokkeveld, Nawapali, Cochabamba, Mighei, Murray, Yamato 82042, Yamato 86720, ALH84034,Nogoya; xenoliths in Jodzie	3, 9, 12 to 26
CI	phyllosilicate	magnetite, olivine, pyroxene, ferrihydrite, pentlandite, pyrrhotite, troilite, cubanite, dolomite, ankerite, aragonite, vaterite, ferroan magnesite, gypsum, hexahydrite, blödite, epsomite, carbonaceous matter	serpentine, montmorillonite, smectite, chlorite, chamosite, saponite	Orgueil, Ivuna, Alais, Yamato-82162	25, 28 to 51
CR	olivine pyroxene	plagioclase, metal, sulfides	saponite, serpentine	Renazzo, EET87770, Essebi, MAC87300	49

* Minerals occur in CAIs. We include them here for comparison.

character. The ferrihydrite adsorbed much S and Ni onto its surfaces, which helps explain the high content of S in CI matrix (Boström & Fredriksson 1966) and the correlation of S and Ni (Kerridge 1977, McSween & Richardson 1977). It has an approximate formula of $5Fe_2O_3 \cdot 9H_2O$ (Chukhrov et al 1973).

Relatively minor phases include magnetite and vein fillings of Mg and Ca sulfates (mainly epsomite and gypsum), together with dolomite, breunnerite, and magnesite as the carbonate phases (DuFresne & Anders 1962, Boström & Fredriksson 1966, Richardson 1978). Magnetite in CIs displays a variety of morphologies; it occurs in framboids, spherulites, plaquettes, and a variety of miscellaneous shapes (Jedwab 1971, Kerridge et al 1979, Tomeoka et al 1989). Isolated carbonates and sulfates provide strong evidence of aqueous alteration on the parent bodies (Boström & Fredriksson 1966, Nagy 1975, Fredriksson & Kerridge 1988).

5.2   Matrix in CMs

CMs contain between 57 and 85 vol% matrix (McSween 1979a), with the remainder consisting of chondrules, CAIs, and isolated mineral fragments. CM matrix is a complex assemblage consisting mainly of phyllosilicates and PCP (see definition below), with lesser amounts of carbonates, sulfides, oxides, hydroxides, silicates, alkali halides, and organic matter (Table 3). The most detailed observations on CM matrix are of Murchison, but Cold Bokkeveld, Nawapali, Cochabamba, Mighei, Murray, Yamato–82042, Yamato–86720, Nogoya, ALH–84034, and CM-like xenoliths in Jodzie have also received careful attention (see Table 3 for references).

Products of aqueous alteration (Table 4) are ubiquitous in CMs and in

---

References:

1. Tomeoka & Buseck (1982a); 2. Cohen et al (1983); 3. Tomeoka & Buseck (1985); 4. Tomeoka & Buseck (1990); 5. Keller & Buseck (1990a,b); 6. Zolensky et al (1989); 7. Keller & Buseck (1991); 8. Zolensky & McSween (1988); 9. Liorca & Brearley (1992); 10. McSween (1977); 11. Wood (1967); 12. Johnson & Prinz (1991); 13. Barber (1977); 14. Barber (1981); 15. Barber et al (1983); 16. Mckee & Moore (1979); 17. Mackinnon & Buseck (1979); 18. Akai (1980); 19. Mackinnon (1980); 20. Mackinnon (1982); 21. Bunch & Chang (1980); 22. Fuchs et al (1973); 23. Ramdohr (1973); 24. Tomeoka & Buseck (1983a,b); 25. McSween & Richardson (1977); 26. Bunch et al (1979); 27. Mackinnon & Zolensky (1984); 28. Bass (1971); 29. Caillére & Rautureau (1974); 30. Fanale & Cannon (1974); 31. Boström & Fredricksson (1966); 32. Kerridge (1976); 33. Zaikowski (1979); 34. Mackinnon (1985); 35. Tomeoka & Buseck (1988); 36. Tomeoka (1990); 37. Chukhrov et al (1973); 38. Herr & Skerra (1969); 39. Wdowiak & Agresti (1984); 40. Madsen et al (1986); 41. Nagy (1975); 42. Fredriksson et al (1980); 43. Richardson (1978); 44. Fredriksson & Kerridge (1988); 45. Jedwab (1971); 46. Tomeoka et al (1989); 47. Kerridge et al (1979); 48. Lewis & Anders (1975); 49. Zolensky (1991); 50. Kerridge (1977); 51 Brearley (1992).

Table 4 Secondary minerals in CC matrix

Group name	Mineral name	Basal spacing (nm)	CC type	Meteorites represented	Morphology	M/Si+Al	References
•Phyllosilicate							
Serpentine	ferroan serpentine $(Mg,Fe)_3Si_2O_5(OH)_4$	0.7–0.73	CO	Lance	platy, ribbon-like tubes,	1.5	1
			CR	Renazzo			8
	cronstedtite $(Fe^{+2},Fe^{+3})_3(Si,Fe^{+3})O_5(OH)_4$	0.7–0.73	CM	Cochabamba	fibers, flat &	1.5	2, 3, 4, 11, 17
	greenalite $(Fe^{+2},Fe^{+3})_{2-3}Si_2O_5(OH)_4$	0.7–0.73	CM	Cold Bokkeveld	corrugated	1–1.5	4, 11
	ferroan antigorite $(Mg,Fe,Mn)_3(Si,Al)O_5(OH)_4$	0.7–0.73	CM	Murchison	sheets, cones	1.5	4, 11, 18
	chrysotile $Mg_3Si_2O_5(OH)_4$	0.7–0.73	CM	Orgueil	tubes, fibers, flat & corrugated sheets	1.5	6
							4, 5
Smectite	montmorillonite $(Na,Ca)_{0.3}(Al,Mg)_2Si_4O_{10}(OH)_2n(H_2O)$	1.0–1.4	CV	Allende	sheets	0.83	8
	saponite	1.0–1.5	CI				13, 14
	$(Ca_{0.5},Na)_{0.3}(Mg,Fe^{+2})_3(Si,Al)_4O_{10}(OH)_2\cdot4H_2O$		CV	Allende, Mokoia		0.83	1, 9 to 12
			CI				6, 15
	sobotkite		CR	Renazzo			7
Mica*	$(Ca_{0.5},K)_{0.3}(Mg_2,Al)(Si_3,Al)O_{10}(OH)_2\cdot5H_2O$		CV				16
	Na-phlogopite	1.0	CV	Mokoia	sheets	0.83	12
	$(Na,K)(Mg,Al,Fe)_3(Si,Al)_4O_{10}(OH)_2$					1	
Chlorite	margarite $Ca(Mg_{0.2},Al_{1.7})(Si_{1.8},Al_{2.2})O_{10}(OH)_2$	2.0	CV	Allende		0.73	19
	clintonite $Ca(Mg_{2.5},Al_{0.3})(Si_{1.6},Al_{2.4})O_{10}(OH)_2$	1.9	CV	Allende		0.95	19
	chamosite $(Fe^{+2},Mg,Fe^{+3})_5Al(Si_3,Al)O_{10}(OH)_8$	1.4	CI		sheets	1.5	15, 20

•Tochilinite	tochilinite ~ $Fe_{1.3}Ni_{0.1}SO_{1.4} \cdot Mg(OH)_2$?	1.07–1.13, 0.54	CM	Murchison	massive, tubes	4, 5, 11, 17, 21 to 30
	haapalaite 4(Fe,Ni)S 3$(Mg,Fe^{+2})(OH)_2$		CM		fibers, flat & corrugated sheets	33
•Hydroxide	brucite $Mg(OH)_2$	0.5	CM		sheets & filled tubes	5, 26, 28, 30, 37
	ferrihydrite $5Fe_2O_3$ $9H_2O$		CI			1, 20, 23, 31, 32, 34 to 36
•Carbonate	calcite & aragonite $(CaCO_3)$		CM,CI	Orgueil		20, 38 to 40
	dolomite $Ca(Mg,Fe^{+2},Mn)(CO_3)_2$		CM,CI			33
	ankerite $Ca(Fe^{+2},Mg,Mn)(CO_3)_2$		CI			33
	vaterite $CaCO_3$		CI			33
	ferroan magnesite $(Mg,Fe^{+2})CO_3$		CI			33
•Sulfate	epsomite $MgSO_4$ $7H_2O$		CI			33
	hexahydrite $MgSO_4$ $6H_2O$					33
	gypsum $CaSO_4$ $2H_2O$					33
	blödite $Na_2Mg(SO_4)_2$ $4H_2O$					33

* The micas may be primary; see text.

References:

1. Keller & Buseck (1990a); 2. Müller et al (1979); 3. Barber (1977); 4. Barber (1981); 5. Mackinnon (1980); 6. Tomeoka & Buseck (1988); 7. Zolensky (1991); 8. Tomeoka & Buseck (1982); 9. Cohen et al (1983); 10. Tomeoka & Buseck (1985); 11. Tomeoka & Buseck (1985); 12. Tomeoka & Buseck (1990); 13. Bass (1971); 14. Mackinnon (1985); 15. Tomeoka (1990); 16. Zolensky et al (1989); 17. Bunch & Chang (1980); 18. Zaikowski (1979); 19. Keller & Buseck (1991); 20. Boström & Fredriksson (1966); 21. Fuchs et al (1973); 22. Ramdohr (1973); 23. McSween & Richardson (1977); 24. Bunch et al (1979); 25. Barber et al (1983); 26. Mackinnon & Buseck (1979); 27. Tomeoka & Buseck (1983a,b); 28. Akai (1980); 29. Mackinnon & Zolensky (1984); 30. Mackinnon (1982); 31. Chukhrov et al (1973); 32. Kerridge (1977); 33. Zolensky & McSween (1988); 34. Herr & Skerra (1969); 35. Wdowiak & Agresti (1984); 36. Madsen et al (1986); 37. Mckee & Moore (1979); 38. Nagy (1975); 39. Fredriksson & Kerridge (1988); 40. DuFresne & Anders (1962).

amounts far greater than those in CVs and COs. Relative to CIs, CMs lack saponite and ferrihydrite; instead, Fe-rich layer silicates such as cronstedtite, ferroan antigorite, and greenalite are the dominant fine-grained phases (Tomeoka & Buseck 1985a). TEM observations show a wide variety of morphologies (Figure 7).

Fuchs et al (1973), as a result of a careful examination of CM matrices, introduced two terms: "PCP" and "spinach." Both have been widely used. "Spinach," a semi-transparent, green, hydrated silicate within chondrules and commonly associated with anhydrous silicates in CMs, is an intergrowth of Fe-rich serpentines (Bunch & Chang 1980, Tomeoka & Buseck 1985a). If it occurs in matrix at all, it does so as a minor component. The situation for PCP, which is the second most abundant material in CM matrix, is more complex.

Identification of the mineral constituents of PCP has a complicated history. It was originally defined by Fuchs et al (1973) as *Poorly Characterized Phase* because its character could not be determined by the then standard methods of mineralogy. Bunch et al (1979) and Bunch & Chang (1980) identified three forms of PCP and suggested that it consists of an Fe-rich serpentine and an Fe-Ni-S-C-rich phase. A series of TEM studies sheds considerable light on its constituent minerals as well as on its origin and development (Mackinnon & Buseck 1979; Mackinnon 1980, 1982; Tomeoka & Buseck 1983a,b,c; Tomeoka & Buseck 1985a). As a result of those studies, Tomeoka & Buseck (1985a) redefined PCP as *Partly Characterized Phases*.

The mineralogical character of PCP was problematical both because of its fine-grained nature and its unusual mineralogy. It now seems widely accepted that it consists of an intimate intergrowth of cronstedtite and tochilinite (see Table 4 for their compositions), both of which are rare terrestrially. Tochilinite, a composite mineral consisting of alternating layers of mackinawite-like sulfide and brucite (Organova et al 1971), was recognized as the identity of a PCP constituent by Mackinnon & Zolensky (1984). Prior to that time, Mackinnon & Buseck (1979) had used transmission electron microscopy to identify alternating layers in PCP from the Murchison CM meteorite; the layers have spacings (~ 7 and 5 Å) that they thought belonged to serpentine- and brucite-type minerals (Figure 8*A*). Tomeoka & Buseck (1983a,b,c) provided more detail regarding the struc-

---

Figure 4 Serpentine (Sr, 7-Å layers)—saponite (Sp, 10-Å to 11-Å layers) intergrowths. (*a*) TEM image showing coherent but disordered intergrowths of the two silicates. (*b*) Selected-area electron-diffraction pattern showing diffuse streaks along [001], with intensity maxima corresponding to ~ 7 Å (Sr) and ~ 12 Å (Sp) spacings. [From Orgueil CI (Tomeoka & Buseck 1988).]

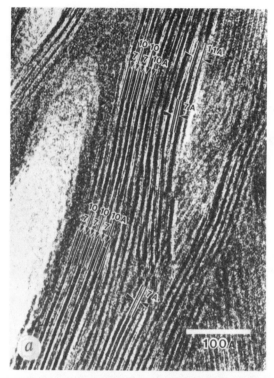

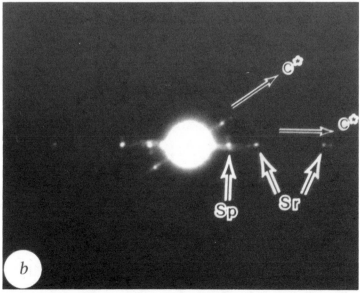

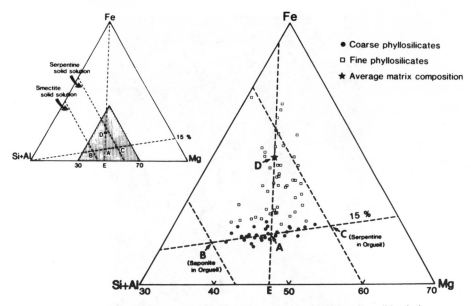

Figure 5   The ideal Mg–Fe composition lines for serpentine and smectite solid solutions. The shaded triangle is enlarged at the right. Line BC represents Fe/(Mg + Fe) = 15 atomic%, D represents the average matrix composition, and A shows the average composition of the matrix phyllosilicates in Orgueil. [From Orgueil CI (Tomeoka & Buseck 1988).]

tures and intergrowths of the two layer types; they also recognized the importance of sulfur in one set of layers. Barber et al (1983) found that PCP has an Fe/S atomic ratio of 1.4 and O/S ratio of 1.3.

A striking characteristic of the tochilinite in PCP is the wide variety of complex and beautiful forms it assumes (Figures 8A–F). Its undulatory character and its curved or even circular lattice fringes result from the misfit between the sulfide and hydroxide layers (Tomeoka & Buseck 1983b, Mackinnon & Zolensky 1984).

The origin and development of PCP was studied by Tomeoka & Buseck (1985a), who observed two major types. Type I occurs in chondrules and aggregates and consists largely of an Fe-Ni-S-O phase, presumably tochilinite; type II, in platelet and acicular fiber shapes, occurs in matrix and consists of tochilinite and cronstedtite. Microprobe analyses show variable S and Si contents, trace Cr and P, and low summations. It forms in cylindrical, corrugated, wavy, and ribbon structures in both ordered and disordered intergrowths. Most common is the ordered sequence with a ~17.8-Å spacing (Tomeoka & Buseck 1985a), with 5.4- , 5.4-, and 7.0-

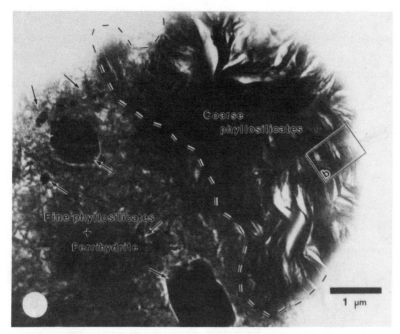

Figure 6 TEM image of ferrihydrite and adjacent phyllosilicates. Large ferrihydrite aggregates (*arrows*) occur together with fine phyllosilicates (left), whereas the coarse phyllosilicates (right) appear free of ferrihydrite. [From Orgueil CI (Tomeoka & Buseck 1988).]

Å spacings, which resembles the SBB sequence of Mackinnon & Buseck (1979; Figure 8*A*). We believe that further work on tochilinite is warranted.

Olsen et al (1988) described several xenoliths (inclusions having characteristics and perhaps sources that differ from their hosts) from the Murchison CM meteorite. These xenoliths have the mineralogical characteristics of other CC types. For example, their xenolith MX1 has alteration minerals (prominent among which is berthierine, a 1:1 Fe-rich layer silicate that is a member of the serpentine-kaolinite group) that bear a close resemblance to those of Lancé (Keller & Buseck 1990a), a CO3 meteorite.

5.3 *Matrix in CVs*

CV meteorites contain abundant chondrules, fragments of olivine and pyroxene, and refractory inclusions. Consequently they contain less matrix than CIs or CMs (35–50 vol%) (McSween 1979a). Typical matrix minerals such as phyllosilicates occur in at least small amounts in most CVs, and they provide important insights into alteration histories. In Allende most phyllosilicates occur in the Ca-, Al-rich inclusions (CAIs) and in some

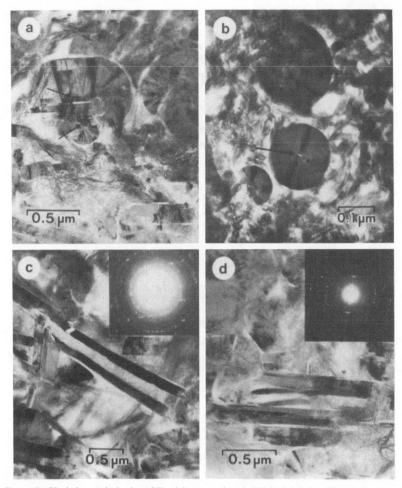

Figure 7 Varied morphologies of Fe-rich serpentines in CMs. (*a*) Spherulites and sectored crystals on platelike nuclei (*arrow*); Nawapali. (*b*) Tubular chrysotile fibers showing a central void (*arrow*); Cochabamba. (*c*) Platy crystals separated by "spongy" material containing small chrysotile-like fibers; Nawapali. (*d*) Boundary between massive partly-spherulitic material (left) and "spongy" material containing small tubular fibers (right); Nawapali. (From Barber 1981.)

chondrules, but in the vast majority of CVs the phyllosilicates also occur in the matrix (Tomeoka & Buseck 1982a, Cohen et al 1983, Fegley & Post 1985, Tomeoka & Buseck 1986, Hashimoto & Grossman 1987, Keller & Thomas 1991). Published observations on CV matrix are from Allende, Bali, Grosnaja, Kaba, Mokoia, and Vigarano (see Table 3 for references).

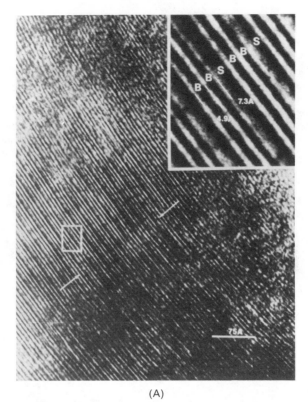

(A)

Figure 8 The varied appearances of tochilinite and its intergrowths in PCP, as viewed by transmission electron microscopy. As discussed in the text, prior to its recognition as tochilinite, this component in PCP was variously called SBB, FESON, or by its major elements. These terms and labels appear in some of the following figures.

(*A*) Ten ordered "SBB" units occur between the arrows. The white rectangle is enlarged in the inset. See text for explanation. [From Murchison CM (Mackinnon & Buseck 1979b).]

(*B*) Presumed tochilinite tube, viewed end-on. (*i*) The concentric layers and cylindrical structure are evident; (*ii*) enlargement of the framed area in (*i*), with the black arrows indicating areas where layers are bent (cf "Povlen" chrysotile). [From Murchison (Tomeoka & Buseck 1985a).]

(*C*) Image showing undulations in the structure and a reversal in the direction of the curvature. Scale bar = 300 Å. [From Murray (Tomeoka & Buseck 1983b).]

(*D*) (*a*) Set of tubes viewed end-on. (*b*), (*c*) Enlargements of the tubes indicated in (*a*); they consist of alternating intergrowths of 5.4- and 7-Å layers. The centers consist of 7-Å layers. [From Murchison (Tomeoka & Buseck 1983a).]

(*E*) Terminations and discontinuities (*arrows*) of the layers in a PCP spherule in an olivine chondrule. [From Mighei (Tomeoka & Buseck 1983b).]

(*F*) 500-kV image of undulations, 7.7-Å transverse spacings (*arrows*), and the corresponding electron-diffraction pattern. [From Murchison (Barber et al 1983).]

(B)

(C)

(D)

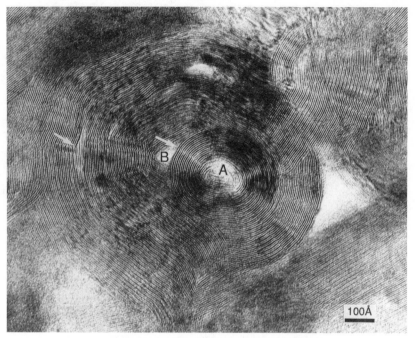

(E)

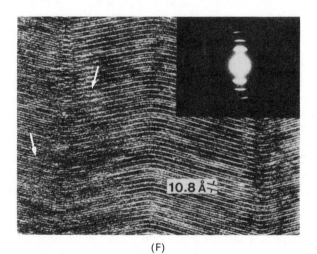

(F)

Mokoia CV matrix has been the topic of a recent study by Tomeoka & Buseck (1990). The main matrix phase is olivine, which is characterized by a wide compositional range (Peck 1983, Tomeoka & Buseck 1990) and so indicates a lack of internal equilibration and consequent homogenization. Additional matrix phases include magnetite, FeNi sulfides, pyroxenes (enstatite, diopside, and fassaite), andradite, and chromite together with lesser spinel, anorthite, and sodalite.

Phyllosilicates are interspersed among these minerals, although they appear to be mainly associated with the olivine. The major phyllosilicate is saponite, similar to that in Orgueil. In places the Mokoia smectite is associated with ferrihydrite, as it also is in Orgueil. However, because in Mokoia the alteration is far less advanced, and so olivine is still present, there is direct observational evidence that the saponite developed from the olivine (Figure 9). It grew along crystal edges as well as interiors, especially along planar defects (Figure 10). Since the metal/Si ratio of olivine (2.0) is appreciably closer to that of serpentine (1.5) than to saponite (0.83, Table 4), it is surprising that saponite is the alteration phase. Tomeoka & Buseck (1990) proposed that the planar features may represent Fe-rich precipitates that segregated from olivine, making the original olivine cation-deficient near these precipitates and thereby promoting saponite development.

The phyllosilicate described in the previous paragraph had been called "LAP" (for *Low-Al Phyllosilicate*; Cohen et al 1983. Cohen also described "HAP" (*High-Al Phyllosilicate*) in CAIs (Table 5). Tomeoka & Buseck (1990) identified it as an intimate intergrowth of serpentine and Na-rich phlogopite and found it in chondrule interstices and mesostasis glass. Although the serpentine-phlogopite intergrowths are not concentrated in matrix, their occurrence raises the possibility that they might be an early stage of matrix formation since they could well be precursors to the components that ultimately comprise much matrix.

Kaba, another CV3 CC, also contains phyllosilicates, although in relatively low abundance (McSween & Richardson 1977, Peck 1984). However, they occur in matrix, CAIs, and altered chondrules (Keller & Buseck 1990b). The phyllosilicates in Kaba matrix occur in coarse- and fine-grained clusters, in isolated packets between olivine grains, and as lamellar replacements of olivine. Only the clusters are large enough for energy-dispersive spectrometry (EDS) analysis; they have the composition of saponite, similar to that in Mokoia, Orgueil, and Yamato–82162 (Zolensky & McSween 1988, Tomeoka et al 1989).

Based on fringe spacings and analytical electron microscopy (AEM) analyses, Keller & Buseck (1990b) assumed that Fe-bearing saponite is the widespread, albeit minor, phyllosilicate that occurs throughout Kaba.

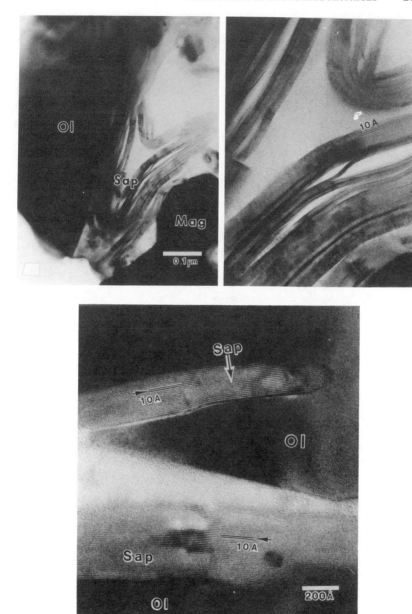

Figure 9   Saponite replacing olivine. (*Top*) The saponite (Sap) layers developed adjacent to the olivine (Ol), close to a large grain of magnetite (mag). (*Bottom*) Saponite layers intruding and presumably replacing the interior of an olivine crystal. [From Mokoia (Tomeoka & Buseck 1990).]

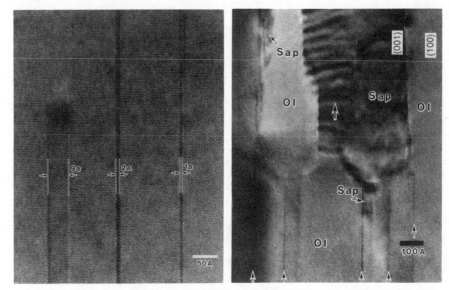

Figure 10   Planar zones of various thicknesses within olivine. (*Left*) From right to left the planes correspond to widths equaling 1, 2, and 9 unit cells along the *a* crystallographic dimension of olivine. (*Right*) Saponite (Sap) with its (001) planes parallel to olivine (Ol) (100). The arrow indicates Moiré fringes, presumably arising from overlapping saponite and olivine crystals. [From Mokoia (Tomeoka & Buseck 1990).]

They concluded that it formed on the parent body, consistent with the general calculations of Prinn & Fegley (1989). The saponite developed from both olivine and enstatite, but it seems to have preferentially attacked the pyroxene in chondrules. The minor amounts of Na, Al, and Si that were needed for the reaction were presumably derived from the mesostasis glass:

$$3\underset{\text{enstatite}}{MgSiO_3} + 0.5\underset{\text{glass}}{NaAlSiO_4} + (n+1)H_2O$$

$$= (Na_{0.5})Mg_3\underset{\text{saponite}}{(Si_{3.5}Al_{0.5})}O_{10}(OH)_2 \cdot nH_2O.$$

As shown by Keller & Buseck (1990b, and Figure 11), the development of saponite from enstatite requires relatively little structural rearrangement, and the diffusion required to introduce the necessary elements was over short paths. The structural rearrangement to alter olivine is more complex, which is perhaps why such alteration is less well developed. The reaction temperature inferred by Keller & Buseck is $< 100°C$, which may be another reason why olivine alteration was inhibited.

The alteration phases in Kaba resemble those in Mokoia (Tomeoka &

Table 5 Acronyms in the literature related to CC matrix

Acronym	Meaning	Possible minerals or materials	References
LAP	Low Aluminum Phyllosilicate	Trioctahedra Ca-Mg smectite	2
HAP	High Aluminum Phyllosilicate	Al-serpentine with mixed layers of Na-, K-rich, phyllosilicates	2
		Al-rich smectite	3
		Coherent intergrowth of Na-rich phlogopite and serpentine	4
PCP	Poorly Characterized Phase	A group of Fe-S-Ni-O- and Fe-S-Ni-Si-O-rich phases	5
		Tochilinite and an unnamed mineral consisting of coherently interstratified serpentine and tochilinite	6, 7, 8, 9, 10, 11
	Partially Characterized Phases	Complex intergrowth of tochilinite, cronstedtite, and submicron particles of magnetite, chromite and a mineral containing Fe, Ni, Cr, and P	12, 13
PAHs	Polycyclic Aromatic Hydrocarbons	Phenanthrene, anthracene, naphthalene, pyrene, fluoranthrene, etc	1, 14

References:
1. Zenobi et al (1989); 2. Cohen et al (1983); 3. Zolensky & McSween (1988); 4. Tomeoka & Buseck (1990); 5. Fuchs et al (1973); 6. Mackinnon & Zolensky (1984); 7. Zolensky et al (1987); 8. Barber et al (1983); 9. Mackinnon (1980); 10. Mackinnon (1982); 11. Mackinnon & Buseck (1979); 12. Tomeoka & Buseck (1983a,b); 13. Tomeoka & Buseck (1985); 14. Wing & Bada (1991).

Buseck 1990), although there are differences in the nature and distribution of the Fe oxides. Also, Fe sulfides are more abundant in Kaba. Alteration in both of these CV3s was in a relatively oxidizing environment, but that in Kaba was perhaps slightly more reducing. Kaba contains neither the Na-phlogopite that occurs in Mokoia nor the mica in Allende (Hashimoto & Grossman 1987, Keller & Buseck 1990b).

The saponite in Kaba is similar in composition to that in the Orgueil and Yamato–82162 CIs (Tomeoka & Buseck 1988, Zolensky et al 1989). However, Kaba is far less altered than any CI, and the saponite is not intergrown with serpentine in the way it is in Orgueil. In Yamato–82162, Zolensky et al (1989) described saponite-ferrihydrite intergrowths similar to those observed in Kaba. The Bali, Vigarano, and Grosnaja CVs contain saponites and Fe-bearing serpentine in the matrices, although they are not as altered as Kaba and Mokoia (Keller & Thomas 1991, Graham & Lee 1992).

5.4 Matrix in COs

Like the CVs, the matrices of most CO chondrites show little evidence of alteration. The main anhydrous matrix minerals are olivine and pyroxene.

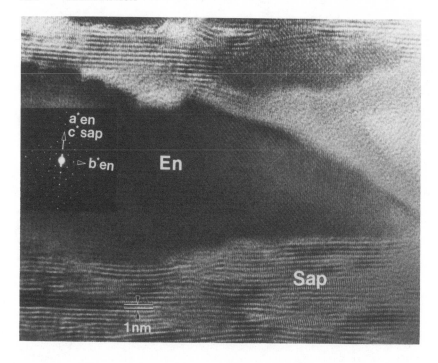

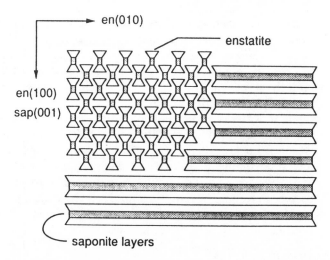

Figure 11 (*Top*) HRTEM image of a saponite-enstatite intergrowth. (*Bottom*) Sketch of an oriented enstatite-saponite intergrowth showing pyroxene I-beams abutting saponite sheets. [From Kaba (Keller & Buseck 1990b).]

Keller & Buseck (1990a) studied three of the 13 known CO falls: Lancé, Warrenton, and Kainsaz. Of these, only Lancé shows evidence of significant matrix alteration, and the data suggest that the maximum post-accretional temperature to which it might have been subjected is 450°C. ALH A–77307, one of the least equilibrated CO3 chondrites, has a rather different matrix mineralogy from other CO3 meteorites; notable is the relatively abundant Si-, Fe-rich amorphous material (Brearley 1990).

The Lancé matrix consists mainly of fragmented Fe-rich olivine, lesser phyllosilicates, poorly crystalline Fe oxides and hydroxyoxides, Ca-rich and Ca-poor clinopyroxenes, metal, and oxides (Keller & Buseck 1990a). An unusual feature of the olivine is that it contains periodically arranged pits a few nanometers in dimension on some of its surfaces (Figure 12); because they are seen in TEM images, which are 2-dimensional, the possibility that these are actually grooves cannot be excluded. Keller & Buseck speculated that these features subsequently formed channels within the

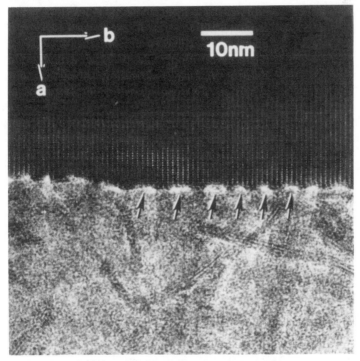

Figure 12 Pits (or grooves; periodic white spots at the crystal margins indicated by *arrows*) in euhedral matrix olivine. Note the serpentine packets in matrix. [From Lancé (Keller & Buseck 1990b).]

olivine, and these, in turn, were sites for both attack by phyllosilicate alteration minerals and fracture during regolith gardening.

The phyllosilicates are either draped around the olivine or parallel to (100) planes of olivine, which suggests limited crystallographic control. Keller & Buseck (1990a) estimated the phyllosilicate composition as $(Mg_{2.2}Fe_{0.8})Si_2O_5(OH)_4$, a variety of serpentine that forms through a hydration reaction with olivine. The Fe released in the process formed the fine-grained Fe oxide, which is either ferrihydrite ($5Fe_2O_3 \cdot 9H_2O$) or ferroxyhyte (δ-FeOOH).

The morphologies of these Fe minerals differ among the few meteorites in which they have been studied in detail. In Lancé (CO3) they occur as finely dispersed, poorly crystalline grains that pervade the sample. In Orgueil they occur in granular masses that appear to be pseudomorphic after framboidal magnetite (Tomeoka & Buseck 1988), and in Kakangari (a unique chondrite) they occur in featherlike masses as well as in rims around troilite (Brearley 1989).

The Kainsaz and Warrenton CO3s are among the least altered CO meteorites, and their anhydrous (unaltered) matrix minerals are essentially the same as those in Lancé. However, in Kainsaz, Keller & Buseck (1990a) observed kamacite, which is one of the first phases to be attacked and consumed during alteration, and sparse graphite.

The anhydrous minerals in CV3 and CO3 chondrites are similar, and it is probable that their prealteration mineralogy differed little from one another. However, as shown by the contrast between Kaba and Lancé (Keller & Buseck 1990a,b), their alteration products differ even though the degree of reaction in both was relatively slight. Lancé contains serpentine and Fe oxides that suggest alteration at higher temperature and more oxidizing conditions; their sparser distribution also suggests less water was available for Lancé alteration.

The matrix mineralogy of Mokoia (CV3) and Lancé (CO3) are similar (Keller & Buseck 1990a). However, the main alteration product of Mokoia olivine is saponite and Fe oxides while in Lancé it is serpentine, presumably reflecting a higher alteration temperature. The phyllosilicates in Kaba resemble those in Mokoia rather than those in Lancé.

5.5   Matrices in CRs and CKs

The proportions of Mg, Fe, and Si to one another in the matrices of the CRs are similar to those in the CM, CO, and CV meteorites, but S is significantly higher than in CVs and COs. Olivine is the dominant phase in the CR matrix. Metal, sulfides, saponite and serpentine are abundant, and the two layer silicates are locally intergrown.

The phyllosilicates and carbonates in CRs differ from those in other meteorite groups (Weisberg & Prinz 1991). Some chlorite-like phases are highly aluminous (10 to 15 wt% Al_2O_3), whereas the serpentines and saponites are not as magnesian as those in CI, CM, and CV chondrites. In distinction to CIs and CMs, the CR carbonates have compositions ranging from pure $CaCO_3$ to containing up to 7.3 wt% FeO, 3.1 wt% MgO, and 1.9 wt% MnO. Unique to the CRs, $CaCO_3$ also occurs as rims surrounding curved, unbroken surfaces of chondrules. Weisberg et al (1991) showed that dark inclusions in CR2 chondrites have textures, compositions, and oxygen isotope ratios similar to the CR matrices, which they interpret as indicating that at least some matrix existed as lithified material prior to the aggregation of CR chondrites.

The phyllosilicates in CR meteorites are significantly finer grained than those in CIs. That, together with the abundant olivine, suggests that aqueous alteration was not pervasive. The lack of tochilinite suggests that the CRs experienced a maximum alteration temperature of $\sim 150°C$. Their mineralogy resembles that of many saponite-class IDPs (Zolensky 1991).

The compositional, textural, and oxygen-isotope data of the CKs are closely related to the CVs and COs, except that all CKs have been significantly metamorphosed, with petrographic grades from 4 to 6; some contain shock features (Kallemeyn et al 1991). Their matrix consists of two types of materials. 90 vol% consists of olivine (50–200 μm), plagioclase (20–100 μm), magnetite and sulfides (several micrometers across)—the olivine grains are unzoned, contain significant (~ 0.5 wt%) NiO, and are appreciably larger than in matrix of the other CC types, as defined in Section 2. 10 vol% consists of fine-grained augite (< 2 μm), low-Ca pyroxene (< 5 μm, $\sim Fs_{27}$), anhedral olivine (< 2 μm, $\sim Fa_{29}$), and subhedral magnetite (< 1 μm, containing $\sim 2\%Cr$, $\sim 1\%Al$) (Brearley et al 1987, Kallemeyn et al 1991).

5.6 Comparison to Interplanetary Dust Particles

Several studies have suggested a relationship between the CCs and the IDPs (e.g. Tomeoka & Buseck 1985b, Rietmeijer & Mackinnon 1985, Bradley & Brownlee 1986). Although most IDPs differ from CCs in the details of their mineralogy and textures, two recent papers provide convincing evidence of a relationship between CCs and certain IDPs. Keller et al (1992) show that IDP number W7013F5 contains Mg-Fe carbonates as well as coherent intergrowths of Mg-Fe serpentine and Fe-bearing saponite on a unit-cell scale, similar to what is observed in CIs. There is also compositional overlap of phyllosilicates, carbonates, and sulfides in W7013F5 with those in the CIs. Bradley & Brownlee (1991) show that the minerals in IDP RB12A44 resemble those in the matrix of certain CMs. It contains tochilinite and, moreover, the tochilinite is intergrown with cronstedtite. The minor phases in RB12A44 also resemble those in CM matrix. Some IDPs are thought to be derived from comets; McSween & Weisman (1989) discussed the possibility that cometary nuclei may be similar in many aspects to CC parent bodies.

6. ALTERATION

6.1 *Alteration Processes and Phases*

6.1.1 GENERAL   In spite of their primitive composition, it has long been perceived that CCs experienced aqueous alteration. The effects of such alteration were first noted by DuFresne & Anders (1962) on the basis of veins containing hydrous sulfates and carbonates. Similar early petrographic observations were made by Nagy et al (1963) and by Boström and Fredriksson (1966). Petrography combined with electron microprobe analyses allowed Richardson (1978), McSween (1979b), and Bunch & Chang (1980) to elaborate on this alteration. Details were provided by subsequent TEM studies (see references in Sections 1 and 5). In the case of CI chondrites, their mineralogy is almost totally the result of such secondary processes, and the matrix is the main repository of these secondary products. Indeed, matrix mineralogy cannot be properly understood without considering alteration. During the past decade much work has provided details, primarily by use of TEM to identify the alteration minerals and, to a lesser extent, also to detail the mechanisms whereby the alteration occurred.

We will explore the alteration by considering matrix in the CCs that are physically most primitive (least altered) and then consider the more heavily altered types. In general, the sequence is CV, CO, CM, and CI, although there are significant variations within each group. It is worth noting that alteration reactions occurred at many temperatures. Thus, there is an extensive literature regarding reactions experienced by minerals in inclusions formed at high ($> \sim 1200$ K) temperatures. These minerals do not, however, form in matrix, and therefore we restrict this discussion to hydrated minerals. Micas are included in Table 4 because they are hydrated and are mineralogically related to the other minerals in the table, although it is probable that they formed before matrix developed (Keller & Buseck 1991).

A thorough electron microprobe study of alteration phases in CM meteorites was provided by Bunch & Chang (1980). Based on compositional and textural evidence, they concluded that alteration was extensive and that it occurred in situ on parent bodies at low temperatures (< 400 K). Their indications of alteration—pseudomorphic replacements, cross-cutting veins, colloform textures, hydrated minerals—have been confirmed by subsequent TEM studies that allowed observation of more details of the minerals and their replacement relations (Tomeoka & Buseck 1985a, Tomeoka et al 1989).

McSween (1979a), as a result of detailed electron microprobe studies of a suite of CMs and CIs, observed that the volume percent of matrix is proportional to the extent of alteration. He therefore concluded that much matrix is secondary, having resulted from the comminution of coarser material. The comminution was produced by regolith gardening, which resulted from impact of other meteorites onto the CC parent bodies. He provides estimates of the amount of matrix in various CCs, and from that it is possible to estimate the degree of processing that occurred to produce a given CC. However, the details of the alteration were not known, although they could potentially provide information about development of matrix and conditions on the parent body.

A series of five alteration steps for CMs were outlined by Tomeoka & Buseck (1985a) and Tomeoka et al (1989) and are illustrated in Figure 13. It is assumed that the initial meteorite contained mainly anhydrous silicates such as olivine and pyroxene and lesser amounts of metal and sulfide, primarily kamacite and troilite. A small amount of carbon was present, but it did not significantly influence the alteration reactions.

• *Step I*: This first low-temperature alteration stage occurred when a S-bearing fluid, presumably generated by decomposing some of the sulfides, attacked the kamacite to produce tochilinite. The presumed presence of hydroxyl reflects the widespread assumption that the fluid contained water.

• *Step II*: More extensive aqueous alteration occurred together with regolith gardening, resulting in greater comminution plus extensive alteration of olivine to phyllosilicates such as serpentine and the development of "spinach" from mesostasis glass. Richardson & McSween (1978) suggested that this alteration greatly weakened the meteorites, thereby facilitating disaggregation (and even breakup). This disaggregation helped produce mixing of the *step-II* products into the matrix.

• *Step III*: Si, released during the alteration of olivine and pyroxene, reacted with the tochilinite to produce magnesian cronstedtite. The result was intimate intergrowths and intercalations of the two phases. This mixture has long been called PCP, and it is widespread in many partly altered CM meteorites.

• *Step IV*: Cronstedtite became the dominant alteration mineral. It formed both by consuming tochilinite and by scavenging Fe from ferroan serpentine, thereby becoming more Fe-rich and resulting in a more magnesian serpentine. The effects of continued regolith gardening were to repeatedly stir the surface of the host parent body, exposing fresh PCP and serpentine as well as olivine and pyroxene to reaction and thus producing more

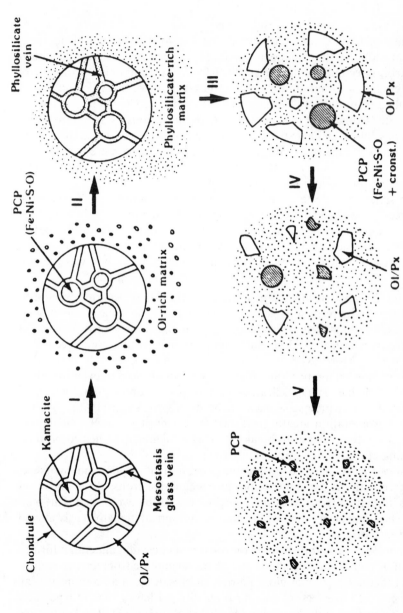

Figure 13 Schematic illustration of the aqueous alteration process that probably occurred in the regolith of the CM parent body. The alteration converted chondrules and fine-grained, olivine-rich matrix to phyllosilicate-rich matrix (modified from Tomeoka & Buseck 1985b).

cronstedtite. The bulk composition of the matrix became richer in Mg. Fe released during this process produced the fine-grained magnetite (perhaps including the framboidal aggregates first described by Kerridge et al 1979) and the Fe-Ni sulfides that contribute to the opaque character of matrix.

• *Step V*: Repeated bombardment and continued comminution combined with alteration resulted in an almost uniformly fine-grained matrix that consists mainly of layer silicates and few, if any, residual coarse-grained fragments.

Some mineral and lithic clasts in polymict achondrites resemble CC matrix material. Brearley & Prinz (1992) propose that, based on oxygen isotopic composition, mineralogy, and bulk chemistry, certain clasts in the Nilpena ureilite are closely related to CIs; they are less altered than is Orgueil and so may serve as the starting material for the aqueous alteration of CI parent bodies. They essentially support the alteration model proposed by Tomeoka & Buseck (1988). Bourcier & Zolensky (1992) used computer models to try to reproduce the mineralogy produced during aqueous alteration. While their models do not work too well for the COs and CVs, the results for the CIs and CMs are encouraging.

6.1.2 MINERALOGICAL AND CHEMICAL CHANGES DURING ALTERATION
Since matrix is a major product of alteration, it follows that both its composition and mineralogy change during the alteration process. Based on broad-beam electron microprobe analyses, McSween (1979a) showed that the Mg content of CM matrix correlates with the amount of matrix (and perhaps with the degree of alteration). The destruction of the refractory anhydrous silicates that presumably resided in the unaltered CCs produced a net transfer of Mg into the matrix, thereby increasing the Mg/Si ratio. The Fe/Si ratio, on the other hand, decreased as the amount of matrix increased, presumably because of the development of iron oxides and sulfides, which McSween excludes from matrix. The combination produces an increase in Mg/Fe in the bulk matrix. The progressive development of tochilinite, serpentine, and cronstedtite as major alteration species is consistent with these trends. Ca was also progressively leached from the matrix and formed carbonate veins in the CIs. McSween (1987) did mass balance calculations from which he inferred the changes in relative proportions of minerals as alteration progressed.

6.2 Water in CCs

The original classification of the CCs was founded in part on their H_2O contents, with average values of 20.1 wt% for type I, 13.4 wt% for type II, and 0.99 wt% for type III CCs (Wiik 1956). These numbers can be compared with the average volume percentages of matrix determined by

McSween (1979a): 99% for CIs, 57–85% for CMs, and 35–50% for COs and CVs. The correlation is evident and is consistent with an increase in water content with degree of alteration. Also, the amount of matrix can be used as an indicator of the extent of alteration.

In spite of this widely accepted relationship, there are no published H_2O measurements on matrix alone (except for CIs, which are essentially only matrix). Many measurements were made by determining total H_2O content, and Van Schmus & Hayes (1974) point out the potentially severe problem of terrestrial contamination, the extent of which is hard to assess. By using H as a variable with Ar in plots of $H/(10^2Si)$ vs $^{36}Ar/(10^{10}Si)$, they were able to separate the CO from CV meteorites.

It seems curious that the amount of H does not increase with degree of alteration within the CM group (McSween 1979a), since many of the presumed alteration phases are hydrated. Moreover, the source of the H_2O that produced the alteration is a major unresolved question. Kerridge & Bunch (1979) discussed possible sources of the H_2O that produced the alteration and speculated that it was one of the earliest recognizable events in the solar system. There are numerous possibilities, none of which have been confirmed. The H_2O may have accreted as ice together with lithic material at the time of parent-body formation. Alternatively, it may have come from cometary ices deposited during impacts, and then later released as the result of heating through the decay of short-lived radionuclides or magnetic induction resulting from interaction with a strong solar plasma (Grimm & McSween 1989) or from impacts onto the surface of the parent body (Lange et al 1985). McSween noted that the amount of retained H_2O seems to have remained approximately constant within CMs, independent of the extent of alteration. He suggests degradation of amino acids in matrix as a possible explanation, in which the H freed by destruction of the amino acids during alteration would be converted to H_2O. Such reactions would presumably produce an increase in the H/C ratio, and there is a suggestion that this does occur. Cronin (1989) noted a correlation between the amino acid content and the amount of PCP in five CMs, which strongly suggests that amino acid synthesis was related to an early stage of aqueous alteration on the parent body.

7. CLOSING THOUGHTS; FUTURE RESEARCH DIRECTIONS

Matrix provides a unique window to the most distant past of our solar system. It is an important component of all chondrites, and it is especially prominent in the CCs; the CIs consist essentially of only matrix. Although an integral and well-studied part of the meteorites, in many ways CC matrix remains enigmatic, primarily because of its fine-grain sizes and intimately mixed nonmatrix materials. For these reasons, study of either individual grains or of bulk matrix presents formidable difficulties. Nonetheless, as shown in the previous sections, significant progress has been made in understanding matrix and the processes that contributed to its development.

It was long thought that matrix is pristine, but it is now widely recognized that it contains a strong imprint of aqueous alteration. The evidence is both chemical and mineralogical. The anhydrous silicates and simple oxides are presumably primary minerals, but there is a possibility that they have been cycled through previous geological or cosmochemical events. The alteration products are among the intriguing and characteristic features of matrix and have been the focus of extensive TEM studies.

The problem of *where* the alteration occurred remains one in which the most likely site—in situ on a parent body rather than in the nebula, prior to incorporation—seems reasonable. However, most evidence to date is textural and thus permissive rather than compelling and so cannot easily be used to disprove alternative sites. The calculations of Prinn & Fegley (1989) indicate that silicate hydration in the nebula would have been kinetically inhibited and therefore improbable.

A contrary view is suggested by Metzler et al (1991, 1992), Metzler & Bischoff (1991), and Graham & Kurat (1991), who reported accretionary dust mantles coating alteration products such as serpentine, tochilinite, calcite, and magnetite. They assume these mantles formed within the solar nebula by adhesion of dusty material onto surfaces of coarse chondritic components and thus conclude that the alteration minerals must have formed in the solar nebula or on small planetesimals.

A plausible question might be whether any matrix is pristine, but this question was answered definitively by the discovery, within the matrix, of minerals having interstellar origins. As discussed in Section 4.3, diamond and SiC have isotopic compositions that point convincingly toward sources outside the solar system. The D/H ratios of amino acids and mono-carboxylic acids extracted from CMs also show interstellar sources.

While isotopic data would be especially helpful for resolving questions regarding the origin and development of matrix, the difficulty of separating "clean" matrix for isotopic analysis is a limiting factor. The ion probe is an important instrument for isotopic analyses of the larger ($> 2 \mu$m) matrix grains, but most of the meteorite matrix is still too fine grained for current instruments.

The origin of organic material has long been an important problem in understanding CCs. One approach is to determine exactly where in the matrix they reside. A potential solution to determining their spatial distribution has appeared within the last few years. The development of L²MS as a high-spatial-resolution analytical technique for large molecules (several hundred amu; Section 4.2) holds promise that in the future precise correlations will be possible between organic species and specific components of CC matrix.

Finally, there is the curious apparent anomaly that the CIs, the meteorites that are chemically the most similar to solar composition, are also the most highly altered. It would appear that alteration occurred in a closed system, but such a supposition is not supported by the depletion of at least some elements such as Na and Ca. The mismatch between the pristine character of CCs as based on chemical vs physical characteristics is intriguing. It is evident that important questions regarding matrix and their host meteorites remain to be solved.

In this review, we have tried to show the wealth of information stored within the fine-grained melange called matrix. It contains secreted within it some of the oldest and most remote known materials, but their study is still difficult and extremely challenging. It is certain that additional research will uncover further clues to this most fascinating and ancient material. Answering questions such as which fraction of matrix is primary and which is the product of alteration and when and where the alteration occurred awaits the development of yet more precise analytical techniques and procedures than exist today.

ACKNOWLEDGMENTS

We thank Drs. H. Y. McSween Jr., A. J. Brearley, R. N. Clayton, J. R. Cronin, and L. Keller for helpful comments on the manuscript and Drs. K. Tomeoka, D. J. Barber, and L. Keller for sending us copies of their figures. This work was supported by NASA grant NAG 9–59.

Literature Cited

Akai, J. 1980. Tubular form of interstratified mineral consisting of a serpentine-like layer plus two brucite-like sheets newly found in the Murchison (C2) meteorite. *Mem. Natl Inst. Polar Res., Spec. Issue* 17: 299–310

Anders, E. 1981. Noble gases in meteorites: evidence for presolar matter and superheavy elements. *Proc. R. Soc. London Ser. A* 374: 207–38

Anders, E. 1987. Local and exotic components of primitive meteorites, and their origin. *Philos. Trans. R. Soc. London Ser. A* 323: 287–304

Anders, E., Ebihara, M., 1982. Solar-system abundances of the elements. *Geochim. Cosmochim. Acta* 46: 2363–80

Anders, E., Grevesse, N., 1989. Abundances of the elements: meteoritic and solar. *Geochim. Cosmochim. Acta* 53: 197–214

Anders, E., Lewis, R. S., Tang, M., Zinner, E. 1988. *Interstellar grains in meteorites: diamond and silicon carbide.* Presented at Interstellar Dust: IAU Symp. 135, NASA-Ames Res. Ctr., Santa Clara, CA

Barber, D. J. 1977. The matrix of C2 and C3 carbonaceous chondrites. *Meteoritics* 12: 172–73

Barber, D. J. 1981. Martix phyllosilicates and associated minerals in C2M carbonaceous chondrites. *Geochim. Cosmochim. Acta* 45: 945–70

Barber, D. J. 1985. Phyllosilicates and other layer-structured materials in stony meteorites. *Clay Minerals* 20: 415–54

Barber, D. J., Bourdillon, A., Freeman, L. A. 1983. Fe-Ni-S-O layer phase in C2M carbonaceous chondrites—a hydrous sulphide? *Nature* 305: 295–97

Basile, B. P., Middleditch, B. S., Oro, J. 1984. Polycyclic aromatic hydrocarbons in the Murchison meteorite. *Org. Geochem.* 5: 211–16

Bass, M. N. 1971. Montmorillonite and serpentine in Orgueil meteorite. *Geochim. Cosmochim. Acta* 35: 139–47

Becker, R., H. Epstein, S. 1982. Carbon, hydrogen and nitrogen isotopes in solvent-extractable organic matter from carbonaceous chondrites. *Geochim. Cosmochim. Acta* 46: 97–103

Begemann, F. 1980. Isotopic anomalies in meteorites. *Rep. Prog. Phys.* 43: 1309–56

Bernatowicz, T., Fraundorf, G., Tang, M., Anders, E., Wopenka, B., et al. 1987. Evidence for interstellar SiC in the Murray carbonaceous meteorite. *Nature* 330: 728–30

Boato, G. 1954. The isotopic composition of hydrogen and carbon in the carbonaceous chondrites. *Geochim. Cosmochim. Acta* 6: 209–20

Boström, K., Fredriksson, K. 1966. Surface conditions of the Orgueil meteorite parent body as indicated by mineral associations. *Smithson. Misc. Collect.* 151: 1–39

Bourcier, W. L., Zolensky, M. E. 1992. Computer modeling of aqueous alteration on carbonaceous chondrite parent bodies. *Lunar Planet. Sci. XXIII*, pp. 143–44. Houston: Lunar Planet. Inst.

Bowring, S. A., Housh, T. B., Isachser, E. E. 1990. The Acasta gneisses: remnant of Earth's early crust. In *Origin of the Earth*, ed. H. E. Newsom, J. H. Jones, pp. 319–40. Oxford: Oxford Univ. Press

Bradley, J. P., Brownlee, D. E. 1986. Cometary particles: thin sectioning and electron beam analysis. *Science* 231: 1542–44

Bradley, J. P., Brownlee, D. E. 1991. An interplanetary dust particle linked directly to type CM meteorites and an asteroidal origin. *Science* 251: 549–52

Brearley, A. J. 1989. Nature and origin of matrix in the unique chondrite, Kakangari. *Geochim. Cosmochim. Acta* 53: 2395–2411

Brearley, A. J. 1990. Matrix mineralogy of the unequilibrated CO3 chondrite ALH A77307: evidence for disequilibrium condensation processes and implications for the origin of chondrite matrices. *Lunar Planet. Sci. XXI*, pp. 125–26. Houston: Lunar Planet. Inst.

Brearley, A. J. 1992. Mineralogy of fine-grained matrix in the Ivuna CI carbonaceous chondrite. *Lunar Planet. Sci. XXIII*, pp. 153–54. Houston: Lunar Planet. Inst.

Brearley, A. J., Prinz, M. 1992. CI chondrite-like clasts in the Nilpena polymict ureilite: implications for aqueous alteration processes in CI chondrites. *Geochim. Cosmochim. Acta* 56: 1373–86

Brearley, A. J., Scott, E. R. D., Mackinnon, I. D. R. 1987. Electron petrography of fine-grained matrix in the Karoonda C4 carbonaceous chondrite. *Meteoritics* 22: 339–40

Bunch, T. E., Chang, S. 1980. Carbonaceous chondrites-II. Carbonaceous chondrite phyllosilicates and light element geochemistry as indicators of parent body processes and surface conditions. *Geochim. Cosmochim. Acta* 44: 1543–77

Bunch, T. E., Chang, S., Frick, U., Neil, J., Moreland, G. 1979. Characterization and significance of carbonaceous chondrite (CM2) xenoliths in the Jodzie howardite. *Geochim. Cosmochim. Acta* 43: 1727–42

Burnett, D. S., Woolum, D. S., Benjamin, T. M., Rogers, P. S. Z., Duffy, C. J., Maggiore, C. 1989. A test of the smoothness of the elemental abundances of carbonaceous chondrites. *Geochim. Cosmochim. Acta* 53: 471–81

Buseck, P. R., Iijima, S. 1974. High resolution electron microscopy of silicates. *Am. Mineral.* 59: 1–21

Buseck, P. R., Iijima, S. 1975. High resolution electron microscopy of enstatite II: geological application. *Am. Mineral.* 60: 771–80

Caillère, S., Rautureau, M. M. 1974. Détermination des silicates phylliteux des météorites carbonées par microscopie et microdiffraction électroniques. *C. R. Acad. Sci. Ser. D* 279: 539–42

Cameron, A. G. W. 1982. Elemental and nuclidic abundances in the solar system. In *Essays in Nuclear Astrophysics*, ed. C. A. Barnes, D. D. Clayton, D. N. Schramm, pp. 23–43. Cambridge: Cambridge Univ. Press

Chang, S., Bunch, T. B. 1986. Clays and organic matter in meteorites. In *Clays Minerals and the Origin of Life*, ed. A. G. Cairns-Smith, H. Hartman, pp. 116–29. Cambridge: Cambridge Univ. Press

Chen, J. H., Wasserburg, G. J. 1981. The isotopic compositions of uranium and lead in Allende inclusions and meteoritic phosphates. *Earth Planet. Sci. Lett.* 52: 1–15

Chukhrov, F. V., Zöyagin, B. B., Gorshkov, A. I., Yermilova, L. P., Balashova, V. V.

1973. O ferrigidrite (ferrihydrite). *Int. Geol. Rev.* 16: 1131–43

Clarke, R. S., Jarosewich, E. Jr., Mason, B., Nelen, J., Gomez, M., Hyde, J. R. 1970. The Allende, Mexico, meteorite shower. *Smithson. Contrib. Earth Sci.* No. 5, 53 pp.

Clayton, R. N. 1986. High temperature isotope effects in the early solar system. In *Stable Isotopes in High Temperature Geological Processes*, vol. 16, ed. J. W. Valley, H. P. Taylor Jr., J. R. O'Neil, pp. 129–40

Clayton, R. N., Mayeda, T. K. 1984. The oxygen isotope record in Murchison and other carbonaceous chondrites. *Earth Planet. Sci. Lett.* 67: 151–61

Clayton, R. N., Mayeda, T. K. 1989. Oxygen isotope classification of carbonaceous chondrites. *Lunar Planet. Sci. XX*, pp. 169–70. Houston: Lunar Planet. Inst.

Clayton, R. N., Onuma, N., Grossman, L., Mayeda, T. K. 1976. A classification of meteorites based on oxygen isotopes. *Earth Planet. Sci. Lett.* 30: 10–18

Clayton, R. N., Onuma, N., Ikeda, Y., Mayeda, T. K., Hutcheon, I. D., et al. 1983. Oxygen isotopic compositions of chondrules in Allende and ordinary chondrites. In *Chondrules and Their Origins*, ed. E. A. King, pp. 37–43. Houston: Lunar Planet. Inst.

Cohen, R. E., Kornacki, A. S., Wood, J. A. 1983. Mineralogy and petrology of chondrules and inclusions in the Mokoia CV3 chondrite. *Geochim. Cosmochim. Acta* 47: 1739–57

Cronin, J. R. 1989. Origin of organic compounds in carbonaceous chondrites. *Adv. Space Res.* 9(2): 59–64

Cronin, J. R., Chang, S. 1992. The organic matter of the Murchison meteorite: molecular and isotopic analyses. In *The Chemistry of Life's Origins*, ed. J. M. Greenberg, V. Pironello. In preparation

Cronin, J. R., Pizzarello, S. 1986. Amino acids of the Murchison meteorite. III. Seven carbon acyclic primary α-amino alkanoic acids. *Geochim. Cosmochim. Acta* 50: 2419–27

Cronin, J. R., Pizzarello, S. 1990. Aliphatic hydrocarbons of the Murchison meteorite. *Geochim. Cosmochim. Acta* 54: 2859–68

Cronin, J. R., Pizzarello, S., Cruikshank, D. P. 1988. Organic matter in carbonaceous chondrites, planetary satellites, asteroids and comets. See Kerridge & Matthews 1988, pp. 819–57

de Vries, M. S., Wendt, H. R., Hunziker, H. 1991. Search for high molecular weight polycyclic aromatic hydrocarbons and fullerenes in carbonaceous meteorites. *Lunar Planet. Sci. XXII*, pp. 315–16.

Houston: Lunar Planet. Inst.

DuFresne, E. R., Anders, E. 1962. On the chemical evolution of the carbonaceous chondrites. *Geochim. Cosmochim. Acta* 26: 1085–1114

Epstein, S., Krishnamurthy, R. V., Cronin, J. R., Pizzarello, S., Yuen, G. U. 1987. Unusual stable isotope ratios in amino acid and carboxylic acid extracts from the Murchison meteorite. *Nature* 326: 477–79

Fanale, F. P., Cannon, W. A. 1974. Surface properties of the Orgueil meteorite: implications for the early history of solar system volatiles. *Geochim. Cosmochim. Acta* 88: 453–70

Fegley, B., Post, J. E. 1985. A refractory inclusion in the Kaba CV3 chondrite: some implications for the origin of spinel-rich objects in chondrites. *Earth Planet. Sci. Lett.* 75: 297–310

Fredriksson, K., Kerridge, J. F. 1988. Carbonates and sulfates in CI chondrites: formation by aqueous activity on the parent body. *Meteoritics* 23: 35–44

Fredriksson, K., Jarosewich, E., Beauchamp, R., Kerridge, J. 1980. Sulfate veins, carbonates, limonite and magnetite: evidence on the late geochemistry of the C–1 regoliths. *Meteoritics* 15: 291–92

Fuchs, L. H., Olsen, E., Jensen, K. J. 1973. Mineralogy, mineral-chemistry, and composition of the Murchison (C2) meteorite. *Smithson. Contrib. Earth Sci.* 10: 1–39

Grady, M. M., Wright, I. P., Fallick, A. E., Pillinger, C. T. 1983. The stable isotopic composition of carbon, nitrogen, and hydrogen in some Yamato meteorites. *Proc. 8th Symp. Antarctic Meteorites*, ed. T. Nagata. Tokyo: Natl. Inst. Polar Res. 30: 292–305

Grady, M. M., Wright, I. P., Awart, P. K., Pillinger, C. T. 1988. The carbon and oxygen isotopic composition of meteoritic carbonates. *Geochim. Cosmochim. Acta* 52: 2855–66

Grady, M. M., Wright, I. P., Pillinger, C. T. 1991. Comparisons between Antarctic and non-Antarctic meteorites based on carbon isotope geochemistry. *Geochim. Cosmochim. Acta* 55: 49–58

Graham, A. L., Kurat, G. 1991. Phyllosilicates in the Yamato–82042 carbonaceous chondrite—primitive or not? *Lunar Planet. Sci. XXII*, pp. 475. Houston: Lunar Planet. Inst.

Graham, A. L., Lee, M. 1992. The matrix mineralogy of the Vigarano (CV3) chondrite. *Lunar Planet. Sci. XXIII*, pp.435. Houston: Lunar Planet. Inst.

Green, H. W., Radcliffe, S. V., Heuer, A. H. 1971. Allende meteorite: a high-voltage electron petrographic study. *Science* 172: 936–39

Greenberg, J. M. 1984. Chemical evolution in space. *Origins of Life* 14: 25–35

Grimm, R. E., McSween, H. Y. Jr. 1989. Water and the thermal evolution of carbonaceous chondrite parent bodies. *Icarus* 82: 244–80

Hahn, J. H., Zenobi, R., Bada, J. L., Zare, R. N. 1988. Application of two-step laser mass spectrometry to cosmogeochemistry: direct analysis of meteorites. *Science* 239: 1523–25

Halbout, J., Robert, F., Javoy, M. 1986. Oxygen and hydrogen isotope relations in water and acid residues of carbonaceous chondrites. *Geochim. Cosmochim. Acta* 50: 1599–1609

Hashimoto, A., Grossman, L. 1987. Alteration of Al-rich inclusions inside amoeboid olivine aggregates in the Allende meteorite. *Geochim. Cosmochim. Acta* 51: 1685–1704

Hayatsu, R., Anders, E. 1981. Organic compounds in meteorites and their origins. In *Cosmo- and Geochemistry*, vol. 99, *Topics in Current Chemistry*, pp. 1–37. Berlin: Springer-Verlag

Hayatsu, R., Matsuoka, S., Scott, R. G., Studier, M., Anders, E. 1977. Origin of organic matter in the early solar system— VII. The organic polymer in carbonaceous chondrites. *Geochim. Cosmochim. Acta* 41: 1325–39

Hayatsu, R., Studier, M. H., Moore, L. P., Anders, E. 1975. Purines and triazines in the Murchison meteorite. *Geochim. Cosmochim. Acta* 39: 471–88

Hayatsu, R., Winans, R. E., Scott, R. G., McBeth, R. L., Moore, L. P., Studier, M. H. 1980. Phenolic ethers in the oranic polymer of the Murchison meteorite. *Science* 207: 1202–4

Hayes, J. M. 1967. Organic constituents of meteorites: a review. *Geochim. Cosmochim. Acta* 31: 1395–440

Herr, W., Skerra, B. 1969. Mössbauer spectroscopy applied to the classification of stone meteorites. In *Meteorite Research*, ed. P. M. Millman, pp. 106–22. Dordrecht: Reidel

Holweger, H. 1977. The solar Na/Ca and S/Ca ratios: a close comparison with carbonaceous chondrites. *Earth Planet. Sci. Lett.* 34: 152–54

Huss, G. R. 1990. Ubiquitous interstellar diamond and SiC in primitive chondrites: abundances reflect metamorphism. *Nature* 347: 159–62

Huss, G. R., Keil, K., Taylor, G. J. 1981. The matrices of unequilibrated ordinary chondrites: implications for the origin and history of chondrites. *Geochim. Cosmochim. Acta* 45: 33–51

Huss, G. R., Lewis, R. S. 1990. Interstellar diamonds and silicon carbide in enstatite chondrites. *Lunar Planet. Sci. XXI*, pp. 542–43. Houston: Lunar Planet. Inst.

Iijima, S., Buseck, P. R. 1975. High resolution electron microscopy of enstatite I: Twinning, polymorphism, and polytypism. *Am. Mineral.* 60: 758–70

Iijima, S., Buseck, P. R. 1978. Experimental study of disordered mica structures by high resolution electron microscopy. *Acta Cryst.* A34: 709–19

Jedwab, J. 1971. La magétite de la météorite d'Orgueil vue au microscope electronique à Balayage. *Icarus* 35: 319–40

Johnson, C. A., Prinz, M. 1991. Carbonate compositions in CM and CI chondrites and Mg-Fe-Mn partitioning during aqueous alteration. *Lunar Planet. Sci. XXII*, pp. 643–44. Houston: Lunar Planet. Inst.

Jungclaus, G. A., Cronin, J. R., Moore, C. B., Yuen, G. U. 1979. Aliphatic amines in the Murchison meteorite. *Nature* 261: 126–28

Jungclaus, G. A., Yuen, G. U., Moore, C. B., Lawless, J. G. 1976. Evidence for the presence of low molecular weight alcohols and carbonyl compounds in the Murchison meteorite. *Meteoritics* 11: 231–37

Kallemeyn, G. W., Rubin, A. E., Wasson, J. T. 1991. The compositional classification of chondrites: V. The Karoonda (CK) group of carbonaceous chondrites. *Geochim. Cosmochim. Acta* 55: 881–92

Keller, L. P., Buseck, P. R. 1990a. Matrix mineralogy of the Lancé CO3 carbonaceous chondrite. *Geochim. Cosmochim. Acta* 54: 1155–63

Keller, L. P., Buseck, P. R. 1990b. Aqueous alteration of the Kaba CV3 carbonaceous chondrite. *Geochim. Cosmochim. Acta* 54: 2113–20

Keller, L. P., Buseck, P. R. 1991. Calcic micas in the Allende meteorite: evidence for hydration reactions in the early solar nebula. *Science* 252: 946–49

Keller, L. P., Thomas, K. L. 1991. Matrix mineralogy of the Bali CV3 carbonaceous chondrite. *Lunar Planet. Sci. XXII*, pp. 705–6. Houston: Lunar Planet. Inst.

Keller, L. P., Thomas, K. L., McKay, D. S. 1992. An interplanetary dust particle with links to CI chondrites. *Geochim. Cosmochim. Acta* 56: 1409–12

Kerridge, J. F. 1976. Major element compositions of phyllosilicates in the Orgueil carbonaceous meteorite. *Earth Planet. Sci. Lett.* 29: 194–200

Kerridge, J. F. 1977. Correlation between nickel and sulfur abundances in Orgueil phyllosilicates. *Geochim. Cosmochim. Acta* 41: 1163–64

Kerridge, J. F. 1985. Carbon, hydrogen and nitrogen in carbonaceous chondrites:

abundances and isotopic compositions. *Geochim. Cosmochim. Acta* 49: 1707–14

Kerridge, J. F., Bunch, T. E. 1979. Aqueous activity on asteroids: evidence from carbonaceous meteorites. In *Asteroids*, ed. R. P. Binzel, T. Gehrels, M. S. Matthews, pp. 745–64. Tucson: Univ. Ariz. Press

Kerridge, J. F., Chang, S., Shipp, R. 1987. Isotopic characterization of kerogen-like material in the Murchison carbonaceous chondrite. *Geochim. Cosmochim. Acta* 51: 2527–40

Kerridge, J. F., Mackay, A. L., Boynton, W. V. 1979. Magnetite in CI carbonaceous meteorites: origin by aqueous activity on a planetesimal surface. *Science* 205: 395–97

Kerridge, J. F., Matthews, M. S., eds. 1988. *Meteorites and the Early Solar System.* Tucson: Univ. Ariz. Press. 1269 pp.

Khare, B. N., Sagan, C. 1973. Red clouds in reducing atmospheres. *Icarus* 20: 311–21

Kolodny, Y., Kerridge, J. F., Kaplan, I. R. 1980. Deuterium in carbonaceous chondrites. *Earth Planet. Sci. Lett.* 46: 149–58

Krishnamurthy, R. V., Epstein, S., Pizzarello, S., Cronin, J. R., Yuen, G. U. 1991. Stable hydrogen and carbon isotope ratios of extractable hydrocarbons in the Murchison meteorite. *Lunar Planet. Sci. XXII*, pp. 757–58. Houston: Lunar Planet. Inst.

Kroto, H. W., Heath, J. R., O'Brien, S. C., Curl, R. F., Smalley, R. E. 1985. C_{60}: Buckminsterfullerene. *Nature* 318: 162–63

Kung, C. C., Clayton, R. N. 1978. Nitrogen abundances and isotopic compositions in stony meteorites. *Earth Planet. Sci. Lett.* 38: 421–35

Lange, M. A., Lambert, P., Ahrens, T. J. 1985. Shock effects on hydrous minerals and implications for carbonaceous chondrites. *Geochim. Cosmochim. Acta* 49: 1715–26

Larimer, J. W., Anders, E. 1970. Chemical fractionations in meteorites—III. Major element fractionations in chondrites. *Geochim. Cosmochim. Acta* 34: 367–87

Lawless, J. G., Zeitman, B., Pereira, W. E., Summons, R. E., Duffield, A. M. 1974. Dicarboxylic acids in the Murchison meteorite. *Nature* 251: 40–41

Lewis, R. S., Anders, E. 1975. Condensation time of the solar nebula from extinct ^{129}I in primitive meteorites. *Proc. Natl. Acad. Sci. USA* 72: 268–73

Lewis, R. S., Anders, E. 1981. Isotopically anomalous xenon in meteorites: a new clue to its origin. *Astrophys. J.* 247: 1122–24

Lewis, R. S., Anders, E., Draine, B. T. 1989. Properties, detectability and origin of interstellar diamonds in meteorites. *Nature* 339: 117–21

Lewis, R. S., Tang, M., Wacker, J. F.,

Anders, E. 1987. Interstellar diamonds in meteorites. *Nature* 326: 160–62

Llorca, J., Brearley, A. J. 1992. Alteration of chondrules in ALH84034, an unusual CM2 carbonaceous chondrite. *Lunar Planet. Sci. XXII*, pp. 793–94. Houston: Lunar Planet. Inst.

Lumpkin, G. R. 1981. Electron microscopy of carbonaceous matter in Allende acid residues. *Proc. Lunar Planet. Sci. Conf.* 12B, pp. 1153–66. Houston: Lunar Planet. Inst.

Mackinnon, I. D. R. 1980. Structures and textures of the Murchison and Mighei carbonaceous chondrite matrices. *Proc. Lunar Planet. Sci. Conf. 11th*, pp. 839–52. Houston: Lunar Planet. Inst.

Mackinnon, I. D. R. 1982. Ordered mixed-layer structures in the Mighei carbonaceous chondrite matrix. *Geochim. Cosmochim. Acta* 46: 479–89

Mackinnon, I. D. R. 1985. Fine-grained phases in carbonaceous chondrites: Alais and Leoville. *Meteoritics* 20: 702–3

Mackinnon, I. D. R., Buseck, P. R. 1979. New phyllosilicate types in a carbonaceous chondrite matrix. *Nature* 280: 219–20

Mackinnon, I. D. R., Zolensky, M. 1984. Proposed structures for poorly characterized phases in C2M carbonaceous chondrite meteorites. *Nature* 309: 240–42

Madsen, M. B., Morup, S., Costa, T. V. V., Knudsen, J. M., Olsen, M. 1986. Superparamagnetic component in the Orgueil meteorite and Mössbauer spectroscopy studies in applied magnetic fields. *Nature* 321: 501–3

Mayeda, T. K., Clayton, R. N., Yanai, K. 1987. Oxygen isotopic compositions of several Antarctic meteorites. *Proc. 11th Symp. Antarctic Meteorites*, ed. K. Yanai, 46: 144–50. Tokyo: Natl. Inst. Polar Res.

Mckee, T. R., Buseck, P. R. 1978. HRTEM observation of stacking and ordered interstratification in rectorite. *Ninth Int. Congr. Electr. Microsc.*, 1: 272–73. Toronto: Microsc. Soc. Can.

Mckee, T. R., Moore, C. B. 1979. Characterization of submicron matrix phyllosilicates from Murray and Nogoya carbonaceous chondrites. *Proc. Lunar Planet. Sci. Conf. 10th*, pp. 921–36. New York: Pergamon

McNaughton, N. J., Hinton, R. W., Pillinger, C. T., Fallick, A. E. 1982. D/H ratios of some ordinary and carbonaceous chondrites. *Meteoritics* 17: 252

McSween, H. Y. Jr. 1977. Chemical and petrographic constraints on the origin of chondrules and inclusions in carbonaceous chondrites. *Geochim. Cosmochim. Acta* 41: 1843–60

McSween, H. Y. Jr. 1979a. Alteration in CM carbonaceous chondrites inferred from modal and chemical variations in matrix. *Geochim. Cosmochim. Acta* 43: 1761–70

McSween, H. Y. Jr. 1979b. Are carbonaceous chondrites primitive or processed? A review. *Rev. Geophys. Space Phys.* 17: 1059–78

McSween, H. Y. Jr. 1987. Aqueous alteration in carbonaceous chondrites: mass balance constraints on matrix mineralogy. *Geochim. Cosmochim. Acta* 51: 2469–77

McSween, H. Y. Jr., Richardson, S. M. 1977. The composition of carbonaceous chondrite matrix. *Geochim. Cosmochim. Acta* 41: 1145–61

McSween, H. Y. Jr., Weissman, P. R. 1989. Cosmochemical implications of the physical processing of cometary nuclei. *Geochim. Cosmochim. Acta* 53: 3263–71

Metzler, K., Bischoff, A. 1991. Evidence for aqueous alteration prior to parent body formation: petrographic observations in CM-chondrites. *Lunar Planet. Sci. XXII*, pp. 893–94. Houston: Lunar Planet. Inst.

Metzler, K., Bischoff, A., Morfill, G. 1991. Accretionary dust mantles in CM chondrites: chemical variations and calculated time scales of formation. *Meteoritics* 26: 372

Metzler, K., Bischoff, A., Stöffer, D. 1992. Accretionary dust mantles in CM chondrites: evidence for solar nebula processes. *Geochim. Cosmochim. Acta* 56: 2873–97

Miller, S. L. 1955. Production of some organic compounds under possible primitive Earth conditions. *J. Am. Chem. Soc.* 77: 2351–61

Müller, W. F., Kurat, G., Kracher, A. 1979. Chemical and crystallographic study of cronstedtite in the matrix of the Cochabamba (CM2) carbonaceous chondrite. *Tschermaks Mineral. Petrogr. Mitt.* 26: 293–304

Nagahara, H. 1984. Matrices of type 3 ordinary chondrites—primitive nebular records. *Geochim. Cosmochim. Acta* 46: 2581–95

Nagy, B. 1975. *Carbonaceous Meteorites.* New York: Elsevier. 747 pp.

Nagy, B., Meinschein, W. G., Hennessy, D. J. 1963. Aqueous low temperature environment of the Orgueil meteorite parent body. *Ann. N. Y. Acad. Sci.* 108: 534–52

Niederer, F. R., Eberhardt, P., Geiss, J. 1985. Carbon isotope abundances in Murchison residue 2C10c. *Meteoritics* 20: 716–18

Olsen, E. J., Davis, A. M., Hutcheon, I. D., Clayton, R. N., Mayeda, T. K., Grossman, L. 1988. Murchison xenoliths. *Geochim. Cosmochim. Acta* 52: 1615–26

Organova, N. I., Genkin, A. D., Drits, V. A., Dmitrik, A. L., Kuzmina, O. V. 1971. Tochilinite: a new sulfide hydroxide of iron and magnesium. *Zap. Vses. Mineral. O.* 4: 477–87

Peck, J. A. 1983. An SEM petrographic study of C3(V) meteorite matrix. *Lunar Planet. Sci. XIV*, pp. 598–99. Houston: Lunar Planet. Inst.

Peck, J. A. 1984. Origin of the variation in properties of CV3 meteorite matrix and matrix clasts. *Lunar Planet. Sci. XV*, pp. 635–36. Houston, Lunar Planet. Inst.

Peltzer, E. T., Bada, J. L. 1978. α-Hydroxycarboxylic acids in the Murchison meteorite. *Nature* 272: 443–44

Peltzer, E. T., Bada, J. L., Schlesinger, G., Miller, S. L. 1984. The chemical conditions on the parent body of the Murchison meteorite: some conclusions based on amino, hydroxy and dicarboxylic acids. *Adv. Space Res.* 4: 69–74

Pierce, L., Buseck, P. R. 1974. Electron imaging of pyrrhotite superstructures. *Science* 186: 1209–12

Prinn, R. G., Fegley, B. 1989. Solar nebula chemistry: origin of planetary satellite, and cometary volatiles. In *Origin and Evolution of Planetary and Satellite Atmospheres*, ed. S. Atreya, J. Pollack, M. S. Matthews, pp. 78–136. Tucson: Univ. Ariz. Press

Ramdohr, P. 1973. *The Opaque Minerals in Stony Meteorites.* New York: Elsevier. 245 pp.

Richardson, S. M. 1978. Vein formation in the C1 carbonaceous chondrites. *Meteoritics* 13: 141–59

Richardson, S. M., McSween, H. Y. Jr. 1978. Textural evidence bearing on the origin of isolated olivine crystals in C2 carbonaceous chondrites. *Earth Planet. Sci. Lett.* 37: 485–91

Rietmeijer, F. J. M., Mackinnon, I. D. R. 1985. Layer silicates in a chondritic porous interplanetary dust particle. *Proc. Lunar Planet. Sci. Conf., 15th, J. Geophys. Res.* 90: D149–55

Robert, F., Epstein, S. 1982. The concentration and isotopic composition of hydrogen, carbon and nitrogen in carbonaceous meteorites. *Geochim. Cosmochim. Acta* 46: 81–95

Rubin, A. E., Kallemeyn, W. G. 1990. Lewis Cliff 85332: a unique carbonaceous chondrite. *Meteoritics* 25: 215–25

Rubin, A. E., Wang, D., Kallemeyn, G. W., Wasson, J. T. 1988. The Ningqiang meteorite: classification and petrology of an anomalous CV chondrite. *Meteoritics* 23: 13–23

Rubin, A. E., Wasson, J. T. 1987. Chondrules, matrix and coarse-grained chondrule rims in the Allende meteorite: origin, interrelationships and possible precursor components. *Geochim. Cosmochim. Acta* 51: 1923–37

Rubin, A. E., Wasson, J. T. 1988. Chondrules and matrix in the Ornans CO3 meteorite: possible precursor components. *Geochim. Cosmochim. Acta* 52: 425–32

Scott, E. R. D., Barber, D. J., Alexander, C. M., Hutchison, R., Peck, J. A. 1988. Primitive material surviving in chondrites: Matrix. See Kerridge & Matthews 1988, pp. 718–45

Scott, E. R. D., Rubin, A. E., Taylor, G. J., Keil, K. 1984. Matrix material in type 3 chondrites—occurrence, heterogeneity and relationship with chondrules. *Geochim. Cosmochim. Acta* 48: 1741–57

Shock, E. L., Schulte, M. D. 1990. Summary and implications of reported amino acid concentrations in the Murchison meteorite. *Geochim. Cosmochim. Acta* 54: 3159–73

Smith, J. W., Kaplan, I. R. 1970. Endogenous carbon in carbonaceous meteorites. *Science* 167: 1367–70

Smith, P. P. K., Buseck, P. R. 1981. Graphitic carbon in the Allende meteorite: a microstructural study. *Science* 212: 322–24

Studier, M. H., Hayatsu, R., Anders, E. 1968. Origin of organic matter in the early solar system—I. Hydrocarbons. *Geochim. Cosmochim. Acta* 32: 151–73

Suess, H. E. 1949. Zur Chemie der Planeten-, und Meteoritenbildung. *Z. Elektrochem.* 53: 237–41

Suess, H. E., Urey, H. C. 1956. Abundances of the elements. *Rev. Mod. Phys.* 28(1): 53–74

Swart, P. K., Grady, M. M., Pillinger, C. T., Lewis, R. S., Anders, E. 1983. Interstellar carbon in meteorites. *Science* 220: 406–10

Tang, M., Anders, E. 1988a. Isotopic anomalies of Ne, Xe, and C in meteorites. II. Interstellar diamond and SiC: carriers of exotic noble gases. *Geochim. Cosmochim. Acta* 52: 1235–44

Tang, M., Anders, E. 1988b. Interstellar silicon carbide: How much older than the solar system? *Astrophys. J. Lett.* 335: L31–34

Tang, M., Anders, E., Hoppe, P., Zinner, E. 1989. Meteoritic silicon carbide and its stellar sources for galactic chemical evolution. *Nature* 339: 351–54

Tang, M., Lewis, R. S., Anders, E., Grady, M. M., Wright, I. P., Pillinger, C. T. 1988. Isotopic anomalies of Ne, Xe, and C in meteorites. I. Separation of carriers by density and chemical resistance. *Geochim. Cosmochim. Acta* 52: 1221–34

Thiemens, M. H. 1988. Heterogeneity in the nebula: evidence from stable isotopes. See Kerridge & Matthews 1988, pp. 899–923

Tielens, A. G., Seab, D. J., Hollenbach, D. J., McKee, C. F. 1987. Shock processing of interstellar dust: diamonds in the sky. *Astrophys. J.* 319: L109–13

Tomeoka, K. 1990. Phyllosilicate veins in a CI meteorite: evidence for aqueous alteration on the parent body. *Nature* 345: 138–40

Tomeoka, K., Buseck, P. R. 1982a. Inter-grown mica and montmorillonite in the Allende carbonaceous chondrite. *Nature* 299: 326–27

Tomeoka, K., Buseck, P. R. 1982b. High-resolution transmission electron microscopy observations of "Poorly characterized phase" in the Mighei C2M carbonaceous chondrite. *Meteoritics* 17: 289–90

Tomeoka, K., Buseck, P. R. 1983a. An exotic Fe-Ni-S-O layered mineral: an improved characterization of the "Poorly Characterized Phase" in C2M carbonaceous chondrites. *Lunar Planet. Sci. XIV*, pp. 789–90. Houston: Lunar Planet. Inst.

Tomeoka, K., Buseck, P. R. 1983b. Unusual microstructures in the C2 carbonaceous chondrites. *41st Annu. Electr. Microsc. Soc. Am.*, pp. 188–89. San Francisco: San Francisco Press

Tomeoka, K., Buseck, P. R. 1983c. A new layered mineral from the Mighei carbonaceous chondrite. *Nature* 306: 354–56

Tomeoka, K., Buseck, P. R. 1985a. Indicators of aqueous alteration in CM carbonaceous chondrites: microtextures of a layered mineral containing Fe, S, O, and Ni. *Geochim. Cosmochim. Acta* 49: 2149–63

Tomeoka, K., Buseck, P. R. 1985b. Hydrated interplanetary dust particle linked with carbonaceous chondrites? *Nature* 314: 338–40

Tomeoka, K., Buseck, P. R. 1986. Phyllo-silicates in the Mokoia CV3 carbonaceous chondrite: petrographic and transmission electron microscope observations. *Lunar Planet. Sci. XVII*, pp. 899–900. Houston: Lunar Planet. Inst.

Tomeoka, K., Buseck, P. R. 1988. Matrix mineralogy of the Orgueil CI carbonaceous chondrite. *Geochim. Cosmochim. Acta* 52: 1627–40

Tomeoka, K., Buseck, P. R. 1990. Phyllo-silicates in the Mokoia CV carbonaceous chondrite: evidence for aqueous alteration in an oxidizing condition. *Geochim. Cosmochim. Acta* 54: 1745–54

Tomeoka, K., McSween, H. Y. Jr., Buseck, P. R. 1989. Mineralogical alteration of CM carbonaceous chondrites: a review. *Proc., NIPR Symp. Antarctic Meteorites* 2: 221–34

Tomeoka, K., Kojima, H., Yanai, K. 1989. Yamato–82162: a new kind of CI carbonaceous chondrite found in Antarctica.

Proc., NIPR. Symp. Antarctic Meteorites 2: 36–54

Van Schmus, W. R. 1969. Mineralogy, petrology and classification of types 3 and 4 carbonaceous chondrite. In Meteorite Research, ed. P. M. Millman, pp. 480–91. Dordrecht: Reidel

Van Schmus, W. R., Hayes, J. M. 1974. Chemical and petrographic correlations among carbonaceous chondrites. Geochim. Cosmochim. Acta 38: 47–64

Van Schmus, W. R., Wood, J. A. 1967. A chemical-petrologic classification for the chondritic meteorites. Geochim. Cosmochim. Acta 31: 737–65

Veblen, D. R., Buseck, P. R. 1979. Serpentine minerals: intergrowths and new combination structures. Science 206(4425): 1398–400

Wasson, J. T. 1974. Meteorites: Classification and Properties. New York: Springer-Verlag. 316 pp.

Wasson, J. T. 1985. Meteorites—Their Record of Early Solar-System History. New York: Freeman. 267 pp.

Wasson, J. T., Kallemeyn, G. W. 1988. Compositions of chondrites. Philos. Trans. R. Soc. London Ser. A 325: 535–44

Wdowiak, T. J., Agresti, D. G. 1984. Presence of a superparamagnetic component in the Orgueil meteorite. Nature 311: 140–42

Weisberg, M. K., Prinz, M. 1991. Aqueous alteration in CR2 chondrites. Lunar Planet. Sci. XXII, pp. 1483–84. Houston: Lunar Planet. Inst.

Weisberg, M. K., Prinz, M., Chatterjee, N., Clayton, R. N., Mayeda, T. K. 1991. Dark inclusions in CR2 Chondrites. Meteoritics 26: 407

Wiik, H. B. 1956. The chemical composition of some stony meteorites. Geochim. Cosmochim. Acta 9: 279–89

Wing, M. R., Bada, J. L. 1991. Geochromatography on the parent body of the carbonaceous chondrite Ivuna. Geochim. Cosmochim. Acta 55: 2937–42

Wood, J. A. 1967. Chondrites: their metallic minerals, thermal histories, and parent planet. Icarus 6: 1–49

Yang, J., S. Epstein. 1983. Interstellar organic matter in meteorites. Geochim. Cosmochim. Acta 47: 2199–2216

Yang, J., Epstein, S. 1984. Relic interstellar grains in meteorites. Nature 311: 544–47

Yuen, G., Blair, N., Des Marais, D. J., Chang, S. 1984. Carbon isotope composition of low molecular weight hydrocarbons and monocarboxylic acids from Murchison meteorite. Nature 307: 252–54

Zaikowski, A. 1979. Infrared spectra of the Orgueil (C–1) chondrite and serpentine minerals. Geochim. Cosmochim. Acta 43: 943–45

Zenobi, R., Philippoz, J. M., Buseck, P. R., Zare, R. N. 1989. Spatially resolved organic analysis of the Allende meteorite. Science 246: 1026–29

Zinner, E., Fahey, A., McKeegan, K. 1986. Ion probe isotopic measurements of refractory phases from CM and CV meteorites. Terra Cognita 6: 129

Zinner, E., Tang, M., Anders, E. 1989. Interstellar SiC in the Murchison and Murray meteorites: isotopic composition of Ne, Xe, Si, C, and N. Geochim. Cosmochim. Acta 53: 3273–90

Zolensky, M. E. 1991. Mineralogy and matrix composition of "CR" chondrites Renazzo and EET 87770, and ungrouped chondrites Essebi and MAC 87300. Meteoritics 26: 414

Zolensky, M. E., Gooding, J. L., Barrett, R. A. 1987. Mineralogical variations within the matrices of CM carbonaceous chondrites. Meteoritics 22: 544–45

Zolensky, M. E., McSween, H. Y. Jr. 1988. Aqueous alteration. See Kerridge & Matthews 1988, pp. 114–43

Zolensky, M. E., Barret, R. A., Prinz, M. 1989. Petrography, mineralogy and matrix composition of Yamato–82162, a new CI2 chondrite. Lunar Planet. Sci. XX, pp. 1253–54. Houston: Lunar Planet. Inst.

Annu. Rev. Earth Planet. Sci. 1993. 21:307–31

ACCRETION AND EROSION IN SUBDUCTION ZONES: The Role of Fluids

Xavier Le Pichon, Pierre Henry, and Siegfried Lallemant

Laboratoire de Géologie de l'Ecole Normale Supérieure, 24 rue Lhomond, 75231 Paris Cedex 05, France

KEY WORDS: fluid flow, accretionary wedges, erosional active margins

INTRODUCTION

Convergence occurs today along about 60,000 km of plate boundaries (Parsons 1981). In this paper, we consider frontal subduction which occurs along about 40,000 km of convergent plate boundaries. Using the kinematic solution of De Mets et al (1990), and including 0.5 km^2 yr^{-1} of marginal basin openings (Otsuki 1989), we find that 2.95 km^2 per year disappear in frontal subduction at an average rate of 7.3 cm yr^{-1}. The upper overriding plate is significantly modified not only through the addition of a large amount of magmatic material (Markhinin 1968) but also through tectonic processes which result either in outgrowth of the margin through addition of sediments (Silver 1969) or in recession through frontal and subcrustal erosion which appears to be linked to sediment subduction (Coats 1962, Gilluly 1963). The various phenomena observed must in some way be related to variations in mechanical coupling between the two plates.

Uyeda & Kanamori (1979) had proposed that the transition from erosion to accretion results from increasing mechanical coupling. The Marianas trench was presented as typical of an erosional margin and the Chile trench of an accreting margin. Ruff & Kanamori (1980) and Kanamori (1986) showed that the magnitude of the subduction earthquakes increases with the convergence velocity and the inverse of the age of the subducted lithosphere. This correlation was qualitatively explained by the control of age on the dip of the plate and of convergence velocity on the

307

amount of friction between the two converging plates. Thus the magnitude of the earthquakes was considered to be a direct measure of the degree of mechanical coupling between plates. Then one would expect erosion and accretion to characterize subduction zones with low and high release of seismic energy respectively.

Yet, a recent systematic survey of accretion and erosion along subduction zones by Von Huene & Scholl (1991) demonstrates that there is no overall correlation between the magnitude of the earthquakes, as described by Kanamori (1986), and the presence or absence of accretion as given by their compilation. Von Huene & Scholl (1991) show that 55% of the frontal subduction margins are actively accreting sediments whereas 45% are nonaccreting. Their detailed study illustrates that the balance between accretion and nonaccretion is governed by the sediment supply. There is no active accretionary wedge where the average trench fill is less than about 1 km thick. The larger the trench fill, the larger the relative importance of accretion versus subduction of sediments. Thus, accretion appears to require the presence of a significant thickness of coarse trench fill. Finally, Von Huene & Scholl (1991) show that frontal as well as basal erosion are frequently present along margins where there is no or little accretion.

It is now realized that our poor understanding of these tectonic processes result from insufficient knowledge of the major tectonic agent in these zones which is the huge release of fluids contained within the oceanic sediments and crust fed into the subduction zones. The release results from mechanical effects (compaction of the pores of the sediments) at relatively low pressures (less than a few hundred MPa) and thermal effects (heating of the sediments and rocks) at higher pressures (up to 30 GPa). Because of the low permeabilities of the sediments and rocks where the fluids are released, very high fluid pressures and consequently very low effective pressures tend to prevail there. These necessarily have important tectonic implications (Von Huene & Lee 1982) because they greatly decrease the energy necessary for motion to occur along décollements and thrusts (Hubbert & Rubey 1959). The presence of high fluid pressures implies high fluid pressure gradients and consequently fluid flow out of these high pressure zones. Thus, it is not surprising that manifestations of fluid flow out of the toes of the upper plates have been increasingly discovered in the last few years. Three special issues (*J. Geophys. Res., Earth Planet. Sci. Lett., Philos. Trans. R. Soc. London*) have recently been devoted to the subject of fluids in subduction zones (Langseth & Moore 1990, Tarney et al 1991, and Kastner & Le Pichon 1992). Von Huene & Scholl (1991) have reviewed the tectonic processes of sediment subduction and subduction erosion and Moore & Vrolijk (1992) have reviewed questions dealing with fluids in accretionary prisms.

No attempt will be made here to repeat these recent reviews. In this paper, instead, we discuss the tectonic implications of the release of fluids from the pores of the accreted and (or) subducted sediments within the outer margin of the upper plate (roughly the part which is less than 20 km thick), using physical models based on the simplest assumptions. Compaction of sediments, assuming steady state dewatering without convection, is an important source of fluids as it produces $1 \text{ km}^3 \text{ yr}^{-1}$ (Von Huene & Scholl 1991) to $1.8 \text{ km}^3 \text{ yr}^{-1}$ (Moore & Vrolijk 1992) in the world subduction zones. It is by far the major source of fluids within the outer margin because the $0.5 \text{ km}^3 \text{ yr}^{-1}$ (Moore & Vrolijk 1992) released by heating from hydrous minerals (Kastner et al 1991) are produced deeper at temperatures exceeding 80°C. As for the fluids released from the oceanic crust, these are produced at temperatures exceeding 600°C (Peacock 1991) at depths of several tens of kilometers; thus they are of no concern for this review.

THE DÉCOLLEMENT AS A ZONE OF HIGH FLUID PRESSURE

Below the leading edge of the overriding plate, a surface of structural disharmony called décollement separates the upper plate from the subducting plate (Figure 1). The décollement is a zone of high shear strain which accommodates the large difference in velocity between the subducting plate and the leading edge of the upper plate. One expects the behavior of this décollement to dominate the tectonics of the plate boundary. Seismic reflection and drilling data show that the décollement separates a highly deformed, generally sedimentary, complex above it from essentially undeformed sediments below (Moore 1989). In simple terms, it can be understood as resulting from the partitioning of the trench sediments between an upper layer, which is accreted, and a lower layer which is subducted.

Because of the high shear strains present along the décollement, one might expect high shear stresses and presumably seismic activity there. Yet the seismicity of this zone to a depth of the décollement of about 15 to 20 km is weak to nonexistent (Yoshii 1979, Hirata et al 1983, Frohlich et al 1982, Byrne et al 1988). If the shear stress were high, one would also expect a high heat flow. However, over the leading edge of the upper plate, detailed heat flow measurements reveal a characteristic decrease landward to a depth of the décollement of 10 to 20 km (Langseth et al 1990, Davis et al 1990, Yamano et al 1992). The magnitude of this decrease is not compatible with a significant shear stress on the décollement. It is only beyond a depth of 10 to 20 km of the décollement that the heat flow flattens

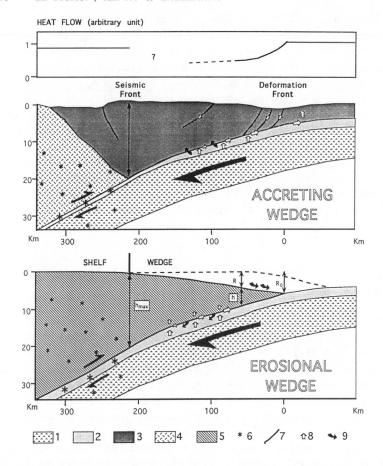

Figure 1 Synthetic cross sections of convergent margins with erosional and accreting wedges. The wedge extends between the trench and the seismic front, and lies on a low friction décollement, lubricated by high pore-fluid pressure. Burial of the lithospheric slab below the wedge results in a heat flow decrease, which is eventually compensated by shear heating landward of the seismic front. 1. Oceanic crust. 2. Material subducted with the oceanic crust. 3. Accreted sediments. 4. Arc or continental crust (backstop). 5. Material of wedge can be anything (arc or continental crust, slope sediments, old accretionary wedge). 6. Earthquake. 7. Major thrust. 8. Fluid flow. 9. Material flow. An accreting wedge forms primarily by offscraping where a sufficient amount of terrigenous sediment is deposited in the trench. Material can be added by underplating below the wedge (*filled arrows*). Fluids coming from compaction of the subducted layer flow along the décollement and are expelled along the frontal thrusts or injected into the trench. Erosion occurs mostly at the base as the décollement migrates upward, but also by collapse at the surface. Fluids under high pressure migrating from the décollement result in basal erosion.

and that a higher shear stress may be present on the décollement (Molnar & England 1990, Westbrook 1991). A third argument indicates that the outer décollement acts as a zone of mechanical decoupling with very low shear stresses. Above the décollement, within accretionary wedges, the principal shear stress (σ_1) is close to horizontal and the tectonic regime is compressional. Below it, σ_1 is vertical and sediments are essentially undeformed (Taira et al 1992). We conclude that, over the leading edge of the plate, the shear stress is quite low within the décollement.

A simple argument demonstrates that a low shear stress on the décollement implies the presence of high fluid pressure there. Because of the relatively small pressure within the wedge, the material within it must be a Coulomb material. Then, the shear stress T acting upon the décollement is (Byerlee 1978)

$$T = \mu(N - P_f),$$

where N is the stress normal to the décollement and P_f the fluid pressure. But the décollement is nearly horizontal so that N is nearly equal to the lithostatic pressure P_t. Call $P_e = P_t - P_f$ the effective pressure and let $\lambda = P_t/P_f$, then

$$T = \mu P_e = \mu(1 - \lambda)P_t. \tag{1}$$

For a depth $z < 10$–20 km, heat flow considerations indicate that T is smaller than about 10 MPa. But μ cannot be very small for likely geological materials (Byerlee 1978). Davis & Von Huene (1987) found $\mu = 0.45$ within the toe of the Aleutian accretionary wedge. Then P_e must be much smaller than P_t and λ must be close to 1. This is the result obtained on the basis of mechanical considerations by Hubbert & Rubey (1959) for large-scale overthrusts. Davis et al (1983), using a Coulomb material at failure throughout an accreting wedge, later showed that the mechanical balance between friction at the base of the wedge and the gravitational force acting on the wedge implied a near lithostatic fluid pressure within the décollement for wedges characterized by the low slopes observed in subduction zones. Their model is now generally called the Coulomb wedge model.

Also note, however, that Molnar & England (1990) concluded that the décollement was characterized by relatively high shear stresses on the basis of a heat flow analysis. But they did not specifically consider the heat flow within the leading part of the wedge. The relatively high shear stress (50 to 100 MPa) on the décollement obtained from their computations does not apply to the leading part of the wedge ($z < 10$–20 km), at least in regions where detailed heat flow data are available as in Barbados

(Langseth et al 1990, Westbrook 1991), Oregon (Davis et al 1990), and Nankai (Yamano et al 1992) as stated above.

Generalizing these observations, we conclude that a high fluid pressure, close to lithostatic, is present within the décollement below the leading part of the wedge whether it is accreting (as in Barbados) or erosional (as in Japan) and that it is this high fluid pressure that prevents high shear stresses and seismicity from existing there. At higher lithostatic pressure ($z > 10$–20 km), high fluid pressures may not be maintained as a rise in heat flow and significant seismic activity indicate the presence of relatively high shear stresses.

COMPACTING SEDIMENTS AS A SOURCE OF FLUIDS

Typically, sediments entering subduction zones have a high porosity. Stacking of thrust packages from the upper part of the entering sediment sequence within the accretionary wedge above the décollement results in rapid compaction and expulsion of most of the interstitial water and in slow growth of the wedge. Rapid underthrusting of the lower part, and in some cases of the totality, of the trench sedimentary column below the décollement feeds large amounts of high porosity sediments, and consequently of interstitial fluid, below the compacting wedge. Knowing the porosity distribution of the entering sediment column, it is possible to compute the expected steady state supply of fluids both above and below the décollement.

Although the vertical porosity distribution of the entering sediments varies according to the nature and history of the column, it can be approximated by an exponential relationship

$$\varphi = \varphi_0 \exp(-az), \tag{2}$$

where the porosity φ is given in terms of the depth z below the seafloor and of two constants, φ_0 and a. φ_0 is the seafloor porosity and is generally high and close to 0.7 whereas a (in km^{-1}) determines the downward decrease of porosity and is of the order of 1 to 1/1.8 (Le Pichon et al 1990b). For our discussion, we assume $a \sim 1/1.5$ (following Le Pichon et al 1991).

The equivalent height of water contained in a vertical column is

$$H_w = (\varphi_0/a)[1 - \exp(-az)].$$

The maximum H_w is φ_0/a which is close to 1 km. For a thickness of sediments of 1 km, H_w is about 500 m and for 3 km, it is 900 m. Increasing the thickness of the entering sedimentary column above 3 km would not increase significantly the amount of interstitial water. For typical entering thicknesses of sediment (1–4 km), the equivalent height of water H_w changes by a factor of less than 2 and is about 700 m $\pm 30\%$. The actual volume of water entering the subduction zone per unit time is $V_w = v_a H_w$, where v_a is equal to the velocity of subduction v_s plus the growth velocity of the wedge. For a mature wedge, $v_a \sim v_s$ (Le Pichon et al 1990b) and thus

$$V_w = v_s H_w \sim 700 v_s.$$

Hence the volume of water entering the subduction zone per unit time does not vary much with the thickness of sediment z provided the entering sediment thickness is 1 km or larger. It is, however, linearly dependent on the velocity of subduction.

However, the partitioning of fluid above and below the décollement changes dramatically depending on the amount of accretion of sediment to the wedge. If there is no accretion, all the fluid is fed below the décollement whereas if a significant thickness is accreted, the amount of fluid fed below the décollement from the nonaccreted lower portion will be several times smaller.

If the porosity distribution with depth is known within the accreting wedge and if there is a kinematic model describing the flow of material through the wedge, it is possible to compute the expected local steady state fluid outflow due to compaction within the wedge (Bray & Karig 1985). Several authors (Langseth et al 1990, Screaton et al 1990, Le Pichon et al 1990b) have attempted to compute this fluid circulation. Expected fluid outflow velocities are proportional to the subduction velocity v_s and to the wedge taper angle α. They never exceed a portion of a millimeter per year. As a consequence the heat flow is not significantly affected. The total expected fluid flux out of the toe of the wedge is a few square meters per year (cubic meters per year per meter of length along the strike of the wedge).

CONDITIONS FOR ACCRETION: FORMATION OF A PROTODÉCOLLEMENT IN THE TRENCH

We now return to the trench to discuss why accretion requires the presence of a significant thickness of coarse terrigenous trench fill, as stated in the introduction. We use a model proposed by Le Pichon & Henry (1992).

We make the simple assumption that the partitioning of the sediments in the trench between upper and lower plates as the sediments enter the subduction zone occurs at a level of least mechanical resistance. As discussed earlier, using the Coulomb law, this mechanical resistance on a horizontal plane such as the future décollement is given by Equation (1).

As μ does not vary significantly with pressure or material (Byerlee 1978), if λ is constant, the mechanical resistance is null at the surface and increases linearly with depth. This is true whatever the value of λ. To obtain a level of least resistance at depth, λ should increase with z more rapidly than P_t. The question we ask then is: What are the conditions that lead to such an increase of λ with z within the trench fill?

Shi & Wang (1985, 1988) have discussed the importance of the different terms in the equation governing the evolution of fluid pressure with depth in the trench fill. We consider here the simplest possible unidimensional equation governing the evolution of fluid pressure within a sedimentary column modified by sedimentation:

$$\alpha \frac{dP_f^*}{dt} - \nabla\left[\frac{k}{\mu}\nabla P_f^*\right] = \alpha\frac{dP_t^*}{dt}, \tag{3}$$

where α is the bulk compressibility (in Pa^{-1}), k is the permeability (in m^2), μ is the viscosity of water (taken to be constant and equal to 10^3 Pa), $P_f^* = P_f - P_{hydrostatic}$ is the excess pore pressure, and $P_t^* = P_{lithostatic} - P_{hydrostatic}$ is the load.

Following Shi & Wang (1985, 1988), we take

$$\alpha = b\varphi/(1-\varphi),$$

where φ is the porosity. Earlier, we used for φ an exponential dependence on z (Equation 2). However, the variable in (2) should not be the depth z but the effective pressure P_e:

$$\varphi = \varphi_0 \exp(-bP_e), \tag{4}$$

where b is a constant. Thus (2) combined with (4) implies that

$$P_e = az/b.$$

Accordingly, (2) implies hydrostatic pressure at the seafloor and progressively higher fluid pressures downward. It is a rough approximation unlikely to be correct in areas of anomalous fluid pressure. However, if we know $\varphi(z)$, one can find P_e through (4). Because relationships between porosity and seismic velocity have been derived for silt, mudstone, and turbidite lithologies (Hamilton 1978), porosities can be derived from seismic velocities and then at least relative variations in P_e can be derived from porosities (Westbrook 1991). These estimates confirm that effective pressures tend to be minimum near the décollement level both in the Barbados (Bangs et al 1990) and in the Nankai trough (Moore et al 1990).

For permeability, either we assume a constant permeability or we take

$$k = k_0(P_0/P_e)^f. \tag{5}$$

The physical properties of a given type of sediment are thus determined from parameters φ_0 and k_0—which correspond respectively to the porosity and permeability of uncompacted sediment at the surface (if $P_0 = 10^5$ Pa), b—the porosity compressibility, and a constant f.

The second term in Equation (3) is the diffusivity term. With a zero permeability, it is null and $dP_f^*/dt = dP_t^*/dt$. Then, fluid pressure is everywhere lithostatic, independent of the sedimentation rate V. But as k is nonzero, the pressure field is determined by the balance between the compressibility term and the diffusivity one. A linear property of this equation is that for a given thickness of the sedimentary column, the pressure field depends only on the ratio V/k_0. However, for a level of least mechanical resistance to be present at depth in the sedimentary column, high pore pressures are not the only requirement. Effective pressure also has to decrease with depth over some part of the sedimentary column. Le Pichon & Henry (1992) have shown that in this model, a decrease of effective pressure with depth is not possible for a homogeneous sedimentation (that is, if parameters φ_0, k_0, b, and f are kept constant). Consequently, if the sedimentation type does not change, the only minimum in mechanical resistance is at the seafloor. On the other hand, a minimum may occur at depth if a high permeability sediment is deposited sufficiently fast on top of a low permeability one.

For the simplified models discussed in this paper, we can consider schematically two types of sediments. Sands and silts have a low compressibility and their permeability changes little with effective pressure P_e as long as there is no cementation. Sands and silts then may be characterized by $f = 0$ in (5) ($k_z = k_0$) with a minimum k_0 of 10^{-17} m^2 (silts), by a relatively low φ_0 of 0.5, and a low constant $b = 0.02$ MPa^{-1}. Clays, on the other hand, have a high compressibility and may be characterized by $f = 1.4$, k_0 of the order of 10^{-16} m^2 (then k is actually of the order of 10^{-19} to 10^{-17} m^2), a relatively high φ_0 of 0.7, and a high constant $b = 0.1$ MPa^{-1} (Shi & Wang 1985, 1988). Accordingly, the permeability of clays decreases rapidly as P_e increases and noncemented sands and silts are much more permeable than clays, except near the surface.

Le Pichon & Henry (1992) have used finite element models to solve Equation (3). The elements were fixed with respect to the grains of sediment as they compacted and sedimentation was simulated by the addition of new elements at hydrostatic pressure at the surface. Figure 2 illustrates the fact that a minimum in mechanical resistance can only occur when a high permeability sediment is deposited rapidly on top of a low permeability one. Curve C1 on Figure 2 corresponds to a rapid sedimentation (1 km/Ma) of clays. The high pore pressures keep P_e at a low value in the whole section, but P_e is always increasing with depth. On the other hand,

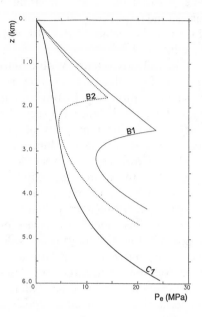

Figure 2 Effective pressure versus depth for different sedimentation models (Le Pichon & Henry 1992). C1 is for a rapid clay-rich sedimentation (1 km/Ma). In B1 and B2 clays are loaded rapidly (1 km/Ma and 2 km/Ma respectively) with silty turbiditic sedimentation $(k = 10^{-17}$ m^2). Fluid pressure rises rapidly in the clayey section but stays low in the more permeable turbidites. Thus, minima in effective pressure and mechanical resistance appear at depth.

curves B1 and B2 could be considered as typical of trenches where coarse turbidites are generally deposited rapidly on top of a finer pelagic layer deposited earlier at a low sedimentation rate (Moore 1989). Thus, an abrupt increase in P_f will in general occur immediately below the turbidites/pelagites transition and a minimum in mechanical resistance will develop at some depth below it. One might expect the protodécollement to develop at this least mechanical resistance level. Consequently, the turbidites and the topmost part of the pelagites will in general be accreted and the décollement will develop below, within the upper portion of pelagites. Note that the décollement does not coincide with the stratigraphic discontinuity.

An example of this succession of turbidites on top of pelagites is given by the Nankai trough in its central portion at drilling site 808 (Taira et al 1992). There, 400 m of low sedimentation rate, hemipelagic clays have been covered by a faster sedimentation rate, coarser, 600-m thick terrigenous layer consisting mostly of distal turbidites. The protodécollement occurs within the upper portion of pelagites as expected and a simple model shows the presence of a level of least mechanical resistance near 800 m (Figure 3).

If, on the other hand, no fast deposition of coarse terrigenous sediments has occurred, one would expect the whole sedimentary column to be

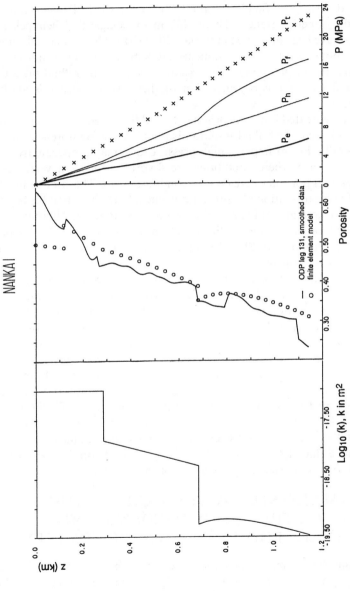

Figure 3 Sedimentation model for the Nankai trough at the location of ODP hole 808 (Le Pichon et al 1992). Sedimentation rate is 60 m/Ma for the hemipagitic clays, and progressively increases to 1.2 km/Ma for the turbidites. Porosity is calculated and compared with data from samples. The calculated effective pore pressure drops to a minimum at the depth where the décollement was drilled (800 m).

subducted because the only minimum in mechanical resistance is at the seafloor. Then, relatively high porosity sediments are rapidly loaded, which leads to high fluid pressures and trenchward flow of fluids. Such a case has been described by Shipley et al (1990) for the Costa Rica section of the Middle America trench. There, 450 m of pelagic and hemipelagic sediments have been deposited in the last 20–25 My. The seismic reflection results demonstrate that the sediments are subducted. Significant compaction occurs during the first four kilometers and a high fluid pressure is indicated by the polarity inversion of the seismic reflections at the décollement.

As the case treated is unidimensional, lateral fluid injection at depth has not been allowed. Such fluid injection is possible if a high-pressure level exists laterally. For example, a high-pressure level may propagate from a portion of trench where conditions are favorable for its development toward another portion where no such conditions exist. This process might account for the formation of a protodécollement in the northern Barbados trench where no minimum of P_e would be expected at depth. Alternatively, the introduction of tectonic loading also leads to a decrease of P_e at depth and has been proposed by Shi & Wang (1985) as the cause of the formation of the protodécollement.

In general, then, the simple model just described accounts for the complete subduction of the thin pelagic layers where no fast and coarse terrigenous sedimentation occurs in the trench. This is typical of the 19,000 km of nonaccreting subduction boundaries described by Von Huene & Scholl (1991) which have an average trench sediment column of only 400 m. On the other hand, large trench sediment columns are generally due to a rapid ($\sim 1 \, km \, Ma^{-1}$) coarse terrigenous sedimentation within the trench. This should lead to accretion of the upper coarse sedimentary fill of the trench as noted by Moore (1989). The larger the trench fill, the more probable it is that this fill is mostly made of geologically recent turbidites. The 25,000 km of accreting trenches have an average sediment thickness that is more than 1 km thick and may reach 5 or 6 km for the largest accretionary wedges such as the Makran or the Aegean.

A SCHEMATIC STEADY STATE CIRCULATION MODEL FOR THE WEDGE-DÉCOLLEMENT-SUBDUCTED SEDIMENTARY LAYER COMPLEX

Let us consider a simple schematic steady state model in which a sedimentary layer is subducted below a wedge which may or may not be actively accretionary but has been built up from trench sediments by accretion. The sediments accreted within the wedge consist of trench fill on top of a small thickness of pelagites. The trench fill is made up of alternating layers of turbiditic coarse (sand to silt) and fine (clay) sediments. The subducted sedimentary layer consists of fine (clay) pelagites overlain by a 20–40 m thick décollement layer of highly sheared and fractured pelagites with a scaly fabric and a near-lithostatic fluid pressure (e.g. Moore 1989, Moore & Vrolijk 1992).

The pressure being near-lithostatic within the décollement, flow into it can only occur from the subducted layer below, not from the wedge above. Thus, to a first approximation, the subducted layer-décollement couple and the wedge form two isolated hydraulic systems. Drilling through the décollement in the Barbados area has confirmed this separation: There is a significant seaward flow of fluid along the décollement and the upward fluid flow within the overlying accretionary wedge appears chemically distinct from the décollement flow (Moore et al 1988). Drilling through the protodécollement below the tip of the Nankai wedge has confirmed that this protodécollement is also overpressured as there is a sharp porosity increase at this level. However clear evidence for present seaward flow along the protodécollement there has not been obtained (Taira et al 1992, Maltman et al 1992). However, the shear strain at the location of this protodécollement is still small since the velocities of the tip of the wedge and of the subducted layer are nearly equal. Thus the permeability of the protodécollement may be too small for the flow to continue to be channelled along it.

Using a simple argument, Henry & Le Pichon (1991) have shown that the absence of significant upward leakage from the décollement into the wedge can only be explained if the permeability along the décollement is much higher than across the wedge. The flux of water along the décollement F_d is of the order of

$$F_d = \frac{K_d}{\mu} e \nabla P / D,$$

where e is the thickness of the décollement, ∇P the excess pore pressure, K_d the permeability along the décollement, and D the depth. The flux of water leaking through the wedge over the length L of the décollement is of the order

$$F_w = \frac{K_w}{\mu} L \nabla P / D,$$

where K_w is the equivalent permeability across the wedge in the vertical direction. If $F_w < F_d$, then

$$K_d / K_w > (L/e)(D/L)^{-1}.$$

As $e \sim 40$ m, $L \sim 40$ km, and $D/L \sim 0.05$, then $K_d/K_w > 2 \times 10^4$. This result had been obtained by Screaton et al (1990) using a finite element hydrological model.

Because the expected equivalent vertical wedge permeability is dominated by the layers of clay, it should be of the order of 10^{-18} m^2 (Shi & Wang 1988) and the permeability along the décollement should be of the order of 10^{-14} to 10^{-13} m^2. Values as high as 10^{-12} m^2 have been measured in scaly fabric material from faults (e.g. Moore & Vrolijk 1992) and these high values are probably due to the presence of a continuous network of small-scale fractures. On the other hand, the expected permeability in the subducted layer has no reason to be that high and consequently no significant seaward flow can occur along it. The only possible flow is upward, into the décollement, provided the effective pressure increases downward. Then, the décollement will be fed with the water released by the compaction of the subducted layer and the seaward flow of water along the décollement will progressively decrease landward.

This is the simplest steady state model of the décollement-subducted layer complex. Henry & Le Pichon (1991) have shown that in this model the décollement will be maintained at a near-lithostatic value with a water flux equal to the flux produced by compaction of the subducted layer if the décollement permeability is not larger than 10^{-14}–10^{-13} m^2 s. However, if the décollement is at a near-lithostatic value, the amount of compaction possible within the subducted layer is small because the vertical pore fluid gradient there cannot be less than hydrostatic. If, however, the décollement permeability increases landward because of increasing shear strain (the difference in grain velocity between the subducted layer and the wedge increases landward), then the pore pressure ratio λ along the décollement will decrease landward to below lithostatic values and efficient compaction of the subducted layer will occur (Henry & Le Pichon 1991).

A simple steady state model which seems to fit the data is schematized in Figure 4. The décollement has a high permeability anisotropy—the horizontal permeability being about three orders of magnitude larger than the vertical one near the toe of the wedge and increasing to four orders of magnitude (10^{-14}–10^{-13} m^2) landward as shear strain increases. As a result, the pore pressure is near lithostatic within the décollement near the toe but progressively decreases to smaller values landward, thus allowing significant compaction of the subducted layer. The expected seaward water velocity along the décollement in this steady state model is of the order of fractions of a meter to meters per year. As shear strain diminishes sharply seaward and as the décollement is transformed into the protodécollement, the permeability drops and the flow of water is channelled through connecting thrust faults in the toe of the wedge or diffused in the trench area.

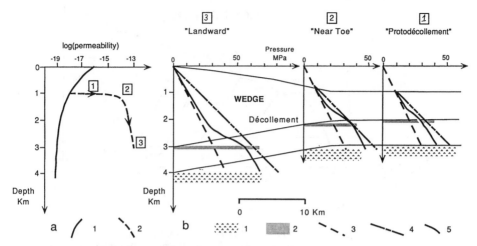

Figure 4 Schematic pressure and permeability distribution in an accretionary wedge (numerical values are only indicative). (*a*) Permeability versus depth: 1. permeability of the sediment in the vertical direction, 2. permeability along the décollement increases rapidly from the protodécollement stage (1) to near the toe (2) where it is about 10,000 times more permeable than surrounding sediments. Permeability may increase further landward (3) as shear increases. (*b*) Pressure versus depth at the three same locations: 1. oceanic crust, 2. décollement, 3. hydrostatic pressure, 4. lithostatic pressure, 5. fluid pressure. Pore pressure is lithostatic in the décollement at the toe and decreases into the trench. Compaction of the subducted sequence and décollement recharge can occur landward only if the décollement pore pressure does not stay lithostatic.

On the landward side, if the permeability increases sufficiently (more than an order of magnitude with respect to the toe), the fluid pressure ratio drops rapidly, the effective pressure increases (Henry & Le Pichon 1991), and the minimum in effective pressure may migrate downward in the subducted layer (Figure 4). This leads to subaccretion with downward migration of the décollement.

Consider now, in this model, the difference between an erosional margin and an accreting margin. In Figure 5, we have assumed that the compacted thickness of the subducted layer (after complete loss of fluid) is the same in both cases (366 m). But in the accreting case, the presubduction thickness in the trench is 466 m for a total trench sediment thickness of 2000 m; in the erosional case, the presubduction thickness is 800 m. As a result, the subducted water volume is 100 m^3/m^2 in the first case and 434 m^3/m^2 in the second case. Because the porosity will not actually decrease to zero but rather to about 10% in the outer part of the subduction zone, the actual water release will be 60 and 400 m^3/m^2 respectively. It is seven

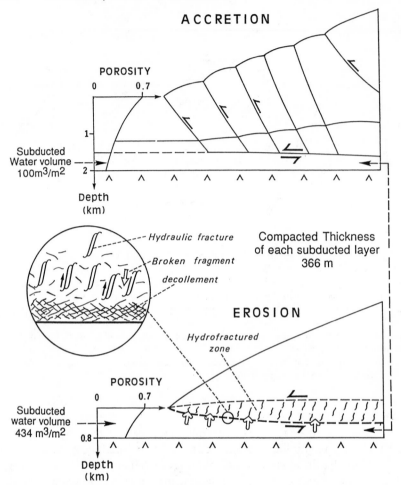

Figure 5 Schematic cross sections of an erosional and an accreting wedge. For a given subduction rate, the flux of sediment grain underthrust below the décollement is typically the same in both cases, but as the décollement is shallower in the erosional case the amount of water available is much larger. Upward migration of these fluids may disrupt rocks above the décollement by hydraulic fracturing. Broken fragments could then be subducted.

times larger for the erosional case than for the accreting case because the subducted layer in the accreting case is already significantly compacted as it is thrust below the wedge.

Consequently, for equivalent compacted layer thicknesses, the water released from the subducted layer is close to an order of magnitude larger for the erosional case than for the accreting case. Using the simple model of Henry & Le Pichon (1991) discussed above, the pore pressure along the décollement is higher and closer to lithostatic much farther from the trench in the erosional case than in the accreting one.

FRONTAL AND BASAL EROSION OF NONACTIVE ACCRETIONARY WEDGES

Erosion of the leading edge of the upper plate in subduction zones was first considered as a widespread phenomenon in the late 1970s (Scholl et al 1977, 1980; Murauchi & Ludwig 1980). In the following years, deep sea drilling coupled with seismic reflection studies established that both frontal and basal erosion contributed to this wastage of upper plate material (Von Huene et al 1982; Aubouin & Von Huene 1985; Von Huene, Suess, et al 1988). Frontal erosion is demonstrated by the proximity to the trench of the paleovolcanic arc and the truncation of old formations near the trench. Basal erosion is demonstrated by large-scale subsidence of the margin and truncation of beds at the décollement.

A quantitative evaluation of the amount of erosion has been made by Von Huene & Lallemand (1990) for the Japan and Peru margins. Their evaluations are based on the amount of subsidence of the wedge which increases from zero (where the thickness of the wedge is now 22–23 km for the Japan trench and 20 km for the Peru trench) to 5–6 km at the trench (Figure 1). Because the present slope of the wedge is approximately constant, the amount of subsidence increases linearly from 0 to 5–6 km. The amount of erosion is larger than the subsidence because the décollement must have moved upward during the erosion as the subducting plate adjusted to the smaller load. However, this upward movement should be relatively small and should increase regularly seaward to a maximum at the trench. The results of Von Huene & Lallemand (1990) on the amount of erosion R can thus be characterized by the following very simple relationship

$$R = R_0(1 - h/h_{max}),$$

where R_0 is the depth at the trench, h the thickness of the wedge, and h_{max} is equal to 20–23 km.

Landward of a wedge thickness of 20–23 km, the topography is generally quite flat, the heat flow increases, and seismicity is present both along the décollement and within the wedge. Thus it is reasonable to assume that the base of the wedge is the site of relatively high shear stresses. The Coulomb wedge model does not apply, possibly because of the increased temperatures along the décollement and the consequent transition to plastic behavior (Davis et al 1983).

Actually, it is unlikely that the transition from a high fluid pressure–low stress décollement to a low fluid pressure–high stress one occurs abruptly. Rather, one would expect a more progressive transition. One indication that such a transition exists is the increase in heat flow on the Barbados wedge, as the wedge thickness increases beyond 10 km (Langseth et al 1990). Similarly, the seismic front in the Japan wedge is located near the part of the wedge that is about 10–15 km thick (Byrne et al 1988). In spite of this last qualification, we conclude that erosion appears to be confined to the Coulomb type wedge and to be maximal at its leading edge, where the stress is minimal.

What is the cause of this erosion? Because of the low stress environment, it is reasonable to assign the process of erosion to the high fluid pressure. Murauchi & Ludwig (1980) proposed that basal erosion occurred through bodily removal of fragments, a kind of piecemeal stoping, due to upward migration of overpressured fluids from the subducted sediments. Von Huene & Lallemand (1990) proposed further that the near-lithostatic fluid pressure within the décollement might enable the formation of a slurry of water and rock fragments moving along it.

Moore (1989) has suggested that a phenomenon similar to the fault valve behavior proposed by Sibson (1990) might occur along the décollement. If the décollement is close to lithostatic, hydraulic fracturing occurs along horizontal veins whenever the fluid pressure exceeds the lithostatic pressure because the tensile strength of the veins is small. The horizontal permeability sharply increases; transient flow occurs and the fluid pressure drops back to sublithostatic value. If the décollement is at a given slightly sublithostatic pore pressure gradient, superlithostatic pressures are more likely to be reached first near the top of the décollement. In addition, the fluid pressure gradient is likely to be maximal just above the décollement (see Figure 4). Thus, the processes of stoping and fluidization will most likely occur on the upper boundary of the décollement and lead to basal erosion of the wedge. But this erosion will only occur if the average pressure is near lithostatic and if a sufficient amount of fluid is available to feed the décollement. We have seen that this is much more likely at greater distances from the toe of the wedge in erosional wedges than in accreting wedges. The decrease in erosion landward may then be attributed to the decreasing fluid pressure and the decreasing amount of water available to feed the décollement.

Another consideration may contribute to explain the absence of erosion within accreting wedges. Using the Coulomb wedge model of Davis et al (1983), the material needs to be at failure throughout the accreting wedge to maintain its critical shape, otherwise the taper will become undercritical. Then, as the material is continuously tectonically reworked, the erosional processes are stifled before they have sufficient time to achieve significant removal of sediments or rocks. Contrarily, in a nonaccreting wedge, there is no need for the material to be maintained at failure within the wedge. Erosion has time to act and progressively increases the taper, making the wedge more and more supercritical (Davis et al 1983). This increasingly steep passive wedge is only modified through increasing erosion as high porosity sediments are being subducted below it and a high-pressure décollement is maintained at its base.

Another mechanism has been shown to result in spectacular frontal erosion: the subduction of large seamounts or ridges (for a detailed review, see Von Huene & Scholl 1991). The mechanism of erosion has generally been considered to be extensive mass wasting due to oversteepening of the wedge slope in the trail of the subducted relief. Lallemand & Le Pichon (1987) have discussed this mechanism in terms of the Coulomb wedge model. However, the Coulomb wedge model generally assumes the presence of a uniform pore pressure gradient λ within the wedge whereas λ increases from near hydrostatic at the seafloor to near lithostatic near the décollement (Figure 4). Let us consider the effect of this downward increase in λ.

The critical slope beyond which failure occurs is (Dahlen 1984)

$$\alpha_c = \arctan\left[\frac{\mu(1-\lambda)}{1-\rho_w/\rho}\right], \tag{6}$$

where ρ_w is the density of seawater and ρ the density of the wedge sediment. In a dry and subaerial wedge, (6) reduces to the well known

$$\alpha_c = \arctan \mu.$$

The numerator of (6) indicates that the effective coefficient of friction is proportional to $(1-\lambda)$ and decreases at increasing pore pressure gradient. The denominator takes into account the presence of seawater. As mass wasting occurs and deeper strata are rapidly exposed, the pore pressure gradient λ increases and thus the critical slope decreases, increasing the amount of slumping. Another consequence of this erosion is a concentration of fluid venting in the scoured area (Orange & Breen 1992). Orange & Breen (1992) also propose that a local estimate of λ can be made in areas of repeated slumping through a measure of the critical slope α_c.

Little attention, however, has been paid to the possibility of basal erosion of the wedge during the subduction of the topographic relief, although Ballance et al (1989) have shown that a large amount of basal erosion was most probably present during the subduction of a seamount in the Tonga Trench. With the model discussed above, this is not surprising. As the seamount is subducted, the décollement is raised by more than 3 kilometers. As a result, there is a very large fluid pressure gradient which drains fluid from the sides as well as from the inside of the wedge. In the case of this seamount, the fluid will be drained toward the top from a width of at least 30 to 40 km. This greatly increased flow will increase the fluid pressure along the décollement to near lithostatic (Henry & Le Pichon 1991) and greatly facilitate basal erosion and upward migration of the décollement. Contrarily, in the trail of the seamount, the flow is drastically reduced or ceases altogether. The fluid pressure drops and the effective pressure increases. Thus, the décollement has a tendency to migrate downward returning to its normal lower position.

This effect accounts for the presence of a large field of mud volcanoes and diapirs seaward of the Barbados accretionary wedge near $14°50'N$ (Le Pichon et al 1990a). There, a huge amount (about 10^6 m^3 yr^{-1}) of low chlorinity fluid is released over a width of a few kilometers (Manon cruise 1992, unpublished results). The fluid has a high content of thermogenic methane which confirms that it originated far below the wedge (Le Pichon et al 1990a). A recent seismic reflection survey (M. Langseth 1992, personal communication) indicates that this diapiric field occurs on top of a 1.5 km high fracture zone ridge which acts as a drain. Such fracture zones, spaced every 50 km (e.g. Brown & Westbrook 1988), progressively sweep across the wedge because they are not parallel to the subduction vector and consequently dominate the drainage pattern of the décollement in this subduction zone.

FLUID FLOW WITHIN THE WEDGE

Up to this point, we have used a local steady state model in which the only source of water is the progressive compaction of the sediments. Although we did mention the probable existence of bursts of transient flow along the décollement attested by the presence of mineral and mud-filled veins (Behrmann et al 1988), we showed that the steady state model can account, at least in a qualitative way, for the major tectonic processes dominated by the massive release of fluids in subduction zones.

However, the predictions of this local steady state model do not match the observations of fluid flow available over accreting wedges (Le Pichon et al 1991, Kastner et al 1991). Numerous vents have been discovered on the seafloor where outflow velocities are six orders of magnitude larger than predicted (100 m yr^{-1} instead of 0.1 mm yr^{-1}; Henry et al 1989, 1992; Carson et al 1990) and corresponding estimated total flows out of the toe of the wedge are one to two orders of magnitude larger than predicted (Le Pichon et al 1991). Estimates of flow along the décollement in the Barbados area, based on thermal models, suggest total flow one to two orders of magnitude larger than predicted by the steady state model (Fisher & Hounslow 1990, Foucher et al 1990).

Kastner et al (1991) have emphasized that the fluids sampled often have Cl^- concentrations lower than seawater. These low chlorinity fluids are often associated with faults or with the décollement and contain large methane concentrations and other geochemical anomalies. In many cases, the methane has been shown to have at least one component that is thermogenic and which originated at depth at temperatures higher than 80°C. The methane is often collected in an intermediate relatively shallow (about 500 m deep) reservoir formed by the hydrate layer at the lower limit of their stability field (Hyndman & Davis 1992).

Thus, the evidence suggests that a large amount of flow within the wedge is channelled along faults and that this flow has chemical signatures indicating a deep and often distant origin. Several possibilities have been proposed to explain the low chlorinity observed [clay membrane filtration, dissociation of gas hydrates, and breakdown of clay minerals (Kastner et al 1992)] but none fully account for the massive flow observed, for example, in the 14°N Barbados diapiric area (Le Pichon et al 1990a). The existence of vents on the seafloor through which flow occurs at velocities of the order of hundreds of meters per year is also difficult to explain. First, it implies extremely high permeabilities (higher than 10^{-11} m^2) in the conduit below the vents (Henry et al 1992) and it is not clear how these permeabilities are obtained. Second, the low temperature of the fluid implies a relatively shallow recharge with seawater and consequently some kind of convection, possibly salinity driven (Henry et al 1992). Finally, the global budget is difficult to account for in a steady state model, even if local convection with seawater dilutes the fluid of deep origin. Because the escape velocity is about 10^6 times larger than the expected velocity for the steady state local model, the vent area should not exceed 10^{-6} of the total area. Henry & Wang (1991) estimated that approximately 4% of the wedge is covered by vents off Oregon. Off Nankai, the percentage is slightly below 1%. Even if one accounts for a 10% dilution by seawater (Henry et al 1992), there is still a 10^3 discrepancy between the model and the observations. One then would have to assume that the explored areas are highly anomalous. But similar discrepancies come from heat flow measurements at the seafloor and temperature anomalies in the drill holes as discussed earlier. Thus, one has to invoke either large-scale transient flows and/or other sources of water such as meteoric water from the adjacent continent (Kastner et al 1991, Le Pichon et al 1991).

SUMMARY AND CONCLUSIONS

Accretion of sediment to the leading edge of the overriding plate appears to occur only where the thickness of sediment in the trench exceeds 500 m to 1 km, whereas nonaccretion and erosion apparently prevail where this thickness is smaller than 500 m. We related the accretion of sediments to the rapid loading of fine pelagic low permeability sediments by coarse terrigenous high permeability sediments. We showed that this stratigraphic succession induces the formation of a level of high fluid pressure and thus of low mechanical resistance within the upper portion of the low permeability sediments along which the décollement will be initiated.

We then used a simple local compaction steady state model to show that basal erosion is much more probable below nonaccreting wedges because of the subduction of high porosity sediments and consequently the greater availability of fluids there. As the amount of fluid subducted is directly proportional to the subduction velocity, erosion is maximum at high subduction rates whereas accretion is favored at low subduction rates. Thus basal erosion is stifled in low subduction rate accreting wedges.

Finally, we discussed the evidence of fluid flow over the toes of accreting wedges which indicates that a large amount of fluid is channelled along faults and that this fluid has a deep origin. The typical low chlorinity observed is not easy to account for because of the large volumes of flow involved. Our conclusion, at this point, is that the steady state local compaction model fails to account in a quantitative way for the fluid circulation within the wedge—although there is no consensus in the literature on this point. We need much better data on in situ permeabilities, which are poorly known, especially within faults, and on fluid pressure distribution and its variation through time. The demonstration of the modulation of the fluid pressure by the earthquake stress cycle, such as proposed by Sibson (1990), would be of great interest and would possibly account for the transient nature of the flow.

Thus, the reader will have realized by now that, although the local compaction–steady state model gives important insights into the role of fluids in the tectonic evolution of subduction zones, we are still quite far from a comprehensive theory of this highly complex field.

ACKNOWLEDGMENTS

We are grateful to Miriam Kastner who suggested that we write such a review. We acknowledge active international cooperation with many colleagues in this fast growing field. We thank the College de France, Ifremer, and CNRS for financial support.

Literature Cited

Aubouin, J., Von Huene, R. 1988. Summary: Leg 84, Middle America Trench transect off Guatamala and Costa Rica. In *Initial Rep. Deep Sea Drilling Proj.*, *84*, ed. R. Von Huene et al, pp. 939–57

Ballance, P. F., Scholl, D. W., Vallier, T. L., Stevenson, A. J., Ryan, H., et al. 1989. Subduction of a late Cretaceous seamount of the Louisville ridge at the Tonga Trench: a mode of normal accelerated tectonic erosion. *Tectonics* 8: 953–62

Bangs, N. L., Westbrook, G. K., Ladd, J. W., Buhl, P. 1990. Seismic velocities from the Barbados ridge complex: indicators of high pore fluid pressure in an accretionary complex. *J. Geophys. Res.* 95: 8767–82

Behrmann, J. H., Brown, K., Moore, J. C., Mascle, A., Taylor, E., et al. 1988. Evolution of structures and fabrics in the Barbados accretionary prism: insights from Leg 110 ODP. *J. Struct. Geol.* 10: 557–91

Bray, C. J., Karig, D. E. 1985. Porosity of sediments in accretionary prisms and some implications for dewatering processes. *J. Geophys. Res.* 90: 768–78

Brown, K., Westbrook, G. K. 1988. Mud diapirism and subcretion in the Barbados ridge accretionary complex: the role of fluids in accretionary processes. *Tectonics* 7: 613–40

Byerlee, J. 1978. Friction of rocks. *Pure Appl. Geophys.* 116: 615–26

Byrne, D. E., Davis, D. M., Sykes, L. R. 1988. Loci and maximum size of thrust earthquakes and the mechanics of the shallow region of subduction zones. *Tectonics* 7: 833–57

Carson, B., Suess, E., Strasser, J. C. 1990. Fluid flow and mass flux determinations at vent sites on the Cascadia margin accretionary prism. *J. Geophys. Res.* 95: 8891–97

Coats, R. R. 1962. Magma type and crustal structure in the Aleutian Arc. In *Crust of the Pacific Basin. Geophys. Monogr. Ser. Vol. 6*, pp. 92–109. Washington, DC: Am. Geophys. Union

Dahlen, F. A. 1984. Noncohesive critical Coulomb wedges: an exact solution. *J. Geophys. Res.* 89: 10,125–33

Davis, D., Suppe, J., Dahlen, F. A. 1983. Mechanics of fold and thrust belts and accretionary wedges. *J. Geophys. Res.* 88: 1153–72

Davis, D. M., Von Huene, R. 1987. Inferences on sediment strength and fault friction from structures at the Aleutian trench. *Geology* 15: 517–22

Davis, E. E., Hyndman, R. D., Willinger, H. 1990. Rates of fluid expulsion across the Northern Cascadia accretionary prism: constraints from new heat flow and multichannel seismic reflection data. *J. Geophys. Res.* 95: 8869–89

De Mets, C., Gordon, R. G., Argus, D. F., Stein, S. 1990. Current plate motions. *Geophys. J. Int.* 101: 425–78

Fisher, A. T., Hounslow, M. W. 1990. Transient fluid flow through the toe of the Barbados accretionary complex: constraints from ocean drilling program Leg 110 heat flow studies and simple models. *J. Geophys. Res.* 95: 8845–59

Foucher, J. P., Le Pichon, S., Lallemant, S., Hobart, M. D., Henry, P., et al. 1990. Heat flow, tectonics, and fluid circulation at the toe of the Barbados Ridge accretionary prism. *J. Geophys. Res.* 95: 8859–68

Frohlich, C., Billington, S., Engdall, E. R., Malahoff, A. 1982. Detection and location of earthquakes in the Central Aleutian subduction zone using island and ocean bottom seismograph stations. *J. Geophys. Res.* 87: 6853–64

Gilluly, J. 1963. Tectonic evolution of the western United States. *Geol. Soc. London Q. J.* 119: 133–74

Hamilton, E. L. 1978. Sound velocity-density in sea-floor sediments and rocks. *J. Acoust. Soc. Am.* 63: 366–77

Henry, P., Le Pichon, X. 1991. Fluid flow along a décollement layer: a model applied to the 16°N section of the Barbados accretionary wedge. *J. Geophys. Res.* 96: 6507–28

Henry, P., Wang, C. Y. 1991. Modeling of fluid flow and pore pressure at the toe of the Oregon and Barbados accretionary wedges. *J. Geophys. Res.* 96: 20,109–30

Henry, P., Lallemant, S., Le Pichon, X., Lallemand, S. 1989. Fluid venting along Japanese trenches: tectonic context and thermal modeling. *Tectonophysics* 160: 277–91

Henry, P., Foucher, J. P., Le Pichon, X., Sibuet, M., Kobayashi, K., et al. 1992. Interpretation of temperature measurements from the Kaiko-Nankai cruise: modeling of fluid flow in clam colonies. *Earth Planet. Sci. Lett.* 109: 355–71

Hirata, N., Kanazawa, T., Suyehiro, K., Shimamura, H. 1983. A seismicity gap beneath the inner wall of the Japan trench as derived by ocean bottom seismographic observations. *Geophys. J. R. Astron. Soc.* 73: 653–69

Hubbert, M. K., Rubey, W. W. 1959. Role of fluid pressure in mechanics of overthrust faulting, 1, Mechanics of fluid-filled solids and its application to overthrust faulting. *Geol. Soc. Am. Bull.* 70: 115–66

Hyndman, R. D., Davis, E. E. 1992. A mechanism for the formation of bottom simulating reflectors by vertical fluid expulsion. *J. Geophys. Res.* 97: 7025–41

Kanamori, H. 1986. Rupture process of subduction-zone earthquakes. *Annu. Rev. Earth Planet. Sci.* 14: 293–322

Kastner, M., Le Pichon, X. 1992. Preface (special issue on fluids in convergent margins). *Earth Planet. Sci. Lett.* 109: I

Kastner, M., Elderfield, H., Martin, J. B. 1991. Fluids in convergent margins: what do we know about their composition, origin, role in diagenesis and importance for oceanic chemical fluxes? *Philos. Trans. R. Soc. London Ser. A* 335: 243–59

Lallemand, S., Le Pichon, X. 1987. Coulomb wedge model applied to subduction of seamounts in the Japan Trench. *Geology* 15: 1065–69

Langseth, M. G., Moore, J. C. 1990. Introduction (special section on the role of fluids in sediment accretion, deformation, diagenesis and metamorphism in subduction zones). *J. Geophys. Res.* 95: 8737–41

Langseth, M. G., Westbrook, G. K., Hobart, M. 1990. Contrasting geothermal regimes of the Barbados Ridge Accretionary Complex. *J. Geophys. Res.* 95: 8829–44

Le Pichon, X., Henry, P. 1992. Erosion and accretion along subduction zones: a model of evolution. *Proc. Kon. Ned. Akad. v. Wetensch.* In press

Le Pichon, X., Foucher, J. P., Boulègue, J., Henry, P., Lallemant, S., et al. 1990a. Mud volcano field seaward of the Barbodos accretionary complex: a submersible study. *J. Geophys. Res.* 95: 8931–45

Le Pichon, X., Henry, P., Lallemant, S. 1990b. Water flow in the Barbados Accretionary Complex. *J. Geophys. Res.* 95: 8945–68

Le Pichon, X., Henry, P., the Kaiko-Nankai Scientific Crew. 1991. Water budgets in accretionary wedges: a comparison. *Philos. Trans. R. Soc. London Ser. A* 335: 315–30

Maltman, A., Byrne, T., Karig, D., Lallemant, S., Leg 131 Shipboard Party. 1992. Structural geological evidence from ODP Leg 131 regarding fluid flow in the Nankai prism, Japan. *Earth Planet. Sci. Lett.* 109: 463–68

Markhinin, E. K. 1968. Volcanism as an agent of formation of the earth's crust. In *The Crust and Upper Mantle of the Pacific Area, Am. Geophys. Union Monogr. 12*, ed. L. Knopoff, C. L. Drake, P. J. Hart, pp. 413–23. Washington, DC: Am. Geophys. Union

Molnar, P., England, P. 1990. Temperatures, heat flux and frictional stress near major

thrust faults. *J. Geophys. Res.* 95: 4833–56

Moore, J. C. 1989. Tectonics and hydrogeology of accretionary prisms: role of the décollement zone. *J. Struct. Geol.* 11: 95–106

Moore, J. C., Vrolijk, P. 1992. Fluids in accretionary prisms. *Rev. Geophys.* 30: 113–35

Moore, J. C., Mascle, A., et al. 1988. Tectonics and hydrogeology of the Northern Barbados Ridge, results from ocean Drilling Program Leg 110. *Geol. Soc. Am. Bull.* 100: 1578–93

Moore, G. F., Shipley, T. H., Stoffa, P. L., Karig, D. E., Taira, A., et al. 1990. Structure of the Nankai Trough accretionary zone from multichannel seismic reflection data. *J. Geophys. Res.* 95: 8753–66

Murauchi, S., Ludwig, W. J. 1980. Crustal structure of the Japan Trench: the effect of subduction of ocean crust. *Initial Rep. Deep Sea Drill. Proj.* 56/57: 463–70

Orange, D. L., Breen, N. A. 1992. The effects of fluid escape on accretionary wedges: 2, Seepage force, slope failure, headless submarine canyons and vents. *J. Geophys. Res.* 97: 9277–95

Otsuki, K. 1989. Empirical relationships among the convergence rate of plates, rollback rate of trench axis and island-arc tectonics: laws of convergence rate of plates. *Tectonophysics* 159: 73–94

Parsons, B. 1981. The rates of plate creation and consumption. *Geophys. J. R. Astron. Soc.* 67: 437–48

Peacock, S. M. 1991. Numerical simulation of subduction zones pressure temperature time paths: constraints on fluid production and arc magmatism. *Philos. Trans. R. Soc. London Ser. A* 335: 341–53

Ruff, L., Kanamori, H. 1980. Seismicity and the subduction process. *Phys. Earth Planet. Inter.* 23: 240–52

Scholl, D. W., Marlow, M. S., Cooper, A. K. 1977. Sediment subduction and offscraping at Pacific margins. In *Island Arcs, Deep Sea Trenches and Back Arc Basins*, ed. M. Talwani, W. C. Pitman III, pp. 199–210, Maurice Ewing, Ser. 1. Washington, DC: Am. Geophys. Union

Scholl, D. W., Von Huene, R., Vallier, T. Z., Howell, D. G. 1980. Sedimentary masses and concepts about tectonic processes at underthrust ocean margins. *Geology* 8: 564–68

Screaton, E. J., Wuthrich, D. R., Dreiss, S. J. 1990. Permeabilities, fluid pressures and flow rates in the Barbados ridge complex. *J. Geophys. Res.* 95: 8997–9007

Shi, Y., Wang, C. Y. 1985. High pore pressure generation in sediments in front of the

Barbados ridge complex. *Geophys. Res. Lett.* 12: 773–76

Shi, Y., Wang, C. Y. 1988. Generation of high pore-pressures in accretionary prisms: inferences from the Barbados subduction complex. *J. Geophys. Res.* 93: 8893–910

Shipley, T. H. S., Stoffa, P. L., Dean, D. T. 1990. Underthrust sediments, fluid migration paths and mud volcanoes associated with the accretionary wedge off Costa-Rica: Middle America Trench. *J. Geophys. Res.* 95: 8743–52

Sibson, R. H. 1990. Conditions for fault valve behaviour. In *Deformation Mechanisms, Rheology and Tectonics*, ed. R. J. Knipe, E. H. Rutter. *Geol. Soc. Spec. Publ.* 54: 15–28

Silver, E. A. 1969. Late Cenozoic underthrusting of the continental margin off northernmost California. *Science* 166: 1265–66

Taira, A., Hill, I., et al. 1992. Sediment deformation and hydrogeology of the Nankai trough accretionary prism: synthesis of shipboard results of ODP leg 131. *Earth Planet. Sci. Lett.* 109: 432–50

Tarney, J., Pickering, K., Knipe, R., Dewey, J. 1991. Preface (special issue on the behaviour and influence of fluids in subductions zones). *Philos. Trans. R. Soc. London Ser. A* 335: V–XI

Uyeda, S., Kanamori, H. 1979. Back arc opening and the mode of subduction. *J. Geophys. Res.* 84: 1049–61

Von Huene, R., Lallemand, S. 1990. Tectonic erosion along the Japan and Peru convergent margins. *Geol. Soc. Am. Bull.* 102: 704–20

Von Huene, R., Lee, H. 1982. The possible significance of pore fluid pressures in subduction zones. In *Studies in Continental Margins*, ed. J. Watkins, C. L. Drake, *Mem. Am. Assoc. Petrol. Geol.* 34: 781–91

Von Huene, R., Scholl, D. V. 1991. Observations at convergent margins concerning sediment subduction, subduction erosion, and the growth of continental crust. *Rev. Geophys.* 29: 279–316

Von Huene, R., Langseth, M., Nasu, N., Okada, H. 1982. A summary of Cenozoic tectonic history along the IPOD trench transect. *Geol. Soc. Am. Bull.* 93: 829–46

Von Huene, R., Suess, E., Leg 112 Shipboard Scientists. 1988. Ocean drilling program leg 112, Peru continental margin: part 1, tectonic history. *Geology* 16: 934–38

Westbrook, G. K. 1991. Geophysical evidence for the role of fluids in accretionary wedge tectonics. *Philos. Trans. R. Soc. London Ser. A* 335: 227–42

Yamano, M., Foucher, J. P., Kinoshita, M., Fisher, A., Hyndman, R. D., ODP Leg 131 Shipboard Scientific Party. 1992. Heat flow and fluid flow regime in the western Nankai accretionary prism. *Earth Planet. Sci. Lett.* 109: 451–62

Yoshii, T. A detailed cross-section of the deep seismic zone between northeastern Honshu, Japan. *Tectonophysics* 55: 349–60

Annu. Rev. Earth Planet. Sci. 1993. 21:333–73

THE SCALING OF IMPACT PROCESSES IN PLANETARY SCIENCES

K. A. Holsapple

Department of Aeronautics and Astronautics, University of Washington FS 10, Seattle, Washington 98195

KEY WORDS: lunar crater scaling, catastrophic disruption

"Make things as simple as possible, but no simpler."

(Attributed to A. Einstein)

1.   INTRODUCTION

Impact processes have shaped the solar system since its early beginnings. They are an inevitable consequence of the myriad of bodies of all sizes traveling through the same regions of space at velocities of tens of kilometers per second, and the gravitational attractions between those bodies. The results of impact processes are readily apparent in almost every image of bodies in our solar system, and are recognized as one of the most important geological processes determining the morphology of those surfaces. The observation of those surfaces is one of the most important remote sensing procedures for understanding the history and evolution of the solar system.

To make use of this unique sensing system requires an understanding and quantification of the processes of impacts. How do the observed shape and size of an impact crater depend on the underlying geology? How do they depend on the size, composition, and velocity of the impactor? What is the size distribution of the remnants of a catastrophic disruption of a body? How can one meaningfully study those processes?

This article reviews "scaling laws" for impact processes, an approach

333

used to study these questions. It presents current approaches, not an historical summary; the older approaches are only briefly mentioned in passing. The interested reader may consult the book by Melosh (1989) for mention of some of the older approaches, as well as for discussions of the mechanics of shock processes and cratering. Here the restriction of space allows only a presentation of the simplest concepts and first order theories that are the most determined; extensions to topics such as atmospheric effects, oblique impacts, time-dependent creep, and modification are only mentioned in passing.

What exactly is meant by the *scaling* of impact events, and by *scaling laws*? By *scaling* we mean to apply some relation, the *scaling law*, to predict the outcome of one event from the results of another, or to predict how the outcome depends on the problem parameters. The parameters that are different between the two events are the variables that are *scaled*. Most often these are the size scale or the velocity scale, but they can also include other parameters including a gravitational field or a material strength.

As an example, consider some easily quantified measure of the outcome of an impact such as the radius R of a crater resulting from an impact into a planet of a hypervelocity body with radius a and velocity U, of a known material. The crater radius depends, among other things, on the impactor size and velocity:

$$R = f(a, U).$$

The fundamental goal of scaling studies is to measure, guess, derive, or otherwise determine the form of such a function, the scaling law, giving the dependence of the outcome of a hypervelocity impact on the size, velocity, or other conditions of the problem.

A number of questions can be raised about such scaling laws. Clearly they result from complex processes involving the balance equations of mass, momentum, and energy of continuum mechanics and the constitutive equations of the materials. The impact processes encompass the gamut of pressures—from many megabars where common solids act like fluids, to near zero where material strength or other retarding actions limit the final crater growth. Must scaling laws be power laws? Why should any such simple algebraic result be expected for such complex phenomena? If not, when are exceptions to be expected, and what form is appropriate in those cases?

Possible approaches to determining such scaling laws include experiments, analytical solutions to the governing equations, or code calculations using those same equations. Each approach has its uses but also its deficiencies, as are now summarized.

1.1   *Experiment of Impacts*

Experiments are severely limited by the shortcomings of existing testing techniques. There are no well characterized techniques to launch competent compact projectiles at speeds in excess of about 8 km/sec, well below the several tens of km/sec of primary interest in the solar system. Further, only gram-sized projectiles can be launched at those speeds. Thus, one cannot test models large enough to be governed by the same physical processes as those for the ubiquitous kilometer-sized craters on solar system bodies. For meter-sized or smaller craters in geological materials the physical processes of cratering or of catastrophic disruption are governed by material strengths of the parent body. In contrast, processes responsible for kilometer-sized craters and bodies are primarily governed by gravitational forces. Experiments in the laboratory do not model the right physics for large craters. The analogy is similar, but even more severe, to attempting to predict the response of a large airplane by conducting tests on small models thrown at hand-launched speeds: The model does not have the right Reynolds number due to deficiencies in both speed and size. Large bodies do not have the simple laminar flow of small ones. For cratering, it is the *Froude number*—the ratio of dynamic to gravitational pressures—that governs the physics of large craters. Experiments typically have a Froude number several decades too large. As a consequence, laboratory experiments have limited usefulness in predicting the outcomes of impacts at planetary scales.

1.2   *Calculations of Impacts*

The physical laws that govern impacts are well known: the balance of mass, momentum, and energy, augmented by the constitutive laws for the materials involved. In theory one could solve those equations with appropriate numerical techniques (e.g. using the finite difference methods of the so-called hydrocodes) to determine how the outcome of an impact process is related to the defining conditions. However we lack the ability to define and correctly model the diverse material behavior over the wide range of conditions involved: Processes can begin with hundreds of megabars of pressure and millions of degrees of temperature, where all materials behave roughly as a gas, and subsequently decay to less than bars of pressure where the models of rock and soil mechanics are necessary. Even if we knew how to model this behavior in principle, we often have very imprecise knowledge of the material in question. Furthermore, from the results of a single or even multiple code runs, it is almost impossible to see the forest for the trees: It is difficult to determine fundamental dependences of the outcome on the parameters of interest, or how they are interrelated. Consequently, while code calculations can be very instructive, one needs to know what to look for, and needs to understand the shortcomings to interpret the results.

1.3   *Theoretical Solutions*

If one cannot model the material behavior sufficiently to expect accurate code calculations of impact processes, then obviously one could not expect to obtain exact analytical solutions to those complicated equations. However, this does not mean that there do not exist solutions in certain limiting and idealized cases. There are such solutions, based on an approximation of the initial phases of the problem as one of a "point source." Those solutions play a significant role in determining key features of all solutions to impact problems. In fact, the primary thesis of this review will be that the existence and applicability of such solutions are the key to the understanding and derivation of scaling laws for impact phenomena. This is true even in those cases where the exact form of the solution is not obtainable: Whenever such a point-source approximation is valid certain power-law forms for scaling laws will follow.

The most well known theoretical point-source solution is for a closely related problem: the propagation in a perfect gas whole-space of the effects of the detonation of a nuclear explosion.[1] That point-source solution was determined by G. I. Taylor (1950) of the United States; and, more completely, by L. I. Sedov (1946) of the Soviet Union. It was obtained by making the approximation that the initial conditions can be described as a point source. Thus, the energy of the nuclear weapon is assumed to be instantaneously deposited in a region of zero extent. While this approximation precludes any meaningful description of the effects very near to the actual nuclear source, it gives a very accurate description for effects that are at distances large compared to the physical dimensions of the device, and at times large compared to the time of the detonation: most of the region of interest. Taylor (1950) used this solution to determine the

---

[1] Problems of the effects of nuclear explosions, and, to a lesser extent the effects of conventional explosives, are physically almost identical to those of hypervelocity impacts. In both cases, there is a deposition of energy and momentum in a very small initial region that subsequently is redistributed in a very large region. That distribution is accomplished by an outgoing shock that decays in time and distance. The flow field behind the shock is adiabatic. Ultimate effects and the remaining signature are determined by the physics of that flow and the physics of the material behavior as the stresses decay back to initial levels. Many of the advances in the understanding of impact physics have resulted from studies and experiments for nuclear explosions.

explosive yield[2] of the first nuclear explosion from observations on the fireball expansion.

This "air-blast" point-source solution is characterized and determined by the total energy of the explosion. The assumption that the correct measure of the point source is this energy is, however, a special assumption which is true because of the assumption of a perfect gas and a whole-space, spherically symmetric problem. For other materials, and for geometries such as half-space problems, there are still point-source approximations, but those solutions are not determined by the *energy* of the source. Instead they are determined by another single scalar measure called a "coupling parameter" by Holsapple (1981) and developed by Holsapple & Schmidt (1987). Almost all the scaling results for both explosions and for impacts are based on such point-source solutions, as will be demonstrated.

2. HISTORICAL APPROACHES

An introduction to current methods and a summary of past approaches is given by considering the most common and simplest scaling law: the prediction for the volume V of the crater that results from the impact of a spherical body of radius a, velocity U, and mass density δ into a planet with surface gravity g, some strength measure Y (some measure with the dimensions of stress), and mass density ρ.

While one can easily imagine other parameters of the problem that may affect the result, this short list suffices for the example. In fact, this is much more general than might be realized. For any material that has rate-independent behavior, all material properties include only combinations of the dimensions of stress and mass density. Thus, every additional material property can be nondimensionalized using the single strength measure Y and the mass density ρ. Then, for scaling size and velocity *in that given material*, those additional material property groups are constant, and the discussion here holds with no loss in generality. The most general material behavior can be considered by including only one additional rate-dependent material property such as a viscosity (see Holsapple & Schmidt 1987).

It will be assumed that the planet can be considered a half-space, and that the impact is normal to the planet surface. Then the volume V is determined by the listed variables:

---

[2] It was about 17 kton of equivalent TNT. Modern nuclear devices typically are in the megaton size. The impact of the bolide of the presumed KT event, with a diameter of perhaps 10 km at 20 km/sec, would have an energy equivalent to over 10^5 gigatons of TNT. That is a factor of 10^4 times the total yield of all nuclear weapons ever built by mankind.

$$V = f[\{a, U, \delta\}, \{\rho, Y\}, g] \tag{1}$$

with a grouping of the variables into those defining the impactor, those defining the material of the planet, and the surface gravity of the planet.

Dimensional analysis is the primary tool used to derive scaling theories. In the case of the Equation (1), there are seven parameters, with the three independent dimensions of mass, length, and time. Therefore, there is a simpler relation among four (seven minus three) dimensionless combinations (groups). The choice of these groups is not unique, and all results are independent of the forms chosen. The form chosen uses the mass m of the impactor as well as its radius:

$$\frac{\rho V}{m} = \bar{f}\left[\frac{ga}{U^2}, \frac{Y}{\rho U^2}, \frac{\rho}{\delta}\right], \quad m = \frac{4\pi}{3}\delta a^3. \tag{2}$$

The dimensionless groups in these relations can be interpreted. On the left is the ratio of the mass of material of the crater to that of the impactor. This is called the *cratering efficiency*, and is often denoted by π_V. The first term inside the function is, to within a numerical factor, the ratio of a lithostatic pressure $\rho g a$ at a characteristic depth equal to one projectile radius to the initial dynamic pressure ρU^2 generated by the impactor. This is the inverse of the definition of the Froude number. It has traditionally[3] been denoted as π_2 (see, for example, Holsapple & Schmidt 1982). Its presence in (2) allows the cratering efficiency to vary (it decreases) as either the size of the impactor, or the gravity level increases. The second term in the function is the ratio of a crustal material strength to the initial dynamic pressure; it will be denoted as π_3. Finally, the last term is the ratio of the mass densities.

If one could completely determine this function by calculation or experiment then the scaling problem would be solved. However, as stated, neither is practical, so further simplifications are pursued.

The ratio of the mass densities is usually about unity (and, since there is no problem matching its value in experiments, it will be omitted for now and the dependence on the remaining variables studied). Strengths of geological materials range from less than bars for soils to more than 10s of kbar for small samples of competent rocks. However, this latter value

---

[3] In many previous cases it has been plotted as $3.22ga/U^2$, where the factor is twice the cube root of $4\pi/3$. This makes it consistent with explosive results, where the specific energy of the device is equivalent to $1/2U^2$ and the $4\pi/3$ factor arises when the cube root of m/δ is replaced by the radius a.

is not characteristic of the large-scale geologies of planetary surfaces because of the presence of faults and cracks. Thus, the effective values of crustal strengths range from less than 1 bar to 1 kbar. In comparison, on the Earth's surface the lithostatic pressure is about 0.3 bars for each meter of depth. These values lead to a natural partition between two size scales of impacts, one appropriate for "strength-dominated" and one for "gravity-dominated" cratering, depending on the size of the event.

If, for the Earth, the impactor is smaller than about meter-sized, then the strength of a soil surface is large compared to the lithostatic pressure and the lithostatic pressure can be ignored. (The maximum size for which strength is large compared to the gravitational pressures depends linearly on the crustal strength and inversely on the surface gravity of a planet.) In this case, Equation (2) becomes

$$\frac{\rho V}{m} = \bar{f}\left[\frac{Y}{\rho U^2}\right]. \tag{3}$$

Consequently, in this "strength regime" the volume of the crater increases linearly with the volume of the impactor, its mass, and—at constant velocity only—its energy. By the same argument, any crater linear dimension such as its diameter d increases linearly with the cube root of the impactor volume and mass: i.e. with the projectile radius. Such a dependence is often called "cube-root" scaling, or "strength scaling." The dependence on the velocity, however, remains undetermined. In principle, experiments to determine the velocity dependence are possible: One can shoot a projectile (at up to 8 km/sec) into appropriate specimens. However, one must then extrapolate to velocities of several 10s of km/sec. Because this extrapolation begins from processes that have little or no melt and extends to processes that have significant melt and vapor near the impact point, the extrapolation is very uncertain. Most experiments of impacts are performed in this strength-dominated regime and at relatively low velocities, and, as a consequence, give limited information on the majority of cases of primary interest.

At the other extreme, when the impactor is kilometer-sized or larger, then the crustal strength is small compared to the lithostatic pressure term. This defines the "gravity regime" for which Equation (2) becomes

$$\frac{\rho V}{m} = \bar{f}\left[\frac{ga}{U^2}\right]. \tag{4}$$

Thus, in the gravity regime the crater volume is not proportional to the impactor volume or mass, nor is it necessarily proportional to its kinetic energy.[4] The dependence on either the size or the velocity, which are related in a particular way, must be determined by experiment or by numerical calculation.

Consider experiments in this case. Problems of interest have variations of the velocity of about a decade, in projectile size of about five decades [spanning laboratory diameters (cm) to solar system diameters (10 km)], and in gravities ranging from near zero to a decade larger than the Earth's. Then the range of inverse Froude numbers of interest in (4) is about 10^{-6} to 10^{-2}, a four decade range. (For very small gravity such as on smaller asteroids, the strength regime governs all impacts.)

To determine the function for increasing π_2, one could perform experiments at a fixed velocity at increasing projectile diameter or one could decrease the velocity at fixed size. However, for velocities below a few km/sec, the impacts are no longer hypervelocity ones; different physics controls the process and additional parameters will occur in (4). Experiments should be at velocities as large as possible. Then the largest possible value for experiments at Earth's gravity and for cm-sized projectiles has an inverse Froude number of about 10^{-6}, and experiments in the actual regime of interest are not feasible.

Only one possibility for experiments in the gravity regime for common soils exists: These involve increasing the gravitational forces. The technique was begun in the late 1970s, both for explosive cratering and for hypervelocity impact cratering, by Schmidt (1977, 1980) and later developed by Schmidt & Holsapple (1978a,b, 1980) who performed experiments on a large geotechnical centrifuge, with gravity increased by a factor approaching three decades. Froude numbers in the gravity regime were thus obtained.

The first experiments of this type were performed in a dry sand. In such cases, the cohesive strength of the material is essentially zero, so that the gravity regime also extends back to much smaller inverse Froude numbers. The results obtained from those experiments were very revealing, and provided the first clues that led to much of the recent theory on scaling. Some results of Schmidt & Holsapple (1980) are reproduced here as Figure 1. (These experiments were for explosive cratering, rather than impact. In this case, the parameters used are the explosive energy per unit mass Q, equivalent to $U^2/2$ for an impact, and the explosive mass W, equivalent to the mass m of the impactor.) While the functional form shown in (4)

---

[4] It is interesting to note that, until the past decade, researchers of nuclear weapons effects always assumed that the strength regime and cube-root scaling, as in Equation (3), applied to the hundred of meter diameter craters produced by nuclear weapons in the megaton yield range. More recently, they have recognized the deficiency of such an assumption for large nuclear craters also. Schmidt et al (1986) give nuclear and conventional explosive cratering estimates using the gravity scaling.

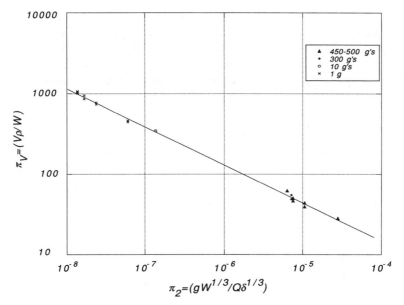

Figure 1 Explosive cratering results in a dry sand at normal and elevated gravity. The power-law fit extends at least four decades in the gravity-scaled size parameter, giving decreasing cratering efficiency with either increasing explosive size or increasing gravity.

predicts that there should be some smooth relation between the groups plotted on the ordinate and abscissa of this figure, the data show a very particular functional form: an exact power law over almost four decades of the abscissa, with a power exponent of -0.47. This power-law form is not predicted by the dimensional analysis leading to (4).

A similar result existed in the literature for impacts into dry sand. Gault & Wedekind (1977) report on a series of one-g and less than one-g experiments at fixed energy. (The interest was in lunar craters. However, the occurrence of the factor ga in the Froude number shows that using a 1 cm projectile at Earth's gravity to simulate, for example, the effect of a 10 m bolide striking the Moon with a gravity 1/6 of that of the Earth requires an *increased* gravity by a factor of 167, not a reduced gravity.) Separately, Oberbeck (1977) reported a series of variable energy impacts at one-g. Schmidt (1980) extended those results to larger inverse Froude numbers by adapting the centrifuge testing methods to impact problems. When combined and plotted as in Figure 1, a power law is observed with a slope of approximately -0.50, as reported by Schmidt (1980). The question was, why were these results all exactly power laws?

Described in the literature was one approach that predicted a power law. An appealing assumption, consistent with the results for the related air-blast problem and with the older literature on explosive cratering, is that the dependence on the impactor size and velocity is on its *kinetic energy* only, irrespective of its separate size and velocity. In this case (1) becomes

$$V = f[\{mU^2\}, \{\rho, Y\}, g]. \tag{5a}$$

The dimensional analysis now leads to a restricted form of (2):

$$\frac{\rho V}{m}\left[\frac{ga}{U^2}\right]^{3/4} = \bar{f}\left\{\left(\frac{Y}{\rho U^2}\right)\left[\frac{ga}{U^2}\right]^{-3/4}\right\}, \tag{5b}$$

which for the strength regime gives a special case of (3):

$$\frac{\rho V}{m} \propto \left(\frac{Y}{\rho U^2}\right)^{-1}, \tag{5c}$$

and for the gravity regime reduces to a special case of (4):

$$\frac{\rho V}{m} \propto \left[\frac{ga}{U^2}\right]^{-3/4}. \tag{5d}$$

In the historical scaling approaches, it was assumed a priori that the energy of the impactor determined the resulting crater size, so that (5c) applied. For example, Shoemaker (1963) used this cube-root law and scaled from a relatively deeply buried nuclear event "Teapot Ess" to estimate the energy of the meteorite that created Meteor Crater in Arizona. This approach was adopted from the explosive cratering literature, in which the form (5c) was introduced by Lampson (1946) in studies of craters from explosions of up to about one ton explosive mass—cases predominantly within the strength regime. In addition, it should be noted that all common chemical explosions have essentially the same specific energy (about 4.2×10^{10} ergs/gm), which is equivalent to a single impact velocity of 2.9 km/sec. Therefore, it was easy to overlook the possibility that there might be a separate specific energy dependence in the data, i.e. to overlook the important difference between (5c) and the more general (3). (Nuclear weapons have specific energies over 6 decades higher.)

In the gravity regime, this "energy-scaling" assumption predicts that there should be a power law, but with exponent $-3/4$ (Equation 5d)— substantially different from the observed slopes of about $-1/2$. Since a linear crater dimension would then vary as the $-1/4$ power of the energy, this is called "quarter-root" scaling: It is the consequence of the energy scaling assumption in the gravity regime.

In the literature of the 1960s pertaining to micrometeorite impacts into the metals of spacecraft, along with those who thought that the energy of an impactor should govern the results, was an opposing camp who promoted the *momentum* of the impactor as its definitive measure. [Later, Holsapple & Schmidt (1982) showed that these are the two limits on possible scaling.] If this momentum, equal to mU, is used in (5a), then one again gets a power law as in (5c) for the gravity regime, but now with a power of $-3/7$. Although closer to the observed results for dry sand, the magnitude of the exponent $(-3/7)$ is *below* that observed.

At about the same time another important experimental result was reported by Gault (1978) for the maximum transient crater formed by hypervelocity impacts into water. In reporting his results, he again assumed a dependence only on the energy of the impactor and reported that a plot of maximum instantaneous volume versus energy gave the exponent of -0.75 as appropriate for energy scaling. However, the data were reinterpreted using the more general dimensionless groups of (4) by Holsapple & Schmidt (1982) who concluded that a substantially better fit was a power law as in (5c) but with an exponent of -0.65.

There were interesting experiments reported at the turn of the century by Worthington & Cole (1897) and Worthington (1908) for (maximum transient) craters formed in water by simply dropping rigid balls into a container of water. The impact velocities were in the range of 1 to 20 m/sec. Theirs and other more recent data were examined by Holsapple & Schmidt (1982) and a fascinating result was observed. The best fit power-law curve through the Gault (1978) data (where the velocities of impact were 6 km/sec) extrapolated to over 6 decades in the Froude number goes right through the center of the low speed data. A power law with exponent of -0.65 holds for over 9 decades of Froude number for impacts into water.

Figure 2 shows these data, as reproduced from Holsapple & Schmidt (1982). Again, there was very powerful experimental evidence for power-law scaling laws for cratering efficiency over multiple decades of the gravity-scaled size.[5]

These observations, as well as others, led to the approach given by Holsapple (1981, 1983) and Holsapple & Schmidt (1987). It was recognized that the assumption that the kinetic energy of the impactor governed all subsequent results is in fact equivalent to a point-source assumption, where

---

[5] The simple water-drop experiments using 1 cm balls with 10 m/sec impact velocity have the same Froude number as a 10 km/sec impact into Earth's ocean of a 10 km bolide!

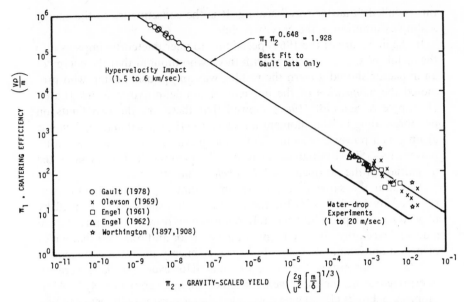

Figure 2   Data for cratering in water for velocities from 1 m/sec to 6 km/sec. A single power-law fit through the hypervelocity data goes right through the low speed data. Thus, a single power-law fit holds for over 8 decades of impactor radius (24 decades of mass), 4 decades in impact velocity, and 8 decades in gravity.

the measure of the impactor is $1/2mU^2$. It is convenient to take a cube root of this, to use the impactor radius, and to drop a numerical factor to get the measure

$$C = \delta^{1/3}aU^{2/3}. \tag{6a}$$

Alternatively, the assumption that the *momentum* dominates all results is equivalent to a different point-source assumption, with the measure

$$C = \delta^{1/3}aU^{1/3}. \tag{6b}$$

While the only well-known actual point-source solution was that for the spherical air-blast problem (for which it was indeed the energy that governed), there were other point-source solutions in related literature where other measures governed. The book by Barenblatt (1979) presents a detailed exposition of a class of problems in mechanics termed "self-similar of the second kind." In distinction to those he calls "of the first kind," those of the second kind are governed by point-source approximations for which certain exponents governing those solutions are not known a priori, but depend on the parameters of the problem and are determined in the course of the solution. A study of those point-source problems and the nature of their solutions leads to the current scaling approaches.

3. POINT-SOURCE SOLUTIONS

In an impact problem, one starts with the assumption that the impactor size a, velocity U, and mass density δ each affect the outcome. Those measures of the impactor give a size scale a, a velocity scale U, a time scale $\tau = a/U$, and a mass scale δa^3. Clearly those three separate measures can separately affect certain aspects of the results. For example, the initial pressure generated for sufficiently high-speed impacts is proportional to the square of the velocity. The amount of melt or vapor may be determined by the velocity and the mass of the projectile. The time to transfer the energy and momentum of the projectile into the target and for the shock to traverse the projectile is proportional to the time scale.

On the other hand, it seems entirely reasonable that these scales can only affect the solution very near to the impact point. For phenomena at a distance of many projectile radii, and at times of many scale times, how much can the details near the impact source affect the results? How much can they affect the final crater volume? Can the source region be approximated as a point source? In an early paper, Rae (1970) assumes so, and considers some aspects of the shock propagation of point-source solutions to the impact problem.

In mathematics a point-source solution is introduced as a limit. A problem very similar to the nuclear air-blast problem mentioned previously provides an example. Consider an infinite region filled with a perfect gas of initial mass density ρ_0, where at the initial instant there is a high pressure spherically symmetric "source" region of radius a and pressure P_1, but where the rest of the infinite region is initially at some low pressure P_0. The sole material property of a perfect gas is the perfect gas constant γ. The initial pressure and size define the initial energy of the problem (internal rather than kinetic in this case),

$$E_1 = \frac{4\pi}{3} \frac{P_1}{\gamma - 1} a^3. \tag{7}$$

Because of that initial spatially discontinuous state, there will be a spherical shock generated which propagates outward with time. Denote by R the position of that shock and by P_f the pressure at the shock which are functions of time t. Behind the shock the pressure p will be a function of radial position r and time t. All depend on the problem parameters. Thus

$$p = p(r, t, a, P_1, P_0, \rho_0, \gamma) \tag{8a}$$

$$R = R(t, a, P_1, P_0, \rho_0, \gamma) \tag{8b}$$

$$P_f = F(t, a, P_1, P_0, \rho_0, \gamma). \tag{8c}$$

These can be written in a nondimensional form as

$$\frac{p}{P_1} = f\left(\frac{r}{a}, \frac{t}{t_1}, \frac{P_0}{P_1}\right) \tag{9a}$$

$$\frac{R}{a} = f\left(\frac{t}{t_1}, \frac{P_0}{P_1}\right) \tag{9b}$$

$$\frac{P_f}{P_1} = f\left(\frac{t}{t_1}, \frac{P_0}{P_1}\right), \tag{9c}$$

where the dependence on the dimensionless perfect gas constant is understood, and the source region length scale a, and a time scale defined as $t_1 = a\sqrt{\rho_0/P_1}$ are used to nondimensionalize. Here we consider only the "strong-shock" regime, where all pressures are very large compared to the pressure P_0. Then the pressure ratio P_0/P_1 can be dropped from further consideration.

For the asymptotic form of these relations when the time t and distance r are large compared to the source scales, the ratios r/a and t/t_1 are large. These ratios also become large in a different limit as the source size and time scale become small, which is the limit leading to the definition of a point source. A point-source problem is then a limit of problems where the initial size scale a and the time scale t_1 go to zero, but where the initial pressure P_1 becomes infinite in some (as yet) undefined way. The fundamental question is whether such a limit exists in a nontrivial way (not become infinite or identically zero); and, if so, how the initial pressure must grow to infinity.

An answer to this last question can be given. Suppose that in the limit the pressure P_1 and the source size a vary in some particular way:

$$\lim_{a \to 0} (P_1 a^\beta) \to \text{constant} \tag{10}$$

for some particular positive value of the exponent β. Then in that limit the sole remaining measure of the source region is the single scalar $P_1 a^\beta$ or, by taking a root, $aP_1^{1/\beta}$. In this case the three variables remaining in (9a) can be recombined into two as

$$\frac{p}{P_1}\left(\frac{r}{a}\right)^{\beta} = f\left[\frac{t}{t_1}\left(\frac{r}{a}\right)^{-(\beta+2)/2}\right],\tag{11}$$

which gives a form that does not become indeterminate in the limit, as a consequence of (10) and because of the definition of the time scale.

Thus, point-source limits may exist when some combined measure of the initial pressure and size is fixed in the limit process.[6] That measure determines all characteristics of the point-source limit.[7] In the limit, Equations (9b) and (9c) can also be recombined to become

$$\frac{R}{a}\left(\frac{t}{t_1}\right)^{-2/(\beta+2)} = \text{constant}\tag{12a}$$

$$\frac{P_f}{P_1}\left(\frac{t}{t_1}\right)^{2\beta/(2+\beta)} = \text{constant}\tag{12b}$$

showing that in this point-source limit the shock front moves outward as a power law, with power $2/(2+\beta)$ and the shock pressure decays with time to the power $-2\beta/(2+\beta)$. The remaining question is, of course, what is the correct value for the exponent β? While one might suppose that the initial energy given in (7) should govern the solution, such an hypothesis (which is in fact correct in this particular spherical, perfect gas case) must be proved. Why could not some other combination of the pressure and size govern?

Assuming that the energy governs the solution, then the total initial energy in (7) must be held constant in the limit process defined in (10), and from (7) it is seen that $\beta = 3$. Then the shock moves outward in time with the power of $2/5$ and the pressure decays in time as the power of $-6/5$. Taken together, these imply that the pressure decays with distance to the power of -3. These results are well known for spherical air-blast problems.

How does one prove the existence of such a limit? How does one obtain the unknown exponent of the source measure? Barenblatt (1979) shows how to determine the unknown exponent for certain examples of self-similar, point-source problems. Generally, it is determined as an eigenvalue

---

[6] The proof of the actual existence of such a limit must be done on a case-by-case basis.

[7] The relation (11) is an example of what are called *self-similar solutions* in the study of solutions to partial differential equations in mathematics. Thus, a point-source limit (in the strong shock regime) is by necessity self-similar. At sufficiently large times, the pressures will decay to where the atmospheric pressure in front of the shock can no longer be ignored. The solution still arises from an initial point source, but it is no longer self-similar. Therefore, while self-similar solutions are point-source solutions, the converse is not true: Point source solutions need not be self-similar, and generally will not be in the far field.

from a nonlinear ordinary differential equation of first order that must satisfy two end conditions. Fortunately, the theoretical determination of the exponent is not needed here; it suffices to know the general form that such point-source solutions must take. The existence of a single scalar measure of the form in (10) is sufficient to determine the form of scaling laws, and relations between different aspects of the same problem. We now return to the impact problem and derive such forms.

4. SCALING LAWS FOR CRATERING

It has been shown above how point-source solutions determine power-law scaling laws in terms of some single combined measure of the source. The converse is also true: Since any power-law form has a combined single measure, that form implies a point-source measure. Based on that, Holsapple (1981) considered a single measure

$$C = aU^\mu \delta^\nu \tag{13}$$

assumed to measure the results in the far field of the impact of a bolide with radius a, velocity U, and mass density δ. This same approach was used by Dienes & Walsh (1970) for impacts into metal targets; they called the approach "late-stage equivalence." The exponents μ and ν remain undetermined for now, but special cases can be noted. If the impactor kinetic energy is the correct measure, as has been assumed for energy scaling in the past (this turns out never to be exactly correct for the problems of interest here), then $\mu = 2/3$ and $\nu = 1/3$. For the momentum assumption, $\mu = 1/3$ and $\nu = 1/3$.

The terminology "coupling parameter" is used since it is the sole measure of the coupling of the energy and momentum of the impactor into the planetary surface. It must then determine all subsequent scaling laws for all phenomena appropriately determined by the far-field solution.[8] This coupling parameter is determined by the two exponents μ and ν. Therefore, all scaling laws of impact processes will involve those exponents in some specific way. Several of the more important examples of the development of those scaling laws are presented here. The reader can refer to Holsapple (1987) and Holsapple & Schmidt (1987) for more complete results and tables of scaling forms.

[8] It is not always clear which effects are sufficiently far from the source to be included. However, code calculations (see Holsapple 1982, 1984) have routinely showed that the solutions approach that of the point-source solution for a distance within one to two impactor radii. (See also O'Keefe & Ahrens 1992b.) For a feature such as the amount of melt or vapor produced, it is more likely that this measure should not be used but instead scaling based on the near-field solution would be better.

4.1 *Crater Volume*

When there is this single measure of the impactor, relation (1) for volume scaling takes the special form

$$V = f[aU^\mu\delta^\nu, \rho, Y, g]. \tag{14}$$

[Holsapple & Schmidt (1987) also included a dependence on a material viscosity, which is useful for cratering in viscous materials, but is probably not required for the majority of applications. The interested reader can consult that reference for the more general results.]

To do a dimensional analysis one must recognize that the dimensions of the coupling parameter depend on the two exponents μ and ν. There are now only two (five minus three) dimensionless groups. Two alternative useful forms obtained are

$$\frac{\rho V}{m}\left(\frac{Y}{\rho U^2}\right)^{3\mu/2}\left(\frac{\rho}{\delta}\right)^{3\nu-1} = F\left[\frac{ga}{U^2}\left(\frac{\rho U^2}{Y}\right)^{(2+\mu)/2}\left(\frac{\rho}{\delta}\right)^{-\nu}\right] \tag{15a}$$

$$\frac{\rho V}{m}\left[\frac{ga}{U^2}\right]^{3\mu/(2+\mu)}\left(\frac{\rho}{\delta}\right)^{(6\nu-2-\mu)/(2+\mu)} = G\left\{\frac{Y}{\rho U^2}\left[\frac{ga}{U^2}\right]^{-2/(2+\mu)}\left(\frac{\rho}{\delta}\right)^{2\nu/(2+\mu)}\right\}. \tag{15b}$$

These forms may look complicated, but the functions now have a *single* variable. They give explicit results in both the strength and the gravity regime limits. In the strength regime gravity can be ignored; the function of the right of (15a) is a constant $F[0]$ so that

$$V \propto \frac{m}{\rho}\left(\frac{\rho U^2}{Y}\right)^{3\mu/2}\left(\frac{\rho}{\delta}\right)^{1-3\nu}. \tag{16a}$$

Similarly, ignoring the strength in (15b) gives for the gravity regime

$$V \propto \frac{m}{\rho}\left[\frac{ga}{U^2}\right]^{-3\mu/(2+\mu)}\left(\frac{\rho}{\delta}\right)^{(2+\mu-6\nu)/(2+\mu)}. \tag{16b}$$

Based on some plausible hypotheses Holsapple & Schmidt (1982) showed that the momentum scaling and the energy scaling give the upper and lower bounds on the scaling exponent μ. Thus, $1/3 \le \mu \le 2/3$. For both limit cases the exponent ν is $1/3$, so this is a likely general value also. Since for all cases of interest the mass density ratio does not deviate from unity by more than a factor of three or so, the terms with that ratio are of the order of unity, and will often be omitted.

All of these results can be illustrated as shown schematically in Figure

3, which shows the cratering efficiency π_V versus the gravity-scaled size parameter π_2 as it would appear on a log-log plot. In the strength regime, the cratering efficiency is constant for increasing impactor size but depends on the velocity. In the gravity regime, the cratering efficiency decreases with increasing impactor size as shown. The exponent of that decrease has been often denoted by $-\alpha$; comparison with (16b) yields

$$\alpha = \frac{3\mu}{2+\mu}. \tag{17}$$

The limits on α are 3/7 to 3/4.

Experiments in an alluvial soil on a centrifuge (Holsapple & Schmidt 1979) for explosions have indicated that the transition between the strength and the gravity regimes typically spans about two decades in the π_2 parameter, over which the gravity lithostatic pressure ranges from about 1/10 the strength to 10 times the strength. Following the approach of Holsapple & Schmidt (1979) a convenient empirical smoothing function to span the transition can be given as

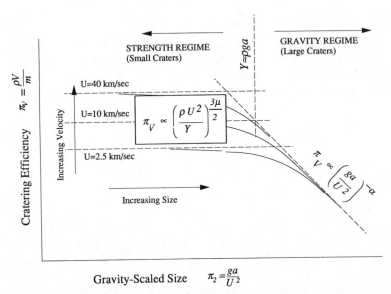

Figure 3 The regimes of cratering for a material with strength. In the strength regime the cratering efficiency depends on the impact velocity, but is independent of gravity-scaled size. For increasing size at a fixed velocity, there is a transition to the gravity regime in which the cratering efficiency has a power law decrease with increasing size. Most experiments in geological materials are by necessity in the strength regime.

$$\pi_V = K_1 \left\{ \pi_2 \left(\frac{\rho}{\delta}\right)^{(6v-2-\mu)/3\mu} + \left[K_2 \pi_3 \left(\frac{\rho}{\delta}\right)^{(6v-2)/3\mu} \right]^{(2+\mu)/2} \right\}^{-3\mu/(2+\mu)}$$

$$\pi_V = \frac{\rho V}{m}, \quad \pi_2 = \frac{ga}{U^2}, \quad \pi_3 = \frac{Y}{\rho U^2}, \tag{18}$$

where the definitions of the dimensionless groups for the cratering efficiency, the gravity-scaled size, and the strength group are again indicated. This has the correct limits in both regimes. The two constants K_1 and K_2 as well as the exponents μ and v can be chosen to match the results in each of the gravity regime and the strength regime. Henceforth, it will be assumed that $v = 1/3$ in all cases, which simplifies the results. [Holsapple & Schmidt (1979) omitted the power on the strength term since it was not recognized that the exponent in the strength regime and that in the gravity regime must both be determined by the common exponent of the point-source solution. Holsapple & Schmidt (1987) later gave a correct form.]

The material strength Y occurs in the numerator of the second term as the product $K_2 Y$ which is then determined from cratering experiments in the strength regime. One can take K_2 equal to unity and then determine an effective strength $\bar{Y}$ to match the strength-dominated, small-scale cratering results. This approach will be used. The notation $\bar{\pi}_3$ is used to denote that choice. Then Equation (19) gets somewhat simpler:

$$\pi_V = K_1 \left[\pi_2 \left(\frac{\delta}{\rho}\right)^{1/3} + \bar{\pi}_3^{(2+\mu)/2} \right]^{-3\mu/(2+\mu)}. \tag{19}$$

This equation is then used for the scaling of crater volume for all materials. The transition between the two limits occurs when the two terms inside the brackets are equal. This occurs when the gravity-scaled size π_2 parameter defined above is equal to $(\bar{Y}/\rho U^2)^{(2+\mu)/2}$.

Schmidt & Housen (1987) present estimates of this form[9] for wet and dry soils and for water. Additionally, estimates can be based on explosive results. Schmidt et al (1986) have presented a complete set of estimates for nuclear and conventional explosives in a variety of geologies and burial depths. It has been determined by Holsapple (1980) that explosive sources buried one to two source radii give the same crater size as an impact event with the same energy and specific energy. Using these data, estimates for impact crater size can be determined. The estimates here are a composite of those two sources. Since the estimates are unknown to within a factor of perhaps two, there is no point in retaining the ratio of the mass densities,

---

[9] Note that the definition of π_2 is different there by a factor of 3.22, see Footnote 3.

and it will be dropped. It could be reintroduced to give estimates of the effect of variations in impactor mass density.

Values for a range of representative geological materials are given in Table 1. Note that the estimates for the strength regime of rocks is based on relatively large craters. It thus requires sufficiently large impacts so that the effective strengths are those of the material on 10 to 100 meter scales, not cm scales. Small laboratory experiments will give much smaller craters than these estimates in such cases.[10] Figures 4 to 7 are plots for these representative materials.

4.2 Ejecta

The methods described above have been applied to the scaling of ejection dynamics and final ejecta blanket profiles by Housen et al (1983). Again, there are important distinctions depending on whether material strength or gravitational forces are the dominant mechanism determining crater size. Laboratory experiments are usually in the strength regime. Of course, once material is ejected from the crater, it is always gravitational forces that determine its ballistic path.

For brevity, only a few of the more important results of ejecta scaling are summarized here; the reader is referred to Housen et al (1983) for the complete analysis (see specifically Table 1 of that reference). Since all scaling aspects are determined by the point-source approximation, they are all given in terms of the single scaling exponent μ, or equally, in terms of the exponent α defined in (17). Those interdependencies have often been violated in strictly empirical approaches to scaling.

Housen et al (1983) give results for the dynamics of the ejecta plume position, velocity, mass, and angle; as well as for the final blanket thickness versus range. In the gravity regime, the crater dimensions and the ejecta are both determined by the coupling parameter and by gravity. When the crater radius R is used to nondimensionalize the ejecta phenomena, the dependence on the coupling parameter cancels out. The results in the gravity regime are listed in Table 2. These results show that in the gravity regime all ejecta blankets are geometrically similar to the crater size. This is not true in the strength regime, where the blanket moves relatively closer in as the crater sizes grow (see Housen et al 1983).

4.3 Crater Depth and Radius

4.3.1 SIMPLE CRATERS   It is the crater radius that is most easily determined from remote observation. Consider first the so-called simple craters

___

[10] Very small craters in competent rocks and other brittle targets are dominated by surface spall effects rather than the excavation mechanisms of the large craters. The photo in figure 2.1 of Melosh (1989) gives a good example for a very small 30 μm crater in glass.

Table 1 Cratering volume estimates for a variety of geological materials

Material	Scaling exponent α	Scaling exponent μ	K_1	$\bar{Y}$ mpa	Strength regime[a]	Gravity regime[a]	Transition impactor diameter[b]
Sand	0.51	0.41	0.24	0	–	$V = 0.14\ m^{0.83}\ G^{-0.51}\ U^{1.02}$	near 0
Dry soil	0.51	0.41	0.24	0.18	$V = 0.04\ m\ U^{1.23}$	$V = 0.14\ m^{0.83}\ G^{-0.51}\ U^{1.02}$	0.2 meters
Wet soil	0.65	0.55	0.20	1.14	$V = 0.05m\ U^{1.65}$	$V = 0.60\ m^{0.783}\ G^{-0.65}\ U^{1.3}$	1.2 meters
Water	0.648	0.55	2.30	0	–	$V = 13.0\ m^{0.783}\ G^{-0.65}\ U^{1.3}$	near 0
Soft rock	0.65	0.55	0.20	7.6	$V = 0.009\ m\ U^{1.65}$	$V = 0.48\ m^{0.783}\ G^{-0.65}\ U^{1.3}$	11 meters
Hard rock	0.60	0.55	0.20	18	$V = 0.005\ m\ U^{1.65}$	$V = 0.48\ m^{0.783}\ G^{-0.65}\ U^{1.3}$	32 meters

[a] Uses mass in kg, velocity in km/sec, gravity in Earth G's, gives volume in m^3.
[b] Impactor diameter at Earth's gravity and impact velocity of 10 km/sec: it is proportional to $g^{-1}U^{-\mu}$.

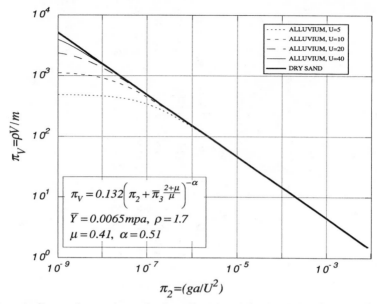

Figure 4   Crater volume estimates for dry soils and sand showing the strength and gravity regimes. Analytical expressions are as shown. At Earth's gravity and an impact velocity of 10 km/sec, the gravity regime holds for impactor radii above about 1 m, with a corresponding crater volume of about 3×10^3 m³.

which are those usual bowl-shaped craters of the laboratory and the smaller lunar and terrestrial craters. There are two radii that can be identified. Experiments usually report the radius R_e measured as the radius of the crater excavation at the original ground surface. Remote observations more often use the rim radius R_r measured to the top of the rim around the crater formed from the uplift and ejecta. An approach as above for either gives a law of the form

$$\pi_R = \left(\frac{\delta}{m}\right)^{1/3} R = K_1 \left[\pi_2 \left(\frac{\delta}{\rho}\right)^{1/3} + \bar{\pi}_3^{(2+\mu)/2} \right]^{-\mu/(2+\mu)}. \tag{20}$$

In the gravity regime this is

$$\left(\frac{\delta}{m}\right)^{1/3} R = K_1 (\pi_2)^{-\alpha/3} \tag{21}$$

(where the mass density ratio raised to a small power has been ignored). Schmidt & Housen (1987) give $K_1 = 0.69$ and $\alpha = 0.51$ for dry soils when

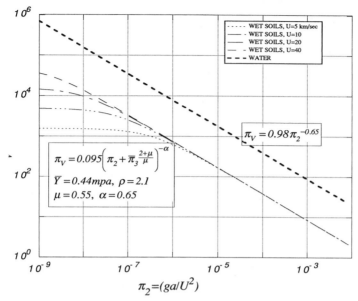

Figure 5   Crater volume estimates for wet soils and water showing the strength (for the soils only) and gravity regimes. Analytical expressions are as shown. At Earth's gravity and an impact velocity of 10 km/sec, the gravity regime for wet soils holds for impactor radii above about 1 m, with a corresponding crater volume of about 2×10^4 m^3.

converted to this form for the excavation radius R_e. This can be expanded to

$$R_e = 7.8G^{-0.17}a^{0.83}U^{0.34} \qquad (22a)$$

using the same $(m, G, \mathrm{km/sec})$ units as in Table 1. The radius to the rim peak around the crater has the same scaling, but with a different coefficient. The measurements of the rim profiles reported by Housen et al (1983) for dry soils give the rim peak at a factor of 1.3 of the excavation radius. In either the strength or gravity regime any linear dimension will follow the same type of scaling law, so the crater shapes are similar at all sizes (but could be different in each regime). Therefore the ratio of rim radius to excavation radius is constant. Thus

$$R_r = 10.14G^{-0.17}a^{0.83}U^{0.34}. \qquad (22b)$$

The reader is referred to Schmidt & Housen (1987) for other estimates of this form for wet and dry soils and for water.

Two crater depths can also be identified: that below the rim peak d_r,

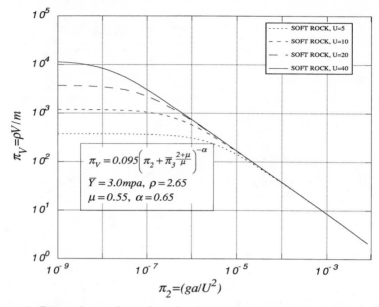

$\pi_V = 0.095\left(\pi_2 + \pi_3 \frac{2+\mu}{\mu}\right)^{-\alpha}$

$\bar{Y} = 3.0mpa, \ \rho = 2.65$

$\mu = 0.55, \ \alpha = 0.65$

Figure 6 Crater volume estimates for soft rocks showing the strength and gravity regimes. Analytical expressions are as shown. At Earth's gravity and an impact velocity of 10 km/sec, the gravity regime holds for impactor radii above about 10 m, with a corresponding crater volume of about 4×10^6 m^3.

and that excavated below the original ground surface d_e—the difference being the rim height h. Experimental craters in an alluvial soil show an aspect ratio R_e/d_e just over 2 for shallow buried explosions in an alluvial soil, and a rim height h/R_e of about 0.07 (Schmidt et al 1986). Pike (1977) shows a constant value of about 2.55 for R_r/d_r and $h/R_r = 0.072$ for lunar craters from a diameter of 0.1 km up to a crater diameter of about 15 km. These are all approximately consistent. Thus, for the depth d, one can simply use (22b) divided by 2.55, and for the rim height h can multiply (22b) by 0.07 to get

$$d_r = 4.0G^{-0.17}a^{0.83}U^{0.34} \tag{22c}$$

for the depth below the rim, and

$$h = 0.71G^{-0.17}a^{0.83}U^{0.34} \tag{22d}$$

for the rim height.

4.3.2 COMPLEX CRATERS For the larger craters, additional phenomena occur. In contrast to the results for the smaller craters, the observed aspect ratio is no longer constant but increases rapidly with increasing crater

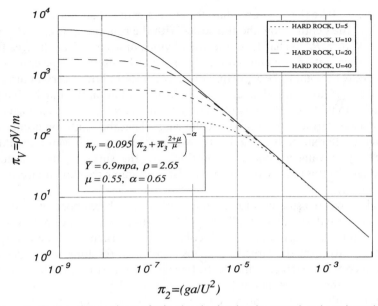

Figure 7 Crater volume estimates for hard rocks showing the strength and gravity regimes. Analytical expressions are as shown. At Earth's gravity and an impact velocity of 10 km/sec, the gravity regime holds for impactor radii above about 20 m, with a corresponding crater volume of about 2×10^7 m³.

Table 2   Scaling of cratering ejecta[a]

Ejecta characteristic	Gravity regime results
Plume Position x versus Time t	$x/R \propto \left(t\sqrt{g/R}\right)^{2\alpha/(\alpha+3)}$
Velocity v versus Position x	$v/\sqrt{gR} \propto \left(x/R\right)^{(\alpha-3)/2\alpha}$
Velocity versus Time	$v/\sqrt{gR} \propto \left(t\sqrt{g/R}\right)^{(\alpha-3)/(\alpha+3)}$
Volume Ejected at Velocity $> v$	$V_e/R^3 \propto \left(v/\sqrt{gR}\right)^{6\alpha/(\alpha-3)}$
Volume Ejected at Position $< x$	$V_e/R^3 \propto \left(x/R\right)^3$
Blanket Thickness B versus Range r	$B/R \propto \left(r/R\right)^{(6+\alpha)/(\alpha-3)}$

[a] From Housen et al 1983.

diameter. Pike (1977) shows that, for craters with a diameter greater than 20 km, the depth below the rim scales with the rim radius as $d_r \propto R_r^{0.3}$ so that $R_r/d_r \propto R_r^{0.7}$. That fact, and the obvious morphological changes for the larger craters, are affirmation of a new mechanism for large craters—generally thought to be a collapse of simple craters above a certain threshold size (15 to 20 km diameter on the Moon, 3 km on the Earth) due to gravitational forces, leading ultimately to the *complex* craters having central peaks, terraced walls, and circular rings. That the transition occurs on several different bodies (the Earth, Moon, Mercury, and Mars; see Pike 1988) at a value inversely proportional to the surface gravity argues strongly for this gravity-driven mechanism (assuming roughly equal crustal strengths).

Thus, when a simple crater is sufficiently large, there apparently occurs a final "gravity-modification" stage of formation. Any shape described by the scaling laws given above will collapse and perhaps, if large enough, oscillate to form the large basins observed, with resulting topologies of slumped walls, terraces, and central peaks. The final resting state of the crater in those cases then will have a much smaller depth and somewhat greater radius than that predicted from (22). However, since this slumping is to first order volume conserving, the volume scaling will not show this effect, and the results above may be applied. (If anything, because of bulking, the crater volume may decrease during these processes, making the predictions given an upper bound. However, over long time periods any low density material may be expected to consolidate back to its initial density by natural processes.)

Thus, for the scaling of the radius and depth the final resting configuration can no longer be simply assumed to depend on the gravity field, since the slumping and final configuration must be determined by some material strength measure. There reappears a dependence on both the gravity (determining the transient crater) and on some strength (determining when the crater is "frozen" into its final configuration). The additional dependence on a material strength will give another transitional regime and the appropriate scaling law for the crater radius may appear as shown schematically in Figure 8.

A simple scaling approach to this problem will be outlined. If the slumping and rebound phenomena are effectively uncoupled from the transient crater stages of the formation, then either of the above Equations (22) for the gravity regime can be used directly for a maximum transient[11]

---

[11] In some cases, it is also important to distinguish between a maximum transient crater and a final resulting simple crater shape. The radii of those two are nearly equal, but the depth of the maximum transient crater is typically a factor of two larger than the final simple crater due to rebound mechanisms. See the measurements by Schmidt & Housen (1987).

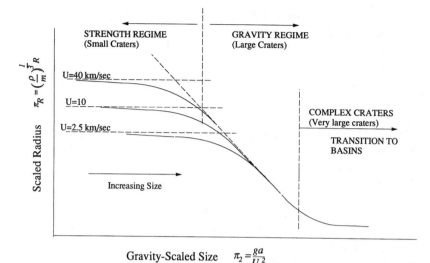

Figure 8 Schematic cratering curves for crater radius, showing an additional "complex crater" regime for very large craters. At Earth's gravity, the strength regime holds for submeter sized craters, the gravity regime for meter to km sizes, and the transition to complex craters begins at diameters of a few kilometers.

crater rim radius R_t. The subsequent crater modification stage does not depend at all on the impactor conditions but only on this transient crater size, on gravity, on the mass density of the material, and on some material strength, so that the final rim radius is given by

$$R = f(R_t, \rho, g, Y) \tag{23}$$

which has the dimensionless form

$$\frac{R}{R_t} = f\left(\frac{\rho g R_t}{Y}\right). \tag{24}$$

Thus, there is a single unknown function to be determined. (Earlier approaches have often assumed a constant function, e.g. $f = 1.3$.) This relation will hold only when the crater radius is greater than some transitional crater radius denoted by R_*. At that transition $R_t = R_*$ and also $R = R_*$ so that the function must be unity when $R_t = R_*$.
    This means

$$R_* \propto \frac{Y}{\rho g} \tag{25}$$

so that the transition radius is proportional to some strength measure and inversely proportional to the gravity. Substituting this back into (25) gives a useful form which eliminates the unknown strength in terms of the observable transition radius:

$$\frac{R}{R_t} = f\left(\frac{R}{R_*}\right). \tag{26}$$

It is tempting to introduce the usual power laws for this function; but, in contrast to the developments above, here there is no obvious theoretical reason for such. It is clear though that the function must be unity for craters below the transition size R_*, and an increasing function above that threshold.

In the past, there have been several authors who considered this problem, most recently Croft (1985). Croft does succumb to the temptation: He assumes as his working hypothesis that the final radius is a power law in the initial energy, with an additional dependence on the velocity. He obtains a relation like

$$\frac{R}{R_t} = \left(\frac{R}{R_*}\right)^\beta \tag{27}$$

as a special form for (26). [A more direct approach would be to simply assume that the final radius depends only on the transient shape as in (24), and assume a power law there. Then one again obtains relation (27).] Croft (1985) examined four different criteria based on extensive lunar observations and chose the exponent β as 0.15. One of those criteria was an assumption of the constancy of the volume in a slumping process, which uses various earlier empirical forms for crater geometry features. Melosh (1989) also presents an argument based on constant volume, together with the simple power-law analytical forms for the ejecta blanket thickness that hold for energy scaling in the far field.

A new analysis (Holsapple, unpublished) is also based on the assumption of constant volume in a slumping mechanism, but with several improvements over the previous analyses. The actual crater shapes and the rim profiles as measured by Schmidt et al (1986) and Schmidt & Housen (1987) were used, rather than a simple analytical power-law model appropriate for the far field only. The crater scaling given here was used for the shape and volume of the simple craters. The morphometric data of Pike (1977) for actual lunar craters gives a basis for determining their volume as a function of the final radius R. Then the extent of the crater slumping can be determined, and a particular function as in (26) determined. The final results of this analysis are as shown in Figure 9.

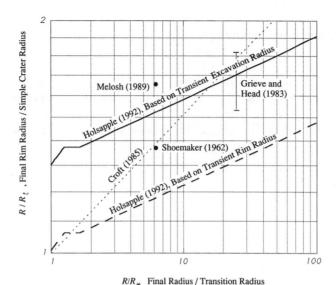

Figure 9 Radius enhancement due to gravity-modification slumping mechanisms based on a constant volume model of the author and observed simple crater profiles.

There are two curves shown, one the ratio of the final rim radius R to the transient *rim* radius R_r and one the ratio of the final rim radius to the transient *excavation* radius R_e. Previous studies have not distinguished between these, and there is a factor of 1.3 difference. Either curve is essentially a power law over most of the domain of interest. These results seem to be as consistent with the large variety of observed features given by Croft (1985) as is his power-law estimate. The Shoemaker (1962) estimate for Copernicus seems to be for the ratio of rim radii, so the present result is definitely below it. The Melosh (1989) estimate for Copernicus is based on a model that has no difference in the two radii. The Grieve & Head (1983) estimate for the 100 km diameter terrestrial crater Manicouagan is most applicable to the ratio using the transient excavation radius, and the present result agrees well.

For analytical calculations, the fit to the functions shown are

$$\frac{R}{R_r} = 1.02\left(\frac{R}{R_*}\right)^{0.079}, \quad \frac{R}{R_e} = 1.32\left(\frac{R}{R_*}\right)^{0.079} \tag{28}$$

for the ratios of the final to either of the transient excavation or rim radius.

Since the ratio of the radius R to the transition radius R_* can be directly observed, this result (28) can be used to determine the original transient radius in terms of the observed final radius R. Then (22) applied to R_t gives the impactor conditions.

For lunar craters the results are as shown in Figure 10. The transition is taken at a crater radius of 8.5 km, the value where the volume given by the scaling of simple craters here equals that determined by the Pike (1977) data. (This plot does not show the strength-dominated craters that would occur for smaller diameters.)

The depth scaling for complex lunar craters was obtained by using the rule

$$\frac{d}{R_*} = 0.313 \left(\frac{R}{R_*}\right)^{0.301} \tag{29}$$

for complex craters—a dimensionless form equivalent for lunar craters with that given by Pike (1977). Note that the final depth is much reduced from the transient; at 100 times the transition size, the depth is reduced by a factor approaching 20 of the predicted simple crater depth (a factor of

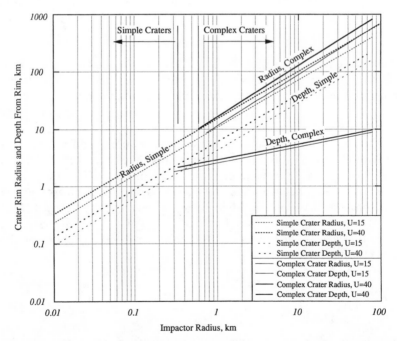

Figure 10 Crater radius and depth for lunar craters showing both the transitory simple crater and the final complex crater sizes.

40 less than the maximum transient crater). The radius is enlarged by a factor of almost 1.5 from the excavation radius, and a factor of about 1.9 from the simple rim radius.

The scaling of these last two sections implies that the impact on the Moon of a 25 km radius impactor traveling at 25 km/sec would first excavate a simple crater with a depth below the rim of about 72 km (using 22c), an excavation radius of 141 km (using 22a), a transient rim radius of 184 km (using 22b), with a previous transient depth of about 150 km. Then, during the gravity modification stage (using Equation 28b), this structure reforms to a depth of "only" 7.3 km and to a radius of 244 km. While an initial excavation to such great depths may seem surprising, code calculations of large impact structures produce similar results (see Roddy et al 1987).

4.4   The Dynamics of Crater Formation

In addition to the final crater, the methods here can also be applied to the actual dynamics of the shock propagation and the crater growth history. The key is the assumption that a single combined coupling parameter measure is appropriate, and that it must determine all aspects of the event. Measurements of observations of any one suffice to give that exponent, and the scaling of any other aspect can be predicted. Dynamic facets include the time of formation (Schmidt & Housen 1987), the shock pressure decay, the transient crater growth history (Holsapple 1984, Holsapple & Schmidt 1987), and others.

Tables 3 and 4 give a variety of results for the dynamics of the crater history. Shown are a general case and six specific special regions that can be identified. (In some cases, the regimes overlap.) Two examples of the use of this table, and a discussion of the various regimes will be given.

4.4.1   SHOCK PRESSURE PROPAGATION AND DECAY   Consider the peak pressure P at the shock wave formed from the impact event. Figure 11 shows a schematic of its decay with range or with time. That pressure can depend on the variables shown in the second column of Table 3, namely ρ, U, a, c, Y, r, t, where c is a sound speed measure of the material. For the initial value at the instant of contact ($r = 0$, $t = 0$), the pressure is typically substantially above any strength measure, so the strength Y can be ignored; and usually the impact velocity is substantially above the wave speed c, so it too can be dropped. [If the impact velocity is not large compared to the sound speed, then the sound speed c must be retained, and the dimensional analysis yields $P/(\rho U^2) = f(c/U)$. The exact value can be obtained by the usual impedance matching approach.] Furthermore,

Table 3 Scales for impact processes for initial values and near-source regions

	General	Initial	Near source
Defined by:		$r = 0,\ t = 0$	$r \approx a$
Variables:	ρ, U, a, c, Y, r, t	ρ, U, a	ρ, U, a, r (or t)
Pressure	$\rho U^2, \rho c^2, Y$	ρU^2	$\rho U^2 f(r/a)$
Velocity	$U, c, \sqrt{Y/\rho}$	U	$U\ f(r/a)$
Time, Durations	$a/U, t$	a/U	$a/U\ f(r/a)$
Position, Length	a, r	a	$a\ f(r/a)$

Table 4 Scaling of cratering events away from source region, as governed by the point-source approximation

	Strong shock	Intermediate	Special intermediate	Weak shock
Defined by:	$P >> \rho c^2,\ r \approx 2-3$	$P \approx \rho c^2$	Shock speed $\approx c$	$P \approx Y$ or $\rho g h$
Variables:	ρ, aU^μ, r (or t)	ρ, aU^μ, c, r (or t)	$\rho c, aU^\mu, r$ (or t)	ρ, aU^μ, c, Y, r (or t)
Pressure	$\rho U^2 (a/r)^{2/\mu}$	$\rho U^2 (a/r)^{2/\mu} f\left((c/U)(a/r)^{-1/\mu}\right)$	$\rho c U (a/r)^{1/\mu}$	f also has $(Y/\rho c^2)$
Velocity	$U(a/r)^{1/\mu}$	$U(a/r)^{1/\mu} f\left((c/U)(a/r)^{-1/\mu}\right)$	$U(a/r)^{1/\mu}$	f also has $(Y/\rho c^2)$
Time, Durations	$(a/U)(r/a)^{(1+\mu)/\mu}$	$(a/U)(r/a)^{(1+\mu)/\mu} f\left((c/U)(a/r)^{-1/\mu}\right)$	$(a/U)(r/a)^{(1+\mu)/\mu}$	f also has $(Y/\rho c^2)$
Position, Length	$a(Ut/a)^{\mu/(1+\mu)}$	$a(Ut/a)^{\mu/(1+\mu)} f\left((c/U)(Ut/a)^{1/(1+\mu)}\right)$	$a(Ut/a)^{\mu/(1+\mu)}$	f also has $(Y/\rho c^2)$

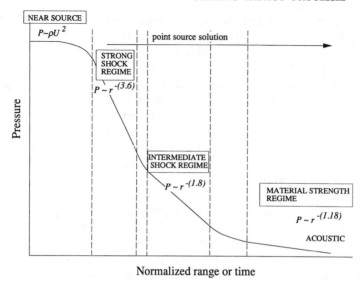

Figure 11   The regimes of the evolution of the peak pressure at the shock for the outgoing shock wave in an impact problem.

the finite curvature of the impactor is not yet a factor, so that the radius a can also be dropped. That leaves only the two variables ρ and U. A dimensional analysis then gives the result shown in the third column of Table 3: $P \propto \rho U^2$. This is indeed the result obtained by a simple one-dimensional "impedance-matching" solution for an impact. (The introduction of a specific Hugoniot result such as a linear shock-particle velocity relation will give the constant of proportionality of this relation again by using the impedance matching approach of shock mechanics.)

This initial shock will then move into the interior of the body and form a nearly hemispherical shock which decays in magnitude as it propagates. The peak pressure at the shock is then a function of position r (or of time t). When r is on the order of the impactor radius a, that radius must be retained in a dimensional analysis which gives (assuming the case when $U \gg c$) $P = \rho U^2 f(r/a)$, as in the fourth column of Table 3. Similar results are obtained for the particle velocity at the shock, for the time duration of the pulse, and its position.

After the shock moves a few impactor radii away from the impact point, the problem transitions to one governed by a point source, as measured by the coupling parameter. In general, the problem can still show three distinct regimes, depending on the magnitude of the pressure. These three possible regimes are called the "strong-shock," the "intermediate," and the "weak shock" regimes. There is often a fourth regime which is labeled as the "special intermediate" in Table 4.

The strong shock regime is defined as one for which the particle velocities are still much larger than the sound speed c. Equally, the pressure $P \gg \rho c^2$. However, there cannot be a separate dependence on the impactor velocity U and its radius a in this regime, since the point-source approximation governs.[12] The second column of Table 4 then shows the results of a dimensional analysis, $P = \rho U^2 (a/r)^{2/\mu}$, giving a power-law decay of shock front pressure with distance. The particle velocity at the shock decays with distance to the power of $1/\mu$.

As that decay progresses, the pressure will ultimately become comparable to the pressure scale ρc^2. Then the sound speed c again reenters the analysis, giving the less specific results shown in the "intermediate regime." However, usually the power-law result for the velocity holds to much greater ranges than that for the pressure, even when the shock velocity is comparable to the sound speed. In that case, the pressure at a shock is linear in the particle velocity, as is apparent from the jump conditions. Thus, since the velocity decays to the power of $1/\mu$, so does the pressure (see the "special intermediate" case in Table 4).

Finally, in the very far field, the pressure will decay to a level comparable to either a material strength Y, or perhaps to some initial pressure P_0, as in an air-blast analysis. Then the peak pressure also depends on this additional stress measure, as shown in the final column of Table 4.

It is important for the reader to remember that the scaling powers of these laws use exactly the same exponent μ as the size scaling. Thus, for nonporous material such as rock or water, the appropriate value is about $\mu = 0.55$. Recent papers by O'Keefe & Ahrens (1992a,b) have verified that fact by a suite of code calculations of the cratering dynamics for generic nonporous materials. (Once one accepts the governance of the point-source solution, then a single calculation at any suitable size and velocity suffices to define all of the point-source solution regime.) The reader can refer to these papers for specific numerical forms for those materials for the scaling laws given here.

4.4.2 DYNAMIC CRATER GROWTH   The scaling of any length scale is shown in the last row of the Table 3. It can be applied to the transient crater growth history (Holsapple 1984, Holsapple & Schmidt 1987). During the

---

[12] The literature is full of errors regarding this point. Some assume that the time scale and rise time of the far-field pulse is a/U, which in fact can only hold very near the impactor, as shown in the appropriate row of Table 2. Away from the impact point, every dependence on the impactor conditions must be in a combined form aU^μ.

initial regime the projectile buries itself in the planet at a constant velocity equal to the particle velocity behind the shock as determined by the jump conditions. Therefore, the projectile/planet interface surface moves at a constant velocity in this "penetration" regime. After a distance of just over one projectile radius the transition to the point-source solution is apparent, and the interface moves according to the power of $\mu/(1+\mu)$ in time, as in the first column of Table 4. That slope governs until just before the crater begins to slow its growth at the maximum transient depth. After maximum depth is achieved, there is a rebound of up to a factor of two to the final observed simple crater depth. Figure 12 (reproduced from Holsapple & Schmidt 1987) clearly demonstrates these regimes and, further, shows the commonalty of the dynamic growth in a wide variety of different materials when scaled according to these laws, and provides numerical values. O'Keefe & Ahrens (1991) clearly identify these same results for a single material in a suite of code calculations at various impactor sizes and

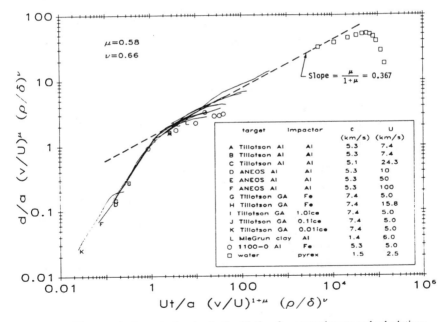

Figure 12 The growth of transient crater depth with time from experiments and calculations for a variety of conditions. In this scaled form all follow a common path, with an early time "penetration" regime and an intermediate time "point-source" regime. The growth is ultimately arrested by either a material strength or the gravity, depending on the conditions.

gravities. O'Keefe & Ahrens (1992b) give many numerical formulas for these specific forms for that material.[13]

5. CATASTROPHIC IMPACTS

The catastrophic disruption of asteroids and other solar system bodies is another inevitable consequence of energetic impacts, and another important application of impact scaling. Theories of collisional fragmentation are at present more speculative than for cratering. Most of our knowledge of such events is based on experimental results. Unfortunately, due again to practical constraints, the experiments are conducted at size and velocity scales that are vastly different from those appropriate to collisions involving asteroids or satellites. Therefore the results must be extrapolated using scaling rules.

Scaling rules that guide the extrapolation of small-scale results can be based on the same concepts as for cratering. The most common scaling method in the previous literature assumes that collisional outcomes (e.g. normalized fragment size and velocity distributions) are determined by the energy of the event divided by the mass of the target body, i.e. by the ratio $Q = E/M$. A specific value Q_* determines the threshold for catastrophic disruption, which is defined as when the largest remaining fragment is 1/2 the size of the original body. This threshold specific energy for target fragmentation is typically assumed to be independent of target size and impact velocity. That approach mirrors the older energy-scaling assumptions for cratering. In addition, it is implicitly based on an assumption that the asteroid strength is independent of size and rate. On the basis of a variety of experimental and theoretical evidence, it now appears that neither of these conditions should hold, thereby casting serious doubt on the validity of using Q as the sole parameter in scaling.

For cratering, the predominant strength measure is the yield function that limits shear stresses, as for example, a Mohr-Coulomb law. In contrast, the disruption of entire bodies is governed more by a tensile fracture strength measure. For geological materials such as rock and for other brittle materials, fracture strength is typically highly strain-rate dependent. In addition, for large bodies gravitational self-compression

---

[13] While they also give results for final crater size, it is difficult to predict final outcomes using numerical approaches. This author believes that the code approaches are better believed for study of the earlier regimes of impact processes, and for relative comparisons of different conditions; not for absolute measures of the final outcome. In particular, codes have extreme difficulty in calculating impact processes for any material with porosity.

deters fragmentation. Both factors must be accounted for in developing scaling theories for catastrophic disruptions.

Holsapple & Housen (1986) and Housen & Holsapple (1990) have considered this problem in detail. The interested reader is referred to these references for the complete analysis and discussion. The situation is similar to the scaling above for cratering. As long as the impactor is relatively small compared to the body being impacted, then the coupling parameter measure can be adopted, greatly simplifying the dependence on the impactor conditions. (While this becomes more of a problem than for cratering, catastrophic disruptions typically occur with the impactor to impacted body diameters in the ratio of about 1:10.)

Again there are two obvious regimes. For smaller bodies gravitational attractions between the parts of the body can be ignored. (Typically these are for asteroids small enough to be nonspherical.) Then one can assume a single measure for the strength of the body. However, as mentioned already, that measure should now be rate-dependent.

The fracture of brittle materials is a consequence of the growth and coalescence of an initial distribution of various sizes of small flaws and cracks. At any material point, all cracks with a length greater than some critical length—which depends on the instantaneous stress level—will be activated and growing. Fracture occurs whenever any crack lengths are able to grow to their inter-crack spacing, and that spacing will determine the dominant fragment sizes.

In a simple constant strain-rate test, the resulting fracture strength based on this model is in accord with the common strain-rate dependent model

$$\sigma_{cr} = S\dot{\varepsilon}^{1/n}, \tag{30}$$

where the exponent n is given by the original crack size distribution, and the coefficient S is a material constant. (The model also predicts fracture strength for more general stress histories, when the strength is not simply proportional to a power of the instantaneous strain rate.) Holsapple & Housen (1986) then used this material measure S for the resisting measure for disruption, together with the coupling parameter as the measure of the input, to determine a variety of scaling results. A further dependence on gravitational forces was allowed, and it dominates for the larger bodies.

One of the more fundamental results is shown as Figure 13, for the threshold specific energy dependence on target body size, as presented in Housen et al (1991). [This figure supersedes the earlier estimates given in Housen & Holsapple (1990); it is based on the theory and recent experiments.] In contrast to the energy-scaling, constant strength approach, the inclusion of a rate dependence for the fracture strength makes the threshold specific energy Q decrease for increasing target size in the strength regime,

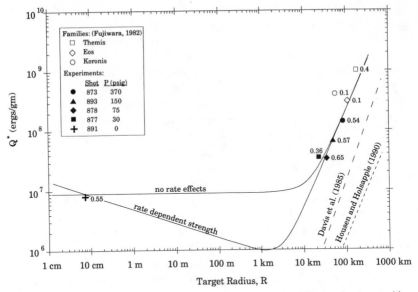

Figure 13 Estimates of the specific energy required to catastrophically break an asteroid. (From Housen et al 1991.)

due to the fact that the larger events have a correspondingly slower stress pulse and thus a weaker governing strength compared to small laboratory events. Then, for bodies more than about 2 km in diameter, the gravitational forces become the dominant factor, and the threshold energy required again increases.

6.   CONCLUDING REMARKS

Much has been learned about scaling over the past decade. Much of that progress is based on the recognition of the point-source approximation as the appropriate simplifying assumption that leads to the power-law results empirically observed; and, as a consequence, the earlier energy-scaling assumptions are replaced with the coupling parameter measure approach. However, there remains much to be learned. The theories presented here are only of first order—a "child's garden of scaling." The specific scaling laws are generally based on two ingredients: a choice of the measure of the "input" (i.e. the coupling parameter measure of the impactor), and a choice of a single measure of a "resistance" (e.g. a single strength or gravity). Many complicating factors exist which are not presently quantified, and will lead to future research.

More study is certainly needed about:

1. Scaling when the crater becomes so small compared to the impactor that the point-source assumption is not appropriate;
2. Atmospheric effects such as might be important on Venus and the Earth;
3. Facets of cratering such as the melt and vapor produced in the near field that are not governed by far-field scaling;
4. Layered and inhomogeneous targets, which must affect the immense transient depths of large craters, or for impacts into the ocean floor;
5. Very small craters where surface tension and viscosity of the material play roles;
6. Impacts on the icy satellites;
7. Subsequent creep mechanisms of materials including ice, and how they change the observed craters;
8. Highly oblique impacts;
9. Very low speed interactions of reaccumulation processes and planetary growth;
10. Dispersed impactors, such as asteroids or comets fractured during atmospheric penetration;
11. The gravity modification stage for complex craters and for the very large basins and ring structures;
12. Impacts of comparable sized bodies;
13. Details of escape mechanisms for near surface material.

ACKNOWLEDGMENTS

The research reported here has been supported by funding from the NASA Planetary Geology and Geophysics Program under grant NAGW-328. I thank my colleagues and critics K. R. Housen and R. M. Schmidt, who were co-investigators and coauthors for many of the results reported.

Literature Cited

Barenblatt, G. I. 1979. *Similarity, Self-similarity, and Intermediate Asymptotics.* New York: Consultants Bureau

Croft, S. K. 1985. The scaling of complex craters. *Proc. Lunar Sci. Conf. 15th. J. Geophys Res.* 90: C828–42 (Suppl.)

Dienes, J. K., Walsh, J. M. 1970. Theory of impact: some general principles and the method of Eulerian codes. In *High-Velocity Impact Phenomena*, ed. R. Kinslow. New York: Academic

Fujiwara, A. 1982. Complete fragmentation of the parent bodies of the Themis, Eos, and Koronis families. *Icarus* 52: 434–43

Gault, D. E. 1978. Experimental impact "craters" formed in water: gravity scaling realized. *Eos Trans. Am. Geophys. Union* 59: 1121 (Abstr.)

Gault, D. E., Wedekind, J. A. 1977. Experimental hypervelocity impact into quartz sand, II, Effects of gravitational acceleration. See Roddy et al 1977, pp. 1231–44

Grieve, R. A. F., Head, J. W. 1983. The Manicouagan impact structure: an analysis of its original dimensions and form.

Proc. Lunar Planet. Sci. Conf. 13th, J. Geophys. Res. 88: A807–18

Holsapple, K. A. 1980. The equivalent depth of burst for impact cratering. J. Geochem. Soc. Meteoritical Soc., Suppl. 14, pp. 2379–2401

Holsapple, K. A. 1981. Coupling parameters in cratering. Eos Trans. Am. Geophys. Union 62(45): 944 (Abstr.)

Holsapple, K. A. 1982. A Comparison of scaling laws for planetary impact cratering: experiments, calculations and theory. Lunar Planet. Sci. XIII, p. 331

Holsapple, K. A. 1983. On the existence and implications of coupling parameters in cratering mechanics. Lunar Planet. Sci. XIV, pp. 319–20

Holsapple, K. A. 1984. On crater dynamics: comparisons of results for different target and impactor conditions. Lunar Planet. Sci. XV, pp. 367–68

Holsapple, K. A. 1987. The scaling of impact phenomenon. Int. J. Impact Eng. 5: 343–55

Holsapple, K. A., Housen, K. R. 1986. Scaling laws for the catastrophic collisions of asteroids. Mem. Soc. Astron. Ital. 57(1): 65–85

Holsapple, K. A., Schmidt, R. M. 1979. A material strength model for apparent crater volume. J. Geochem. Soc. Meteoritical Soc., Suppl. 13, pp. 2757–77

Holsapple, K. A., Schmidt, R. M. 1982. On the scaling of crater dimensions 2. Impact processes. J. Geophy. Res. 87(B3): 1849–70

Holsapple, K. A., Schmidt, R. M. 1987. Point-source solutions and coupling parameters in cratering mechanics. J. Geophys. Res. 92(B7): 6350–76

Housen, K. R., Holsapple, K. A. 1990. On the fragmentation of asteroids and planetary satellites. Icarus 84: 226–53

Housen, K. R., Schmidt, R. M., Holsapple, K. A. 1991. Laboratory simulations of large scale fragmentation events. Icarus 94: 180–90

Housen, K. R., Schmidt, R. M., Holsapple, K. A. 1983. Crater ejection scaling laws: fundamental forms based on dimensional analysis. J. Geophys. Res. 88: (B3): 2485–99

Lampson, C. W. 1946. Explosions in earth. In Effects of Impact and Explosion V.1, Chap. 3, p. 110. Washington, DC: Off. Sci. Res. Dev.

Melosh, H. J. 1989. Impact Cratering, A Geological Process. New York: Oxford Univ. Press

Oberbeck, V. R. 1977. Application of high explosion cratering data to planetary problems. See Roddy et al 1977, pp. 45–66

O'Keefe, J. D., Ahrens, T. J. 1991. Tsunamis from giant impacts. Lunar Planet. Sci. XXII, pp. 997–98

O'Keefe, J. D., Ahrens, T. J. 1992a. Melting and shock weakening effects on impact crater morphology. Lunar Planet. Sci. XXIII, pp. 1017–18

O'Keefe, J. D., Ahrens, T. J. 1992b. Planetary cratering mechanics. Submitted

Pike, R. J. 1977. Size-dependence in the shape of fresh impact craters on the moon. See Roddy et al 1977, pp. 489–509

Pike, R. J. 1988. Geomorphology of impact craters on Mercury. In Mercury, ed. F. Vilas, C. R. Chapman, M. Shapley-Matthews, pp. 165–273. Tucson: Univ. Ariz. Press

Rae, W. J. 1970. Analytical studies of impact-generated shock propagation: survey and new results. In High-Velocity Impact Phenomena, ed. R. Kinslow. New York: Academic

Roddy, D. J., Schuster, S. H., Rosenblatt, M., Grant, L. B., Hassig, P. J., Kreyenhagen, K. N. 1987. Computer simulations of large asteroid impacts into oceanic and continental sites—preliminary results on atmospheric, cratering and ejecta dynamics. Int. J. Impact Eng. 5: 525–41

Roddy, D. J., Pepin, R. O., Merrill, R. B., eds. 1977. Impact and Explosive Cratering, New York: Pergamon

Schmidt, R. M. 1977. A centrifuge cratering experiment: development of a gravity-scaled yield parameter. See Roddy et al 1977, p. 1261–78

Schmidt, R. M. 1980. Meteor Crater: energy of formation—implications of centrifuge scaling. Proc. Lunar Sci. Conf. 11th, pp. 2099–2128

Schmidt, R. M., Holsapple, K. A. 1978a. A gravity-scaled energy parameter relating impact and explosive crater size. Eos Trans. Am. Geophys. Union 59(12): 1121 (Abstr.)

Schmidt, R. M., Holsapple, K. A. 1978b. Centrifuge cratering experiments in dry granular soils. DNA Rep. 4568F. Washington, DC: Defense Nucl. Agency

Schmidt, R. M., Holsapple, K. A. 1980. Theory and experiments on centrifuge cratering. J. Geophys. Res. 85: 235–52

Schmidt, R. M., Holsapple, K. A. 1982. Estimates of crater size for large-body impact: gravity-scaling results. Geol. Soc. Am. Spec. Pap. No. 190

Schmidt, R. M., Holsapple, K. A., Housen, K. R. 1986. Gravity effects in cratering. DNA Rep. DNA-TR–86–182. Washington, DC: Defense Nucl. Agency

Schmidt, R. M., Housen, K. R. 1987. Some recent advances in the scaling of impact and explosion cratering. *Int. J. Impact Eng.* 5: 543–60

Sedov, L. I. 1946. *Appl. Math. Mech. Leningrad* 10(2): 241

Shoemaker, E. M. 1962. Interpretation of lunar craters. In *Physics and Astronomy of the Moon*, ed. Z. Kopal, pp. 283–359. New York: Academic

Shoemaker, E. M. 1963.. Impact mechanics at Meteor Crater, Arizona. In *The Moon, Meteorites and Comets*, ed. B. M. Middlehurst, G. P. Kuiper, pp. 301–6, Chicago:

Univ. Chicago Press

Taylor G. I. 1950. The formation of a blast wave by a very intense explosion: II. The atomic explosion of 1945. *Proc. R. Soc. London Ser. A* 201: 175–86

Worthington, A. M. 1908. *A Study of Splashes*. London: Longmans & Green

Worthington, A. M., Cole, R. S. 1897. Impact with a liquid surface, studied by the aid of instantaneous photography. *Philos. Trans. R. Soc. London Ser. A* 193: 137–48

Annu. Rev. Earth Planet. Sci. 1993. 21:375–406

PROGRESS IN THE EXPERIMENTAL STUDY OF SEISMIC WAVE ATTENUATION

Ian Jackson

Research School of Earth Sciences, Australian National University, GPO Box 4, Canberra ACT 2601, Australia

KEY WORDS: rock anelasticity, internal friction, ultramafic rocks, creep, shear modulus

INTRODUCTION

Seismological Motivation

New impetus for the laboratory study of seismic wave attenuation was provided during the 1960s and 1970s by the publication of models for the variation of attenuation with depth in the Earth's mantle (e.g. Anderson et al 1965, Anderson & Hart 1978, Sailor & Dziewonski 1978). It was clearly demonstrated that attenuation is concentrated in shear, and that zones of relatively high attenuation and low shear wave speed are features of both the upper and the lowermost mantle. The character and lateral variability of the upper mantle zone of high attenuation and low velocity have been explored in numerous studies, particularly of the dispersion of surface waves. Such studies indicate that this zone is most highly developed beneath young oceanic crust and other regions of recent tectonic activity, where the shear wave speed V_S may be reduced by as much as 10% relative to the overlying high-velocity "lid" (e.g. Forsyth 1975) and the internal friction Q_G^{-1} for shear waves may exceed 0.1 (Chan et al 1989). Observations of the attenuation of body waves, surface waves, and free oscillations, of the response of the Earth to lunar tides, and of perturbations to the Earth's rotation (the Chandler wobble) provide constraints on the frequency or period dependence of Q_G^{-1} for oscillation periods T_o between 10^{-2} and 10^8 s. Such analyses suggest that Q_G^{-1} varies approximately as

375

0084–6597/93/0515–0375$02.00

T_o^z with α in the range 0.15–0.4 (Anderson & Minster 1979, Molodenskiy & Zharkov 1982).

These intriguing observations have defied definitive interpretation because of the lack of directly relevant data for representative mantle materials under controlled laboratory conditions that approach those of seismic wave propagation. The production of basaltic magma by partial melting of peridotite beneath mid-ocean ridges suggested that the observed low velocities and high attenuation in the upper mantle might be attributed to the presence of a small fraction of melt (e.g. Anderson & Sammis 1970), although it has since become widely recognized that high subsolidus conditions might suffice (Goetze 1977a, Shankland et al 1981, Minster & Anderson 1981). The temptation to associate these anomalous seismological features of the upper mantle with low resistance to shear on the much longer timescales of glacial advance and retreat, seamount loading, and plate-scale mantle convection is superficially attractive and has proved irresistible, but is similarly without secure foundation (e.g. Jackson 1991).

The Experimental Challenge

Interpretation of these seismological observations in terms of temperature, grainsize, dislocation density, and the presence or absence of melt obviously requires the conduct of laboratory experiments on appropriate materials (especially ultramafic rocks) under controlled conditions. In this context Goetze (1977b) noted that "the direct laboratory measurement of either sonic velocity or attenuation under conditions . . . pertinent to a discussion of the low velocity zone has not so far been accomplished for any silicates." He went on to explain why there was such a dearth of relevant experimental observations—pointing out that such experiments need to be conducted at low strain amplitudes in order to sample the regime of linear, amplitude-independent mechanical behavior, and at seismic frequencies which are many orders of magnitude lower than those of the widely used ultrasonic wave propagation methods. Furthermore, it was noted that the simultaneous application of pressures of a few hundred MPa (a few kilobars) and temperatures above about 1000°C is needed in order to suppress thermal microcracking, and to allow the exploration of the impact of the mobility of thermally activated crystal defects including point defects and dislocations, and of melt. During the past two decades, considerable progress has been made in the development of such experimental capability.

Scope of this Review

This review will focus on these developments and on the insights beginning to emerge into the anelastic behavior of ultramafic rocks at high subsolidus temperatures and beyond. For further information on the phenomenology of anelasticity, laboratory techniques, and a survey of experimental results mainly for metals, the reader is referred to Nowick & Berry (1972). The application of mainly relatively high-frequency resonance methods in the study of the role of volatiles in the anelasticity of sedimentary rocks is covered by Vassiliou et al (1984). Jackson (1986) and Peselnick & Liu (1987) reviewed low frequency experimental methodologies for the study of rock anelasticity with particular emphasis in the former study on the role of pressure at room temperature. The defect microdynamics ultimately responsible for solid-state anelastic relaxation in mantle materials are the subject of a recent review by Karato & Spetzler (1990).

THE IMPORTANCE OF LINEARITY

Theory: The Creep Function, the Superposition Principle, and the Dynamic Compliance

In the absence of directly relevant experimental data, Goetze (1971, 1977b) sought to construct bounds on the expected seismic wave attenuation, using information obtained in compressive creep tests at much smaller strains than those typical of steady-state experimental rock deformation. In a creep test, the response is measured to the sudden application of a steady stress. To be more precise, the *creep function* $J(t)$ is defined (e.g. Nowick & Berry 1972) as the strain resulting from the application of a stress given by the Heaviside function, namely

$$\sigma(t) = H(t) = 0, \text{ for } t < 0 \tag{1}$$
$$= 1, \text{ for } t \geq 0.$$

In the context of wave propagation, however, it is the response to a sinusoidally time-varying stress $\sigma(t) = \sigma_0 \exp(i\omega t)$ which determines both the phase speed and the attenuation. If the relationship between stress and strain and their respective time derivatives is linear, then the resulting strain will also vary sinusoidally with time. Only for the special case of genuinely elastic behavior are these stress and strain sinusoids in phase. More generally, i.e. for anelastic or linearly viscoelastic rheology, the strain lags behind the applied stress resulting in the dissipation of strain energy. The *dynamic compliance* $J^*(\omega)$ is defined as the ratio of the instantaneous strain $\varepsilon(t) = \varepsilon_0 \exp i(\omega t - \delta)$ to the instantaneous stress, i.e.

$$J^*(\omega) = \varepsilon(t)/\sigma(t) = J_1(\omega) - iJ_2(\omega) = |J| \exp(-i\delta). \tag{2}$$

The real and negative imaginary parts of the complex dynamic compliance $J^*(\omega)$ may be combined to yield various more commonly employed measures of the departure from perfect elasticity. Thus the *quality factor* Q, defined by analogy with dissipation in electric circuits, is given in terms of the ratio of the strain energy dissipated per cycle to the maximum energy stored, and is related to the *internal friction* $\tan \delta$, by

$$Q^{-1} = J_2/J_1 = \tan \delta. \tag{3}$$

Expressions for the phase speed V and attenuation coefficient α for a plane traveling wave (e.g. Nowick & Berry 1972, Jackson 1986) may be obtained through the substitution of the appropriate expression for the complex propagation constant,

$$k^*(\omega) = \omega[\rho J^*(\omega)]^{1/2}, \tag{4}$$

into the equation for the particle displacement associated with the traveling wave:

$$u(x, t) = u_0 \exp i[\omega t - k^*(\omega)x] = u_0 \exp(-\alpha x) \exp i\omega(t - x/V) \tag{5}$$

with

$$\alpha(\omega) \approx \pi Q^{-1}/\lambda = \omega Q^{-1}/2V$$

and

$$V(\omega) \approx [\rho J_1(\omega)]^{-1/2},$$

with the approximations applying for $Q^{-1} \ll 1$.

The connection between the creep function $J(t)$ and the dynamic compliance $J^*(\omega)$ is readily established as follows. Any prior history of stress application $\sigma(t')$ may be approximated arbitrarily well by a series of infinitesimal increments of magnitude $\dot{\sigma}(t')dt'$ applied suddenly during the time interval $(t', t' + dt')$. By definition of the creep function, the corresponding contribution to the time-dependent strain is

$$d\varepsilon(t) = \dot{\sigma}(t')J(t - t')dt'. \tag{6}$$

Provided again that the stress-strain relationship is linear, the total response at time t may be calculated as the sum of the responses to the individual increments of stress application—a result known as the Boltzmann superposition principle. Thus,

$$\varepsilon(t) = \int_{-\infty}^{t} \dot{\sigma}(t')J(t - t')dt'. \tag{7}$$

For the special case of most interest where stress varies sinusoidally with time, we obtain after changing the variable to $\xi = t - t'$

$$\varepsilon(t) = i\omega\sigma(t) \int_0^\infty J(\xi) \exp(-i\omega\xi) \, d\xi, \tag{8}$$

yielding for the dynamic compliance

$$J^*(\omega) = \varepsilon(t)/\sigma(t) = i\omega \int_0^\infty J(\xi) \exp(-i\omega\xi) \, d\xi, \tag{9}$$

where the integral is essentially the Fourier transform of the creep function.

Finally, it is important to note in this brief survey of linear theory that the frequency dependences of the real and imaginary parts of the complex compliance, or of its reciprocal—the modulus [e.g. $G^*(\omega) = 1/J^*(\omega)$], are linked through the Kronig-Kramers relationships (e.g. Brennan & Smylie 1981). For the situation commonly encountered both seismologically and experimentally where the internal friction $Q^{-1} \ll 1$ and varies with oscillation period $T_o = 2\pi/\omega$ as

$$Q^{-1}(T_o) = Q_0^{-1} T_o^\alpha, \tag{10}$$

this relation adopts the form

$$G(T_o)/G(T_o^{\text{ref}}) = 1 - \tan[\pi(1-\alpha)/2][Q^{-1}(T_o) - Q^{-1}(T_o^{\text{ref}})]. \tag{11}$$

Access to the Linear Regime

In an attempt to constrain seismic wave attenuation in rocks at high subsolidus temperatures through the theory outlined above, Goetze (1971; Goetze & Brace 1972) conducted compressive creep tests at 600 MPa confining pressure, temperatures of 500–1000°C, and relatively low deviatoric stresses for a variety of rock types. Inelastic strains of 10^{-4} to 10^{-3} developed over periods of ~ 3000 s in response to deviatoric stresses of 60 to 90 MPa. The inelastic strain, only part of which proved to be recoverable following removal of the applied stress, varied with time t, stress σ, and temperature T, approximately as

$$\varepsilon(t) = A\sigma^n t^m \exp(-mE^*/RT). \tag{12}$$

The exponents n and m varied with rock type within the ranges 1.7–2.0 and 0.35–0.49, respectively. Unfortunately, the nonlinear rheology severely complicates the estimation of $J^*(\omega)$ since the superposition principle is not strictly applicable. Accordingly, the only reasonably robust conclusion to emerge from this analysis was the suggestion (Goetze 1977b) that large values of Q^{-1} might be anticipated even under subsolidus conditions. The promulgation of this finding (Goetze 1977a) provided much of the motivation for the further experimental work which was to follow.

By the late 1970s the viability of low-frequency experimentation within the realm of linear, amplitude-independent rheology had been convincingly demonstrated, especially through the work of Brennan & Stacey (1977; Brennan 1981). These authors reported quantitative observations based on capacitance micrometry of the forced torsional oscillation of hollow rock cylinders at strain amplitudes in the range 10^{-8}–10^{-6}, under ambient laboratory conditions. The measured shear modulus and internal friction were shown to be independent of strain amplitude. Moreover, the dispersion (frequency dependence) of the modulus was shown to be consistent with the measured internal friction through the Kronig-Kramers relations of linear theory. In compressive creep tests performed on peridotite at high temperatures (to 1300°C) and atmospheric pressure, Berckhemer et al (1979) observed no departure from linearity for inductively measured strains below 6×10^{-5}.

In the wake of these developments, rock anelasticity/viscoelasticity was being studied during the 1980s at low frequencies and within the linear regime through the observation of forced torsional oscillations both at high temperature (Berckhemer et al 1982, Gueguen et al 1989, Getting et al 1991) and also at high pressure (Jackson et al 1984, Jackson & Paterson 1987). Very recently, the capacity to perform such measurements under the *simultaneous* application of high pressure and temperature has been demonstrated (Jackson et al 1992)—finally realizing Goetze's dream.

QUASI-STATIC METHODS: CREEP TESTS

Strengths and Weaknesses

As we have seen above, the dynamic compliance $J^*(\omega)$, which describes the dispersion and attenuation of traveling waves, is calculable in principle from the creep function $J(t)$, provided that the rheology is linear. The creep test is therefore at least superficially very attractive, since the frequency dependence of J^* is ostensibly derivable from a single creep test spanning an appropriate time interval. Alternatively, development of a similar insight would require the conduct of a number of forced oscillation tests at different oscillation periods. More importantly, the creep test in which strain is monitored both during the period of steady stress application, *and also after its removal*, provides for the identification of the recoverable (anelastic) and permanent (viscous) contributions to the total inelastic strain. Such information is not normally available from forced oscillation experiments because of the symmetrical history of positive and negative stress application.

In practice, there are some disadvantages to the extraction of $J^*(\omega)$ from creep records. Not the least of these is the common presence of significant background drift which will introduce bias into the apparent viscosity at long times, leading to corresponding bias (of opposite sign) in the inferred values of Q^{-1}. There are also complications associated with the estimation of the infinite integral in Equation (9) from a discretely sampled record of finite length (Jackson 1993).

Experimental Procedure

Since many of the examples of torsional mode creep tests and forced sinusoidal oscillation experiments used to illustrate this review are drawn from the work of Jackson et al (1992), a brief description of their apparatus is appropriate (Figure 1). An elastic standard, machined from steel of known elastic modulus and negligible internal friction, and a compound specimen assembly described below, are connected mechanically in series and are therefore exposed to the same torque. The twist induced by the applied torque is measured at stations above and below the elastic standard by pairs of sensitive capacitance displacement transducers—yielding displacement signals $d_1(t)$ and $d_2(t)$, respectively. The displacement associated with the twist of the elastic standard is given by the difference $d_2(t) - d_1(t)$, which provides a measure of the amplitude, and for the forced oscillation experiments of the phase, of the applied torque. The signal $d_1(t)$ measures the response of the specimen assembly (Figure 2) which usually consists of a cylindrical rock specimen of 15 mm diameter and 150 mm length mounted within the hot zone of the internal furnace between two torsion rods. These are of 99.8% alumina ceramic in the regions of strong temperature gradient, and elsewhere of steel. The specimen and ceramic parts of the torsion rods are housed within a thin-walled iron sleeve (jacket) which is sealed with O-rings against the steel parts of the torsion rods at either end. In order to provide for the elimination of the contribution to the overall compliance from that part of the assembly located outside of the furnace hot zone (see section entitled Calibration of Forced Oscillation Apparatus), tests are also performed on a "reference" assembly in which the rock specimen is replaced by a dimensionally equivalent cylinder of the same high-grade polycrystalline alumina used to span the regions of strong temperature gradient (Figure 2).

This reference assembly has been employed recently to conduct some exploratory torsional creep tests as follows. Creep tests of total duration $5T$ are conducted such that within five successive time intervals T, the torque assumes steady values of 0, $+L$, 0, $-L$, and 0, respectively. The displacements measured for the reference assembly of 99.8% purity polycrystalline alumina (labeled "specimen") and for the elastic standard in such a test conducted at 1200°C and 300 MPa are shown in Figure 3. The time-dependent distortion of the specimen evident in the second and fourth segments of this record is clear evidence of inelastic behavior. Partial

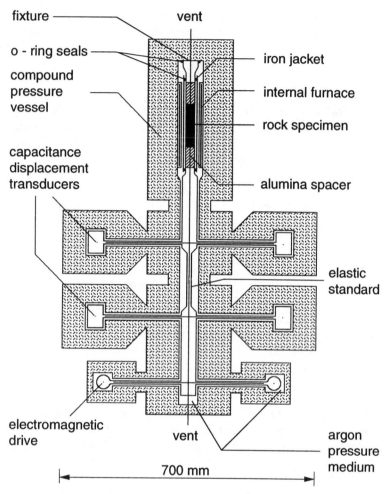

fixture

vent

o - ring seals

iron jacket

compound
pressure
vessel

internal furnace

rock specimen

capacitance
displacement
transducers

alumina spacer

elastic
standard

electromagnetic
drive

vent

argon
pressure
medium

700 mm

Figure 1 A schematic cross-section of the apparatus developed for the study of anelasticity under conditions of high pressure and temperature through the observation of the forced torsional oscillation of cylindrical rock specimens. [Reproduced from Jackson et al (1992) with permission of the Royal Astronomical Society.]

recovery of this deformation following removal of the torque is evident in the third and fifth segments of the record, thereby allowing quantification of the relative magnitudes of the anelastic and viscous contributions. Finally, following the removal of the negative torque, there is a close approach to the original condition—as expected even for viscoelastic material subject to a symmetrical history of stress application.

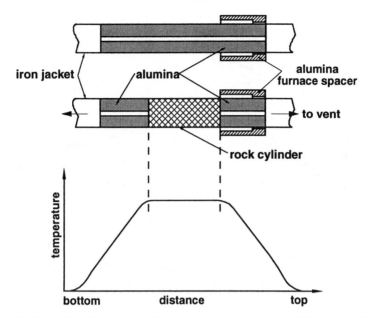

Figure 2 Illustration of the rock-bearing and alumina reference assemblies employed in the comparative experiments described in the text. The establishment of a hot zone of appropriate length requires that a close-fitting cylindrical alumina sleeve be inserted between the specimen assembly and the core of the wire-wound furnace in order to prevent excessive convective heat loss from the top of the furnace.

Analysis of Creep Records

If the behavior is linear and described by a torsional creep function $J(\xi)$, then the response to the history of torque application recorded in Figure 3 is calculable through superposition of the responses to the application of steady torques $+L$, $-L$, $-L$, and $+L$ at $\xi = T$, $2T$, $3T$, and $4T$ respectively. Thus the instantaneous value of the angular twist of the specimen $\phi(t)$ is

$$\phi(t) = L[J(t-T) - J(t-2T) - J(t-3T) + J(t-4T)]. \tag{13}$$

It is simpler in the first instance, however, to focus attention on the second segment for which there is no history of prior nonzero stress application. Since the specimen assembly and elastic standard are subject to the same torque, and the latter is of known torsional compliance, it is straightforward to convert the measured displacement $d_1(t)$ associated with the twist of the former (dominated by that part of the jacketed

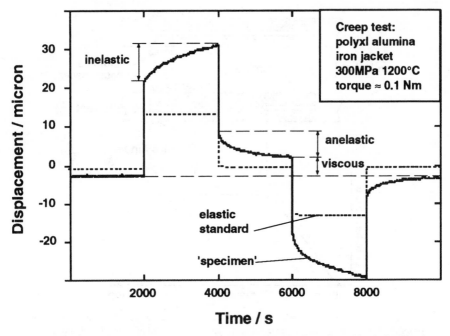

Figure 3 Displacement versus time records representing the distortion of both the "specimen" and the elastic standard resulting from a creep test in which the torque assumed values of 0, +L, 0, −L, and 0, during successive 2000 s time intervals. The label "specimen" denotes the reference assembly comprised of polycrystalline alumina within a thin-walled iron jacket in series with steel torsion rods (see Figure 1). The peak strain at the periphery of the alumina ceramic in the furnace hot zone is less than 5×10^{-6}.

alumina specimen within the hot zone of the furnace) into an instantaneous torsional compliance, i.e. the angular twist per unit torque (Figure 4).

Jackson (1993) has recently demonstrated that this time-dependent torsional compliance may be adequately represented by the Andrade model commonly used to fit transient creep data (e.g. Goetze 1971, Poirier 1985) whereby

$$J_A(t) = J_u + \beta t^m + t/\eta, \tag{14}$$

with exponent m usually in the range 1/3–1/2. For given m, the values of the coefficients J_u, β, and $1/\eta$ for the Andrade model that best fits the observed compliance may be determined by the method of (linear) least squares. The misfit is plotted for two representative models ($m = 1/3$ and

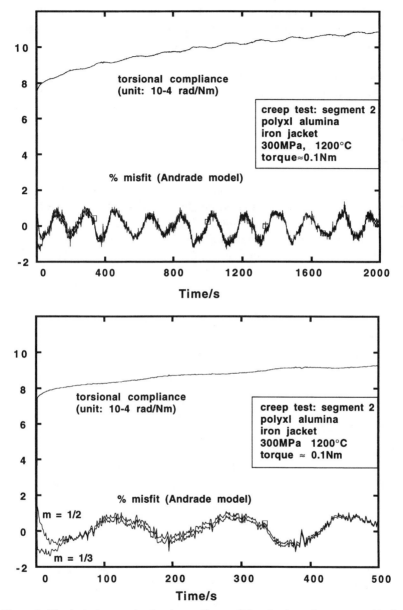

Figure 4 The instantaneous torsional compliance of the alumina reference assembly for segment #2 of the creep test of Figure 3 along with the misfit for two representative Andrade models—with exponents $m = \frac{1}{3}$ and $m = \frac{1}{2}$. The enlargement (*lower panel*) shows more clearly the misfit for $t < 50$ s. The fluctuation in compliance of about 200 s period is thought to result from small oscillations of temperature about the control point.

$m = 1/2)$ in Figure 4 along with the observed compliance. The RMS misfit is almost entirely attributable to the ~ 200 s oscillation associated with temperature fluctuations caused by imperfect furnace temperature control.

Despite some reservations concerning its lack of physical transparency, the Andrade model provides a useful and parametrically economical interpolation formula which is here demonstrably superior (Jackson 1993) to the simple Burgers model with a single anelastic relaxation time τ such that

$$J_B(\xi) = J_u + \delta J[1 - \exp(-\xi/\tau)] + \xi/\eta. \tag{15}$$

The dynamic compliance $J^*(\omega)$ may be calculated from the Andrade fit to the creep record through a hybrid numerical/analytic approach presented and tested by Jackson (1993). For this purpose, the creep function is described by the Andrade model $J_A(\xi)$ fitted to the experimental observations obtained on the interval $(0, T)$. Extrapolation beyond $\xi = T$ is based upon the close approach to linearly viscous behavior at sufficiently long times such that

$$J(\xi) = J_A(T) + (\xi - T)/\eta_{eff}, \tag{16}$$

with η_{eff} chosen so that both J and $dJ/d\xi$ are continuous at $\xi = T$. Under these circumstances, the infinite integral in Equation (9) may be evaluated numerically on the interval $(0, T)$ in terms of the discrete Fourier transform (DFT) of $J_A(\xi)$, and analytically on the interval (T, ∞), yielding as an estimate of the dynamic compliance:

$$J^*(\omega_k) = (i2\pi k/\sqrt{N}) \, \mathrm{DFT}(J_A) + J_A(T) - i/\eta_{eff}\omega_k,$$

where

$$\mathrm{DFT}(J) = (1/\sqrt{N}) \sum_{j=0}^{N-1} J(\xi_j) \exp(-i2\pi jk)$$

and

$$\omega_k = 2\pi k/T. \tag{17}$$

The Andrade model J_A fitted to the creep record of length T is sampled at N equally spaced times $\xi_j = jT/N$. Comparison of fully analytic results for the Burgers model (Equation 15) with those obtained using this more versatile hybrid numerical/analytic method (Equation 17) indicates that the sampling frequency $f_S = N/T$ must be carefully chosen if $Q^{-1} = J_2(\omega)/J_1(\omega)$ (Equation 3) is to be reliably calculated (Jackson 1992). For example, for $T = 2000$ s, f_S should exceed 5 Hz if the analytic results for the internal friction are to be reproduced within 10% for oscillation periods $T_o = 2\pi/\omega$ greater than 100 s. At shorter oscillation periods and lower sampling frequencies, the DFT sum is a poor approximation to the integral, and aliasing effects contaminate the result (Jackson 1993). In Figure 5 the internal friction estimated in this way from the Andrade fits to the creep record of Figure 4 is compared with the apparent internal friction derived from complementary studies of forced sinusoidal oscillations. Although there is some sensitivity to the value of the exponent m, a close consistency is evident between the results obtained from the forced oscillations and the creep tests.

This demonstration of the consistency at long periods between Q^{-1} measured directly in forced oscillation experiments and computed from creep records underscores the essential complementarity of the two methods within the realm of linear behavior. The latter has the major advantage of allowing the distinction to be drawn between anelastic and viscous strains, but is no substitute in practice for forced oscillation experiments at relatively short periods.

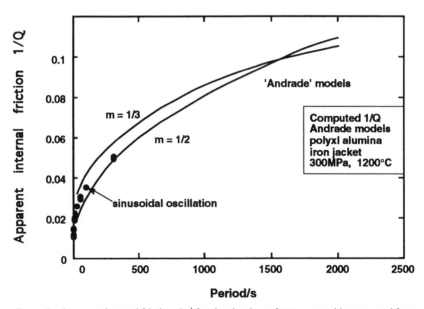

Figure 5   Apparent internal friction Q^{-1} for the alumina reference assembly computed from the $m = \frac{1}{3}$ and $m = \frac{1}{2}$ Andrade models for the creep function, compared with the results of the forced oscillation studies. Consistency between the computed and observed values is a consequence of linearity of the stress-strain relationship.

DYNAMIC METHODS

Ultrasonic Wave Propagation

The amplitude of the particle displacement associated with a plane wave propagating through a linearly inelastic medium decays with distance according to Equation (5), i.e.

$$u(x) = u_0 \exp(-\alpha x) \approx u_0(\omega) \exp(-\omega Q^{-1} x/2V). \tag{18}$$

If that part of the frequency dependence of $u(x)$ attributable to wave propagation through the inelastic material can be isolated through the construction of an appropriate ratio, essentially $u(x)/u_0$, then

$$\partial \ln[u(x)/u_0(\omega)]/\partial \omega = -Q^{-1} x/2V \tag{19}$$

which allows Q^{-1} to be estimated from the spectrum of amplitude ratios.

This method has been employed in pioneering experiments by Sato et al (1989) on fine-grained ultramafic rocks at high subsolidus and trans-solidus temperatures and confining pressures to 700 MPa. Within the precision of the measurement, the compressional mode Q^{-1} is independent of frequency within the range ~ 100–900 kHz, and is a function only of homologous temperature T/T_M, where T_M is the solidus temperature at the appropriate pressure.

The principal disadvantage of this method is the interval of five orders of magnitude between the ultrasonic frequencies employed in laboratory wave propagation experiments and those of seismic waves. Nevertheless, Sato et al have argued that there exists a simple grainsize-frequency scaling law (based on the assumption of a grain boundary relaxation mechanism) which allows their results to be applied directly in the interpretation of seismological data. However, in the absence of compelling support for the grain boundary model, such as demonstration of the expected variation of Q^{-1} with grainsize, this extrapolation is very hazardous.

Resonance Methods: Free Oscillations

Resonance methods have been particularly widely used in the study (especially at relatively low temperatures) of anelasticity in nongeologic materials such as metals and ceramics. The amplitude and phase of $J^*(\omega)$ are determined, respectively, from the resonant frequency, and from either the width of the resonance peak or from the rate of decay of free oscillations (e.g. Nowick & Berry 1972, Jackson 1986). In the absence of external inertia, fundamental resonance frequencies for laboratory specimens of convenient size are of order 1–20 kHz. The application of such bar resonance methods has yielded some results of considerable geophysical significance relating particularly to the influence of confining pressure, volatiles, and thermal cycling upon internal friction in rocks (see reviews by Vassiliou et al 1984 and Jackson 1986). "Pendulums," in which the anelastic restoring force works against substantial external inertia, provide access to lower resonance frequencies, typically 1–100 Hz (Woirgard & Gueguen 1978).

The application of resonance methods, in which the frequency is not readily varied through a wide range, has often involved the variation of temperature at fixed frequency as a means of exploring the mechanical relaxation spectrum, i.e. $J^*(\omega\tau)$. For a thermally activated relaxation mechanism, τ is the time required for diffusion of the appropriate species across some characteristic distance l, and therefore varies with temperature as

$$\tau = l^2/D = \tau_0 \exp(E^*/RT), \tag{20}$$

where D is the relevant diffusivity. The disadvantage of this approach is that microstructural changes consequent upon changing temperature, especially at high subsolidus temperatures, are likely to change the mechanical relaxation spectrum while it is being explored (Woirgard et al 1980, Gueguen et al 1981).

Subresonance Methods: Forced Oscillations

There are several major advantages to the study of subresonant forced oscillations. The dynamic compliance $J^*(\omega) = \varepsilon(t)/\sigma(t)$ is most directly determined by measurement of the relative amplitudes and phase of a sinusoidally fluctuating stress and the resultant strain. Moreover, the mechanical relaxation spectrum can be explored noninvasively by varying the frequency ω at fixed T and τ. Only relatively recently, however, has it become possible to measure with sufficient precision the generally small phase angle δ between stress and strain, at the low strain amplitudes of the linear regime. Torsional geometry has proved especially practicable because of the opportunity it provides for mechanical advantage associated with off-axis location of sensitive displacement transducers employing inductive, capacitive, or optical interferometric measurement strategies (Woirgard et al 1977, Brennan & Stacey 1977, Brennan 1981, Berckhemer et al 1982, Jackson et al 1984, Getting et al 1990).

Ideally such studies of forced torsional oscillation yield two digitally sampled displacement versus time signals which are, respectively, the responses of an elastic standard and of the "unknown" specimen to the same sinusoidal torque—acquired after an empirically determined delay designed to achieve steady state (Jackson & Paterson 1987). A typical example of such a pair of records from an experiment on a polycrystalline alumina reference assembly at 300 MPa, 1200°C, and a maximum strain amplitude of $\sim 2 \times 10^{-6}$ is shown in Figure 6. Such records, although essentially sinusoids, involve low-level harmonic distortion due to slight nonlinearity of the conversion of electric current to torque in the electromagnetic drivers and noise of both mechanical and electrical origin. (The latter comprises harmonics of the 50 Hz mains frequency attributable to the phase angle firing strategy employed in furnace power control.) Fourier analysis of the records provides robust estimates of the relative amplitudes and phase—the latter consistently within $\sim 5\%$ for phase lags of 100–3000 microcycles (1–20 milliradians) which are typical of high temperature conditions. Figure 7 shows the variation of amplitude with integer frequency for the Fourier transform of the displacement-time record $d_1(t)$ of Figure 6.

CALIBRATION OF A FORCED OSCILLATION APPARATUS: POTENTIAL EXTRANEOUS SOURCES OF APPARENT ANELASTICITY

The next step in establishing the integrity of such a forced oscillation experiment is the documentation of any contamination of the relative

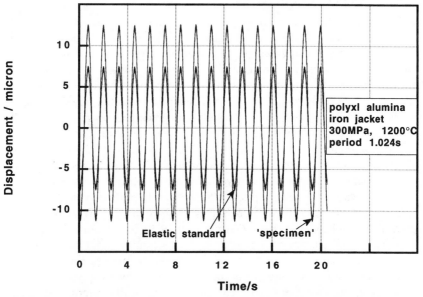

Figure 6 The displacements associated with the distortion of the elastic standard and the alumina reference assembly (labeled "specimen") for a forced sinusoidal oscillation experiment at 1.024 s period. The displacement versus time record is sampled at 128 samples per oscillation period for exactly 16 successive periods.

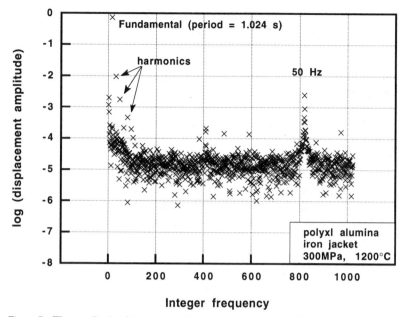

Figure 7 The amplitude of the discrete Fourier transform of the displacement versus time record for the alumina reference assembly of Figure 8, showing a signal-to-noise ratio approaching 10^4 at the fundamental forcing frequency. Harmonics resulting from non-linearity in the conversion of electric current into torque in the electromagnetic drivers are also evident, as is interference at the mains frequency 50 Hz and its harmonics consequent upon the switching of large furnace currents.

amplitudes and phase of the stress and strain sinusoids arising from influences extraneous to the mechanical behavior of the specimen under test. Potential sources of difficulty include inelastic behavior of the load-bearing members (e.g. torsion rods) through which the stress is transmitted to the specimen, imperfect mechanical coupling between these components, and any mechanical interaction between the specimen assembly and its immediate environment (e.g. gas pressure medium, close-fitting furnace components).

Interfaces

It has been demonstrated, for example, that the behavior of interfaces between load-bearing pistons and the specimen is the major source of systematic error in the measurement of the relative phase of the stress and strain sinusoids in extensional mode forced oscillation experiments conducted at ambient pressure and temperature (Liu & Peselnick 1983). Preparation of slightly concave polished surfaces was found to reduce the error in relative phase measurement to 5×10^{-4} radians, equivalent to an apparent Q^{-1} of the same magnitude. In apparatus already described (Figure 1) for high pressure study of shear mode anelasticity through observation of forced torsional oscillations, a normal stress equal to the confining pressure acts across all such interfaces. A "null experiment" was conducted in which the usual rock specimen and ceramic torsion rods (Figure 2) were replaced by a cylinder of the same low Q^{-1} steel used throughout the remainder of the specimen assembly (Jackson et al 1984). At pressures > 50 MPa, the ratio of the amplitudes of the two displacement versus time sinusoids, determined from the Fourier-transformed data, becomes pressure-independent with a value in excellent agreement with the amplitude ratio (3.14) calculated from the geometry of the specimen assembly (Figure 8; Jackson & Paterson 1987). Furthermore, the phase angle between the two sinusoids closely approaches zero (as expected in the absence of extraneous effects) for oscillation periods greater than 10 s.

Interaction with the Pressure Medium

The phase leads, and indeed the slight perturbation to the displacement amplitude ratio evident at shorter periods (Figure 8), are now quantitatively understood in terms of the interaction between the torsional assembly and the argon pressure medium. Flow of the pressure medium between the closely spaced steel plates of the capacitance displacement transducers in response to the imposed oscillatory motion exerts torques on the specimen assembly at each of the displacement measurement stations. These torques and the resulting perturbations to the displacement amplitudes and relative phase are calculable from the known geometry and physical properties of argon (Jackson & Paterson 1987). Correction for these effects yields a near zero phase angle throughout the 1–100 s range of oscillation periods corresponding to an apparent Q^{-1} in the null experiment of only 1.2×10^{-4} (Figure 8; Jackson & Paterson 1987). The effective radius R of the transducer plate and its encircling earthed guard ring is the only adjustable parameter in this analysis. Subsequent experience with the specimen assembly reconfigured for high temperature operation yielded a smaller value for the effective plate radius (Jackson et al 1992). It was concluded that the correction calculated in this way includes a component related to slight variation in the phase of the stress along the length of the specimen assembly in forced oscillation at short periods (see also Kampfmann & Berckhemer 1985).

Torsion Rods, Jackets, and Furnace Components

For forced oscillation experiments at high temperature, the establishment of a low-loss background against which the inelastic behavior of the rock specimen can be measured, is less straightforward. Electromagnetic drivers which generate the applied stress, the elastic standard, and the displacement transducers must be maintained at room temperature, and are therefore inevitably connected to the specimen located within the furnace hot zone by load-bearing members, commonly high-grade alumina ceramics, which span regions of strong temperature gradient (e.g. Figures 1 and 2).

In torsional oscillation experiments performed under conditions of *both* high temperature and high pressure, the need to enclose the specimen and associated ceramic and steel torsion rods within a thin-walled metal jacket, in order to exclude the pressure medium from the pore space of the specimen, creates extra difficulties. Iron, the most suitable material for this purpose, undergoes a ferromagnetic → paramagnetic transition at 770°C and a phase transformation from the body-centered to the face-centered cubic structure near 910°C. Relatively high values of Q^{-1} are associated with each of these transitions (Wolfenden 1978, Garber & Kharitonova 1968). Moreover, both fcc iron and any binder phase distributed along alumina grain boundaries within the ceramic components will experience relatively high homologous temperatures T/T_M during experimentation at $T \geq 1000°C$, and inelastic behavior is to be anticipated. Finally, the successful operation of internal furnaces within a gas-medium high-pressure apparatus (Paterson 1970) requires the insertion of a close-fitting cylindrical sleeve between the jacketed specimen and the furnace core to prevent excessive convective heat loss from the top of the furnace. Any friction between the oscillating specimen assembly and the stationary furnace will be manifest as apparent internal friction.

In the experiments of Jackson et al (1992) under high *T-P* conditions, the contribution of the composite torsion rods—of alumina and steel, partially jacketed in iron—towards the overall compliance of the specimen assembly (Figure 1) was eliminated by conducting pairs of comparative experiments in which the rock specimen itself was replaced by a dimensionally equivalent cylinder of the high-grade alumina ceramic material used in the regions of temperature gradient (Figure 2). The difference between the observed torsional compliances of the rock-bearing and alumina reference assemblies is then the difference at the hot zone temperature T between the torsional compliances of the rock and substitute alumina specimens jacketed in iron. Conversion of the result of such comparative experiments into an absolute determination of the complex shear modulus $G^*(\omega)$ [i.e. $1/J^*(\omega) = G(\omega) \exp(i\delta)$] of the rock specimen therefore requires knowledge of the corresponding properties of the alumina ceramic and of the iron jacket material.

The behavior of the reference assembly was found to be imperfectly

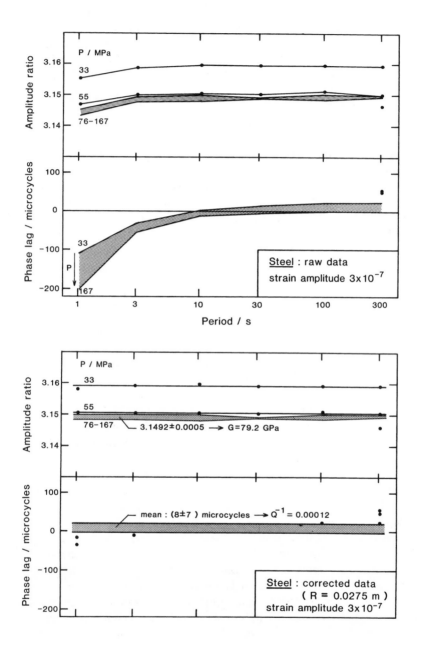

elastic. Within the temperature interval 600–1000°C, the apparent internal friction, indicated by the phase angle between the two displacement sinusoids, increases consistently with increasing temperature but remains small relative to that of the rock-bearing assembly even at 1000°C (Jackson et al 1992, Figure 2). Under these circumstances, $G^*(\omega)$ was reasonably calculated on the assumption that the alumina ceramic *itself* behaves perfectly elastically at these temperatures, with the observed dissipation being more plausibly attributed to the iron jacket and/or to minor but highly reproducible frictional interaction between the specimen assembly and the furnace. The value of $G^*(\omega)$ so determined for the rock specimen is insensitive to the anelasticity of the iron jacket because of the high degree of compensation effected by calculation of the difference in torsional compliance between the two iron-jacketed assemblies.

Subsequent exploratory experiments at 1100°C and 1200°C have revealed a dramatic increase in the apparent internal friction of the reference assembly—such that it can no longer be treated essentially as a low-level subtractable background undeserving of detailed attention. In order to document more thoroughly the source of the apparent internal friction, both forced sinusoidal oscillation and torsional creep tests (described above under "Quasi-Static Methods") are underway, not only on the polycrystalline alumina reference assembly, but also on assemblies containing specimens of single-crystal aluminum oxide or polycrystalline iron. Particularly interesting are indications that only part of the inelastic deformation of the reference assembly (Figure 3) is recoverable, and therefore genuinely anelastic in character. There is also strong evidence that neither the interfaces between the steel and alumina components nor frictional interaction with the furnace contribute significantly to the apparent internal friction.

Rather it is the inelastic behavior of alumina and iron at these temperatures which must be carefully investigated. Flexural pendulum measurements of Q_E^{-1} on a similar 99.7% polycrystalline alumina at 12 Hz and unreported strain amplitude, reveal a roughly 10-fold increase of Q^{-1} from the apparent background level of about 0.003 below 1200°C to

---

Figure 8   Results of a "null" experiment in which the rock specimen was replaced by a dimensionally equivalent cylinder of the same low Q^{-1} steel used throughout the remainder of the specimen assembly. The ratio of displacements measured at the two stations and their relative phase are shown both before (*raw*) and after (*corrected*) correction as explained in the text for the interaction between the transducer plates and the gas pressure medium. The corrected quantities closely approach the values expected for the amplitude ratio (3.14) and phase angle (zero) in the absence of extraneous influences on the apparent anelasticity (reproduced with permission from Jackson & Paterson 1987).

0.03 at 1300°C (Turnbaugh & Norton 1968). In marked contrast, Q^{-1} for single-crystal aluminum oxide showed no increase above background in experiments to 1600°C. The results of these and other experiments on a variety of relatively pure polycrystalline aluminas suggested that the dramatic increase of Q^{-1} at high temperature invariably observed in the polycrystalline materials reflects stress relaxation at grain boundaries intensified, at least in the more impure aluminas, by the presence of a glassy grain boundary phase (Turnbaugh & Norton 1968). The apparent internal friction Q_G^{-1} deriving from forced torsional oscillation experiments on 99.8% polycrystalline alumina at 300 MPa and 1200°C (Figure 5) varies as the oscillation period raised to the power α (Equation 10) with $\alpha = 0.26$. Extrapolation of this variation to the higher frequency (12 Hz) of Turnbaugh & Norton's measurements yields an estimated Q_G^{-1} of 0.006, which would result in a flexural mode Q^{-1} of 0.005 if the dissipation is entirely associated with the shear mode (e.g. Berckhemer et al 1979, equation 27)—not inconsistent with the findings of Turnbaugh & Norton. It is therefore plausible to attribute much of the increase in internal friction of the reference assembly beyond 1000°C to anelasticity/viscoelasticity of the 99.8% polycrystalline alumina, although the contribution of the iron jacket also remains to be clarified. It is anticipated that these experiments in progress will soon provide a firm basis for the quantitative interpretation of high-pressure torsional oscillation experiments already conducted on Åheim dunite beyond 1000°C.

THE ANELASTICITY OF ULTRAMAFIC ROCKS: EXPERIMENTAL RESULTS

Demonstration of Linearity

The importance of the characterization of rock anelasticity/viscoelasticity within the linear regime is a major theme of this review. Perhaps the most compelling experimental evidence which can be adduced for linearity is the strain amplitude independence of the dynamic compliance $J^*(\omega)$, or equivalently of the modulus and phase of its reciprocal, here given by $G^*(\omega) = 1/J^*(\omega) = G(\omega)\exp(i\delta)$, with $\tan\delta = Q^{-1}$. In tests conducted at room temperature, it is possible to operate forced torsional oscillation experiments with sufficient signal-to-noise ratio to allow precise determination of $G^*(\omega)$ over a wide range of relatively low strain amplitudes. The example shown in Figure 9 provides such evidence of linearity for strain amplitudes below 10^{-6}. The lower signal-to-noise ratios typical of high temperature environments tend to restrict attention to the upper part of the experimentally accessible range displayed in Figure 9. Nevertheless, it is usually possible to demonstrate at least that the measured modulus

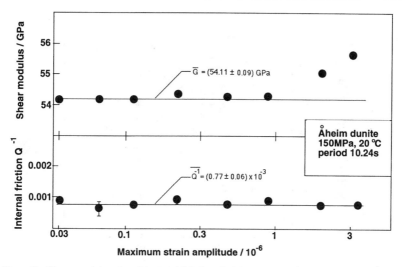

Figure 9 Shear modulus G and internal friction Q^{-1} for an oven-dried specimen of Åheim dunite as functions of the strain amplitude measured at the periphery of the cylindrical specimen. The amplitude independence of G and Q^{-1} for strain amplitudes below 10^{-6} provides compelling evidence of linearity of the stress-strain relationship. [Reproduced from Jackson et al (1992) with permission of the Royal Astronomical Society.]

and internal friction are insensitive to a 2- to 4-fold change in strain amplitude (Berckhemer et al 1979, 1982; Jackson et al 1992).

Arguably more diagnostic of linearity at high temperature, in the presence of pronounced internal friction and associated modulus dispersion, is consistency with the Kronig-Kramers relationships of linear theory (see "The Importance of Linearity: Theory" above). For example, in Figure 10, the measured frequency dependence of the shear modulus compares well with the dispersion calculated from the measured frequency-dependent internal friction through the Kronig-Kramers relation (Equation 11). Such internal consistency provides strong support for the contention, based on strain amplitude independence of G and Q^{-1}, that the stress-strain relationship is indeed linear. The reconciliation of the dynamic compliance $J^*(\omega)$ calculated from the measured creep function $J(t)$ via Equation (9), with that derived from forced oscillation experiments (e.g. Figure 5), can be used to provide further testimony to the linearity of experimentally observed mechanical behavior.

Typical Results at High Temperature

Measurements have thus far been reported at high temperature for a variety of natural and synthetic ultramafic rocks and for single crystals of

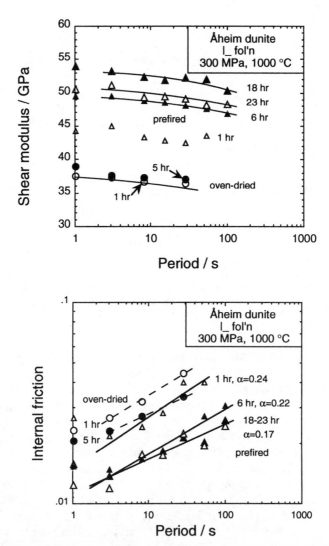

Figure 10   Shear modulus G and internal friction Q^{-1} for oven-dried and prefired specimens of Åheim dunite as functions of oscillation period and the duration of exposure to the prevailing conditions—300 MPa and 1000°C. The lines superimposed upon the Q^{-1} data represent the best-fitting power law models (Equation 10) labeled with the value of the exponent α. The curves superimposed upon the $G(T_o)$ data represent the dispersion calculated from the power law model of the frequency dependence of the internal friction through the Kronig-Kramers relation of linear theory. [Reproduced from Jackson et al (1992) with permission of the Royal Astronomical Society.]

natural olivine and synthetic forsterite. Most directly comparable among the available data are those derived from the forced oscillation experiments of Berckhemer et al (1982) and Jackson et al (1992) on Åheim dunite, and similar experiments by Gueguen et al (1989) and Getting et al (1991) on single-crystal olivines. Of these, only the study of Jackson et al was conducted under high pressure conditions. The results obtained by these latter authors at 300 MPa and 1000°C (Figure 10) best constrain the subsolidus behavior of dunite and also illustrate some of the complexity. For both the oven-dried and prefired specimens (the latter having previously been treated for 24 hr at 1200°C under controlled CO/CO_2 atmosphere), the internal friction measured at fixed oscillation period decreases with increasing duration of the exposure to the prevailing conditions of pressure and temperature. This temporal variation of Q^{-1} has tentatively been associated with the gradual decay of stress concentrations expected to arise at sites of contact between microscopically rough grain surfaces, as a consequence of the anisotropy of thermal expansivity and compressibility. It is planned to test this hypothesis by conduct of similar experiments on aggregates of cubic crystallites in which such stresses are not expected, and on aggregates of weaker materials in which the stresses of thermal origin should decay more rapidly. In the meantime, the values of Q^{-1} reached after many hours of exposure to the imposed P-T conditions are regarded as the most representative, and will form the basis of the following discussion. Similarly, the results obtained for the prefired specimen will be emphasized because they are free from the complication of temporally and spatially variable water pressure consequent upon progressive in situ dehydration during the conduct of the forced oscillation experiment. For the prefired specimen, initially of considerable porosity following prior heat treatment and dehydration (Jackson et al 1992), a consistent trend towards higher shear modulus with increasing exposure to the P-T conditions is observed amongst the most robust determinations. (The results obtained after a 23 hr exposure pertain to a lower strain amplitude at which the calibration of the displacement transducers is less precise and therefore the magnitude of the shear modulus is less secure.) These observations along with permanent dimensional changes (shrinkage) and light microscopic evidence of the closure of much of the initial porosity recorded in post mortem examination, and the very subdued temperature dependence of the shear modulus (Figure 11), all suggest that the reduction of porosity, by sintering at the higher temperatures, has a significant effect on the observed mechanical properties.

The internal friction is found to vary with oscillation period T_o approximately as T_o^α (Equation 10) with exponent α in the range 1/6 to 1/4. At temperatures greater than 1000°C, and at long periods, a trend towards

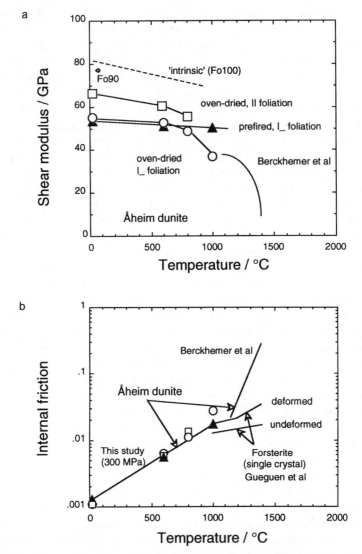

Figure 11 The temperature dependence of (*a*) shear modulus and (*b*) internal friction in Åheim dunite and single-crystal forsterite form forced torsional oscillation studies (Berckhemer et al 1982, Gueguen et al 1989, Jackson et al 1992). All results pertain to an oscillation period of 8.2 s. The suppression of crack porosity results in higher moduli especially for the anhydrous prefired specimen in the high pressure study of Jackson et al. The values of Q^{-1} measured in all three studies converge near $T = 1100°C$ and $Q^{-1} = 0.02$, but display very different temperature sensitivities attesting to the operation of several distinct relaxation mechanisms. [Reproduced from Jackson et al (1992) with permission of the Royal Astronomical Society.]

larger values of α is observed which might herald the transition from anelastic to linearly viscoelastic behavior for which $\alpha = 1$ is expected (i.e. $Q^{-1} = J_2/J_1 \sim 1/\eta\omega \sim T_o/\eta$). However, an exponent $\alpha = 1/4$ provides an adequate description of the results obtained by Berckhemer et al (1982) also for Åheim dunite but at atmospheric pressure and mainly supersolidus temperatures. As noted previously, the appreciable dispersion encountered (amounting to 5% modulus relaxation between 1 and 100 s periods at 300 MPa and 1000°C) is quantitatively consistent with the measured internal friction through the Kronig-Kramers relation of linear theory.

The temperature dependence of the shear modulus for Åheim dunite is sketched in Figure 11a. At room temperature, the moduli measured by Jackson et al for both oven-dried and prefired specimens are low relative to the value of 78 GPa expected for a monominerallic Fo_{90} olivine aggregate of zero porosity. For the oven-dried material, this discrepancy and the evident anisotropy are ascribed to the presence of several volume % of highly aligned platelets of the hydrous layer silicate mineral clinochlore which has very low elastic resistance to shear on its basal plane. The similarly low moduli for the prefired specimens reflect the presence of porosity formed during dehydration, especially lens-shaped voids within the fine-grained aggregate of anhydrous minerals that replace clinochlore. These voids, aligned parallel to the basal plane of the parent clinochlore, mimic its highly anisotropic elastic properties (Jackson et al 1992).

The broken line labeled "intrinsic" represents the temperature dependence of the shear modulus calculated for a polycrystalline forsterite aggregate of zero porosity from high-frequency (1 MHz) resonance measurements of the single-crystal elastic constants (Suzuki et al 1983). The shear modulus measured at 8 s period and 300 MPa by Jackson et al (1992) is much less temperature sensitive, especially for the prefired specimen of Åheim dunite, presumably as the result of the progressive elimination of porosity. The measurements of Berckhemer et al (1982; Kampfmann & Berckhemer 1985) at atmospheric pressure were made as temperature was progressively reduced following fusion bonding of the specimen to alumina torsion rods under axial compressive stress at 1300–1500°C. The strong temperature sensitivity of the modulus between 1400°C and 1250°C is attributed to decreasing melt fraction as the temperature is lowered. Below the solidus at about 1250°C, microcracking caused by intergranular stresses of thermal origin becomes progressively more important. The creation of substantial crack porosity is responsible for the decline in the temperature sensitivity of the shear modulus as temperature falls towards 1100°C (Kampfmann & Berckhemer 1985). Indeed the measured modulus is essentially temperature independent for the interval 1100–1200°C, and some 25% lower than the trend of the data for the prefired specimen of

Jackson et al (1992), showing clearly the positive influence of confining pressure in suppressing crack porosity in the latter study.

The internal friction data for the various specimens studied by Jackson et al define a remarkably consistent trend, whereby Q^{-1} at 8 s period increases approximately 10-fold between 20°C and 1000°C. This trend connects reasonably well with the results of Berckhemer et al for Åheim dunite at higher temperatures and those of Gueguen et al (1989) for predeformed single crystal forsterite, although the latter two studies define, respectively, much greater and somewhat lesser temperature sensitivities. The internal friction recently measured on single-crystal San Carlos olivine at 1500 K (1227°C) by Getting et al (1991) varies with oscillation period as $T_o^{0.2}$, in general accord with the results of the other studies. The Q^{-1} of approximately 0.012 at 8 s period plots close to the trend of Gueguen et al (1989) for undeformed single-crystal forsterite (Figure 11). It should be noted, however, that possible contributions from ceramic torsion rods to the internal friction measured by Getting et al are still being evaluated (personal communication; see relevant section under Calibration of Forced Torsional Oscillation Apparatus). The wide divergence between the Berckhemer et al data for the dunite and the much lower internal friction for predeformed single-crystal forsterite in the temperature interval 1200–1400°C suggests an increasingly important role for intergranular (melt-related?) relaxation as the dunite is exposed to progressively higher temperatures beyond 1200°C.

Anelastic Relaxation Mechanisms

It is not yet possible with any certainty to identify, at the microscopic level, the relaxation mechanisms responsible for the observed anelasticity. However, some useful constraints are beginning to emerge. The fact that very similar internal friction was observed by Jackson et al (1992) for oven-dried and prefired specimens of Åheim dunite suggests that the hydrous phases in the former and their dehydration products in the latter do not exert a significant influence. Rather it would appear that the anelastic relaxation must be concentrated in the olivine aggregate, either within the olivine grains per se, or at olivine grain boundaries. The intragranular alternative derives support from the observation of comparable values of Q^{-1} at 1000°C for both Åheim dunite and predeformed specimens of single-crystal forsterite (Figures 11 and 12).

There is another characteristic common to all of the published studies of the anelasticity of ultramafic materials at relatively high temperature and low frequency. Internal friction peaks of the Debye type associated with a relaxation mechanism having a unique relaxation time, or a narrow distribution of relaxation times, are generally not observed. Instead, it is

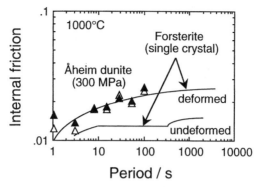

Figure 12 A more detailed comparison at 1000°C of the variation of Q^{-1} with oscillation period for prefired Åheim dunite at 300 MPa (Jackson et al 1992) and both undeformed and predeformed specimens of single-crystal forsterite (Gueguen et al 1989). The close similarity between these results for the dunite and the deformed single crystal suggests that the dunite anelasticity is associated primarily with intragranular, probably dislocation-related, relaxation mechanisms. [Reproduced from Jackson et al (1992) with permssion of the Royal Astronomical Society.]

found that Q^{-1} increases monotonically with increasing oscillation period T_o and temperature T according to the expression

$$Q^{-1}(T_o, T) = A[T_o \exp(-E^*/RT)]^\alpha, \tag{21}$$

with $0 < \alpha \ll 1$, over wide ranges of T_o and T. The observation of such behavior requires that there must exist a broad distribution of relaxation times τ of the form $D(\tau) \sim \tau^{(\alpha-1)}$ for τ within an interval (τ_m, τ_M) such that $\tau_m \ll T_o \ll \tau_M$, and zero elsewhere (Minster & Anderson 1981). For thermally activated processes it is expected that the relaxation time τ will vary with temperature according to Equation (20), so that the required distribution of relaxation times might involve an appropriately wide distribution of preexponential factors or activation energies or both.

The sensitivity of Q^{-1} for single-crystal forsterite to prior deformation (Figures 11 and 12) suggests that the mobility of dislocations might be responsible for the observed anelastic relaxation (Gueguen et al 1989). Interestingly, however, no temporal evolution of Q^{-1} was reported in their study performed under conditions (1000–1400°C) that might have been expected to be conducive to recovery of the dislocation microstructure— in contrast to the recent experience of Getting et al (1991). Clearly, more extensive high resolution studies of the anelasticity, both of olivine crystals and of olivine-rich rocks, of systematically manipulated microstructure are required to resolve the details.

Finally, it must be anticipated that over very wide ranges of temperature and oscillation period, a variety of relaxation mechanisms will be operative. The failure of Equation (24) to fit the data of Gueguen et al over the entire range of oscillation periods and temperature, and the very different temperature sensitivities among the three studies compared in Figure 11, indicate that several distinct mechanisms (probably including both dislocation- and melt-related processes) must be contributing to the variability of Q^{-1} under the widely varying experimental conditions. Under these circumstances, the assumption of a grain boundary relaxation model by Sato et al (1989) for the extrapolation of their ultrasonic attenuation data to seismic frequencies must be viewed with considerable suspicion.

THE PROMISE OF THE FUTURE

After two decades of persistent development of appropriate experimental methodologies, a new age is dawning in the experimental study of seismic wave dispersion and attenuation in rocks. Quantitative experiments of the type envisaged by Goetze so long ago can now be conducted at seismic frequencies, under conditions of high temperature and pressure, within the linear regime. Careful documentation and systematic variation of the microstructure of both single-crystal and polycrystalline materials being tested promises to give us a detailed mechanistic understanding of the observed anelasticity/viscoelasticity, which will provide the basis for informed interpretation of seismological models.

ACKNOWLEDGMENTS

The assistance of G. R. Horwood and Z. Guziak in the conduct of the experiments, of Z. Guziak in the drafting of the illustrations, and of M. Walsh in the preparation of the manuscript is gratefully acknowledged. The manuscript was significantly improved in response to suggestions from B. L. N. Kennett.

Literature Cited

Anderson, D. L., Ben-Menahem, A., Archambeau, C. B. 1965. Attenuation of seismic energy in the upper mantle. *J. Geophys. Res.* 70: 1441–48

Anderson, D. L., Hart, R. S. 1978. Q of the Earth. *J. Geophys. Res.* 83: 5869–82

Anderson, D. L., Minster, J. B. 1979. The frequency dependence of Q in the Earth and implications for mantle rheology and Chandler wobble. *Geophys. J. R. Astron. Soc.* 58: 431–40

Anderson, D. L., Sammis, C. G. 1970. Partial melting in the upper mantle. *Phys. Earth Planet. Inter.* 3: 41–50

Berckhemer, H., Auer, F., Drisler, J. 1979. High temperature anelasticity and elasticity of mantle peridotite. *Phys. Earth Planet. Inter.* 20: 48–59

Berckhemer, H., Kampfmann, W., Aulbach, E., Schmeling, H. 1982. Shear modulus and Q of forsterite and dunite near partial melting from forced oscillation experiments. *Phys. Earth Planet. Inter.* 29: 30–41

Brennan, B. J. 1981. Linear viscoelastic behaviour in rocks. In *Anelasticity in the Earth*, ed. F. D. Stacey, M. S. Paterson, A. Nicolas, Geodyn. Ser. 4: 13–22. Washington: Am. Geophys. Union

Brennan, B. J., Smylie, D. E. 1981. Linear viscoelasticity and dispersion in seismic wave propagation. *Rev. Geophys. Space Phys.* 19: 233–46

Brennan, B. J., Stacey, F. D. 1977. Frequency dependence of elasticity of rock—test of seismic velocity dispersion. *Nature* 268: 220–22

Chan, W. W., Sacks, I. S., Morrow, R. 1989. Anelasticity of the Iceland Plateau from surface wave analysis. *J. Geophys. Res.* 94: 5675–88

Forsyth, D. W. 1975. The early structural evolution and anisotropy of the oceanic upper mantle. *Geophys. J. R. Astron. Soc.* 43: 103–62

Garber, R. I., Kharitonova, Z. F. 1968. Some features of the internal friction of iron on the $\alpha \rightarrow \gamma$ transition. *Phys. Met. Mettalogr.* 26: 115–20

Getting, I. C., Paffenholz, J., Spetzler, H. A. 1990. Measuring attenuation in geological materials at seismic frequencies and amplitudes, In *The Brittle-Ductile Transition in Rocks, Geophys. Monogr. Ser.*, ed. A. G. Duba, W. B. Durham, J. W. Handin, H. F. Wang, 56: 239–43. Washington: Am. Geophys. Union

Getting, I. C., Spetzler, H. A., Karato, S., Hanson, D. R. 1991. Shear attenuation in olivine. *Eos Trans. Am. Geophys. Union* 72: 451

Goetze, C. 1971. High temperature rheology of Westerley granite. *J. Geophys. Res.* 76: 1223–30

Goetze, C. 1977a. A brief summary of our present day understanding of the effect of volatiles and partial melt on the mechanical properties of the upper mantle. In *High-Pressure Research: Applications in Geophysics*, ed. M. H. Manghnani, S. Akimoto, pp. 3–23. New York: Academic

Goetze, C. 1977b. Bounds on the subsolidus attenuation for four rock types at simultaneous high temperature and pressure. *Tectonophysics* 42: T1–T5

Goetze, C., Brace, W. F. 1972. Laboratory observations of high-temperature rheology of rocks. *Tectonophysics* 13: 583–600

Gueguen, Y., Darot, M., Mazot, P., Woirgard, J. 1989. Q^{-1} of forsterite single crystals. *Phys. Earth Planet. Inter.* 55: 254–58

Gueguen, Y., Woirgard, J., Darot, M. 1981. Attenuation mechanisms and anelasticity in the upper mantle. In *Anelasticity in the Earth*, ed. F. D. Stacey, M. S. Paterson, A. Nicolas, Geodyn. Ser. 4: 86–94. Washington: Am. Geophys. Union

Jackson, I. 1986 The laboratory study of seismic wave attenuation. In *Mineral and Rock Deformation—Laboratory Studies, Geophys. Monogr. Ser.* 36: 11–23, ed. B. E. Hobbs, H. C. Heard. Washington: Am. Geophys. Union.

Jackson, I. 1991 The petrophysical basis for the interpretation of seismological models for the continental lithosphere. In *The Australian Lithosphere*, ed. B. Drummond, Geol. Soc. Aust. Spec. Publ. 17: 81–114

Jackson, I. 1993. Dynamic compliance from torsional creep and forced oscillation tests: an experimental demonstration of linear viscoelasticity. *Geophys. Res. Lett.* Submitted

Jackson, I., Paterson, M. S. 1987 Shear modulus and internal friction of calcite rocks at seismic frequencies : pressure, frequency and grainsize dependence. *Phys. Earth Planet. Inter.* 45: 349–67

Jackson, I., Paterson, M. S., FitzGerald, J. D. 1992. Seismic wave dispersion and attenuation in Åheim dunite: an experimental study. *Geophys. J. Int.* 108: 517–34

Jackson, I., Paterson, M. S., Niesler, H., Waterford, R. M. 1984. Rock anelasticity measurements at high pressure, low strain amplitude and seismic frequency. *Geophys. Res. Lett.* 11: 1235–38

Kampfmann, W., Berckhemer, H. 1985. High temperature experiments on the elastic and anelastic behaviour of magmatic rocks. *Phys. Earth Planet. Inter.* 40: 223–47

Karato, S., Spetzler, H. A. 1990. Defect microdynamics in mineals and solid state mechanisms of sesimic wave attenuation and velocity dispersion in the mantle. *Rev. Geophys.* 28: 399–421

Liu, H.-P. Peselnick, L. 1983. Investigation of internal friction in fused quartz, steel, plexiglass and Westerley Granite from 0.01 to 1.00 Hz at 10^{-8} to 10^{-7} strain amplitude. *J. Geophys. Res.* 88: 2367–79

Minster, J. B., Anderson, D. L. 1981 A model of dislocation-controlled rheology for the mantle. *Phil. Trans. R. Soc. London Ser. A* 299: 319–56

Molodenskiy, S. M., Zharkov, V. N. 1982. Chandler wobble and the frequency dependence of Q_μ of the Earth's mantle. *Izv. Earth Phys.* 18: 245–54

Nowick, A. S., Berry, B. S. 1972. *Anelastic Relaxation in Crystalline Solids*. New York: Academic. 677pp.

Paterson, M. S. 1970. A high-pressure, high-temperature apparatus for rock deformation. *Int. J. Rock Mech. Min. Sci.* 7: 517–26

Peselnick, L., Liu, H.-P. 1987 Laboratory measurement of internal friction in rocks and minerals at seismic frequencies. In *Methods of Experimental Physics* 24A: 349–69. New York: Academic

Poirier, J.-P. 1985. *Creep of Crystals.* Cambridge: Cambridge Univ. Press. 260 pp.

Sailor, R. V., Dziewonski, A. M. 1978. Measurements and interpretation of normal mode attenuation. *Geophys. J. R. Astron. Soc.* 53: 559–81

Sato, H., Sacks, I. S., Murase, T., Muncill, G., Fukuyama, H. 1989. Q_p-melting temperature relation in peridotite at high pressure and temperature: attenuation mechanism and implications for the mechanical properties of the upper mantle. *J. Geophys. Res.* 94: 10,647–61

Shankland, T. J., O'Connell, R. J., Waff, H. S. 1981. Geophysical constraints on partial melt in the upper mantle. *Rev. Geophys. Space Phys.* 19: 394–406

Suzuki, I., Anderson, O. L., Sumino. 1983. Elastic properties of a single-crystal forsterite Mg_2SiO_4 up to 1200 K. *Phys. Chem. Minerals* 10: 38–46

Turnbaugh, J. E., Norton, F. H. 1968. Low-frequency grain-boundary relaxation in alumina. *J. Am. Ceram. Soc.* 51: 344–48

Vassiliou, M., Salvado, C. A., Tittman, B. R. 1984. Seismic Attenuation. In *CRC Handbook of Physical Properties of Rocks*, ed. R. S. Carmichael, Vol. III: 295–328. Boca Raton: CRC Press

Woirgard, J., Gueguen, Y. 1978 Elastic modulus and internal friction in enstatite, forsterite and peridotite at seismic frequencies and high temperatures. *Phys. Earth Planet. Inter.* 17: 140–46

Woirgard, J., Gerland, M. Riviére, A. 1980. High temperature and low frequency damping of aluminium single crystals. In *Internal Friction and Ultrasonic Attenuation in Solids*, ed. C. C. Smith. Oxford: Pergamon

Woirgard, J., Sarrazin, Y., Chaumet, H. 1977. Apparatus for the measurement of internal friction as a function of frequency between 10^{-5} and 10 Hz. *Rev. Sci. Instrum.* 48: 1322–25

Wolfenden, A. 1978. Internal friction in polycrystalline iron near the Curie point. *Z. Metallkunde* 69: 308–11

Annu. Rev. Earth Planet. Sci. 1993. 21:407–26

THE GLOBAL METHANE CYCLE

Martin Wahlen

Scripps Institution of Oceanography, University of California San Diego, La Jolla, California 92093

KEY WORDS: trace gases, biogeochemical cycles, biosphere-atmosphere exchange, isotopic species of methane

INTRODUCTION

Methane (CH_4) is the most abundant organic species in the Earth's atmosphere. It is a greenhouse gas, as are water vapor (H_2O), carbon dioxide (CO_2), nitrous oxide (N_2O), ozone (O_3), and the chlorofluorocarbon compounds. It absorbs long wave radiation emitted from the Earth's surface in the 4–100 μm atmospheric window and therefore affects atmospheric temperature directly (Lacis et al 1981, Ramanathan 1988, Hansen et al 1988). It is chemically reactive, and influences the abundance of ozone in the troposphere and in the stratosphere (Johnston 1984), and it is a major source of stratospheric water (Ehhalt 1979, Pollock et al 1980). Methane thus affects temperature indirectly through its chemical interactions.

Systematic measurements of the global tropospheric CH_4 mixing ratios since 1978 reveal a steady increase with time by about 1% per year, due to anthropogenic activities (Blake & Rowland 1988, Steele et al 1987). The history of CH_4 atmospheric mixing ratios has been reconstructed from measurements of air occluded in ice cores for glacial and interglacial times, as well as during the more recent 200 years through the industrial era (Chappellaz et al 1990; Pearman et al 1986; Etheridge et al 1988; Stauffer et al 1985; Craig & Chou 1982; Rasmussen & Khalil 1981a, 1981c; Robbins et al 1973). Large natural and anthropogenically influenced variations are observed over different time scales.

In the troposphere CH_4 is oxidized to CO and ultimately to CO_2 and H_2O. This oxidation reaction sequence is initiated by the hyroxyl (OH) radical. This constitutes the major sink for CH_4. The atmospheric lifetime for methane is 8–12 yr. Methane emitted from inundated anoxic environ-

407

ments can be substantially reduced by bacterial methane oxidation in oxic layers above. A small sink can be attributed to bacterial oxidation on relatively dry soils (Born et al 1990), which relies on consumption of atmospheric methane. Some methane is exported to the stratosphere. In the lower stratosphere the same oxidation reaction as in the troposphere occurs, initiated by OH; at higher altitudes reactions with excited oxygen atoms [$O(^1D)$] and with chlorine atoms lead to mixing ratio profiles that decrease with altitude.

Methane is produced by bacteria under anaerobic conditions in wet environments such as wetlands, swamps, bogs, fens, tundra, rice fields, and landfills. It is also produced in the stomachs of ruminants (cattle and other cud-chewing mammals), and possibly by termites. Most of this biogenic methane is generated by two major bacterial pathways, namely by fermentation of acetate and by reduction of CO_2 with H_2 (Whiticar et al 1986, Wolin & Miller 1987, Cicerone & Oremland 1988). Other sources of CH_4 are from leakage of natural gas upon drilling and distribution, and from coal mining. A further source is from biomass burning where CH_4 is a product of incomplete combustion (Crutzen et al 1979). The annual production rates and the magnitude of the different sources and sinks are still somewhat uncertain.

ATMOSPHERIC DISTRIBUTION OF METHANE

The Recent Atmospheric Record

The presence of methane in the Earth's atmosphere was discovered by Migeotte (1948), from infrared absorption features in the solar spectrum. Hutchinson (1954), Koyama (1964), Fink et al (1965), Lamontagne et al (1974), Ehhalt (1974), Ehhalt & Schmidt (1978), and Ehhalt (1979) gave the first accounts on atmospheric methane and attempted to delineate, in principle, its sources and sinks.

Systematic worldwide time series measurements of the troposheric CH_4 mixing ratio from different latitudes started in 1978. These early measurements established that the global average CH_4 mixing ratio has been increasing approximately linearly by about 1% per year over the past decade and a half (Rasmussen & Khalil 1981b, 1981c, 1984; Steele et al 1987; Blake & Rowland 1988). Figure 1 (from Blake & Rowland 1988) illustrates this trend. Today's average global mixing ratio is about 1750 ppb, corresponding to a global atmospheric inventory of about 4900 Tg of CH_4. This increase is attributed to the anthropogenically-affected methane sources (see separate section) such as those from rice production, ruminants, and biomass burning. In part, such an increase may be due to the decreasing oxidative capacity of the atmosphere, i.e. if the abundances

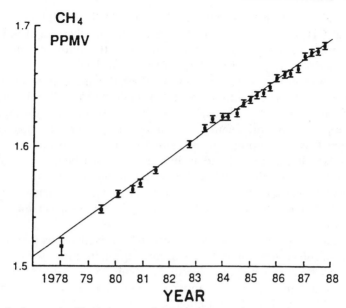

Figure 1   Average worldwide increase of the atmospheric methane mixing ratio after Blake & Rowland (1988). (Reproduced with the permission of *Science*, American Association for the Advancement of Science.)

of CO, CH_4, and nonmethane hydrocarbons increase with time, then the amount of atmospheric OH decreases as it is consumed in oxidation reactions with these trace gases (Thompson 1992; see also Methane Sinks). Other data (Steele et al 1992) indicate a substantial slowing of the global accumulation rate since 1983. The reason for this is not clear.

These studies also show that there is a marked interhemispheric gradient in the tropospheric mixing ratio of methane, with the ratio in the northern hemisphere being about 80–100 ppb or 5% higher than in the southern hemisphere on average. This reflects the larger sources of methane in the northern hemisphere (approximately a factor of three over the southern hemisphere) which, given the 8–12 year atmospheric methane lifetime and an interhemispheric atmospheric air exchange coefficient of $1/1$–$1/1.3 \, yr^{-1}$, results in the observed interhemispheric gradient of the atmospheric mixing ratios of methane.

Time series measurements in remote locations at different latitudes also reveal a seasonal cycle in the atmospheric methane mixing ratios (Rasmussen & Khalil 1981b, Khalil & Rasmussen 1983, Steele et al 1987,

Quay et al 1991, Fung et al 1991, Steele et al 1992). The amplitudes of the seasonal CH_4 mixing ratios are strongest in the northern high latitudes (30–40 ppb) and diminish toward the equator. In the southern hemisphere they are smaller (10–20 ppb) and quite constant with latitude. The findings are illustrated in Figure 2 (from Fung et al 1991) which contains the data of Steele et al (1987) and additional data from the NOAA/CMDL network for the years 1984–1987. Atmospheric mixing ratios are lowest during local summer and fall in both hemispheres, and higher during the remainder of the year. This is interpreted to be caused by methane oxidation by OH radicals which are more abundant in summer. Other factors may play a role: Vertical atmospheric mixing is stronger in summer than in winter, as evidenced by seasonal variations in measurements of the chemically inert ^{85}Kr (M. Wahlen et al, unpublished data; P. Povinec, personal communication).

In the stratosphere the chemical destruction of methane by OH, $O(^1D)$, and Cl becomes evident in profiles of decreasing mixing ratios with altitude, as vertical mixing is highly reduced. At about 25 km altitude the CH_4 mixing ratios are reduced by about a factor of two over tropospheric values (Ehhalt & Heidt 1973, Ehhalt & Thonnissen 1980).

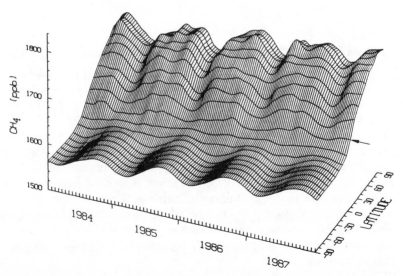

Figure 2 Temporal and latitudinal variation of the tropospheric methane mixing ratio (from Fung et al 1991). (Reproduced with the permission of the American Geophysical Union.)

Reanalyses of solar absorption spectra taken in 1951 at Jungfraujoch (Switzerland) and at Kitt Peak in 1981 bridge the records between ice-core observations and direct atmospheric observations (Rinsland et al 1985). The results were 1.14 ppm (April 1951) and 1.58 ppm (February 1981), respectively.

The Longer Term Record

The atmospheric methane mixing ratios over various periods in the past, up to 160,000 years ago, have been reconstructed from the analyses of air trapped in bubbles of deep ice cores, using wet extraction techniques and flame ionization gas chromatography. These studies reveal substantial variations of the atmospheric mixing ratios through time.

Ice core work compellingly demonstrates that the atmospheric CH_4 mixing ratio has more than doubled from preindustrial times to today (Craig & Chou 1982, Rasmussen & Khalil 1984, Stauffer et al 1985, Pearman et al 1986, Etheridge et al 1988, Pearman & Fraser 1988). This is attributed to anthropogenically influenced sources such as methane from rice production, animals, and from biomass burning. Figure 3 illustrates the increase of atmospheric CH_4 mixing ratios from about 750 ppb to today's value over the last two centuries (Houghton et al 1990).

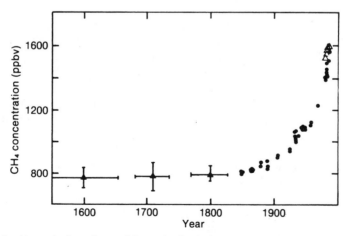

Figure 3 Atmospheric methane mixing ratios for the last several hundred years as deduced from measurements in air occluded in ice cores, and direct atmospheric observations over the last two decades (from Houghton et al 1990). (Reproduced with permission of the University Press, Cambridge.)

Longer term records are from Antarctica and Greenland ice cores (Raynaud et al 1988, Stauffer et al 1988, Barnola et al 1987, Chappellaz et al 1990). They reveal that the atmospheric CH_4 mixing ratios varied by a factor of two between the last glacial maximum (about 350 ppb) and the Holocene (about 650 ppb). The same variations are observed for the penultimate glacial and the previous warm period. There are large oscillations in the methane mixing ratio during the period of the last glacial maximum. These oscillations, ranging from glacial low values to almost Holocene high values, seem to positively correlate with temperature variations, as displayed in Figure 4. Temperature variations are derived from measurements of D/H in the ice, and a change between glacial and interglacial times of 6–8°C is indicated for Antarctica. Thus there seems to be a fundamental link between temperature and atmospheric CH_4, which seems to react to temperature change quickly. This is most likely related to changes in the global methane source strength, and/or to changes in the atmospheric OH mixing ratios. Preindustrial and Holocene methane levels can be reasonably explained by stripping anthropogenically influenced sources, according to population growth. To maintain glacial time CH_4 mixing ratios when the high to mid-northern latitudes were covered with ice would, however, require increased fluxes (over today's) from natural wetlands at low latitudes, which could be accomplished by increased precipitation. This pattern is somewhat supplemented by the observation that precipitation rates were higher at low latitudes in glacial times than in interglacial times—opposite to what is observed in polar regions. Evidence comes from observations in low latitude ice cores (L. Thompson, personal communication) where concentrations of sulphates and nitrates and the abundance of particulates were found to be lower during glacial times. Methane hydrate destabilization upon the decay of the glacial sheets is another possible explanation for increased atmospheric CH_4 during warm times (Kvenvolden 1988), but experimental evidence is lacking.

Spectral analyses of the Vostok CH_4 and temperature records reveal the presence of all major Milankovitch periodicities, namely approximately 100, 41, 23, and 19 kyr. These findings illustrate the important role of methane in the past, present, and future greenhouse effect. Presently, climate forcing by methane is considered to be about 30% of that by CO_2 (Raynaud et al 1988, Chapellaz et al 1990, Lorius et al 1990).

METHANE SINKS

Atmospheric Chemistry

Methane is predominantly destroyed in the troposphere by oxidation reactions initiated by the OH radical:

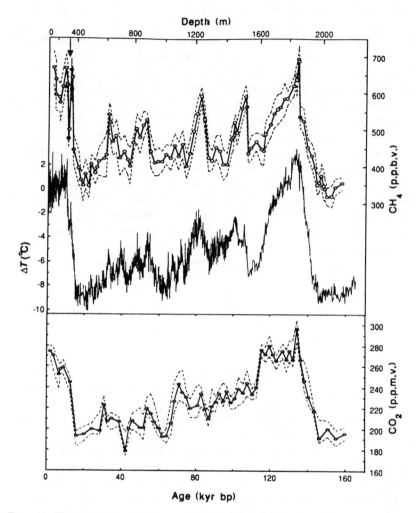

Figure 4 The long term (160,00 years BP) atmospheric CH_4 record as deduced from measurements of the mixing ratio in the air occluded in ice from the Vostok ice core (*top*); CO_2 mixing ratios in the same core (*bottom*); temperature variation as deduced from measurements of D/H in the ice of the core (from Chappellaz et al 1990). (Reproduced with permission from *Science*, American Association for the Advancement of Science.)

$$CH_4 + OH \Rightarrow CH_3 + H_2O.$$

The reaction sequence eventually leads to formaldehyde, then CO, and ultimately to CO_2 and H_2O, consuming additional OH radicals (Ravishankara 1988). The methane chemistry is closely linked with that of CO, as methane oxidation produces as much as 30% of the atmospheric CO. Moreover, atmospheric NO_x abundances influence the CH_4 oxidation chemistry, and the methane oxidation chain can either produce or consume ozone (Thompson & Cicerone 1986). In the presence of adequate NO_x, methane oxidation produces O_3. When NO_x is low, methane oxidation consumes O_3. The oxidation of methane, CO, and nonmethane hydrocarbons in principle controls the tropospheric levels of OH. Thus increasing sources of methane (and possibly CO and nonmethane hydrocarbons) could deplete atmospheric OH, and by feedback, contribute to their atmospheric mixing ratio increase. There is no plausible way to experimentally distinguish between increasing CH_4 sources and decreasing OH, unless the global distribution of OH abundances can be more precisely determined and monitored. Recently two new techniques have been compared to measure the number density of OH in the lower troposphere (Mount & Eisele 1992) with sensitivities accurate enough to test photochemical model predictions. These initial measurements indicate OH number densities below those predicted by photochemical models.

One approach to constructing global OH fields by photochemical models uses compounds for which the source strength and the destruction reaction rate constant is well known. Such a compound is methylchloroform (Mayer et al 1982, Prinn et al 1987) for which the release rates and the oxidation reaction constant with OH are fairly well known. Scaling the methane destruction and atmospheric lifetime results in the values cited above. A recent redetermination of the $CH_4 + OH$ reaction constant by Vaghjiani & Ravishankara (1991) indicates a lower rate constant by about 20%, which would increase the CH_4 atmospheric lifetime or lower the global source strength, respectively. However, more recently a redetermination of the rate constant for the reaction of OH with methylchloroform found a lower value by 15% (Talukdar et al 1992). As this reaction rate constant together with the methylchloroform release rate is used to construct the OH fields, the above two reaction rate constant redeterminations cancel the effect on the global methane source/sink rate and lifetime, and effectively restore the previous values.

Some tropospheric methane (5–15%) is mixed into the stratosphere where it is destroyed by OH, $O(^1D)$, and Cl. Reaction with Cl is important, as it sequesters ozone-destroying Cl into relatively unreactive HCl. Furthermore stratospheric CH_4 oxidation is a substantial source of stratospheric H_2O (Pollock et al 1980).

Bacterial Methane Oxidation

Bacterial oxidation of methane under aerobic conditions, occurring above anaerobic methane production zones in wet environments, influences the net flux of methane emitted to the atmosphere (Harriss et al 1982, Keller et al 1983, King et al 1989, Whalen & Reeburgh 1990a). This is particularly evident in lakes and in the oceans where methane produced in anaerobic sediments cannot escape to the atmosphere due to aerobic methane oxidation in the oxic water column (Oremland et al 1987). Direct uptake and consumption of atmospheric CH_4 on relatively dry soils have been observed by Keller et al (1983, 1990), Whalen & Reeburgh (1990a), Born et al (1990), and Striegl et al (1992). Born et al (1990) estimate that this sink globally amounts to about 1–15% of the chemical sink in the atmosphere.

METHANE SOURCES

Numerous studies have investigated the strengths of various methane sources contributing to the atmospheric inventory. This has been done in a number of ways: by experimental measurements of methane emission rates from various ecosystems and methane producing processes, by investigations of the isotopic composition of methane sources and of atmospheric methane, by biostatistical analyses, and by modeling of the global methane mixing ratio distributions using Global Circulation Models. The results of some of the more recent investigations are summarized in Table 1. These investigations show that there are basically four major CH_4 sources of about equal size contributing to the atmospheric inventory. These are: methane from natural wetlands and tundra, from ruminants, from rice production, and from fossil methane (natural gas and from coal mining). Smaller sources are from biomass burning and from landfills. Termites were once thought to be a major source (Zimmerman et al 1982) but this source is now considered a minor one. Additional minor sources are from freshwater and the oceans. Methane from clathrate destabilization is a potential source which has yet to be quantified (Kvenvolden 1988). Table 1 summarizes some recent methane budgets based on various techniques.

Wetlands, Soils, and Tundra

Extensive efforts have been made to characterize methane fluxes from wet and inundated environments, where methane is produced under anaerobic conditions by bacteria mainly via acetate fermentation and CO_2 reduction (see e.g. Keller et al 1986; Sebacher et al 1986; Crill et al 1988a,

Table 1 Sources and sinks of methane

	Annual release and range (Tg CH$_4^-$/yr)			
	Cicerone & Oremland 1988	Wahlen et al 1989	Houghton et al 1990	Fung et al 1991 (scenario 7)
Source				
Natural wetlands (bogs, swamps, tundra, etc)	115 100-200	147 (incl. landfills)	115 100-200	115
Rice paddies	110 60-170	136	110 25-170	100
Enteric fermentation (animals)	80 65-100	119	80 65-100	80
Fossil methane:				
Gas drilling, venting, transmission	45 25- 50	123	45 25- 50	40
Coal mining	35 25- 45		35 19- 50	35
Biomass burning	55 50-100	55	40 20- 80	55
Landfills	40 30- 70		40 20- 70	40
Termites	40 10-100		40 10-100	20
Oceans and fresh waters	15 6- 45		15 6- 45	10
Hydrate destabilization	5? 0-100		5 0-100	5
Total	540 400-640	580	525 290-965	510
Sink				
Reaction with OH in atmosphere	500 405-595		500 400-600	450
Removal by soils			30 15- 45	10
Atmospheric increase	40- 46	55	44 40- 48	

1988b; Burke et al 1988; Bartlett et al 1988; Quay et al 1988; Whalen & Reeburgh 1988, 1990a, 1990b; Moore et al 1990; Roulet et al 1992a, 1992b). Methane flux measurements for a given ecosystem scatter widely over several orders of magnitude, both in time and space. They exhibit seasonal variations, but there are no other clear correlations with any parameters (such as temperature or precipitation) which could be used to obtain reliable global scale extrapolations. Global extrapolations for the methane flux from these sources therefore remain somewhat uncertain. Matthews & Fung (1987) and Aselmann & Crutzen (1989) have presented global extrapolations for these sources using average flux values and global distribution of these ecosystems, as well as net primary productivity. The resulting annual CH_4 production, approximately 110 Tg yr^{-1} is within the range indicated in Table 1.

Methane production from tundra would be expected to increase with global warming. Warming would lower permafrost levels and make large stores of carbon available for bacterial methane production.

Rice Fields

Rice fields which are flooded for extended periods of the growing season provide, like natural wetlands, anaerobic conditions for bacterial methane production. Methane fluxes from rice fields have been investigated by Cicerone et al 1981; Holzapfel-Pschorn & Seiler 1986; Schuetz et al 1989, 1990; Schuetz & Seiler 1989; and Sass et al 1990. Methane fluxes seem to vary during the growing season, and are affected by environmental factors and agricultural practices. Different fertilization procedures seem to affect the methane production as well. Global extrapolations are difficult, given the limited data base. Experimental data from the major rice growing countries (China and Southeast Asia) are lacking. Matthews et al (1991) compiled a global data base for methane by rice production with an estimated source strength of about 100 Tg yr^{-1}. Methane production from rice cultivation is expected to increase with world population growth and therefore will influence the atmospheric CH_4 inventory in the future. Holzapfel-Pschorn & Seiler (1986) estimated that this source has been increasing worldwide by about 1.6% per year since 1940.

Animals

Methane is produced by enteric fermentation in the stomachs of ruminants. Methane production by domestic and wild ruminants is by the CO_2 reduction pathway (Wolin & Miller 1987). Crutzen et al (1986) and Lerner et al (1988) have investigated this source on a global scale. The annual CH_4 production from this source is estimated at about 80 Tg yr^{-1}. This CH_4 source is thought to contribute to the atmospheric mixing ratio increase over the recent past as domestic ruminants likely scale with population growth.

Fossil Methane

Fossil methane is being released to the atmosphere in the drilling and exploration of oil and natural gas. Losses furthermore occur in transmission and distribution of natural gas. Fossil methane is also released from coal mining. There is some discrepancy between fossil methane releases accounted for by statistical analyses and those derived from the analyses of $^{14}CH_4$ which is most suitable to estimate this methane source (see Isotopic Tracing of the Methane Cycle).

Landfills

Anaerobic conditions in soil-covered landfills favor biogenic methane and CO_2 production which is vented to the atmosphere. Biogenic methane production from biodegradable carbon lasts for a long time (about two decades). Bingemer & Crutzen (1987) evaluated the production from waste and its degradable carbon content. They estimated the global methane production rate from this source to be 30–70 Tg yr^{-1}, and found the annual methane emission rate to be somewhat lower due to methane oxidation.

Biomass Burning

Biomass burning produces CH_4 in incomplete combustion. Crutzen et al (1979) realized this to be a substantial source for atmospheric methane. Crutzen & Andreae (1990) estimate that biomass burning in the tropics is contributing 11–53 Tg of methane per year to the atmospheric inventory. Craig et al (1988), Cicerone & Oremland (1988), and Quay et al (1991) estimate the biomass burning contribution to be about 50 Tg yr^{-1}.

ISOTOPIC TRACING OF THE METHANE CYCLE

Rust (1981) and Stevens & Rust (1982) first proposed the isotopic approach to a CH_4 budget using ^{13}C. This approach was carried further to include ^{14}C and deuterium in methane (Burke et al 1988; Wahlen et al 1989, 1990; Quay et al 1991; Manning et al 1990; Levin et al 1992). The isotopic composition is influenced by many steps in the carbon cycle: Atmospheric CO_2 is converted to organic matter, organic matter is broken down into decomposition products, and methane is generated from these products by bacteria in a number of ways. Methane generation also includes aspects of the hydrologic cycle as hydrogen derived from water is incorporated into the final product methane. All these transformations will introduce isotopic fractionations which are influenced by a number of factors. While most of the details are not known, it appears that the isotopic composition of methane from a particular source is determined by the composition of the substrate material and the isotopic fractionation of the methane generating pathway.

In principle therefore it is possible to derive a global methane budget from the comparison of the isotopic composition of methane from the various sources to that of the atmospheric inventory. This requires the knowledge of the isotopic fractionation in the atmospheric destruction reaction for methane. This fractionation was determined variously in the laboratory for ^{13}C (Rust & Stevens 1980, Davidson et al 1987, Cantrell et al 1990) and was found to be from 3 to 10 permil. Substantially larger fractionation has been observed from measurements in the lower strato-sphere by Wahlen et al (1990).

Carbon–14 can be used to assess the contributions by fossil methane to the atmospheric inventory. Fossil CH_4 from natural gas exploration and distribution, from coal mining, and possibly from seepage of natural gas reservoirs on land and near shore is free of ^{14}C, while biogenic methane contains more or less contemporary ^{14}C of about 120 pMC (percent modern carbon). Via ^{14}C analyses, the biogenic contribution has been determined to be $21 \pm 3\%$ of the annual input to the atmosphere by Wahlen et al (1989), 17–25% by Manning et al (1990), and $16 \pm 12\%$ by Quay et al (1991). There is some complication due to a substantial anthropogenic contribution to the atmospheric $^{14}CH_4$ inventory from emissions from pressurized light water reactors (Kunz 1985) which adds to the uncertainty. Figure 5 shows the reconstruction of the atmospheric ^{14}C inventory from dated samples including the turnover of ^{14}C by nuclear atmospheric input from bomb testing, with and without the nuclear reactor contribution. Statistical analyses of fossil methane releases (Cicerone & Oremland 1988, Barns & Edmonds 1990, Fung et al 1991) are lower than those observed by ^{14}C. The discrepancy may be explained by releases of ^{14}C-depleted methane from ecosystems where old carbon is being processed. It is also possible that additional fossil CH_4 is released through seepage from natural gas reservoirs on land and near shore.

The stable isotopic composition of methane sources and of atmospheric methane have been investigated by Schoell (1980), Stevens & Rust (1982), Stevens & Engelkemeir (1988), Tyler (1986), Tyler et al (1988), Chanton et al (1988), Wahlen et al (1989), and Quay et al (1991). The atmospheric ^{13}C concentration is about -47 permil (PDB). Biogenic methane is largely

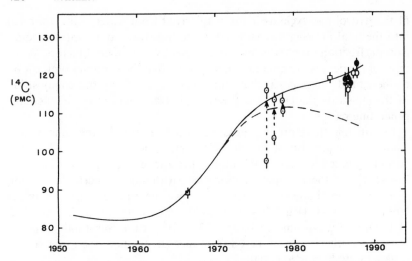

Figure 5 Reconstruction of the turnover of atmospheric bomb-produced ^{14}C into ^{14}CH$_4$ in the atmosphere (from Wahlen et al 1989). ^{14}CH$_4$ is expressed in pMC, (percent Modern Carbon). Upward corrections are for urban samples which contain excess CH$_4$ of fossil origin. (Reproduced with permission of *Science*, American Association for the Advancement of Science.)

depleted in ^{13}C due to fractionation. There are seasonal variations in the atmospheric ^{13}C records which are attributed to the OH destruction (Quay et al 1991). Furthermore there is a hemispheric difference in δ^{13}C reflecting a heavy southern hemispheric source from biomass burning (Quay et al 1991).

Figure 6 shows the power of the isotopic approach (from Wahlen 1992, Wahlen et al 1990). The isotopic composition of the various methane sources are quite distinct. There are more sources than isotopes, thus the system is underdetermined. Nevertheless, strong contraints on the global methane budget can be obtained from the isotopic composition.

CLIMATIC IMPACT OF CH$_4$

The effect of the combined radiatively important trace gases to the terrestrial greenhouse forcing has been estimated to be 45–60% by Lorius et al (1990) from analyses of the gas composition in ice cores. Rodhe (1990) has studied the relative role for the different gases. The findings are summarized in Table 2, which shows the relative importance of CH$_4$. The

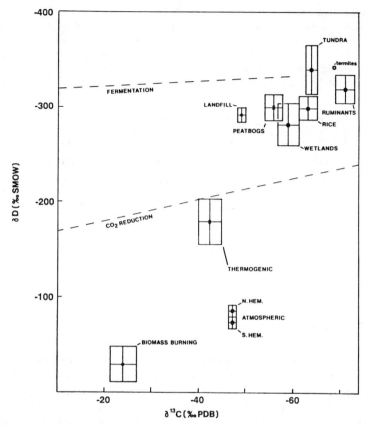

Figure 6 Systematics among the stable isotope composition of methane from various sources and from atmospheric methane (Wahlen et al, unpublished data). Boxes show the mean values for sets of samples for ^{13}C and D in methane.

first and second columns show the abundances and the observed annual increase in the atmospheric mixing ratio of the various gases. Column 3 shows the decay time against additional emission to the atmosphere. These times are essentially exponential, except for CO_2, due to the carbon chemistry in the ocean. Column 4 illustrates the relative importance to the greenhouse effect on a molecular basis. The last column indicates the relative contribution to the terrestrial greenhouse effect by the various trace gases, indicating that methane contributes about 25% of that of CO_2.

Table 2 Trace gas contributions to the greenhouse effect†

Species	Mixing ratio (ppb)	Rate of increase (% per year)	Decay time (yr)	Relative contribution (to CO_2) on a per mole basis (Mol^{-1})	Contribution to greenhouse effect (%)
CO_2	353 x 10³	0.5	120	1	60
CH_4	1.7 x 10³	1	10	25	15
N_2O	310	0.2	150	200	5
O_3 [a]	10–50	0.5	0.1	2000	8
CFC–11 [b]	0.28	4	65	12000	4
CFC–12 [b]	0.48	4	120	15000	8

† After Rodhe (1990).
[a] In the troposphere.
[b] Chlorofluorocarbons.

SUMMARY

Methane is a radiatively and chemically important constituent of the Earth's atmosphere, and it substantially contributes to the terrestrial greenhouse effect. Atmospheric mixing ratios of methane are increasing rapidly today, and large natural fluctuations, by a factor of two, have occurred in different climatic periods in the past. Over the recent 200 years, atmospheric methane mixing ratios have increased more than a factor of two over preindustrial values due to anthropogenic influence. While the gross features of the global methane budget are known, the details about individual source strengths are still poorly understood.

ACKNOWLEDGMENTS

I thank D. Mastroianni for help with the references and figures.

Literature Cited

Aselmann, I., Crutzen, P. J. 1989. Global distribution of natural freshwater wetlands and rice paddies, their net primary productivity, seasonality and possible methane emissions. *J. Atmos. Chem.* 8: 307–58

Barnola, J. M., Raynaud, D., Korotkevich, Y. S., Lorius, C. 1987. Vostok ice core

provides 160,000-year record of atmospheric CO_2. *Nature* 329: 408–14

Barns, D. W., Edmonds, J. A. 1990. An evaluation of the relationship between the production and use of energy and atmospheric methane emissions. *Rep. TRO47* Carbon Dioxide Res. Prog., U.S. Dept. Energy, Washington, DC (Available as

DOE/NBB–0088*P*, Natl. Tech. Inf. Serv., Springfield, Va., 1990)

Bartlett, K. B., Crill, P. M., Sebacher, D. I., Harriss, R. C., Wilson, J. O., et al. 1988. Methane flux from the central Amazonian floodplain. *J. Geophys. Res.* 93: 1571–82

Bingemer, H. G., Crutzen, P. J. 1987. The production of methane from solid wastes. *J. Geophys. Res.* 92: 2181–87

Blake, D. R., Rowland, F. S. 1988. Continuing worldwide increase in tropospheric methane, 1978–1987. *Science* 239: 1129–31

Born, M., Doerr, H., Levin, I. 1990. Methane consumption in aerated soils of the temperate zone. *Tellus* 42B: 2–8

Burke, R. A., Barber, T. R., Sackett, W. M. 1988. Methane flux and stable hydrogen and carbon composition of sedimentary methane from the Florida everglades. *Global Biogeochem. Cycles* 2: 329–40

Cantrell, C. A., Shetter, R. E., McDaniel, A. H., Calvert, J. G., Davidson, J. A., et al. 1990. Carbon kinetic isotope effect in the oxidation of methane by the hydroxyl radical. *J. Geophys. Res.* 95: 22,455–62

Chanton, J. P., Pauly, G. G., Martens, C. S., Blair, N. E., Dacey, J. W. H. 1988. Carbon isotopic composition of methane in Florida everglades soils and fractionation during its transport to the troposphere. *Global Biogeochem. Cycles* 2: 245–52

Chappellaz, J., Barnola, J. M., Raynaud, D., Korotkevich, Y. S., Lorius, C. 1990. Ice-core record of atmospheric methane over the past 160,000 years. *Nature* 345: 127–31

Cicerone R. J., Oremland, R. S. 1988. Biogeochemical aspects of atmospheric methane. *Global Biogeochem. Cycles* 2: 299–327

Cicerone, R. J., Shetter, J. D., Delwiche, C. 1981. Seasonal variations of methane flux from a California rice paddy. *J. Geophys. Res.* 88: 11,022–24

Craig, H., Chou, C. C. 1982. Methane: the record in polar ice cores. *Geophys. Res. Lett.* 9: 1221–24

Craig, H., Chou, C. C., Welhan, J. A., Stevens, C. M., Engelkemeir, A. 1988. The isotopic composition of methane in polar ice cores. *Science* 242: 1535–39

Crill, P. M., Bartlett, K. B., Harriss, R. C., Gorham, E., Verry, E. S., et al. 1988a. Methane flux from Minnesota peatlands. *Global Biogeochem. Cycles* 2: 371–84

Crill, P. M., Bartlett, K. B., Wilson, J. O., Sebacher, D. I., Harriss, R. C., et al. 1988b. Tropospheric methane from an Amazonian floodplain lake. *J. Geophys. Res.* 93: 1564–70

Crutzen, P. J., Andreae, M. O. 1990. Biomass burning in the tropics: impact on

atmospheric chemistry and biogeochemical cycles. *Science* 250: 1669–78

Crutzen, P. J., Aselmann, I., Seiler, W. 1986. Methane production by domestic animals, wild ruminants, other herbivorous fauna, and humans. *Tellus* 38B: 271–84

Crutzen, P. J., Heidt, L. E., Krasnec, J. P., Pollock, W. H., Seiler, W. 1979. Biomass burning as a source of atmospheric gases CO, H_2, N_2O, NO, CH_3Cl and COS. *Nature* 282: 253–56

Davidson, J. A., Cantrell, C. A., Tyler, S. C., Shetter, R. E., Cicerone, R. J., et al. 1987. Carbon kinetic isotope effect in the reaction of CH_4 with HO. *J. Geophys. Res.* 92: 2195–99

Ehhalt, D. H. 1974. The atmospheric cycle of methane. *Tellus* 26: 58–70

Ehhalt, D. H. 1979. Der atmosphaerische Kreislauf von Methan. *Naturwissenschaften* 66: 307–11

Ehhalt, D. H., Heidt, L. E. 1973. Vertical profiles of CH_4 in the troposphere and stratosphere. *J. Geophys. Res.* 78: 5265–71

Ehhalt, D. H., Schmidt, U. 1978. Sources and sinks of atmospheric methane. *Pure Appl. Geophys.* 116: 452–64

Ehhalt, D. H., Tonnissen, A. 1980. Hydrogen and carbon compounds in the stratosphere. *Proc. Nato Advanced Study Inst. on Atmos. Ozone: Its Variation and Human Influences*, ed. M. Nicolet, A. C. Aikin, pp. 129–51. Washington, DC: US Dept. Transp.

Etheridge, D. M., Pearman, G. I., de Silva, F. 1988. Atmospheric trace-gas variations as revealed by air trapped in an ice core from Law Dome, Antarctica. *Ann. Glaciology* 10: 28–33

Fink, U., Rank, D. H., Wiggins, T. Q. 1965. Abundance of methane in the Earth's atmosphere. *Tech. Rep. on Off. Naval Res. contract NONR*–656(12), *NR* 014–401, Penn. State Univ., Univ. Park, Pa

Fung, I., John, J., Lerner, J., Matthews, E., Prather, M., et al. 1991. Three-dimensional model synthesis of the global methane cycle. *J. Geophys. Res.* 96: 13,033–65

Hansen, J., Fung, I., Lacis, A., Rind, D., Lebedeff, S., et al. 1988. Global climate changes as forecast by Goddard Institute for Space Studies three-dimensional model. *J. Geophys. Res.* 93: 9341–64

Harriss, R. C., Sebacher, D. I., Day, F. P. 1982. Methane flux in the Great Dismal Swamp. *Nature* 297: 673–74

Holzapfel-Pschorn, A., Seiler, W. 1986. Methane emmission during a cultivation period from an Italian rice paddy. *J. Geophys. Res.* 91: 11,803–14

Houghton, J. T., Jenkins, G. J., Ephraums, J. J., eds. 1990. *Climate Change: The IPCC*

Scientific Assessment. Cambridge: Cambridge Univ. Press. 337 pp.

Hutchinson, G. E. 1954. The biochemistry of the terrestrial atmosphere. In The Solar System, II. The Earth as a Planet, ed. G. P. Kuiper, pp. 371–433. Chicago: Chicago Press

Johnston, H. S. 1984. Human effects on the global atmosphere. Annu. Rev. Phys. Chem. 35: 481–505

Keller, M., Goreau, T. J., Wofsy, S. C., Kaplan, W. A., McElroy, M. B. 1983. Production of nitrous oxide and consumption of methane by forest soils. Geophys. Res. Lett. 10: 1156–59

Keller, M., Kaplan, W. A., Wofsy, S. C. 1986. Emissions of N_2O, CH_4 and CO_2 from tropical forest soils. J. Geophys. Res. 91: 11,791–802

Keller, M., Mitre, M. E., Stallard, R. F. 1990. Consumption of atmospheric methane in soils of central Panama: effects of agricultural development. Global Biogeochem. Cycles 4: 21–28

Khalil, M. A. K., Rasmussen, R. A. 1983. Sources, sinks, and seasonal cycles of atmospheric methane. J. Geophys. Res. 88: 5131–44

King, S. L., Quay, P. D., Lansdown, J. M. 1989. The $^{13}C/^{12}C$ kinetic isotope effect for soil oxidation of methane at ambient atmospheric concentrations. J. Geophys. Res. 94: 18,273–77

Koyama, T. 1964. Biogeochemical studies on lake sediments and paddy soils and the production of atmospheric methane and hydrogen. In Recent Researches in the Field of Hydrosphere, Atmosphere, and Nuclear Geochemistry, ed. Y. Miyake, T. Koyama, pp. 143–77. Tokyo: Murucen

Kunz, C. 1985. Carbon–14 discharges at three light-water reactors. Health Phys. 49: 25–35

Kvenvolden, K. A. 1988. Methane hydrates and global climate. Global Biogeochem. Cycles 2: 221–29

Lacis, A., Hansen, J., Lee, P., Mitchell, T., Lebedeff, S. 1981. Greenhouse effect of trace gases, 1970–1980. Geophys. Res. Lett. 8: 1035–38

Lamontagne, R. A., Swinnerton, J. W., Linnenboom, V. J. 1974. C_1-C_4 hydrocarbons in the North and South Pacific. Tellus 26: 71–77

Lerner, J., Matthews, E., Fung, I. 1988. Methane emission from animals: a global high-resolution data base. Global Biogeochem. Cycles 2: 139–56

Levin, I., Bosinger, R., Bonani, G., Francey, R., Kromer, B., et al. 1992. Radiocarbon in atmospheric carbon dioxide and methane: global distributions and trends. In Radiocarbon after Four Decades: An Interdisciplinary Perspective, ed. R. E. Taylor, A. Long, R. S. Kra, pp. 503–18. New York: Springer-Verlag

Lorius, C., Jouzel, J., Raynaud, D., Hansen, J., Le Treut, H. 1990. The ice-core record: climate sensitivity and future greenhouse warming. Nature 347: 139–45

Manning, M. R., Lowe, D. C., Melhuish, W. H., Sparks, R. J., Brenninkmeijer, C. A. M., et al. 1990. The use of radiocarbon measurements in atmospheric studies. Radiocarbon 32: 37–58

Matthews, E., Fung, I. 1987. Methane emissions from natural wetlands: global distribution, area, and environmental characteristics of sources. Global Biogeochem. Cycles 1: 61–86

Matthews, E., Fung, I., Lerner, J. 1991. Methane emission from rice cultivation: geographic and seasonal distribution of cultivated areas and emissions. Global Biogeochem. Cycles 5: 3–24

Mayer, E. W., Blake, D. R., Tyler, S. C., Makide, Y., Montague, D. C., et al. 1982. Methane: interhemispheric concentration gradient and atmospheric residence time. Proc. Natl. Acad. Sci. USA 79: 1366–70

Migeotte, M. V. 1948. Spectroscopic evidence of methane in the Earth's atmosphere. Phys. Rev. 73: 519–20

Moore, T., Roulet, N., Knowles, R. 1990. Spatial and temporal variations of methane flux from subarctic/northern boreal fens. Global Biogeochem. Cycles 4: 29–46

Mount, G., Eisele, F. 1992. An intercomparison of tropospheric OH measurements at Fritz Peak Observatory, Colorado. Science 256: 1187–90

Oremland, R. S., Miller, L. J., Whiticar, M. J. 1987. Sources and flux of natural gas from Mono Lake, California. Geochim. Cosmochim. Acta 51: 2915–29

Pearman, G. I., Etheridge, D., de Silva, F., Fraser, P. J. 1986. Evidence of changing concentrations of atmospheric CO_2, N_2O and CH_4 from air bubbles in Antarctic ice. Nature 320: 248–50

Pearman, G. I., Fraser, P. J. 1988. Sources of increased methane. Nature 332: 489–90

Pollock, W., Heidt, L. E., Lueb, R., Ehhalt, D. 1980. Measurement of stratospheric water vapor by cryogenic collection. J. Geophys. Res. 85: 5555–68

Prinn, R. D., Cunnold, R., Rasmussen, R., Simmonds, P., Alyea, F., et al. 1987. Atmospheric trends in methyl chloroform and the global average for the hydroxyl radical. Science 238: 946–50

Quay, P. D., King, S. L., Lansdown, J. M., Wilbur, D. O. 1988. Isotopic composition of methane released from wetlands: implications for the increase in atmospheric methane. *Global Biogeochem. Cycles* 2: 385–97

Quay, P. D., King, S. L., Stutsman, J., Wilbur, D. O., Steele, L. P., et al. 1991. Carbon isotopic composition of CH_4: fossil and biomass burning source strengths. *Global Biogeochem. Cycles* 5: 25–47

Ramanathan, V. 1988. The greenhouse theory of climate change: a test by an inadvertent global experiment. *Science* 240: 293–99

Rasmussen, R. A., Khalil, M. A. K. 1981a. Differences in the concentrations of atmospheric trace gases in and above the tropical boundary layer. *Pure Appl. Geophys.* 119: 990–97

Rasmussen, R. A., Khalil, M. A. K. 1981b. Atmospheric methane (CH_4): trends and seasonal cycles. *J. Geophys. Res.* 86: 9826–32

Rasmussen, R. A., Khalil, M. A. K. 1981c. Increase in the concentration of atmospheric methane. *Atmos. Environ.* 15: 883–86

Rasmussen, R. A., Khalil, M. A. K. 1984. Atmospheric methane in the recent and ancient atmospheres: concentrations, trends, and interhemispheric gradient. *J. Geophys. Res.* 89: 11,599–605

Ravishankara, A. R. 1988. Kinetics of radical reactions in the atmospheric oxidation of CH_4. *Annu. Rev. Phys. Chem.* 39: 367–94

Raynaud, D., Chappellaz, J., Barnola, J. M., Korotkevich, Y. S., Lorius, C. 1988. Climatic and CH_4 cycle implications of glacial-interglacial CH_4 change in the Vostok ice core. *Nature* 333: 655–57

Rinsland, C. P., Levine, J. S., Miles, T. 1985. Concentration of methane in the troposphere deduced from 1951 infrared solar spectra. *Nature* 330: 245–49

Robbins, R. C., Cavanagh, L. A., Salas, L. J., Robinson, E. 1973. Analysis of ancient atmospheres. *J. Geophys. Res.* 78: 5341–44

Rodhe, H. 1990. A comparison of the contribution of various gases to the greenhouse effect. *Science* 248: 1217–19

Roulet, N. T., Ash, R., Moore, T. R. 1992a. Low boreal wetlands as a source of atmospheric methane. *J. Geophys. Res.* 97/D4: 3739–49

Roulet, N., Moore, T., Bubier, J., Lafleur, P. 1992b. Northern fens: methane flux and climate change. *Tellus* 44B: 100–5

Rust, F. E. 1981. Ruminant methane $\delta(^{13}C/^{12}C)$ values: relation to atmospheric methane. *Science* 211: 1044–46

Rust, F., Stevens, C. M. 1980. Carbon kinetic effect in the oxidation of methane by hydroxyl. *Int. J. Chem. Kinet.* 12: 371–77

Sass, R. L., Fisher, F. M., Harcombe, P. A.,

Turner, F. T. 1990. Methane production and emission in a Texas rice field. *Global Biogeochem. Cycles* 4: 47–68

Schoell, M. 1980. The hydrogen and carbon isotopic composition from natural gases of various origin. *Geochim. Cosmochim. Acta* 44: 649–61

Schuetz, H., Holzapfel-Pschorn, A., Conrad, R., Rennenberg, H., Seiler, W. 1989. A 3-year continuous record on the influence of daytime, season, and fertilizer treatment on methane emission rates from an Italian rice paddy. *J. Geophys. Res.* 94: 16,405–16

Schuetz, H., Seiler, W. 1989. Methane flux measurements: methods and results. In *Exchange of Trace Gases between Terrestrial Ecosystems and the Atmosphere*, ed. M. O. Andreae, D. S. Schimel, pp. 209–28. New York: Wiley

Schuetz, H., Seiler, W., Rennenberg, H. 1990. Soil and land use related sources and sinks of methane in the context of the global methane budget. In *Soils and the Greenhouse Effect*, ed. A. F. Bouwman, pp. 269–85. New York: Wiley

Sebacher, D. I., Harriss, R. C., Bartlett, K. B., Sebacher, S. M., Grice, S. S. 1986. Atmospheric methane sources: Alaskan tundra bogs, an alpine fen, and a subarctic boreal marsh. *Tellus* 38B: 1–10

Stauffer, B., Fischer, G., Neftel, A., Oeschger, H. 1985. Increase of atmospheric methane recorded in Antarctic ice core. *Science* 229: 1386–88

Stauffer, B., Lochbronner, E., Oeschger, H., Schwander, J. 1988. Methane concentration in the glacial atmosphere was only half that of the preindustrial Holocene. *Nature* 332: 812–14

Steele, L. P., Duglokencky, E. J., Lang, P. M., Tans, P. P., Martin, R. C., et al. 1992. Slowing of the global accumulation of atmospheric methane during the 1980s. *Nature* 358: 313–16.

Steele, L. P., Fraser, P. J., Rasmussen, R. A., Khalil, M. A. K., Conway, T. J., et al. 1987. The global distribution of methane in the troposphere. *J. Atmos. Chem.* 5: 125–71

Stevens, C. M., Engelkemeir, A. 1988. Stable carbon isotopic composition of methane from some natural and anthropogenic sources. *J. Geophys. Res.* 93: 725–33

Stevens, C. M., Rust, F. E. 1982. The carbon isotopic composition of atmospheric methane. *J. Geophys. Res.* 87: 4879–82

Striegl, R. G., McConnaughey, T. A., Thorstenson, D. C., Weeks, E. P., Woodward, J. C. 1992. Consumption of atmospheric methane by dersert soils. *Nature* 357: 145–47

Talukdar, R. K., Mellouki, A., Schmoltner,

A., Watson, T., Montzka, S., et al. 1992. Kinetics of the OH reaction with methyl chloroform and its atmospheric implication. *Science* 257: 227–30

Thompson, A. M. 1992. The oxidizing capacity of the Earth's atmosphere: probable past and future changes. *Science* 256: 1157–65

Thompson, A. M., Cicerone, R. J. 1986. Possible perturbations to atmospheric CO_2, CH_4, and OH. *J. Geophys. Res.* 91: 10,853–64

Tyler, S. C. 1986. Stable carbon isotope ratios in atmospheric methane and some of its sources. *J. Geophys. Res.* 91: 13,232–38

Tyler, S. C., Zimmerman, P. R., Cumberbatch, C., Greenberg, J. P., Westberg, C., et al. 1988. Measurements and interpretation of $\delta^{13}C$ of methane from termites, rice paddies, and wetlands in Kenya. *Global Biogeochem. Cycles* 2: 341–55

Vaghjiani, G.L., Ravishankara, A.R. 1991. New measurement of the rate coefficient for the reaction of OH with methane. *Nature* 350: 406–9

Wahlen, M., Deck, B., Henry, R., Tanaka, N., Shemesh, A., et al. 1990. Profiles of $\delta^{13}C$ and δD of CH_4 from the lower stratosphere. *Eos, Trans. Am. Geophys. Union* 70(43): 1017

Wahlen, M., Tanaka, N., Henry, R., Deck, B., Broecker, W., et al. 1990. ^{13}C, D and ^{14}C in methane. In *Rep. to Congress and the EPA on NASA Upper Atmos. Res. Prog.*, pp. 324–25. Washington, DC: NASA

Wahlen, M., Tanaka, N., Henry, R., Deck, B., Zeglen, J., et al. 1989. Carbon–14 in methane sources and in atmospheric methane: the contribution from fossil carbon. *Science* 245: 286–90

Whalen, S. C., Reeburgh, W. S. 1988. A methane flux time series for tundra environments. *Global Biogeochem. Cycles* 2: 399–409

Whalen, S. C., Reeburgh, W. S. 1990a. Consumption of atmospheric methane by tundra soils. *Nature* 346: 160–62

Whalen, S. C., Reeburgh, W. S. 1990b. A methane flux transect along the trans-Alaska pipeline haul road. *Tellus* 42B: 237–49

Whiticar, M. J., Faber, E., Schoell, M. 1986. Biogenic methane formation in marine and freshwater environments: CO_2 reduction vs. acetate fermentation—isotope evidence. *Geochim. Cosmochim. Acta* 50: 693–709

Wolin, M. J., Miller, T. L. 1987. Bioconversion of organic carbon to CH_4 and CO_2. *Geomicrobiol. J.* 5: 239–59

Zimmerman, P. R., Greenberg, J. P., Wandiga, S. O., Crutzen, P. J. 1982. Termites: a potentially large source of atmospheric methane, carbon dioxide, and molecular hydrogen. *Science* 218: 563–65

Annu. Rev. Earth Planet. Sci. 1993. 21:427–52

TERRESTRIAL VOLCANISM IN SPACE AND TIME[1]

Tom Simkin

Smithsonian Institution, NHB Mail Stop 119, Washington, DC 20560

KEY WORDS: eruption, hazard, climate, tectonics

INTRODUCTION

Volcanoes are powerful agents of change. Sometimes the changes are broadly destructive, as in the 1991 eruption of Mount Pinatubo, in the Philippines. Sometimes they cost tens of thousands of lives, as in the tragic 1985 eruption of Colombia's Nevado del Ruiz. And sometimes they are even beneficial, as in the replenishment of nutrients by ashfall on Java's fertile fields, or the inventive use of geothermal energy by most of Iceland's population. The 1980 eruption of Mount St. Helens, in the northwestern United States, heightened volcano consciousness in much of the western world, and the decade since has seen dramatic increases not only in volcanological research but also in interdisciplinary studies. Extraterrestrial volcanism has received much attention, and on this planet several narrowly averted catastrophes have energized the air transport industry to find ways of keeping modern jumbo jets away from volcanic ash clouds.

The main interdisciplinary interest, however, has focused on the global climatic impact of major eruptions. Aerosols formed from volcanic gases are distributed throughout the lower stratosphere after major eruptions, and act as a filter, or "sunscreen," to reduce the amount of solar radiation reaching the Earth's surface. This natural agent of global change has been called upon to explain phenomena ranging from June snowfalls in New England to mass extinctions in the geologic record.

Most studies of volcanism—from hazard reduction efforts to global

---

427

climatic change—need information on the rate and frequency of volcanic eruptions in space and time. This paper attempts to review what we have learned, with primary emphasis on the historical record. This record is short, by geological standards, but is a rich and important source of information on the low-to-middle range of volcanism's full spectrum. It provides our best basis for extrapolating the observations and measurements of contemporary eruptions to better understand those of both the past and the future.

This review draws on two Smithsonian resources: (*a*) our reporting of current volcanic activity around the world, and (*b*) our computerized, retrospective database of historic and other dated volcanism over the past 10,000 years (Holocene time). The first began in 1968 as the Center for Short-Lived Phenomena, but was moved, reorganized, and renamed as the Scientific Event Alert Network (SEAN) in 1975. Further concentration on volcanology led to its being renamed the Global Volcanism Network (GVN) in 1990, and it continues today with a monthly *GVN Bulletin* and other outlets. The database began in 1971, was published 10 years later as *Volcanoes of the World* (Simkin et al 1981), and a second edition will soon be in press.

VOLCANOES

Volcanoes are not all symmetrical, snow-capped cones. They range from huge piles of lava, rising up to 10 km above the ocean floor, through small cinder cones, to tranquil lakes on flat (or even inward-dipping) landscapes. They reflect the varied processes that shaped them, and with a wide range of plumbing systems, supply rates, and magma types it is not surprising that a wide range of features results. Furthermore, volcanic processes include many that destroy and modify earlier constructions, adding to the variety of forms that fall under the name "volcano."

Subsurface magma, normally less dense than surrounding rock, rises buoyantly toward the surface and understandably exploits local fracture systems in its path upward. Where long, linear fractures are present, a "curtain of fire" fissure eruption may result. Alternatively, magma may follow the line of weakness formed where two vertical fracture planes intersect, resulting in a pipe-like vertical conduit feeding a symmetrical cone. Many gradations and complexities exist, even changing in the course of a single eruption. The supply of magma to the surface is commonly irregular, even sporadic, and if the supply ceases for too long the next batch will be forced to find another route to the surface because the old conduit will have cooled and hardened. Thus fields of numerous small cones are formed, or (again depending on supply rate and plumbing systems) large mountains dotted by smaller cones.

Perhaps the most influential factor in shaping volcano landforms, though, is the manner in which gas exits the magma. As magma nears the surface, the attendant decrease in pressure permits exsolution of dissolved gases which then drive the eruption vertically (the only direction in which it is free to expand). Both gas content and viscosity vary widely in magmas, and the more gas-rich, viscous, and lower temperature compositions common to continental margin magmas tend to explode violently, fragmenting the liquid into countless small particles which cool swiftly as volcanic glass. This process forms small cones of ash, layer upon layer, or (depending on the violence and longevity of the explosions) thin, widely dispersed ash layers. In contrast, gas separates more easily from the less viscous, hotter, and less gas-rich magmas of the oceanic islands, and lava flows smoothly from vents to form gentle slopes as on Hawaii. All gradations of products exist—from gentle flows, through spatter vents, to scoria, cinder, and ash—with resulting volcanic shapes depending on the gas content and its manner of departing the magma. Often the most violent explosive eruptions are followed by slow, "toothpaste tube"-like extrusions of viscous, degassed magma to form steep-sided domes in the summit vent.

The wide variety of initial constructional shapes is further expanded by additional processes during a volcano's lifetime. For example, lateral shift in the plumbing system of a simple strata-volcano turns it into an asymmetric compound volcano. Destructional processes such as major slope failure (as at Mt. St. Helens in 1980) or caldera collapse (as at Crater Lake, 6,800 years ago) may dramatically change a volcano's shape, and erosion is a relentless modifier in most parts of the world. Such varied processes lead inevitably to the enormous diversity of volcanic forms on the planet (Williams & McBirney 1979, Wilson et al 1987).

VARIATION THROUGH SPACE

Figure 1 shows the distribution of about 1300 terrestrial volcanoes; roughly half have been historically active and all are believed active in the last 10,000 years. Some are scattered, but most are concentrated in linear belts. These belts number about 20 and have accounted for >94% of known historic eruptions. The belts total 32,000 km in length and (assuming a generous 100 km belt width) cover less than 0.6% of the Earth's surface. Most are adjacent to deep oceanic trenches, and it is clear that subduction of one plate beneath another (where oceanic crust is consumed at converging plate margins) causes most of the volcanism that we see. The complementary tectonic process of rifting (the creation of new crust at diverging plate margins) takes place largely on the deep ocean floor, unwitnessed by humans. The left half of the inset in Figure 1 shows the small proportion of the world's historically documented eruptions that have taken place in the ocean rift environment (and in the intraplate environment). It is possible, however, to use relative plate motions to calculate the budget of new lava reaching the Earth's surface each year. This approach was pioneered by the late Kazuaki Nakamura (1974) and has been carried on by Joy Crisp (1984) whose data are shown in the right half of the inset in Figure 1. Clearly mid-oceanic ridge volcanism dominates the planet and, although most of it poses no threat to humans, this dominance must be remembered for global perspective.

Number of Volcanoes

How many active volcanoes are there in the world? This common question can only be answered by knowing which are considered "active." At least 15 volcanoes are almost certainly erupting as this is being written (ongoing eruptions can last tens to thousands of years, e.g. Stromboli); 62 erupted in 1991, 165 in the decade 1980–1989, 538 have had historically documented eruptions; over 1300 have erupted in the past 10,000 years, and some estimates of seafloor volcanoes exceed a million. Take your pick! But remember that the dormant intervals between major eruptions at a single volcano may be thousands of years long and the historic record very short.

Although the recency of activity required to define "active" is important in answering the number question, so is the definition of "volcano." The broad definition used in our compilations allows the record of a single large plumbing system to be viewed as a whole, but requires careful work in field and lab to establish the integrity of the group's common magmatic link. The problem is particularly difficult in Iceland, where eruptions separated by many tens of kilometers along a single rift may share the same magmatic system. Perhaps the most honest answer to the number question is that we do not really know, but that there are at least a thousand identified magma systems—on land alone—likely to erupt in the future.

Tectonic Setting

Although differences between broad volcano groups were recognized by early volcanologists, the conceptual development of plate tectonics in recent decades has provided a framework for such studies (Clapperton 1977). Much attention has been devoted to understanding the features common to specific tectonic settings and how they relate to processes of magma origin, modification, transport, and eruption.

SUBDUCTION   The most common eruptions that we see, and by far the

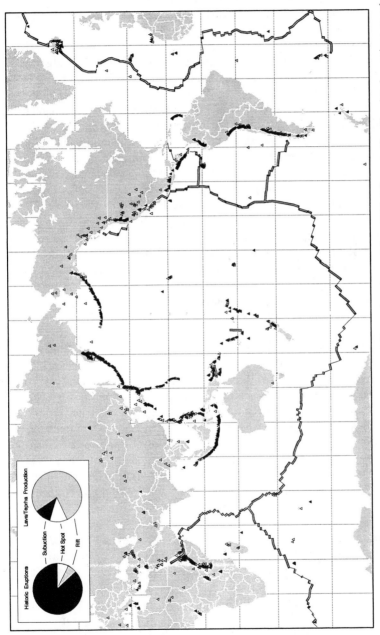

Figure 1 Volcanoes of the world shown by small triangles (filled if eruptions known since 1900 AD, open if earlier dated eruptions or other evidence for eruptions in the last 10,000 years). Mid-oceanic ridge system shown by double black line (from Simkin et al 1989). Pie diagram in inset-left shows proportion of documented historic eruptions from subduction zones (*black*), mid-ocean ridges (*stipple*), and hotspot settings (*white*). Pie diagram in inset-right shows proportion of annual magma budget in the same settings (with same symbols) to show the dramatic contrast between the volcanism that we see and that we don't. Data for inset-right from Crisp (1984) and remainder from Smithsonian.

most dangerous to human populations, are fron volcanoes overlying the world's subduction zones. Because of their explosive nature, these eruptions are also most likely to affect our climate.

As an oceanic plate descends into the hotter, higher pressure interior, it is dehydrated and fluids from it rise into the overlying mantle wedge. In this complex region of shear and melting, new magma is born and rises in buoyant diapirs, but this rise is not steady. There may be many stops along the way, with early-crystallizing minerals separating from the magma, and melting of surrounding rocks further changing its composition by contamination. This slow, complex, and varied route to the surface results, not surprisingly, in the eruption of complexly varied materials at different times and places. These products are reviewed in Gill (1981) and Grove & Kinzler (1986); aspects of the process are discussed by Tatsumi & others (1983, 1991), Plank & Langmuir (1988), Hildreth & Moorbath (1988), and Morris (1991).

In addition to compositional diversity and varied supply rates, a wide variation in the physical nature of subduction further contributes to the broad range of subduction-zone volcanism observed. Segmentation of volcanic chains into discrete linear belts 100–300 km long, commonly offset by zones of more intense seismicity, has been recognized in Central America (Stoiber & Carr 1973, Carr 1984), the Cascades (Hughes et al 1980, Guffanti & Weaver 1988), the Aleutians (Kay et al 1982), and the Andes (Hall & Wood 1985). The geometry of subduction greatly influences the resulting volcanism. In the Andes, for example, volcanism is weak or absent above downgoing slabs that are inclined at only a gentle angle to the horizontal (Barazangi & Isacks 1976, Sacks 1984). Simkin & Siebert (1984) attempted to quantify the volcanic vigor of arc segments around the world, in a contribution to what Uyeda (1981) has called "comparative subductology." We found that relatively weak volcanism also correlated with thin overriding plates, with young downgoing slabs, with decreased angle between the direction of the downgoing slab and the bearing of the arc itself, and with low aseismic slip rates (where great earthquakes indicate strong coupling between converging plates).

MID-OCEANIC RIDGES   The eruptions that we normally do *not* see are those along the 70,000-km oceanic rift system. Here magma is relatively homogeneous basalt, originating in the uppermost mantle (Yoder 1976, 1979) and arriving at the surface in relatively primitive condition. Its extrusion is generally under 1–4 km of water, both shielding it from view and inhibiting volatile release. The result is gentle, effusive eruptions.

The Neo-Volcanic Zone is narrow (often only 1–2 km wide) and volcanoes range, with decreasing spreading rate, from flat fissure eruptions, through elongate shields, to discontinuous chains of central volcanoes constructed of fresh pillow lavas (Macdonald 1982, Solomon & Toomey 1992). Symmetrical central volcanoes are also found outside the rift itself, particularly near large offsets in the ridge system and near topographic highs of the faster spreading ridges; they outnumber subduction-zone and hotspot volcanoes by several orders of magnitude (Batiza 1989, Smith 1991). Hints of the behavior of ocean-rift volcanoes can be found in Iceland, one of the few places where the world's rift system emerges above sea level (Steinthorsson 1989). Here the historic record suggests that, although crustal spreading is probably steady, the rift zone's response to it is episodic: rift eruptions (and intrusions) over several years cause a few meters of separation at intervals of perhaps a hundred years, rather than a regular few centimeters of separation *every* year.

HOTSPOTS AND FLOOD BASALTS   Hotspot volcanoes are believed to be formed by long-lived plumes, anchored deep in the mantle and feeding magma to the surface through the overlying plate. Hawaii is the best-known example of hotspot volcanism, and the variable volumes of the progressively older volcanoes to the northwest testify to the correspondingly variable magma supply as they have been carried in that direction by the Pacific plate during the last 70 million years. Steady feeding of relatively consistent basaltic magma to irregular, shallow chambers yields a delicately balanced system ready to erupt over long time periods and therefore vulnerable to triggering influences such as earthquakes and fortnightly tides. Careful study of historical eruptions on Hawaii shows them to be largely random phenomena, although large-volume eruptions tend to be followed by longer reposes (Klein 1982). Distinctive stages in the long-term development of Hawaiian volcanoes have been recognized, the last of which takes place several million years after the volcano has been separated by plate motion from the feeding plume (Wright 1989).

Stages in the development of the plume itself are believed by many to begin with the first batch of magma rising buoyantly through the mantle to reach the surface as a giant outpouring of flood basalts (Sleep 1992). Flood basalt provinces, such as India's Deccan plateau (60–65 Ma) and the Columbia plateau in the northwestern U.S. (around 15 Ma), have produced hundreds of thousands of km^3 of lava and individual flows that covered thousands of km^2 in just a few days. Such catastrophic episodes are fortunately rare in the planet's history.

INTRACONTINENTAL   This setting has much in common with the two described above in that some intracontinental volcanoes result from rifting and others from hotspots. Rosendahl (1987) has reviewed the former, with special reference to East Africa, and Sleep (1992) reviewed the latter along with oceanic hotspots. Because of the complexities of the thick, siliceous continental crust that is inevitably involved in these eruptions, they are the most violent and least well understood forms of volcanism on the planet. Smith & Luedke (1984) discuss nonsubduction volcanism in the western U. S., emphasizing the relatively long life of these centers (perhaps 5 million years, or ten times the active life of many subduction-zone volcanoes). Pulses of plate motion are thought to trigger magma formation and rise into the complex, extensional environment of the western U.S., and 60 loci are identified as potential eruption sites in the future. Another regional review of intraplate volcanism covers eastern Australia and New Zealand (Johnson 1989a, 1989b), where the overall eruption rate has been slow but apparently continuous throughout the past 95 million years.

ERUPTIONS

Looking at the distribution of volcanism through time again raises questions of definition: Is an individual explosion an "eruption"? or should the word be used (as we tend to use it) for clearly linked events that may be separated by hours, days, or even months of surface quiet? Arrival of volcanic products at the Earth's surface is normally termed an eruption, but the historical record is often unclear. Steam venting is commonly only a byproduct of subsurface heat, and is rarely dignified by the word "eruption," but steam may be all that was seen by a sea captain reporting an "erupting" volcano 200 years ago. Volcanic heat is also the prime source of "phreatic," or "steam-blast," explosions in which heat and water combine to fragment preexisting volcanic materials and hurl the resulting ash into the sky. These events can be quite violent, however, and are rightfully called eruptions, even without the direct involvement of new magma (since few volcanologists are willing to tell a citizen that it is not *really* an eruption because the ash raining down on him contains no fresh volcanic glass).

Gas is an unequivocal volcanic product, but the gentle, continuous emission of fumarolic gases—common to most volcanoes—is not considered an "eruption." Occasionally, though, large quantities of gas are suddenly vented—as in the fatal eruption of the Dieng volcano complex on Java in 1979. In 1984 and 1986, the African nation of Cameroon suffered catastrophic expulsions of CO_2 resulting in many fatalities by suffocation. The origin of these events is controversial, but the evidence suggests that they resulted not from an eruption, but from the sudden overturn of crater lake water, with catastrophic exsolution of volcanic gas that had gradually accumulated in the bottom waters over the course of many years (Sigurdsson et al 1987).

Types and Processes

The great range of eruption types has been mentioned above. The physical processes involved have been reviewed in this series by Wilson et al (1987) and individual eruption types are described in more detail in texts such as Williams & McBirney (1979), Fisher & Schmincke (1984, for explosive eruptions), and Macdonald (1972, for effusive eruptions). Walker (1980) has discussed the various measures of "bigness" of eruptions; volume of products is the most widely used by volcanologists. Eruptions of equal volume, however, may release quite different amounts of gas, with comparably different effects on climate. The larger eruptions inevitably receive the greatest share of scientific (and media) attention, but the steady pulse of smaller events, such as those that make up most of the 10-year global record of McClelland et al (1989), add up to significant contributions. In assessments of annual volcanic SO_2 production, for example, the steady, unspectacular contributions of the world's fumaroles and small, ongoing eruptions are found to outweigh those of the less frequent larger eruptions (Stoiber et al 1987). In any case, the range of eruption types is large, and there are many that, in Scrope's words of 130 years ago (Scrope 1862, p. 14), "exhibit an infinite variety of phases intermediate between the extremes of vivacity and sluggishness."

Durations

Clearly some eruptions last for a *very* long time, like Stromboli's 2500+ year continuing show. At the time of this writing (mid–1992) the following 15 volcanoes have been erupting through the last 17 years (the reporting span of SEAN/GVN) and are likely to remain active for some time: Stromboli; Erta ale, Ethiopia; Manam, Langila, and Bagana, Papua New Guinea; Yasur, Vanuatu; Semeru and Dukono, Indonesia; Suwanose-jima and Sakura-jima, Japan; Santa Maria and Pacaya, Guatemala; Arenal, Costa Rica; Sangay, Ecuador; and Erebus, Antarctica. However, other eruptions end swiftly, with 10% of the eruptions for which we have accurate durations having lasted no longer than a single day. As shown in Figure 2, most end in less than 100 days and few last longer than 1000 days (3 years). The median duration is about 7 weeks.

History's largest explosive eruption, Tambora in 1815, provides sobering lessons about eruption duration and about the wide range of behavior in a single eruption. Mild eruptive activity over three fulls years was followed by a dramatic eruption, with cloud heights estimated at 33 km— but this was not its end. After a lull of five days, the culminating eruption

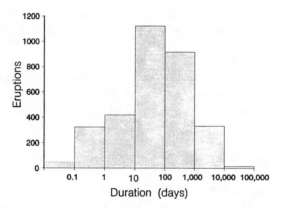

Figure 2 Eruption duration histogram. 3069 eruptions for which stop (as well as start) date is known. The data for ≤0.1 days (≤2½ hours) are dashed because relatively few durations listed as ≤1 day carry sufficient time information to be more specific.

reached heights estimated at 44 km and caused three days of total darkness over an area covering 500 km from the volcano (Sigurdsson & Carey 1989). Knowing when an eruption will end is no easier than predicting its beginning.

VARIATION THROUGH TIME

Historical Patterns

Has global volcanism changed through time? A look at the number of volcanoes active per year, over the last 500 years, shows a dramatic increase, but one that is closely related to increases in the world's human population and communication. We believe that this is an increased *reporting* of eruptions, rather than increased *frequency* of global volcanism: more observers, in wider geographic distribution, with better communication, and broader publication. The past 200 years (Figure 3) show this generally increasing trend along with some major "peaks and valleys" which suggest global pulsations. A closer look at the two largest valleys, however, shows that they coincide with the two World Wars, when people (including editors) were preoccupied with other things. Many more eruptions were probably witnessed during those times, but reports do not survive in the scientific literature.

If these apparent drops in global volcanism are caused by decreased human attention to volcanoes, then it is reasonable to expect that increased attention after major, newsworthy eruptions should result in higher-than-average numbers of volcanoes being reported in the historic literature. The

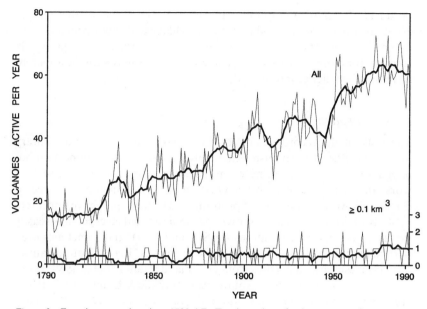

Figure 3   Eruption reporting since 1790 AD. Total number of volcanoes erupting per year (*thin upper line*) and 10-year running mean of same data (*thick upper line*). Lower lines show only the annual number of volcanoes producing large eruptions (≥ 0.1 km^3 of tephra or magma) and scale is enlarged (on right axis). Thick lower line again shows 10-year running mean.

1902 disasters at Mont Pelee, St. Vincent, and Santa Maria (Figure 3) were highly newsworthy events. They represent a genuine pulse in Caribbean volcanism, but we believe that the higher numbers in following years (and following Krakatau in 1883) result from increased human interest in volcanism. People reported events that they might not otherwise have reported and editors were more likely to print those reports. We felt so confident of our interpretation that we predicted, in our 1981 book *Volcanoes of the World*, a similar increased reporting of volcanism after the 1980 eruption of Mount St. Helens. This did not happen (the prediction game is difficult to play) but we would like to think, by way of partial explanation, that we are finally approaching reasonably comprehensive reporting of global volcanism. Note (in Figure 3) that the number of erupting volcanoes has leveled off between 55 and 70 per year during the past 3 decades.

Further evidence that the historical increase in global volcanism is more apparent than real comes from the lower plot of Figure 3. Here only the larger eruptions (generating at least 0.1 km^3 of tephra, the fragmental products of explosive eruptions) are plotted. The effects of these larger events are often regional, and therefore less likely to escape documentation in remote areas. The frequency of these events has remained impressively constant for more than a century, and contrasts strongly with the apparent increase of smaller eruptions with time.

New Approaches

The bulk of the record described above has come from contemporary monitoring, the sifting of historical records, and the terribly important but unglamorous (and increasingly unfunded) combination of careful field work and radiocarbon or other Quaternary dating methods. In recent decades, however, a number of promising new techniques have appeared.

Study of deep ice cores has shown that some annual layers contain easily measured acidity concentrations (H_2SO_4 and HCl), and sometimes even particulate matter from major volcanic eruptions. Records of volcanism have been established in cores from Greenland (Hammer et al 1980, Hammer 1984, Lyons et al 1990), Antarctica (Legrande & Delmas 1987, Palais et al 1990, Delmas et al 1992), and both (Langway et al 1988). Relatively small eruptions from high latitudes leave strong records in nearby polar ice cores, but 95% of the world's known Holocene volcanoes lie more than 30° from the poles, and ice cores clearly show very large events from low latitudes. An example of the technique's value is a major eruption in 1259 AD: it has not been found in the historical record, but its effects are clearly shown in ice cores from both polar regions.

Tree-ring dating has also been effectively used to refine dates for major eruptions (e.g. LaMarche & Hirschboeck 1984, Baillie 1991), although only climatic impact is dated and other evidence must be found to tie such an impact to volcanism. Tree-ring study close to a specific volcano, however, can yield valuable information on its history (e.g. Yamaguchi 1985).

Statistical techniques, despite problems of incompleteness in the volcanic record, have been increasingly used in attempts to learn from that record. Wickman (1966, 1976) applied stochastic principles to the histories of 29 volcanoes, and found 5 that showed simple Poisson distributions but more that did not. This approach has been extended by De la Cruz-Reyna (1991). The most careful recent work has been by Stothers, who used time-series analysis on the Smithsonian data set to show that no month-to-month seasonal variation could be found on a global basis (Stothers 1989a), but that two weak periodicities, at around 11 and 80 years, are probably statistically significant (Stothers 1989b). Both correlate with cycles of solar activity; slightly increased volcanism occurring around times of solar minima. Using the same data set, Zemtsov & Tron (1985) found a 22-year cycle significant, while Guttorp & Thompson (1991) found only a 40-year cycle in the record of Icelandic volcanoes. Gudmundsson & Saemundsson (1980), however, found no statistically significant temporal patterns in the history of Icelandic eruptions. Fractal analysis has recently been applied to the record of four basaltic volcanoes (Dubois & Cheminee 1991) and pattern recognition to one (Mulargia et al 1991). Statistical analysis is being applied to the much longer geologic record of the proposed nuclear waste repository in Nevada (Ho et al 1991).

Satellite technology has aided subaerial volcanology in recent decades, with images which allow mapping of remote, difficult areas such as the central Andes (de Silva & Francis 1991). The satellites also monitor volcanic activity in real time. For more than a decade, weather satellites have routinely provided valuable information on the timing, size, and direction of volcanic plumes (Matson 1984, Holasek & Rose 1991, Casadevall 1993), but perhaps the most exciting application has been Arlin Kreuger's use of the TOMS instrument (Total Ozone Mapping Spectrometer) on the polar-orbiting *Nimbus* 7 satellite to provide quantitative mapping of SO_2 dispersal from erupting volcanoes. This instrument, launched in 1978, has provided valuable data on major eruptions over an important volcanological period, showing that El Chichón in 1982, although producing magma volumes quite similar to Mount St. Helens in 1980, vented dramatically more SO_2 to the stratosphere and had a comparably larger impact on global climate. TOMS measurements of Pinatubo in 1991 show 3 times as much SO_2 production as El Chichón, and emphasize its importance as a natural experiment in the climatic effect of volcanism (Bluth et al 1992). Tracking of the Pinatubo aerosol by satellite and other instruments (McCormick 1992) clearly demonstrates the value of modern technology in understanding the volcano-climate relationship.

Additional monitoring, prompted apparently by national defense purposes, has also been very beneficial to volcanology. An airborne stratospheric sampling program in 1981–1991 gathered valuable sulfate measurements of eruptions (Sedlacek et al 1983), a heat-sensing satellite allowed precise minute-by-minute timing of the development of Mount St. Helens' eruptive cloud in 1980 (Moore & Rice 1984), and submarine acoustic networks have provided some of the only historical data available on the deep-ocean volcanism that dominates Earth's magma budget (Norris & Johnson 1969, Talandier 1989). These examples have been uncommon openings for scientists into largely classified defense monitoring. However, previously classified listening arrays, such as SOSUS, are now being used to locate submarine volcanism, and aircraft-deployed sonobuoys have also been pressed into service (Vogt et al 1990). As the outlook for world peace brightens, we can hope that more of this technology will be declassified to permit a better understanding of our planet.

Volcanic Cycles and Episodicity

In the global historical record, the most active year known was 1974, when 5 of 7 western Papua New Guinean volcanoes erupted (Cooke et al 1976). Similar temporal concentrations of volcanism around a single plate margin are known from 1902 (Caribbean plate, noted above) and 1835 [when Darwin (1840) observed an unusual number of Andean volcanoes active around the time of the famous Concepción earthquake]. These may well result from irregularities in local plate motion, but we see no global periodicity or episodicity that rises above the socio-economic effects on reporting discussed above. As in 1835, regional tectonic earthquakes, and perhaps aseismic plate lurches, appear to have triggered synchronous eruptions. A necessary precondition for tectonic triggering of eruptions is the length of time involved in the cooling and crystallization of large magma chambers. During this time, volatile concentration increases inexorably, and it is reasonable to expect that many of the world's thousand or more chambers are poised and ready to erupt at any given time. As "the straw that broke the camel's back," even earth tides are enough to trigger eruptions.

The search for order in individual volcanic histories—with obvious significance to forecasting new eruptions—has been part of most field/ petrologic studies of active volcanoes. Many have revealed only daunting complexity, but others, particularly vigorous volcanoes with centuries-long historical records, have shown clear cycles. Baratta (1897) and Alfano & Friedlaender (1929) recognized cycles in the long and richly detailed record of Vesuvius. At Mayon, Newhall (1979) observed that basaltic magma is apparently added periodically to shallow reservoirs where fractional crystallization produces successively more andesitic lavas (and more volatile, explosive eruptions), over 75–100-year periods. At Colima, Luhr & Carmichael (1990) showed that large ashflow eruptions usher in a new cycle that begins with crater dome building for 50 or more years and continues with an equal period of intermittent eruptions of andesitic block lava before terminating in another Pelean-type eruption. Luhr & Carmichael point out, however, that many different styles and intensities of eruption are involved, and stochastic modeling (assuming independent, homogeneous phenomena) cannot be realistically applied to Colima.

Study of the more siliceous, longer lived, continental volcanoes has been more difficult. Starting in 1979, however, a series of important papers

(Smith 1979, Hildreth 1981, Spera & Crisp 1981, Lipman 1984, Shaw 1985) have synthesized large numbers of individual studies and brought better understanding to these dangerous systems. Rates of magma supply and production, repose intervals, and individual eruption volumes are clearly related. "Eruption occurs when the input rate of primitive magma, over and above rates of leakage of primitive and fractionated magma, is sufficient to generate a fractionated volume capable of sustaining ash-flow eruptions when, at the same time, conditions of extensional fracture are met" (Shaw 1985, p. 11,277). These relationships are shown graphically in Figure 12 of Smith (1979). Mafic magma is supplied at a rate of about 10^6 m^3/year, and a variety of magmas are produced from the tops of differentiating chambers at a rate of about 10^6 m^3/year, resulting (if other influences are small or balance each other) in a volume of accumulated magma that is roughly proportional to the time interval since the last eruption.

Although such relationships illuminate the periodicity of individual volcanic systems, they imply no synchroneity of eruptions from multiple volcanoes. They do, however, bring home the point that magma chambers spend substantial portions of their lives in near-eruptive states and are thus vulnerable to external triggering influences such as major earthquakes, landslides, or earth tides. Dzurisin (1980) has shown a weak but significant correlation between fortnightly earth tides and the historic eruptions of Kilauea (but none with neighboring Mauna Loa), and earthquakes have triggered eruptions from Kilauea (1974) to Mount St. Helens (1980). Even the crustal flexure of rapid deglaciation has been invoked as an eruption trigger (Grove 1976, Wallmann et al 1988, Bogaard & Schmincke 1990).

Regional/Global Episodicity

To explore broader episodicity of volcanic belts, and possible synchroneity of global activity, it is necessary to enlarge our framework of both space and time: to change scale from individual historic eruptions to larger pulsations of volcanism. An example is regional study of explosive vol-canism over several million years (e.g. Machida 1990). Another is bathy-metric study of the Hawaiian hotspot trace, shown by Shaw et al (1980) to have produced major edifices (most which are now seamounts) during episodic vigorous phases separated by less vigorous phases as the Pacific plate has moved relentlessly to the northwest. However, any such studies inevitably confront the twin problems of inadequate mapping in many parts of the world and increasing uncertainty as we go back in time.

Gilluly (1969, p. 2303) summarizes well the geologic problems of inter-preting the older record: "As with optical perspective, it is clear that a feature distant in time must be larger than one close at hand if it is to appear as distinctly." This is "not simply because younger rocks come to bury the older, but also because the younger have been largely derived by the cannibalizing of the older." Nevertheless, Gilluly (1965) reviewed magmatic activity in the western U.S. and concluded that, for most of Phanerozoic time, volcanism has been "episodic but on the whole quasi-continuous." After early development of plate tectonics, Gilluly (1973) called upon the immense inertia of the plates to support his contention that magmatism, although locally episodic, has proceeded steadily through time in the Cordillera as a whole.

Working from a far larger data set, Armstrong & Ward (1991, p. 13,203) conclude that "although magmatism is always locally episodic, the episodes blur into a continuum of magmatic activity when the whole Cordilleran region is viewed during later Cenozoic time."

The Oligocene maximum was followed by a general decline in Cascade volcanism since 35 Ma, but there is some evidence of a Pliocene lull followed by accelerated volcanism during the past 2–3 Ma. The Cascades arc was established in its present location by 35 Ma and has "remained a persistent magmatic feature for the rest of Cenozoic time" (Armstrong & Ward 1991). In the past 10 Ma its magmatic front has migrated eastward in central Oregon and jumped westward in British Columbia. Additional aspects of episodic volcanism in this well-studied region are discussed in Armstrong & Ward (1991), in McBirney (1978), and in a volume edited by Lipman & Glanzer (1991).

While the extended continental record of volcanism is complicated (and often lost) by erosion, orogeny, glaciation, and more, the deep-sea floor offers the hope of a relatively continuous record. Cores of deep sea sediments contain far-traveled tephra from numerous explosive eruptions, and their study (Kennett et al 1977; Kennett 1981, 1989) has provided evidence of episodicity in many regions. Past changes in wind patterns or plate motions cause complications, as do both diagenesis and bioturbation, but study of many Pacific cores indicates episodes of intense volcanism (e.g. past 2 Ma, 17–14 Ma) separated by long quiescent periods (e.g. Late Miocene). There is strong evidence for widespread synchronism of these episodes over large areas of the Pacific and similarities with both mean spreading rate and global climate change. Vogt (1979, 1986) believes that both hotspot and island arc volcanism exhibit some episodic behavior on scales of 1–10 Ma and that this behavior is to an extent globally synchronous. He closes his 1986 paper, however, with the caution that "If hotspots, plate speed, island arc volcanism, and mountain building all sing in the same global chorus, it will be difficult to isolate from the geologic record the environmental impact of volcanism alone."

Magnitude and Frequency of Volcanism

How much do eruptions vary in size? Like earthquakes, the frequency of eruptions decreases with increasing size: There are a lot of small ones, fewer medium-sized ones, and not so many large ones. We have borrowed from the seismologists' magnitude-frequency plots, using as our best measure of magnitude the VEI (Volcanic Explosivity Index) of Newhall & Self (1982). This closely parallels the volume of erupted tephra, which increases by an order of magnitude with each VEI larger integer. In Figure 4, the number of eruptions at and above each VEI increment is shown for various time intervals in the recent past: 30, 200, 1000, and 2 million years. We have used longer time intervals for data from successively larger VEI groups because the smaller events have been adequately reported only in recent decades whereas the larger events, being rarer, need a longer time interval for representative counts. This is a somewhat subjective process, but the intervals are chosen in the hope that they are long enough to be representative yet short enough to have reasonably accurate reporting. As discussed with Figure 3, larger eruptions are more likely to have been accurately reported over longer periods, on a global basis, than smaller eruptions.

Another way of interpreting this plot is that subaerial eruptions producing at least 10^6 m^3 of tephra (VEI 2) take place at an average rate of once every few weeks somewhere on Earth. Those producing $\geqslant 10^7$ m^3, like the VEI 3 Ruiz event that generated such fatal mudflows in 1985, take place several times a year. Eruptions the size of St. Helens in 1980 (10^9 m^3, or 1 km^3 and VEI 5) occur perhaps once a decade, and those like Krakatau 1883 (> 10 km^3) on the order of once a century. The historic record, however, fails to prepare us for the much larger eruptions of the past. The Toba eruption, only 74,000 years ago in Indonesia, produced nearly 10,000 times more magma (2800 km^3, Chesner & Rose 1991) than the St. Helens eruption. Even larger eruptions are known from the older geologic record.

Simple extrapolation of the Holocene record in Figure 4 is not warranted above VEI 7, however, and that value, we must remember, is based on a single data point. The shape of the magnitude-frequency relationship at this end is unclear, but Decker (1990) has attempted to improve it by plotting data from Newhall & Dzurisin's (1988) compilation of caldera unrest during the last 2 million years. The data set is incomplete, as the authors themselves are quick to acknowledge, and Decker is forced to make some large assumptions, but his extension is probably closer to the truth than extrapolation of the historical data. Decker's calculations suggest 22 VEI 7 events every 10,000 years and 2 VEI 8 events (Yellowstone-sized) every 100,000 years.

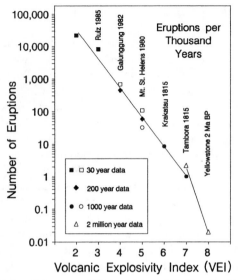

Figure 4 Magnitude and frequency of Holocene eruptions. Cumulative number of eruptions during various intervals (going back in time from 1 January 1992) for each VEI class and normalized to "per 1000 years." Eruptions of low explosivity (VEI 0 and 1) are not shown, and a logarithmic scale is used to allow resolution of the lower frequencies found with the larger eruptions. VEI scale is also logarithmic since each integer represents a factor of 10 increase in volume of tephra erupted. Best-fit line is determined by an exponential regression model for VEI 2–7 using data points that are filled. Data points from Decker (1990) for VEI 7 and 8 are shown by open triangles. Well-known eruptions are labeled to illustrate each VEI class.

HUGE ERUPTIONS AND THEIR EFFECTS

It is both important and easy to say that prehistoric eruptions have dwarfed the largest known in human history, but it is difficult to say how big they have been or how often they occur. Considerable interest attends these questions, though, not only because of the likelihood of their recurrence in the future but also because of their impact on prehistoric climate and biota (Rampino et al 1988, Simarski 1992). An eruption at Yellowstone— the Huckleberry Ridge Tuff—about 2 million years ago produced 2500 km³ of magma that cooled as a single unit (Christiansen 1984). Volcanic ash from this eruption was distributed over at least a 16-state area (including Los Angeles, Tucson, El Paso, and Des Moines) with compacted thicknesses of 0.2 m at distances of 1500 km from the source (Izett & Wilcox 1982). Its impact must have been profound. A much more recent eruption from Yellowstone, 0.66 Ma, was nearly as large, and numerous Oligocene eruptions in the western U. S. were even larger (G. A. Izett

1992, personal communication). Alteration and erosion of distant ash layers makes older records even more difficult to interpret, but individual eruption volumes of 10^4 km^3 have not been recognized.

It may, however, be useful to look at the known effect of large historic eruptions in the context of Figure 4. The 1980 eruption of Mount St. Helens (VEI 5, or 1/decade) was a *local* catastrophe with devastation over hundreds of square kilometers. The 1883 eruption of Krakatau (VEI 6, or 1/century) was a *regional* catastrophe, with 36,000 dead, some as far as India. Another order of magnitude larger, the 1815 eruption of Tambora (VEI 7, or 1/milennia) caused global cooling, with summer snowstorms and crop failures on the opposite side of the globe (Stommel & Stommel 1983). A New England farmer in 1816 would have been justified in calling it a *global* catastrophe. The effects of a VEI 8 event, such as Yellowstone (2 Ma), must have been global by any definition. These effects deserve more attention. Although unlikely as a cause for the planet's largest mass extinctions, explosive volcanism has surely caused smaller catastrophes and may have contributed—through the triggering effect of impact shock waves on poised magma chambers—to larger catastrophes. It is worth remembering how little was known of impacts 25 years ago, and reminding ourselves that there is much yet to learn about huge eruptions.

VOLCANIC HAZARD

To conclude with some comments on volcanic hazard, Figure 5 shows the volcanic death toll for the years since 1800. The total number of fatalities— indirect as well as direct—are plotted on a logarithmic axis and all eruptions with > 1000 fatalities are labeled. Large death tolls continue through the record, from Tambora in 1815 to Ruiz in 1985. Many of the numbers are uncertain, and there is no obvious pattern to the larger figures, but the frequency of fatal eruptions clearly increases with time. Human populations grow dramatically while volcanism shows no sign of slowing down, ensuring that volcanic hazard will be a continuing problem for the future.

One of the certain contributors to large volcanic death tolls is the fact that the repose interval between eruptions is very much longer than the historic records in many parts of the world. Figure 6 shows the relationship between eruption interval and the proportion of historic eruptions in each VEI class. Most eruptions of low explosivity follow repose intervals of only 1 to 10 years, but with increasing time intervals the explosivity increases significantly. Long repose periods precede large eruptions, and in regions with short historic records the human population is commonly unprepared. The results are often tragic.

Humans have not been successful in controlling volcanoes in a sub-

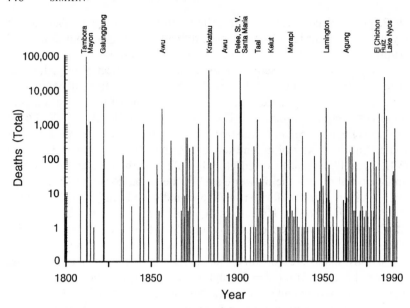

Figure 5 Volcanic death toll since 1800 AD. The total number of fatalities—indirect as well as direct—is plotted on a logarithmic axis and all eruptions with >1000 fatalities are labeled. Many of these data are from Blong (1984). (Updated from McClelland et al 1989.)

stantive way and are not likely to be so in the future. We can, however, develop a more successful coexistence with volcanoes, neither fatalistically accepting their actions nor relying on last-minute attempts at control. To do this, we must emphasize understanding as the first element of prediction, and improve our human response to eruptions when they occur (Fiske 1988, Decker 1986, Latter 1989, Tilling 1989). This requires understanding of the huge eruptions of the geologic past as well as the more frequent and more familiar events of our own lifetimes. With limited resources, volcanology should concentrate efforts on a limited number of dangerous volcanoes, but it would be a mistake to overemphasize the historic record in selecting them. For at least two centuries we have had an average of one or two eruptions per year from volcanoes with *no* previous historic activity. Other volcanoes have erupted after hundreds of years of quiet. Because their danger was not widely recognized, and because unusually violent eruptions are known to follow unusually long periods of quiet, these events included some of history's worst natural disasters. Table 1 lists the most violent eruptions of the last 2 centuries—those with a VEI ≥5 (or ≥1 km³ of tephra). Notice that 12 of 16 were the first historic

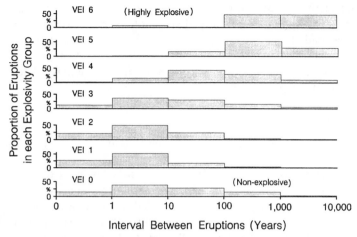

Figure 6 Explosivity and time intervals between eruptions. For each VEI unit, eruptions are grouped by time interval from start of previous eruption. The number of eruptions in VEI groups 0 to 6 are, respectively, 465, 654, 2992, 728, 195, 48, 15. (Updated from Simkin et al 1981.)

Table 1 Largest explosive eruptions of the 19th and 20th centuries[a]

Year	Volcano	First historic?	Deaths
1991	Cerro Hudson (Chile)	no	0
1991	Pinatubo (Philippines)	yes	>740
1982	El Chichón (Mexico)	yes	2,000
1980	Mount St. Helens (USA)	no	57
1956	Bezymianny (Kamchatka)	yes	0
1932	Cerro Azul/Quizapu (Chile)	no	0
1912	Novarupta/Katmai (Alaska)	yes	2
1907	Ksudach (Kamchatka)	yes	0
1902	Santa María (Guatemala)	yes	>5,000
1886	Tarawera (New Zealand)	yes	>150
1883	Krakatau (Indonesia)	no	36,417
1875	Askja (Iceland)	yes	0
1854	Shiveluch (Kamchatka)	yes	0
1835	Cosiguina (Nicaragua)	yes	5–10
1822	Galunggung (Indonesia)	yes	4,011
1815	Tambora (Indonesia)	yes	92,000

[a] VEI $\geq$ 5, or tephra volume ≥ 1 km^3. All but four were the first historic eruption known from the volcano, and the high death tolls (in heavily populated regions) reflect that fact. (Updated from McClelland et al 1989.)

eruption at that volcano (and for others, like Krakatau, earlier activity was not widely known). The study of volcanology tends to concentrate on volcanoes made famous by eruptions that were either destructively large or photogenically long-lasting, but we must not neglect the nonfamous, thickly vegetated volcanoes that have had no historic activity. They may be the most dangerous of all.

ACKNOWLEDGMENTS

This paper is an outgrowth of a symposium held at the 28th International Geological Congress in 1989.

Lindsay McClelland and Lee Siebert have worked with me for nearly 20 years, and their able efforts in gathering contemporary and older eruption information (respectively) have been essential to the data and ideas presented here. Roland Pool translated jumbled numbers into crisp computer graphics, and Jim Luhr, Lee Siebert, Bob Smith, Chris Newhall, and Peter Vogt helped improve the text. All are warmly thanked (and absolved of any blame). Smithsonian funding of the Global Volcanism Program and NASA grant 1438-VCIP-1675 are also gratefully acknowledged.

Literature Cited

Alfano, G. B., Friedlaender, I. 1929. *La Storia del Vesuvio Illustrata da Documenti Coevi.* Ulm: Karl Hohn. 172 pp.

Armstrong, R. L., Ward, P. 1991. Evolving geographic patterns of Cenozoic magmatism in the North American Cordillera: the temporal and spatial association of magmatism and metamorphic core complexes. *J. Geophys. Res.* 96: 13,201–24

Baillie, M. G. L. 1991. Marking in marker dates—toward an archaeology with historical precision. *World Archaeol.* 23: 433–44

Baratta, M. 1897. *Il Vesuvio e le sue Eruzioni.* Rome: Soc. Editrice Dante Alighieri. 203 pp.

Barazangi, M., Isacks, B. L. 1976. Spatial distribution of earthquakes and subduction of the Nazca plate beneath South America. *Geology* 4: 686–92

Batiza, R. 1989. Origin and evolution of oceanic volcanoes near East Pacific Rise. *Washington, DC: 28th Int. Geol. Cong. Abstr. Vol.* 1/100–1

Blong, R.J., 1984. *Volcanic Hazards.* Sydney: Academic. 427 pp.

Bluth, G. J. S., Doiron, S. D., Schnetzler, C. C., Krueger, A. J., Walter, L. S. 1992. Global tracking of the SO₂ clouds from the June, 1991 Mount Pinatubo eruptions. *Geophys. Res. Lett.* 19: 151–55

Bogaard, P. V. D., Schmincke, H.-U. 1990. The 700,000-year eruption and paleoclimate record of the East Eifel Volcanic Field—warm climate because of eruptions, or eruptions because of warm climate? *Abstr. vol. Int. Volcanol. Congr., Mainz (FRG), IAVCEI (Int. Assoc. Volcanol. Chem. Earth's Inter.)*

Boyd, F. R., ed. 1984. *Explosive Volcanism: Inception, Evolution, and Hazards.* Washington, DC: Natl. Acad. 176 pp.

Carr, M. J, 1984. Symmetrical and segmented variation of physical and geochemical characteristics of the Central American volcanic front. *J. Volcanol. Geotherm. Res.* 20: 231–52

Casadevall, T. J., ed. 1993. Proc. Vol., First Int. Symp. on Volcanic Ash and Aviation Safety. *US Geol. Surv. Bull.* In press.

Chesner, C. A., Rose, W. I. 1991. Stratigraphy of the Toba Tuffs and the evolution of the Toba Caldera Complex, Sumatra, Indonesia. *Bull. Volcanol.* 53: 343–56

Christiansen, R. L., 1984. Yellowstone magmatic evolution: its bearing on understanding large-volume explosive volcanism. See Boyd 1984, pp. 84–95

Clapperton, C. N. 1977. Volcanism in space and time. *Prog. Phys. Geogr.* 1: 375–411

Cooke, R. J. S., McKee, C. O., Dent, V. F.,

Wallace, D. A. 1976. Striking sequence of volcanic eruptions in the Bismarck Volcanic Arc, Papua New Guinea, in 1972–75. In *Volcanism in Australasia*, ed. R. W. Johnson, pp. 149–72. Amsterdam: Elsevier

Crisp, J., 1984. Rates of magma emplacement and volcanic output.*J. Volcanol. Geotherm. Res.* 20: 177–211

Darwin, C., 1840. On the connexion of certain volcanic phenomena in South America; and on the formation of mountain chains and volcanos, as the effect of the same power by which continents are elevated. *Trans. Geol. Soc. London* 5(2nd Ser.): 601–31

Decker, R. W. 1986. Forecasting volcanic eruptions. *Annu. Rev. Earth Planet. Sci.* 14: 267–91

Decker, R. W. 1990. How often does a Minoan eruption occur? In *Thera and the Aegean World III*, ed. D. A. Hardy 2: 444–52. London: Thera Found.

De la Cruz-Reyna, S. 1991. Poisson-distributed patterns of volcanic activity. *Bull. Volcanol.* 54: 57–67

Delmas, R. J., Kirchner, S., Palais, J. M., Petit, J. R. 1992. 1000 years of explosive volcanism recorded at South Pole. *Tellus* 44B: 335–50

de Silva, S. L., Francis, P. W. 1991. *Volcanoes of the Central Andes*. Berlin: Springer-Verlag. 216 pp.

Dubois, J., Cheminee, J. L. 1991. Fractal analysis of eruptive activity of some basaltic volcanoes. *J. Volcanol. Geotherm. Res.* 45: 197–208

Dzurisin, D. 1980. Influence of fortnightly earth tides at Kilauea volcano, Hawaii. *Geophys. Res. Lett.* 7: 925–28

Fisher, R. V., Schmincke, H.-U. 1984. *Pyroclastic Rocks*. Berlin: Springer-Verlag. 472 pp.

Fiske, R. S. 1988. Volcanoes and society—challenges of coexistence. *Proc. Kagoshima Int. Conf. on Volcanoes*, pp. 14–21. Kagoshima Prefectural Gov.

Gill, J. B. 1981. *Orogenic Andesites and Plate Tectonics*. Berlin: Springer-Verlag. 390 pp.

Gilluly, J. 1965. Volcanism, tectonism, and plutonism in the western United States. *Geol. Soc. Am. Spec. Pap.* 80, 69 pp.

Gilluly, J. 1969. Geologic perspective and the completeness of the geologic record. *Geol. Soc. Am. Bull.* 80: 2303–12

Gilluly, J. 1973. Steady plate motion and episodic orogeny and magmatism. *Geol. Soc. Am. Bull.* 84: 499–515

Grove, E. W. 1976. Deglaciation—a possible triggering mechanism for recent volcanism. In *Andean and Antarctic Volcanology Problems. Santiago: Int. Assoc. Volcanol. Symp.*, pp. 88–97

Grove, T. L., Kinzler, R. J. 1986. Petrogenesis of Andesites. *Annu. Rev. Earth Planet. Sci.* 14: 417–54

Gudmundsson, G., Saemundsson, K. 1980. Statistical analysis of damaging earthquakes and volcanic eruptions in Iceland from 1550–1978. *J. Geophys.* 47: 99–109

Guffanti, M., Weaver, C. S. 1988. Distribution of late Cenozoic volcanic vents in the Cascade range: volcanic arc segmentation and tectonic considerations. *J. Geophys. Res.* 93: 6513–29

Guttorp, P., Thompson, M. L. 1991. Estimating second-order parameters of volcanicity from historical data. *J. Am. Stat. Assoc.* 86: 578–83

Hall, M. L., Wood, C. A. 1985. Volcano-tectonic segmentation of the northern Andes. *Geology* 13: 203–7

Hammer, C. H. 1984. Traces of Icelandic eruptions in the Greenland ice sheet. *Jökull* 34: 51–65

Hammer, C. H., Clausen, H. B., Dansgaard, W. 1980. Greenland ice sheet evidence of post-glacial volcanism and its climatic impact. *Nature* 288: 230–35

Hildreth, W. 1981. Gradients in silicic magma chambers: implications for lithospheric magmatism. *J. Geophys. Res.* 86: 10,153–92

Hildreth, W., Moorbath, S. 1988. Crustal contributions to arc magmatism in the Andes of Central Chile. *Contrib. Mineral. Petrol.* 98: 455–89

Ho, C.-H., Smith, E. I., Feuerbach, D. L., Naumann, T. R. 1991. Eruptive probability calculation for the Yucca Mountain site, USA: statistical estimation of recurrence rates. *Bull. Volcanol.* 54: 50–56

Holasek, R. E., Rose, W. I. 1991. Anatomy of 1986 Augustine volcano eruptions as recorded by multispectral image processing of digital AVHRR weather satellite data. *Bull. Volcanol.* 53: 420–35

Hughes, J. M., Stoiber, R. E., Carr, M. J. 1980. Segmentation of the Cascade volcanic chain. *Geology* 13: 203–7

Izett, G. A., Wilcox, R. E. 1982. Map showing localities and inferred distributions of the Huckelberry Ridge, Mesa Falls, and Lava Creek ash beds (Pearlette family ash beds) of Pliocene and Pleistocene age in the western United States and southern Canada. *US Geol. Surv. Misc. Invest. Ser. Map* I–1325

Johnson, R. W. 1989a. Time-space relationships in intraplate volcanism of eastern Australia and New Zealand. *Washington, DC: 28th Int. Geol. Congr. Abstr. Vol.* 2/134

Johnson, R. W., ed. 1989b. *Intraplate Volcanism in Eastern Australia and New Zea-*

land. Cambridge: Cambridge Univ. Press. 408 pp.

Kay, S. M., Kay, R. W., Citron, G. P. 1982. Tectonic controls on tholeiitic and calc-alkaline magmatism in the Aleutian arc. *J. Geophys. Res.* 87: 4051–72

Kennett, J. P. 1981. Marine tephro-chronology. In *The Sea, Vol. 7, The Oceanic Lithosphere*, ed. C. Emiliani, pp. 1373–435. New York: Wiley

Kennett, J. P. 1989. Neogene history of explosive volcanism: the deep-sea tephra record. *Washington, DC: 28th Int. Geol. Congr. Abstr. Vol.* 2/175–6

Kennett, J. P., McBirney, A. R., Thunnell, R. C. 1977. Episodes of Cenozoic volcanism in the circum-Pacific region. *J. Volcanol. Geotherm. Res.* 2: 145–63

Klein, F. W. 1982. Patterns of historical eruptions at Hawaiian volcanoes. *J. Volcanol. Geotherm. Res.* 12: 1–35

LaMarche, V. C., Hirschboeck, K. K. 1984. Frost rings as records of major volcanic eruptions. *Nature* 307: 121–26

Langway, C. C., Clausen, H. B., Hammer, C. U. 1988. An interhemispheric volcanic time-marker in ice cores from Greenland and Antarctica. *Ann. Glaciol.* 10: 102–8

Latter, J. H., ed. 1989. *Volcanic Hazards: Assessment and Monitoring*. Berlin: Springer-Verlag. 625 pp.

Legrande, M., Delmas, R. J. 1987. A 220-year continuous record of volcanic H_2SO_4 in the Antarctic ice sheet. *Nature* 327: 671–76

Lipman, P. W. 1984. The roots of ash flow calderas in western North America: windows into the tops of granitic batholiths. *J. Geophys. Res.* 89: 8801–41

Lipman, P. W., Glazner, A. F. 1991. Introduction to middle Tertiary Cordilleran volcanism: magma sources and relation to regional tectonics. *J. Geophys. Res.* 96: 13,193–99 (and following pp. thru 13,735)

Luhr, J. F., Carmichael, I. S. E. 1990. Petrologic monitoring of cyclical eruptive activity at Volcan Colima, Mexico. *J. Volcanol. Geotherm. Res.* 42: 235–60

Lyons, W. B., Mayewski, P. A., Spencer, M. J., Twickler, M. S., Graedel, T. E. 1990. A northern hemisphere volcanic chemistry record (1869–1984) and climatic implications using a south Greenland ice core. *Ann. Glaciol.* 14: 176–82

Macdonald, G. A. 1972. *Volcanoes*. Englewood Cliffs, NJ: Prentice Hall. 510 pp.

Macdonald, K. C. 1982. Mid-ocean ridges: fine scale tectonic, volcanic and hydrothermal processes within the plate boundary zone. *Annu. Rev. Earth Planet. Sci.* 10: 155–90

Machida, H. 1990. Frequency and magnitude of catastrophic explosive volcanism in the Japan region during the past 130 ka: implications for human occupance of volcanic regions. *Geol. Soc. Aust. Symp. Proc.* 1: 27–36

Matson, M. 1984. The 1982 El Chichón volcano eruptions—a satellite perspective. *J. Volcanol. Geotherm. Res.* 23: 1–10

McBirney, A. R. 1978. Volcanic evolution of the Cascade Range. *Annu. Rev. Earth Planet. Sci.* 6: 437–56

McClelland, L, Simkin, T., Summers, M., Nielsen, E., Stein, T. C., 1989. *Global Volcanism 1975–1985*. Englewood Cliffs, NJ: Prentice Hall, and Washington, DC: Am. Geophys. Union 655 pp.

McCormick, M. P. 1992. Initial assessment of the stratospheric and climatic impact of the 1991 Mount Pinatubo eruption: prologue. *Geophys. Res. Lett.* 19: 149. (See also pp. 149–218 of this special section of GRL devoted to this subject, ed. M. P . McCormick.)

Moore, J. G., Rice, C. J. 1984. Chronology and character of the May 18, 1980, explosive eruptions of Mount St. Helens. See Boyd 1984, pp. 133–42

Morris, J. L. 1991. Applications of cosmogenic [10]Be to problems in the earth sciences. *Annu. Rev. Earth Planet. Sci.* 19: 313–50

Mulargia, F., Gasparini, P., Marzocchi, W. 1991. Pattern recognition applied to volcanic activity: identification of the precursory patterns to Etna recent flank eruptions and periods of rest. *J. Volcanol. Geotherm. Res.* 45: 187–96

Nakamura, K., 1974. Preliminary estimate of global volcanic production rate. In *Utilization of Volcanic Energy*, ed. J. Colp, A. S. Furimoto, pp. 273–86. Hilo: Univ. Hawaii & Sandia Corp.

Newhall, C. G. 1979. Temporal variation in the lavas of Mayon volcano, Philippines. *J. Volcanol. Geotherm. Res.* 6: 61–83

Newhall, C. G., Dzurisin, D. 1988. Historical unrest at large calderas of the world. *US Geol. Surv. Bull.* 1855: 1108 pp.

Newhall, C. G., Self, S. 1982. The Volcanic Explosivity Index (VEI): an estimate of explosive magnitude for historical volcanism. *J. Geophys. Res. (Oceans & Atmos.)* 87: 1231–38

Norris, R. A., Johnson, R. H. 1969. Submarine volcanic eruptions recently located in the Pacific by SOFAR hydrophones. *J. Geophys. Res.* 74: 650–64

Palais, J. M., Kirchner, S., Delmas, R. J. 1990. Identification of some global volcanic horizons by major element analysis of fine ash in Antarctic ice. *Ann. Glaciol.* 14: 216–20

Plank, T., Langmuir, C. H. 1988. An evaluation of the global variations in the major element chemistry of arc basalts. *Earth Planet. Sci. Lett.* 90: 349–70

Rampino, M. R., Self, S., Stothers, R. B. 1988. Volcanic winters. *Annu. Rev. Earth Planet. Sci.* 16: 73–99

Rosendahl, B. R. 1987. Architecture of continental rifts with special reference to East Africa. *Annu. Rev. Earth Planet. Sci.* 15: 445–503

Sacks, I. S. 1984. Subduction geometry and magma genesis. See Boyd 1984, pp. 34–46

Scrope, G. P. 1862. *Volcanoes: The Character of Their Phenomena, Their Share in the Structure and Composition of the Surface of the Globe and Their Relation to its Internal Forces.* London: Longman, Green, Longmans, Roberts. 490 pp. 2nd ed.

Sedlacek, W. A., Mroz, E. J., Lazrus, A. L., Gandrud, B. W. 1983. A decade of stratospheric sulfate measurements compared with observations of volcanic eruptions. *J. Geophys. Res.* 88: 3741–76

Shaw, H. R. 1985. Links between magmatectonic rate balances, plutonism, and volcanism. *J. Geophys. Res.* 90: 11,275–88

Shaw, H. R., Jackson, E. D., Bargar, K. E. 1980. Volcanic periodicity along the Hawaiian-Emperor chain. *Am. J. Sci.* 280-A: 667–708

Sigurdsson, H., Carey, S. 1989. Plinian and co-ignimbrite tephra fall from the 1815 eruption of Tambora volcano. *Bull. Volcanol.* 51: 243–70

Sigurdsson, H., Devine, J. D., Tchoua, F. M., et al. 1987. Origin of the lethal gas burst from Lake Monoun, Cameroon. *J. Volcanol. Geotherm. Res.* 31: 1–16

Simarski, L. T. 1992. *Volcanism and Climate Change.* Washington, DC: Am. Geophys. Union Spec. Rep. 27 pp.

Simkin, T. 1988. Terrestrial volcanic eruptions in time and space. *Proc. Kagoshima Int. Conf. on Volcanoes,* pp. 412–15. Kagoshima Prefectural Gov.

Simkin, T., Siebert, L., 1984. Explosive eruptions in space and time: durations, intervals, and a comparison of the world's active volcanic belts. See Boyd 1984, pp. 110–21

Simkin, T., Siebert, L., McClelland, L., Bridge, D., Newhall, C., Latter, J. H. 1981. *Volcanoes of the World: A Regional Directory, Gazetteer, and Chronology of Volcanism During the Last 10,000 Years.* Stroudsburg, Penn: Hutchinson Ross. 240 pp.

Simkin, T., Tilling, R. I., Taggart, J. N., Jones, W. J., Spall, H. 1989. *This Dynamic Planet.* US Geol. Surv. Map (2nd ed. in press)

Sleep, N. H. 1992. Hotspot volcanism and mantle plumes. *Annu. Rev. Earth Planet. Sci.* 20: 19–43

Smith, D. K. 1991. Seamount abundances and size distributions, and their geographic variations. *CRC: Rev. Aquat. Sci.* 5: 197–210

Smith, R. L. 1979. Ash-flow magmatism. *Geol. Soc. Am. Spec. Pap.* 180: 5–27

Smith, R. L., Luedke, R. G., 1984. Potentially active volcanic lineaments and loci in western conterminous United States. See Boyd 1984, pp. 96–109

Solomon, S. C., Toomey, D. R. 1992. The structure of mid-ocean ridges. *Annu. Rev. Earth Planet. Sci.* 20: 329–64

Spera, F. J., Crisp, J. A. 1981. Eruption volume, periodicity, and caldera area: relationships and inferences on development of compositional zonation in silicic magma chambers. *J. Volcanol. Geotherm. Res.* 11: 169–87

Steinthorsson, S. 1989. Mode and rate of volcanism in zones of crustal accretion, as exemplified by Iceland. *Washington, DC: 28th Int. Geol. Congr. Abstr. Vol.* 3/174

Stoiber, R. E., Carr, M. J. 1973. Quaternary volcanic and tectonic segmentation of Central America. *Bull. Volcanol.* 37: 304–25

Stoiber, R. E., Williams, S. N., Huebert, B. J. 1987. Annual contribution of sulfur dioxide to the atmosphere by volcanoes. *J. Volcanol. Geotherm. Res.* 33: 1–8

Stommel, H., Stommel, E. 1983. *Volcano Weather.* Newport, RI: Seven Seas. 177 pp.

Stothers, R. B. 1989a. Seasonal variations of volcanic eruption frequencies. *Geophys. Res. Lett.* 16: 453–55

Stothers, R. B. 1989b. Volcanic eruptions and solar activity. *J. Geophys. Res.* 94: 17,371–81

Talandier, J. 1989. Submarine volcanic activity: detection, monitoring, and interpretation. *Eos, Trans. Am. Geophys. Union* 70: 561, 568–69

Tatsumi, Y., Sakuyama, M., Fukuyama, H., Kushiro, I. 1983. Generation of arc basalt magmas and thermal structure of the mantle wedge in subduction zones. *J. Geophys. Res.* 88: 5815–25

Tatsumi, Y., Murasaki, M. Arsadi, E. M., Nobda, S. 1991. Geochemistry of Quaternary lavas from NE Sulawesi: transfer of subduction componenets into the mantle wedge. *Contrib. Mineral. Petrol.* 107: 137–49

Tilling, R. I., ed. 1989. *Volcanic Hazards.* Washington, DC: Am. Geophys. Union. 123 pp.

Uyeda, S. 1981. Subduction zones and back-arc basins—a review. *Geol. Rundsch.* 70: 552–69

Vogt, P. R. 1974. Volcanic spacing, fractures, and the thickness of the lithosphere. *Earth Planet. Sci. Lett.* 21: 235–52

Vogt, P. R. 1979. Global magmatic episodes: new evidence and implications for the "steady state" mid-oceanic ridge. *Geology* 7: 93–98

Vogt, P. R. 1986. Global episodicity of volcanism. In *Environmental Impact of Volcanism. Norman D. Watkins Symp—Abstr. Vol.*, pp. 104–6 , ed. J. M. Palais. Kingston: Univ. Rhode Island

Vogt, P. R., Nishimura, C., Jarvis, J. 1990. Prospecting for active volcanism: a promising role for long-range aircraft along the Mid-Oceanic Ridge. *Eos, Trans. Am. Geophys. Union* 71: 773–74

Walker, G. P. L. 1980. The Taupo pumice: product of the most powerful known (ultraplinian) eruption. *J. Volcanol. Geotherm. Res.* 8: 69–94

Wallmann, P. C., Mahood, G. A., Pollard, D. D. 1988. Mechanical models for correlation of ring-fracture eruptions at Pantelleria, Strait of Sicily, with glacial sea-level drawdown. *Bull. Volcanol.* 50: 327–39

Wickman, F. E. 1966. Repose period patterns of volcanoes. *Ark. Mineral. Geol.* 4(7–11): 291–367 (in 5 parts)

Wickman, F. E. 1976. Markov models of repose-period patterns of volcanoes. In *Random Processes in Geology*, ed. D. F. Merriam, pp.135–61 Berlin: Springer-Verlag

Williams, H., McBirney, A. R. 1979. *Volcanology*. San Francisco: Freeman & Cooper. 397 pp.

Wilson, L., Pinkerton, H., Macdonald, R. 1987. Physical processes in volcanic eruptions. *Annu. Rev. Earth Planet. Sci.* 15: 73–95

Wright, T. L. 1989. Hot-spot volcanism in space and time: Hawaiian example. *Washington, DC*: 28th Int. Geol. Congr. *Abstr. Vol.* 3/384–5

Yamaguchi, D. K. 1985. Tree-ring evidence for a two-year interval between recent prehistoric explosive eruptions of Mt. St. Helens. *Geology* 13: 554–57

Yoder, H. S. 1976. *Generation of Basaltic Magma*. Washington, DC: Natl. Acad. 265 pp.

Yoder, H. S., ed. 1979. *The Evolution of the Igneous Rocks, Fiftieth Anniversary Perspectives*. Princeton, NJ: Princeton Univ. Press. 588 pp.

Zemtsov, A. N., Tron, A. A. 1985 A statistical analysis of catalogs of volcanic eruptions. *Dokl. Akad. Nauk SSSR* 285: 582–85

Annu. Rev. Earth Planet. Sci. 1993. 21:453–85

PRECAMBRIAN HISTORY OF THE WEST AUSTRALIAN CRATON AND ADJACENT OROGENS

John S. Myers

Geological Survey of Western Australia, 100 Plain Street, East Perth, Western Australia 6004, Australia

KEY WORDS: tectonics, continental growth and fragmentation

INTRODUCTION

This review is a synthesis of the Precambrian geology of the West Australian Craton and its fringing orogens. It describes the evidence that has recently been reinvestigated and reinterpreted to reveal a long dynamic history of repeated aggregation and dispersal of crustal fragments.

Cratons are large composite geological units of continental crust that have survived intact for significant periods of geological time. They are generally tabular bodies with great areal extent (millions of square kilometers) and lithospheric thickness (200–400 km). Orogens are narrow steep zones of tectonic and metamorphic activity associated with the deposition of sediments and the intrusion of plutonic rocks. They occur within or along the margins of continents and are generally a few hundred kilometers wide, thousands of kilometers long, and extend to the base of the lithosphere. Most orogens are initially associated with compressive tectonics, crustal thickening, and the transient formation of mountain belts.

Cratons and orogens may survive for very long periods of time. But as part of a thin crust on a dynamic mantle they are inherently mobile and travel as rafts for great distances around the surface of the Earth. They may be torn apart, dispersed, and reassembled with other continental fragments. The boundaries of these crustal fragments, once established, remain zones of crustal weakness that may be reactivated as intracontinental mobile belts or lines along which cratons are assembled and dismembered (cf Hoffman 1988).

453

0084-6597/93/0515-0453$02.00

This review provides examples of the transient nature of cratons and the reactivation of orogens. The West Australian Craton (Figure 1*b*) formed during the early Proterozoic (2000–1800 Ma) by the collision and combination of the Pilbara and Yilgarn Cratons which are fragments of formerly more extensive Archean continents. The collision zone is marked by the Capricorn Orogen. The West Australian Craton was itself fragmented by rifting between 1600 and 1300 Ma, and then involved in collision and amalgamation with other continental fragments between 1300 and 1100 Ma along the Albany-Fraser and Rudall-Musgrave Orogens (Figure 1*c*). The eastern and western margins of the resulting Austral-Antarctic Craton were zones of renewed tectonic and magmatic activity between 750 and 600 Ma (Figure 1*d*) that reflect further crustal amalgamation and the assembly of this portion of the Gondwana supercontinent.

The Precambrian history outlined here contrasts with most previous interpretations of the geology. This part of Gondwana was thought to have remained virtually intact between 1800 and 450 Ma (Veevers & McElhinny 1976, Lindsay et al 1987, Idnurm & Giddings 1988). Belts of folding with ages within that range, metamorphism, magmatism, and sedimentary deposition were interpreted as intracontinental zones of crustal weakness dominated by vertical and extensional tectonic processes (Rutland 1973, Gee 1979, Etheridge et al 1987). The Archean cratons were thought to have been dominated by similar tectonic processes, with greenstone belts formed in continental rift zones (Groves et al 1987), and deformation and metamorphism caused by rising granite diapirs (Glikson 1979, Hickman 1983, Campbell & Hill 1988).

These essentially fixist models of vertical and extensional tectonics within a single intact continent contrast with the interpretation outlined below of predominantly horizontal tectonic processes associated with the repeated fragmentation and reassembly of small continents.

Figure 1 Evolution of the main Precambrian tectonic units of Western Australia. (*a*) 2780–2600 Ma: rifting of the already ancient Pilbara Craton and extrusion of flood basalt (Fortescue Group); formation of the Yilgarn Craton by amalgamation of microcontinents and volcanic arcs. N—fragment of microcontinent formed between 3730 and 3300 Ma; A—microcontinent(s) with major component formed at about 3000 Ma; B—mainly volcanic arcs generated and amalgamated with minor microcontinents between 2780 and 2600 Ma. Arrows indicate crustal convergence or rifting. (*b*) 2000–1800 Ma: Pilbara and Yilgarn Archean Cratons joined by the suture zone of the Capricorn Orogen to form the West Australian Craton. The exposed parts of the cratons are marked by denser ornament. The West Australian Craton was formerly more extensive; it is now bounded by faults that truncate the rocks and structures of the craton. (*c*) 1300–1100 Ma: West Australian Craton united along the Albany-Fraser, Rudall-Musgrave, and proto-Pinjarra(?) Orogens with other continents to form part of an Austral-Antarctic supercontinent. (*d*) 750–600 Ma: Orogenic activity in the Paterson and Pinjarra Orogens within the Austral-Antarctic Craton.

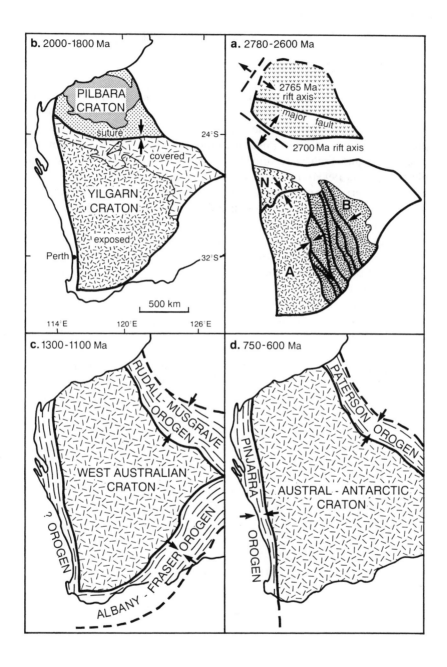

The time frame used in this review is based on U-Pb isotopic ages of zircons unless otherwise indicated. All ages are approximations.

YILGARN CRATON

The Yilgarn Craton (Figure 2) largely consists of volcanic and sedimentary rocks (greenstones) and granites that formed between 3000 and 2600 Ma and were metamorphosed at low-grade. In the southwest, granites and greenstones were metamorphosed in granulite facies at 2640 Ma (Nemchin et al 1993) and then were intruded by high-level late Archean granite plutons (Myers 1990a). High-grade gneiss complexes older than the granite-greenstone terrains are exposed along the western margin of the craton.

Rocks and structures are superficially similar over large parts of the craton, but in detail they can be subdivided into units with distinct geological histories which range in age from 3730 to 2550 Ma. These units form a number of tectonostratigraphic terranes (Figure 2) that, regardless of their age, all indicate intense tectonic, volcanic, plutonic, and metamorphic activity between 2780 and 2630 Ma. This is interpreted as a major episode of plate tectonic activity which swept together and amalgamated a number of diverse crustal fragments—including volcanic arcs, back arc basins, and microcontinents—to form the Yilgarn Craton at this time (Figure 1a).

This conclusion contrasts with most previous interpretations of the Yilgarn Craton. During the 1960s and 1970s the greenstones were considered to represent remnants of primordial crust that was deformed during the diapiric emplacement of steep-sided granite batholiths (Glikson 1979). In the 1980s most models interpreted the greenstones and granites as products of rifting within older continental crust (Gee et al 1981; Groves & Batt 1984; Hallberg 1986; Blake & Groves 1987; Groves et al 1985, 1987). Groves et al (1978) and Blake & Groves (1987) concluded that there was no evidence of subduction-related processes in either the granites or greenstones. However Barley & McNaughton (1988) interpreted the Norseman-Wiluna belt (approximately the Kalgoorlie, Gindalbi, and Jubilee Terranes of Figure 2) in terms of a westward-dipping subduction zone, and this was further elaborated by Barley et al (1989). In contrast, Campbell & Hill (1988) and Hill et al (1991) proposed that the rocks and

---

Figure 2 Main tectonostratigraphic terranes of the Yilgarn Craton: B—Barlee, G—Gindalbie, K—Kalgoorlie, K_u—Kurnalpi, L—Laverton, M—Murchison, N—Narryer, P—Pinjin, Y—Yellowdine. SW—southwest Yilgarn composite terrane, N_c—Narryer Terrane involved in the Capricorn Orogen.

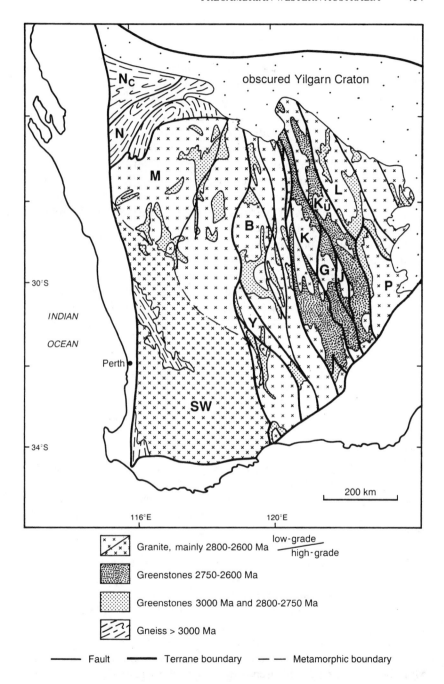

Granite, mainly 2800-2600 Ma low-grade / high-grade

Greenstones 2750-2600 Ma

Greenstones 3000 Ma and 2800-2750 Ma

Gneiss > 3000 Ma

——— Fault   ■■■ Terrane boundary   — — Metamorphic boundary

structures in the southern part of this belt reflected plume-tectonics, and had nothing to do with plate tectonics.

The Yilgarn Craton was formerly divided into the Eastern Goldfields, Southern Cross and Murchison Provinces, and the Western Gneiss Terrane (Gee et al 1981), but these units have not stood up to subsequent more detailed mapping and are best abandoned. The Western Gneiss Terrane was found to contain gneisses with diverse crustal histories, some of which are integral parts of the granite-greenstone terranes (Myers 1990a). The granite-greenstone provinces were found to comprise a number of distinct tectonostratigraphic terranes (Figure 2; Myers 1990c).

The terranes can be grouped into four superterranes, here called the West Yilgarn, West Central Yilgarn, East Central Yilgarn, and East Yilgarn Superterranes. The West Yilgarn Superterrane (Narryer and Murchison Terranes and southwest composite terrane, units A and N of Figure 1a) contains substantial portions of continental crust older than 2800 Ma. It is truncated to the east by faults which join it to the West Central Yilgarn Superterrane (Barlee and Yellowdine Terranes) which contains 2800 and 2700 Ma greenstones. In the adjacent East Central Yilgarn Superterrane (Kalgoorlie, Gindalbie, Jubilee, and Kurnalpi Terranes) all the greenstones formed at about 2700 Ma. The East Yilgarn Superterrane (Laverton and Pinjin Terranes) is less well known but includes high-grade gneiss, granite, and 2800 Ma greenstones.

The geologic histories of three terranes which have recently been studied in detail are summarized below as examples of the diversity of the crustal components within the craton and the nature of the processes by which these components were combined to form the Yilgarn Craton.

Narryer Terrane

The Narryer Terrane (Figure 2) (Myers 1988, 1990b) forms the north-western part of the craton and is the oldest known component. It consists of high-grade gneiss derived from granite and minor amounts of granodiorite, tonalite, layered basic intrusions, and metasedimentary rocks. Meeberrie Gneiss, which formed from 3730 Ma tonalite and 3680 to 3600 Ma granite, and the Manfred Complex of 3730 Ma layered anorthosite, gabbro, and ultramafic rocks, are the oldest known rock units. After deformation and metamorphism these rocks were intruded by 3400 Ma granite which forms the Dugel Gneiss, and 3300 Ma granite. This second plutonic episode culminated in high-grade metamorphism at 3300 Ma marked by melt patches, veins, and overgrowths on zircons.

Sedimentary rocks—now mainly quartzite, pelite, and BIF (banded iron formation)—were deposited on the older rocks between 3100 and 2700 Ma. These contain detrital zircons ranging back to 4270 Ma which are the oldest known fragments of terrestrial rocks (Froude et al 1983, Compston & Pidgeon 1986). A third plutonic episode is marked by the widespread emplacement of granite sheets and plutons between 2780 and 2630 Ma. This was accompanied by intense deformation and high-grade metamorphism associated with collision of the Narryer and Murchison Terranes and subduction of intervening oceanic crust.

The boundary between the Narryer and Murchison Terranes is marked by the Yalgar Fault. It is a zone of intense repeated deformation that truncates the structures in both terranes. Sheets of late Archean granite were emplaced in and alongside the fault zone and are partly converted to mylonite. Within the Narryer Terrane, fold interference structures comprise D1 recumbent folds and thrusts refolded successively by upright folds with east-west (D2) and then north-south (D3) axes which also fold the Yalgar Fault.

Murchison Terrane

The Murchison Terrane mainly consists of granite and greenstones metamorphosed in greenschist facies (Figure 2). Deeper levels of the terrane are exposed in the west and reveal a heterogeneous granitic gneiss complex (Murgoo Gneiss Complex) (Myers 1990a,b), formerly included with the Narryer Gneiss Complex in the Western Gneiss Terrane by Gee et al (1981). Most of the Murchison Terrane, except the eastern and southern boundaries, has recently been mapped and described in detail by Watkins & Hickman (1990) whose work is summarized below.

The granite-greenstone component of the terrane comprises two major sequences called the Luke Creek and Mount Farmer Groups. The lower part of the older, 3000 Ma (Pidgeon & Wilde 1990), Luke Creek Group consists of two sequences of submarine tholeiitic and high-Mg basalt lava flows overlain by BIF, which formed during two episodes of lava-plain volcanism. Most of the lavas are massive and contain very few pillow lava structures. The upper part of the Luke Creek Group consists of interlayered basalt and silicic volcanic and volcaniclastic rocks. There are fundamental differences in the geochemistry of the volcanic rocks in the northern and southern parts of the Luke Creek Group which suggest that the volcanics were derived from different upper mantle sources.

The Luke Creek Group was intruded by sheets of granite at 2900 Ma and, together with the granite, was deformed in a subhorizontal tectonic regime, D1, which converted the granite into gneiss. The younger 2800 Ma Mount Farmer Group unconformably overlies the Luke Creek Group and consists of remnants of nine distinct volcanic centers and one epiclastic sedimentary basin, all of local extent.

Thick sheets of monzogranite were intruded into the volcanic rocks and gneissose monzogranite between at least 2690 and 2680 Ma. They are compositionally similar to the older gneissose monzogranite and were probably derived by similar processes from similar lower crustal source rocks.

All these rocks were deformed and repeatedly folded together with the adjacent Narryer Terrane between 2680 and 2640 Ma. Fold interference structures were formed by older subhorizontal D1 thrusts and fold limbs refolded successively by upright folds with east-west (D2) and then north-south (D3) axes (Myers & Watkins 1985). The rocks were recrystallized in greenschist or low amphibolite facies except in the southwest and west where the regional metamorphic grade is in amphibolite facies.

The last stages of ductile deformation (D4) in the Murchison Terrane formed major north-south shear zones. These shear zones formed in a continuation of the regional D3 stress regime and spanned the emplacement of 2630 Ma plutons of granite, granodiorite and tonalite, and the peak of regional metamorphism. The plutons form two distinct groups with different compositions, which suggests that they were derived from different crustal sources: older siliceous crust in the south and younger mafic crust in the north. This also appears to reflect the two distinct kinds of lithosphere that were already juxtaposed to form the basement of the Murchison Terrane before the eruption of the Luke Creek Group of volcanic rocks (Watkins et al 1991).

Kalgoorlie Terrane

The Kalgoorlie Terrane (Figure 2) is a fault-bounded unit of granite and greenstones with a greenstone sequence and deformation history that are distinct from adjacent mapable granite-greenstone units (Swager et al 1990). It is one of a number of superficially similar terranes that have recently been delineated in the eastern part of the Yilgarn Craton (Myers 1990c).

The Kalgoorlie Terrane comprises a stratigraphic sequence of volcanic rocks that formed between 2700 and 2690 Ma (Swager et al 1990). A lower basalt unit comprised of high-Mg basalt which passes upwards into tholeiite is overlain by a unit of komatiite flows. This in turn is overlain by an upper unit of tholeiitic and high-Mg basalt, followed by a unit of silicic volcanic and sedimentary rocks. The silicic volcanic rocks are predominantly dacitic but range from rhyolite to andesite, and include lava and pyroclastic flows and waterlain tuff. They are interbedded with siltstone and sandstone. The upper part of this unit is dominated by clastic sedimentary rocks largely derived from silicic volcanic centers.

The greenstone sequence was stacked by thrusts and recumbent folds (D1). These structures were deformed by upright folds with NNW-SSE axes (D2), and then modified by transcurrent faulting and associated *en echelon* folds (D3). All three episodes of deformation occurred between 2680 and 2650 Ma. They led to the pronounced NNW-SSE tectonic grain of the region, and deformed and reactivated the terrane boundary faults.

The Ida and Mount Monger Faults which form the western and eastern boundaries of the Kalgoorlie Terrane were zones of repeated intense deformation. Major faults and shear zones within the Kalgoorlie Terrane subdivide the terrane into tectonic domains which contain various parts of the regional greenstone sequence.

Monzogranite was emplaced during three episodes (Witt & Swager 1989) at 2680, 2660, and 2620–2600 Ma (Campbell & Hill 1988, Hill et al 1991), mainly during the later stages of deformation. The granites are generally little deformed and weakly foliated, but were converted into gneiss in narrow belts of intense deformation.

The rocks are generally in amphibolite facies in the western part of the terrane and greenschist facies in the east.

PILBARA CRATON

The Pilbara Craton (Figure 1*b*) comprises a basement of 3600 to 2800 Ma granite-greenstone terrain unconformably overlain by a cover sequence called the Mount Bruce Supergroup which was deposited between 2765 and 2400 Ma in a region called the Hamersley Basin.

Basement

The Pilbara granite-greenstone terrain consists of ovoid outcrops of granitoid rocks, mainly granite, granodiorite, and tonalite, separated by narrow belts of steeply dipping greenstones. These form three distinct stratigraphic units called the Warrawoona, Gorge Creek, and Whim Creek Groups (Figure 3) (Hickman 1983).

The 3500 to 3450 Ma Warrawoona Group comprises basalt with minor components of komatiite, dacitic and rhyolitic tuff, agglomerate, and sedimentary rocks (Hickman 1983, Thorpe et al 1992). It is deformed and recrystallized in greenschist facies but the deformation is generally not intense and primary depositional features are widely preserved. Sedimentary rocks include supra- and intertidal facies, with evaporites and stromatolites (Lowe 1980, Walter et al 1980) and contain fossil cyanobacteria which provide some of the oldest evidence of life on Earth (Awramik et al 1983, Schopf & Packer 1987).

The Warrawoona Group is unconformably overlain by the 3300 to 3000 Ma Gorge Creek Group of sandstone with minor conglomerate, banded-iron formation, and basalt. This is unconformably overlain in the west by

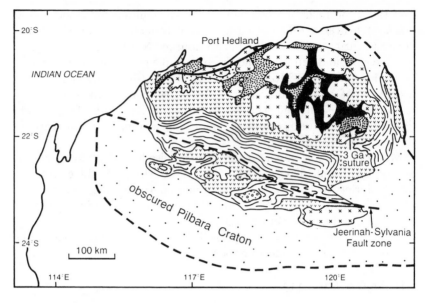

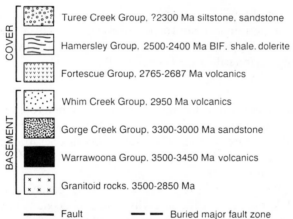

Figure 3   Main geological units of the Pilbara Craton.

the 2950 Ma Whim Creek Group of basalt and dacite, volcaniclastics, and terrigenous sedimentary rocks. Barley (1987) and Krapez & Barley (1987) consider that the Whim Creek Group developed as a rift or pull-apart structure betweeen strike-slip faults.

The granitoid rocks show the same age range as the greenstones, and many are thought to be the intrusive equivalents of the volcanic rocks (Bickle et al 1983, Hickman 1983, Williams & Collins 1990). The known ages cluster into distinct groups at 3500 to 3400 Ma, 3300 Ma, and 3050 to 2850 Ma (Bickle et al 1993). A high pressure kyanite-sillimanite metamorphic event occurred between 3300 and 3230 Ma, and was followed by lower pressure andalusite-sillimanite metamorphic events at 2950 and 2900 to 2840 Ma [evidence from $^{40}Ar/^{39}Ar$ step-heating analyses of amphiboles by Wijbrans & McDougall (1987)].

The most prominent tectonic structures are now steep and have been interpreted by Hickman (1983) as the result of granite diapirism, although Bickle et al (1980) found evidence of early flat-lying structures that predated the steep structures. Tyler et al (1992) recognized a suture in the eastern part of the granite-greenstone terrain (Figure 3) that joined two terranes at 3000 Ma, but the regional tectonics and tectonic evolution of the Pilbara granite-greenstone terrain is still little known.

Post-tectonic tin-bearing granite plutons were emplaced by stoping at 2850 Ma (Blockley 1980, Bickle et al 1989).

Cover

The Mount Bruce Supergroup comprises three conformable rock units called the Fortescue, Hamersley, and Turee Creek Groups (Figure 3) (Trendall 1979).

The Fortescue Group consists of basaltic lava and pyroclastic rocks and a smaller amount of silicic volcanic rocks and non-volcanogenic sedimentary rocks (Blake 1993, Thorne & Trendall 1994). The sequence accumulated between 2765 and 2687 Ma (Arndt et al 1991) and records three distinct tectonic episodes:

1. NNE trending rifts (Figure 1a). In the northeast, the lowest basalts are overlain by the Hardey Formation comprising shallow water, continental fluvial, alluvial fan, and lacustrine deposits. Deposition of the Hardey Formation was controlled by a system of NNE-trending normal faults with downthrows to the west (Blake 1993). A swarm of basaltic dikes called the Black Range Dike Suite is parallel to these extensional faults, but the exact age of these dikes is unknown. The asymmetry of the faults and sedimentation suggest that a major axis of rifting, which could have become a continental margin, lay to the northwest (Figure 1a) (Blake 1993).

   By contrast, the Hardey Formation in the southwest consists of fluvial and deltaic deposits. The sedimentary facies, thickness data, and palaeocurrents indicate the presence of a WNW-trending basement ridge to the northeast, and Thorne & Trendall (1994) infer that a major WNW-ESE trending fault (Figure 1a) called the Jeerinah-Sylvania Fault (Figure 3) split the Pilbara Craton into two regions with different tectonic histories. The fault acted as a transfer fault during deposition of the Hardey Formation at 2765 Ma, and subsequently remained an important regional structure.

2. WNW trending rifts. The Hardey Formation was suceeded by a major episode of basaltic volcanism. There was a major facies and thickness change across the Jeerinah-Sylvania Fault. To the north, 1.5–2 km of subaerial and shallow marine basalts (Kylena and Maddina Basalt) and volcaniclastics and carbonates (Tumbiana Formation) were deposited on a flat coastal plain and shallow marine shelf. To the south of the fault, a 3.5 km thick sequence of pillow lava, volcaniclastics, spinifex-textured basalt, and komatiite accumulated in a deeper marine shelf setting. Blake & Groves (1987), Tyler (1991), and Thorne & Trendall (1994) suggest that these rocks were deposited during major rifting parallel to the Jeerinah-Sylvania Fault, along or to the south of the present southern margin of the Pilbara Craton (Figure 1a).

3. Post-rifting subsidence. The uppermost unit of the Fortescue Group (the Jeerinah Formation) consists of marine sedimentary rocks, mainly shale and mafic and siliceous volcanics which indicate a marine trans-gression across the Pilbara from south to north. Thorne & Trendall (1994) suggest that it was deposited on a cooling and subsiding trailing continental margin, a regime that continued during the deposition of the ensuing Hamersley Group.

The Hamersley Group is a finely stratified sequence of banded iron formations, shale, chert, dolomite, and tuff, 2.5 km thick. It contains numerous sills of dolerite (forming up to 1 km of the thickness of the Weeli Wolli Formation) and a giant sill of rhyolite (Woongarra Rhyolite). It was deposited in shallow marine conditions on either the margin of a subdued continent from which there was no significant drainage or an isolated submerged plateau.

The banded iron formations (Marra Mamba, Brockman, and Boolgeeda Formations) are finely layered with both microbands generally about 1 mm thick, mesobands on the scale of centimeters, and macrobands involv-ing layers of shale meters thick (Trendall & Blockley 1970). The micro-bands were interpreted as varves produced by seasonal, climatic controlled variations in biogenic activity and the deposits were thought to have accumulated in a barred basin with restricted oceanic connection to the northwest. Horwitz & Smith (1978), Ewers & Morris (1981), and Morris & Horwitz (1983) considered that the banded iron formations were deposited on a shelf open to the ocean to the west. They envisaged that the iron was transported by periodic upwelling of deep ocean currents.

Tyler (1991) suggested that the Jeerinah-Sylvania Fault (Figure 3) continued to influence deposition of the Hamersley Group. He noted that only the lowermost of the banded iron formations (Marra Mamba Formation) is seen in the northern part of the Hamersley Basin, and the overlying Carawine Dolomite was deposited in shallow water. In contrast, in the southern part of the basin, banded iron formations are thicker and more extensive and were deposited in deeper water.

Below the uppermost Boolgeeda Iron Formation there is a 400 m thick sill of rhyolite called the Woongarra Rhyolite (A. F. Trendall 1992, personal communication) with an age of 2440 Ma (Pidgeon & Horwitz 1991). This intrusion could be related to the same magmatic event that led to the massive emplacement of dolerite sills in the underlying Weeli Wolli Formation. The Woongarra Rhyolite could be a product of crustal melting associated with the emplacement and ponding of basic magma in the crust. It could represent an episode of extension of the Pilbara Craton or reflect the passage of the craton over a mantle plume.

The Turee Creek Group conformably overlies the Boolgeeda Iron Formation along the southern margin of the Pilbara Craton (Figure 3). It is mainly composed of mudstone, siltstone, and immature sandstone, and is locally 5 km thick (Trendall 1990). The lower part of the Group was deposited in deep water but the upper part contains mature sandstone, conglomerate, stromatolitic carbonate, and basalt, and appears to have been deposited in shallow water during renewed tectonic activity (Trendall 1983). Horwitz (1982) considered that the sediments were derived from a ridge to the south, perhaps along the continental margin.

In the lower part of the Turee Creek Group, the Meteorite Bore Member of the Kungarra Formation consists of boulders and pebbles of sandstone and silicic volcanic rocks (probably Woongarra Rhyolite) randomly scattered in siltstone (Trendall 1990). Some of the pebbles are striated, grooved, and faceted which suggests that this deposit is of glacial origin (Trendall 1976).

THE CAPRICORN OROGEN

The Capricorn Orogen is a 300 km wide belt of early Proterozoic sedimentary deposition, deformation, metamorphism, and plutonic intrusions that lies between the Pilbara and Yilgarn Cratons (Figure 4). It was defined by Gee (1979, p. 352) as a zone of "geosynclinal sedimentation, metamorphism, basement reworking and granitoid emplacement." The Pilbara and Yilgarn Cratons were thought to have been part of a single

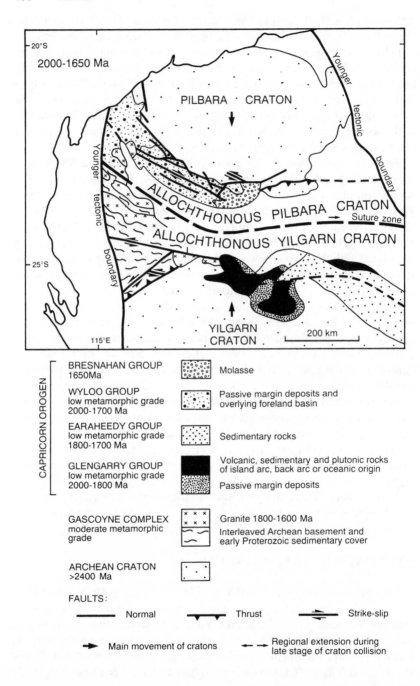

20°S

2000-1650 Ma

Younger tectonic boundary

PILBARA · CRATON

ALLOCHTHONOUS PILBARA CRATON
Suture zone

ALLOCHTHONOUS YILGARN CRATON

Younger tectonic boundary

25°S

YILGARN
CRATON

200 km

115°E

CAPRICORN OROGEN

BRESNAHAN GROUP
1650Ma

WYLOO GROUP
low metamorphic grade
2000-1700 Ma

EARAHEEDY GROUP
low metamorphic grade
1800-1700 Ma

GLENGARRY GROUP
low metamorphic grade
2000-1800 Ma

GASCOYNE COMPLEX
moderate metamorphic
grade

ARCHEAN CRATON
>2400 Ma

Molasse

Passive margin deposits and
overlying foreland basin

Sedimentary rocks

Volcanic, sedimentary and plutonic rocks
of island arc, back arc or oceanic origin

Passive margin deposits

Granite 1800-1600 Ma

Interleaved Archean basement and
early Proterozoic sedimentary cover

FAULTS:

———— Normal   Thrust   Strike-slip

→ Main movement of cratons   Regional extension during
late stage of craton collision

entity that once extended under the geosynclinal deposits of the Capricorn Orogen (Horwitz & Smith 1978, Williams 1986). Gee (1979, p. 361) considered that the orogen formed in "a totally ensialic setting" (i.e. intracontinental) and found "no evidence for subduction or any other manifestation of plate tectonics."

Reappraisal of the evidence indicates that the Archean geology of each craton is completely different (Myers 1990c), and that the Capricorn Orogen reflects the collision and amalgamation of the Pilbara and Yilgarn Cratons (Myers 1989, 1990c,d; Tyler & Thorne 1990; Thorne & Seymour 1991; Tyler 1991). These authors demonstrate that the collision was oblique, with the southeastern part of the Pilbara Craton impinging first against the middle of the surviving northern margin of the Yilgarn Craton. The isotopic timing of events in the Capricorn Orogen is not well known.

Pilbara Foreland

The southern margin of the Pilbara Craton became the foreland of the Capricorn Orogen. The edge of the craton was extensively disrupted by thrusts that carried slices of the craton northwards (Figure 5b). In this displaced (allochthonous) belt the craton is largely buried by sedimentary rocks that formed both during and after the Capricorn Orogeny (Figure 4). Sedimentary rocks that accumulated during the orogeny form the Wyloo Group and occur both on the displaced margin and on the adjacent in situ (autochthonous) part of the craton.

The oldest known ductile structures related to the Capricorn Orogen are dome and basin folds that formed south of the Jeerinah-Sylvania Fault (Figure 3). They predate the Wyloo Group and are considered by Tyler & Thorne (1990) and Tyler (1991) to be associated with the early stages of ocean closure.

These structures are overlain by sandstone and conglomerate (Beasley River Quartzite) which make up the lowest unit of the Wyloo Group (Figure 4). The sediments were deposited in shallow marine and deltaic environments during intermittent faulting on the southern margin of the Pilbara Craton (Thorne & Seymour 1991). Their deposition was followed by eruption of the Cheela Springs Basalt which Thorne & Seymour (1991) relate to flexuring of the continental margin during the final stages of ocean basin closure.

A northward-facing fold-and-thrust belt then developed on the southeast margin of the Pilbara Craton (Figure 5b), probably during oblique

---

Figure 4 Main features of the Capricorn Orogen resulting from collision of Pilbara and Yilgarn Cratons between 2000 and 1650 Ma.

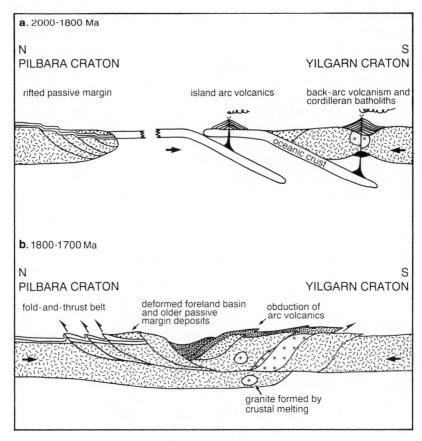

a. 2000-1800 Ma

N   S
PILBARA CRATON   YILGARN CRATON

rifted passive margin   island arc volcanics   back-arc volcanism and
                                                          cordilleran batholiths

oceanic crust

b. 1800-1700 Ma

N   S
PILBARA CRATON   YILGARN CRATON

fold-and-thrust belt   deformed foreland basin   obduction of
                          and older passive   arc volcanics
                          margin deposits

granite formed by
crustal melting

Figure 5 Cross-sections illustrating the plate tectonic evolution of the Capricorn Orogen
which joined the Pilbara and Yilgarn Cratons.

collision with arcs lying in advance of the Yilgarn Craton (Tyler & Thorne
1990). Archean basement in the southeastern part of the Pilbara Craton
was sliced up by ductile shear zones and thrust northwards over the cover
sequences (Figures 4 & 5*b*) (Tyler 1991).

A suite of WNW-trending dolerite dikes was intruded, followed by
uplift and erosion. Detritus accumulated in deltas and shallow marine
environments and formed sandstone and conglomerate of the Mount
McGrath Formation which unconformably overlies the Cheela Springs
Basalt (Tyler & Thorne 1990). Deposition of the succeeding Duck Creek

Dolomite began in deep water and shallowed during progradation of the carbonate shelf (Thorne & Seymour 1991).

Extension and collapse of the carbonate shelf led to the local eruption of basaltic lava (June Hill Volcanics). A spatially associated silicic volcanic rock has given an age of 1843 Ma (Pidgeon & Horwitz 1991). The volcanic eruptions were followed by the development of a deep water basin along the southern margin of the Pilbara Craton. The trough was filled by submarine fan deposits (Ashburton Formation—the uppermost unit of the Wyloo Group). The first major fan spread westwards from uplifted Archean basement to the east (Figure 4), followed by a second fan derived from the Yilgarn Craton to the south (Thorne & Seymour 1991). Intensification of the collision led to recumbent folding and northward-directed thrusting of the Ashburton Formation. The associated metamorphism increases in grade towards the southwest where greater thrust stacking, uplift, and erosion has exposed deeper levels in the northern part of the Gascoyne Complex (Figure 4) (Williams 1986).

During the last phase of the collision, progressive westward closure of the oblique continental margins led to large-scale dextral wrench faulting parallel to the margin of the Pilbara Craton. Corresponding sinistral strike-slip movements occurred in the northwest part of the Yilgarn Craton, as the crustal wedge forming the western part of the Capricorn Orogen was extruded westward.

Yilgarn Hinterland

On the northern margin of the Yilgarn Craton, early Proterozoic shelf deposits form the lower part of the Glengarry Group of the Nabberu Basin (Figure 4) (Gee 1979). They are overlain by a sequence of basaltic volcanic rocks (Narracoota Volcanics) and volcanic derived greywackes (Thaduna Greywacke), of oceanic affinity (Hynes & Gee 1986). The latter have been interpreted by Myers (1990c) as obducted island-arc, back-arc basin, or oceanic crust (Figure 5b). To the west, a thrust sheet of deformed and recrystallized gabbro and ultramafic rocks (Trillbar Complex) was transported southward onto the Archean Narryer Terrane of the Yilgarn Craton (Myers 1989). It is overlain by thrusts carrying interleaved continental shelf deposits and arc volcanics and greywacke.

The Archean basement itself was sliced up by major ductile shear zones and transported southward (Figure 4). Dolerite dikes were emplaced before, during, and after the formation of these shear zones. Many dolerites contain garnet in undeformed igneous textures as well as in schistose portions, and mineral compositions indicate that these rocks were deformed at a deep crustal level (Muhling 1988). Temperature-pressure evidence from coronas in the metadolerites suggest a subsequent episode of isothermal uplift of more than 10 km (Muhling 1990). This may reflect relatively rapid tectonic stacking and uplift of hot crust from a great depth. These events were accompanied by the intrusion of granite sheets. The high-grade rocks are cut by dolerite dikes and shear zones with mylonite formed in greenschist facies associated with southward transport of tectonic slices of Archean basement rocks.

The southernmost major shear zone (Errabiddy Fault) is a zone of mylonitized Archean gneiss, 5–10 km thick, which marks the boundary between displaced Archean basement to the north, and largely in situ Yilgarn Craton to the south. To the north of the Errabiddy Fault, zones of intense deformation up to tens of kilometers wide form an anastomosing network, enclosing lenticular zones of relatively little-deformed Archean rocks. They also disrupt high-level granite plutons which are most abundant in a 75 km wide zone parallel to the former margins of the Pilbara and Yilgarn Cratons (Figure 4). The plutons may be parts of cordilleran-type batholiths which formed above a subduction zone before the cratons were amalgamated (Figure 5a). They are cut by muscovite-tourmaline-rich granite and leucogranite which may have been derived by crustal melting during advanced stages of the continental collision (Figure 5b) (Williams 1986). The inferred southward-dipping suture between the Pilbara and Yilgarn Cratons may lie just to the north of, and parallel to, this belt of elongate batholiths (Figure 4).

BANGEMALL BASIN

The eroded rocks of the Capricorn Orogen are unconformably overlain by the Bangemall Group (Muhling & Brakel 1985, Williams 1990a). Deposition was initially associated with the reactivation of steeply dipping faults in the basement, and alluvial fans were deposited on the flanks of horst blocks. These deposits were overlain by a transgressive marine sequence of pyritic black shale, sandstone, and stromatolitic dolomite, followed by chert thought to have formed in hypersaline water (Discovery Chert). This is overlain by a shallow marine platform sequence of dolomite, shale, and sandstone followed by deeper-water turbidites. All these rocks form the Edmund Subgroup (Figure 6). They were intruded by dolerite sills and dikes, and then folded. Belts of tight upright folds formed in narrow zones parallel to older east-west faults, separated by wide zones of open folds. The age of the Edmund Subgroup is poorly known but is thought to be between 1650 and 1400 Ma (Williams 1992). Zircons from a rhyolite flow thought to be part of the Edmund Subgroup have given an age of 1650 Ma (David Nelson 1992, personal communication).

The folded rocks of the Edmund Subgroup are unconformably overlain

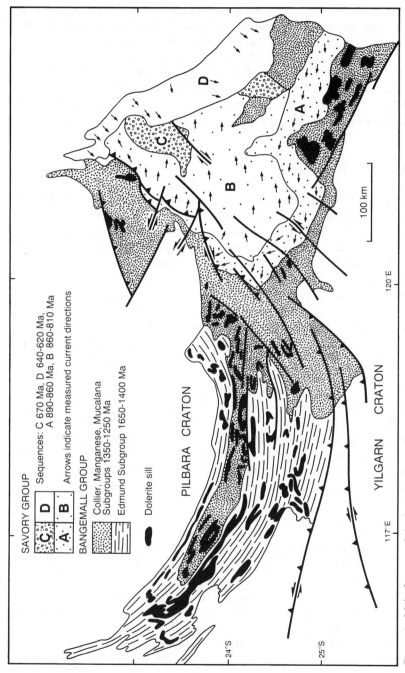

Figure 6 Main features of the Bangemall and Savory Basins which formed on the Capricorn Orogen between 1650 and 620 Ma, illustrating intermittent tectonic reactivation of this zone of crustal weakness.

by the Collier, Manganese, and Mucalana Subgroups of marine sandstone, shale, and stromatolitic dolomite (Figure 6) (Muhling & Brakel 1985, Williams 1990a). The age of these deposits is poorly known, but glauconite from a sandstone within the Manganese Subgroup has given a Rb-Sr age of 1330 to 1260 Ma (Williams 1992). These rocks were folded about east-west axes, and in the east they were refolded by tight NW-SE folds with northeasterly-dipping axial surfaces related to the 1300 to 1250 Ma Rudall-Musgrave Orogeny (Figure 1c) (Williams 1992).

SAVORY BASIN

The rocks of the Bangemall Basin are unconformably overlain in the east by the Savory Group in a region called the Savory Basin (Williams 1992). These deposits reflect four distinct episodes of tectonic activity. Unless otherwise indicated, isotopic ages are based on stratigraphic correlation by Williams (1992) with dated units in the Amadeus Basin to the east in central Australia (Lindsay & Korsch 1991).

The oldest rocks (Figure 6, sequence A) are sandstones . These make up a large northwesterly flowing fluvial and deltaic system which prograded onto a shallow marine shelf between 890 and 860 Ma (Williams 1992) . They formed in an extensional tectonic regime associated with reactivation of east-west fault systems and the local emplacement of dolerite and basalt.

These sandstones are overlain by fan-delta conglomerate (Coondra Formation) and braided stream sandstone (Spearhole Formation) derived from the west (Figure 6, sequence B). They reflect the reactivation of old NE-SW faults that led to uplift and erosion of part of the Pilbara Craton. They pass upwards and eastwards into shallow marine, lagoonal, tidal flat, and evaporite deposits (Mundadjini Formation). Williams (1992) correlates this sequence on the basis of stratigraphy and stromatolites with rocks in the Amadeus Basin that formed between 860 and 812 Ma (Lindsay & Korsch 1991).

An episode of compressive deformation reactivated NE-SW faults across the Savory Basin. The faults extend westwards across the central part of the Bangemall Basin and swing into E-W faults that were initiated during the Capricorn Orogeny on the margin of the Yilgarn Craton (Figure 6). The reactivated faults indicate both transcurrent and reverse movements with thrust transport towards both the Pilbara and Yilgarn Cratons. Within the Savory Basin faulting was especially prominent in the northwest and was accompanied by upright folds parallel to the faults (Blake fault-and-fold belt, Williams 1992). The glaciogenic marine Boondawari Formation (Figure 6, sequence C) was deposited during this episode and intruded by dolerite sills—a sample of which has given a whole rock Rb-Sr isochron age of 640 Ma.

These rocks and structures are unconformably overlain by sandstone and conglomerate of the McFadden, Tchukardine, and Woora Woora Formations of the Savory Group (Figure 6, sequence D) which were associated with tectonic activity in the Paterson Orogen (Figure 1d). The clastic debris was deposited in the Wells Foreland Basin which developed in advance of southwest-directed thrusts on the southwest foreland of the Paterson Orogen between 640 and 620 Ma (Williams 1992). These deposits were subsequently caught up in the advancing front of deformation and were folded and cut by thrusts and faults.

PATERSON OROGEN

The Paterson Orogen truncates the northeast margin of the Austral-Antarctic Craton and is dominated by late Precambrian structures (Figure 1d) (Myers 1990c). It contains components ranging from at least 1980 to 600 Ma and reflects late Proterozoic reactivation of a mid-Proterozoic orogen—here refered to as the Rudall-Musgrave Orogen—which resulted from the collision and amalgamation of the West Australian Craton with continental crust to the northeast (Figure 1c).

The orogen contains two different crystalline basement complexes (Rudall and Musgrave Complexes), which were both involved in mid-Proterozoic orogenesis (Figure 1c, Rudall-Musgrave Orogen). These are unconformably overlain by different sequences of sedimentary and volcanic rocks which predate the Paterson Orogeny (Yeneena Group and Bentley Supergroup), and sedimentary foreland basin deposits related to the orogeny (Figure 7).

Basement

The Rudall Complex comprises two interlayered units of metamorphic rocks with different metamorphic and tectonic histories (Chin et al 1980). The oldest unit consists of granitic gneiss and paragneiss (gneiss formed largely from sedimentary rocks) and the younger unit comprises quartzite and schist. Clarke (1991) called the older paragneiss the Yandagooge Formation. It was derived from greywacke and arkose, with minor BIF, chert, and calcsilicate rock, and occurs as inclusions within some of the granitic gneiss (Chin et al 1980, Clarke 1991). Two samples of granitic gneiss have given ages of 1980 and 1760 Ma (David Nelson 1991, personal communication).

The Yandagooge Formation and granitic gneiss were deformed and foliated (D1) before deposition of, or juxtaposition with, the younger unit

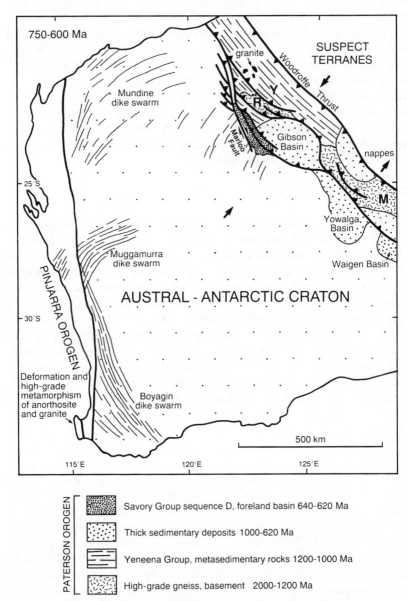

Figure 7 Main components of the Paterson Orogen involved in tectonic activity between 750 and 600 Ma; contemporaneous dike swarms emplaced into the Austral-Antarctic Craton, and plutonic intrusions, deformation, and metamorphism in the Pinjarra Orogen.

of quartzite and schist, called the Tjingkulatjatjarra Formation by Clarke (1991). The latter consists of quartzite and quarz-mica schist with minor BIF, marble, and graphite schist. It was deformed with the older rocks into recumbent tight or isoclinal folds with axes trending ENE. This episode of deformation (D2), called the Watrara Orogeny by Clarke (1991), was accompanied by syntectonic metamorphism in prograde amphibolite facies and granite intrusion. Bodies of granitoid rocks and serpentinite, pyroxenite, and gabbro were intruded after D1 and predate, or were contemporaneous with, D2. These rocks and structures were refolded by D3 folds with upright or northeast-dipping axial surfaces and southeast trend, considered by Clarke (1991) to predate the Yeneena Group. Samples of granitic gneiss have given a whole rock Rb-Sr isochron age of 1330 Ma (Chin & de Laeter 1981).

The Musgrave Complex comprises granitic gneiss and layered "gran-ulites" thought by Daniels (1974, 1975) to be of metasedimentary origin, but which may largely represent strongly deformed heterogeneous plutonic rocks. These rocks were repeatedly deformed and recrystallized in granulite facies, and then intruded by thick sheets of basic magma and further deformed whilst still at a deep crustal level at about 1200 Ma (Goode & Moore 1975, Goode 1978). This magma formed a number of layered intrusions (Giles Complex) consisting of gabbro, norite, troctolite, anor-thosite, and pyroxenite (Daniels 1974, 1975).

Cover

The Yeneena Group lies unconformably on both the Rudall Complex and the easternmost part of the Manganese Subgroup (Williams 1990b). It was therefore deposited after the juxtaposition of the Rudall Complex with the West Australian Craton. It comprises three distinct tectonic units bounded by NW-SE faults which are parallel with the trend of the Paterson Orogen (Figure 7). The faults thrust the rock units southwestward and some moved dextrally. Williams (1990b) suggested that they may reflect reactivation of faults formerly active during deposition of the Yeneena Group.

The western and central units comprise shallow marine sandstone, silt-stone, and local conglomerate and stromatolitic carbonates, interpreted by Williams (1990b) as shallow-water shelf deposits in which sediments were derived from the west. The rocks in the western unit form open folds that strike southeast and are metamorphosed at very low grade. In the central unit the rocks are more deformed and metamorphosed and the amount of deformation and metamorphism increases eastwards and down-wards. The unit is dominated by large-scale tight asymmetric folds with axial surfaces dipping northeastward.

The uppermost formation in the western and central units (Isdell Formation) is the lowest exposed part of the Yeneena Group in the eastern unit. It is overlain by a sequence of sandstone, siltstone, shale, and dolomite, interpreted by Williams (1990b) as a marine sequence deposited in a slowly subsiding basin or shelf. The structure is characterized by dome and basin fold interference structures elongated parallel with the orogen. Upright folds refold older thrusts and related folds, and the rocks recrystallized at medium metamorphic grade. These structures were successively intruded by dolerite sills and dikes, monzogranite and syenogranite plutons, and a NNW trending dolerite dike swarm. Samples of granite have given a Rb-Sr whole rock isochron age of 600 Ma and biotite age of 570 Ma (de Laeter et al 1977). Goellnicht et al (1991) obtained a Pb-Pb age of 690 Ma from the same granite and considered that the granite magmas were mantle derived.

The Musgrave Complex is unconformably overlain by volcanic and sedimentary rocks called the Bentley Supergroup (Daniels 1974). They comprise basalt, rhyolite, sandstone, conglomerate, and dolomite, and are deformed and metamorphosed in greenschist facies. Spatially associated cauldron subsidence complexes of granite were thought by Daniels (1974) to be coeval with the volcanics which have given a Rb-Sr isochron of 1060 ± 140 Ma (Compston & Nesbitt 1967). These rocks were unconformably overlain by sandstone (Townsend Quartzite) prior to deformation related to the Paterson Orogeny.

Paterson Orogeny

The main tectonic features of the Paterson Orogeny are thrusts and folds, associated with greenschist facies metamorphism. The Rudall Complex was tectonically interleaved and folded together with the Yeneena Group. In the southwest, folds within the Yeneena Group are overturned towards the southwest and the rocks were transported southwestward over foreland basin deposits (Savory Basin sequence D, and Gibson, Yowalga, and Waigen Basins in Figure 7). In the northeast of the orogen, the Musgrave Complex and its volcanic and sedimentary cover were transported northeastward on the Woodroffe Thrust over a major northward-facing nappe complex called the Petermann Ranges Nappe (Figure 7) (Wells et al 1970). Dolerite dikes (Mundine dike swarm, Figure 7) were intruded across the northern part of the Austral-Antarctic Craton and cut foreland basin deposits in the Savory Basin.

PINJARRA OROGEN

The Proterozoic Pinjarra Orogen truncates the western margin of the West Australian Craton (Figures 1c,d). It became a rift valley during episodes that culminated in the Phanerozoic separation of India and Australia (430–130 Ma) and was buried by 10–15 km of sedimentary rocks which form the Perth Basin. Therefore the Precambrian geological history of the orogen is poorly known. Precambrian rocks are exposed in three small horsts (Leeuwin, Mullingarra, and Northampton complexes) and are also known from boreholes through the sedimentary cover.

The Mullingarra and Northampton complexes comprise strongly deformed granitoid and sedimentary rocks recrystallized in amphibolite or granulite facies. Most Sm-Nd (T_{DM}) model ages from these complexes, and from boreholes in the Perth Basin, range from 2060 to 2030 Ma (recalculated by I. R. Fletcher 1988, personal communication, from Fletcher et al 1985). Granulite facies gneissses from the Northampton Complex have given a whole rock Rb-Sr isochron of 1020 ± 146 Ma (Richards et al 1985). Similar Rb-Sr ages of 1100 to 1050 Ma have been obtained from micas in both the Pinjarra Orogen and the adjacent Yilgarn Craton (Compston & Arriens 1968).

The Leeuwin Complex is largely derived from a layered intrusion of anorthosite, leucogabbro, and gabbro that was intruded by granite and strongly deformed and metamorphosed in granulite facies (Myers 1990e). Zircons from a sample of granite gneiss have given a U-Pb age of 570–550 Ma (Wilde & Murphy 1990).

The Orogen is bounded to the east by the Darling Fault Zone, a 30 km wide belt within the margin of the craton containing shear zones, mylonite, and phyllonite. These structures increase in intensity westward towards the Darling Fault which bounds the Phanerozoic rift. Bretan (1985) recognized three Precambrian episodes of tectonic activity in the fault zone: dextral transcurrent movements, followed by sinistral transcurrent movements, cut by mylonites perhaps related to late Precambrian compression and steep reverse movements.

Two major swarms of dolerite dikes (Muggamurra and Boyagin dike swarms, Figure 7) are probably related to late Precambrian tectonic activity along the orogen (Myers 1990c).

The scant evidence collectively suggests that the rocks of the Pinjarra Orogen reflect three episodes of major orogenic activity at about 2000 Ma, 1100 Ma, and 750–550 Ma. The latter involves the adjacent craton but the locations of the previous episodes in relation to the craton are unknown.

ALBANY-FRASER OROGEN

The Albany-Fraser Orogen truncates the southern part of the West Australian Craton (Figure 1c). The main tectonic and metamorphic features formed between 1300 and 1100 Ma during an episode of continental collision between the West Australian Craton and continental crust to the south and southeast, now in Antarctica and below the Tertiary Eucla Basin of southern Australia. The orogen is characterized by high-grade granitoid gneiss, granite intrusions, and major thrusting of lower and middle crustal rocks northwestward onto the margin of the West Australian Craton.

The orogen is longitudinally divided into two parts: the Biranup Complex of mainly late Archean and early Proterozoic gneiss, and the Nornalup Complex dominated by mid-Proterozoic granite (Figure 8) (Myers 1990f). The Archean rocks of the Biranup Complex have a different Archean history from those of the adjacent craton (Myers 1993), suggesting that they were part of a different continental plate. They were not significantly deformed during the Archean and were intruded by granite, granodiorite, and tonalite plutons between 1700 and 1600 Ma (David Nelson 1991–1992, personal communication). The rocks were not substantially deformed or metamorphosed at this time, and this may indicate that the Biranup Complex was amalgamated with the West Australian Craton by transcurrent movements.

The age of this amalgamation is unknown but both the margin of the West Australian Craton and these different Archean and early Proterozoic rocks were intruded by a dense swarm of dolerite dikes (Gnowangerup dike swarm) between 1600 and 1300 Ma. The dikes are well preserved in a 100 km wide belt along the edge of the craton (Figure 8). Towards the margin of the craton the dikes are increasingly recrystallized in greenschist facies. Within the orogen the dikes are intensely deformed and recrystallized and were largely reduced to thin layers and lenses parallel to the main foliation of the gneisses.

Low-grade metasedimentary rocks (Stirling Range Formation, Mount Barren Group, Woodline Beds) occur on the margin of the West Australian Craton (Figure 8). They largely comprise psammitic with subordinate pelitic rocks and are deformed by northward-verging folds and thrusts. They are cut by dolerite dikes but these dikes appear to be less abundant than in the adjacent Gnowangerup dike swarm. The spatial distribution of these metasedimentary rocks and the dikes along the margin of the orogen suggest that their formation was related to tectonic activity along the margin of the West Australian Craton. The dike swarm may reflect an episode of rifting prior to the mid-Proterozoic orogenic activity, and the metasedimentary rocks may have formed in association with early stages of this orogenesis.

A major intrusion of gabbro called the Fraser Complex (Figure 8) (Myers 1985) was intruded and crystallized at a deep crustal level at 1300 Ma (Fletcher et al 1991). It was tectonically interleaved with meta-

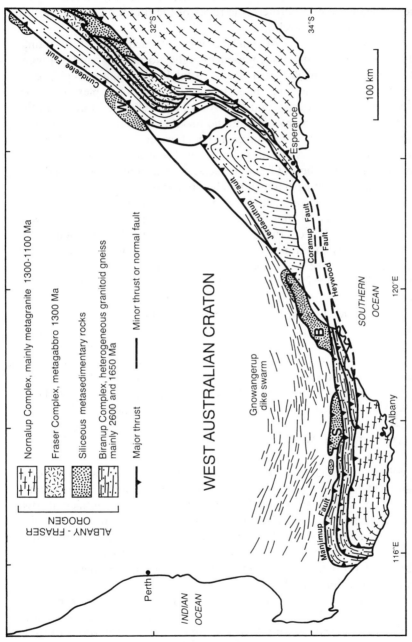

Figure 8 Main features of the 1300–1100 Ma Albany-Fraser Orogen which joined the West Australian Craton with continental crust to the southeast. Metasedimentary rocks: S—Stirling Range Formation, B—Mount Barren Group, W—Woodline Beds.

sedimentary rocks, mainly quartzite, and heterogeneously deformed with the Archean and early Proterozoic rocks of the Biranup Complex whilst still at a deep crustal level.

The Biranup and Fraser Complexes were intruded by sheets of granite which formed the first major component of the Nornalup Complex (Figure 8). They were emplaced at a deep crustal level during an episode of thrusting. Deformation was most intense in a belt of duplex structures bounded by the Coramup and Heywood Faults (Figure 8) in which the granite and early Proterozoic plutonic rocks were converted to mylonite and gneiss. The rocks were transported northwestward over folded duplex structures involving Archean gneiss, early Proterozoic gneiss, and mid-Proterozoic metagabbro. Granulite facies metamorphism outlasted deformation and formed partial melt patches and granoblastic textures.

To the east and south, above the Heywood Fault, the granites are much less deformed. After recrystallization in granulite facies they were transported to a higher crustal level, probably along the Heywood Fault, and intruded by the granite sheets and plutons that make up the second major component of the Nornalup Complex. These younger granites were recrystallized in amphibolite facies whilst heterogeneous deformation continued with declining intensity. The sequence of structures and fabrics indicate a gradual change from ductile amphibolite facies conditions to brittle greenschist facies conditions. Transport dirctions continued to be northwestward with dextral movement between the orogen and the craton.

U-Pb zircon ages indicate that the early granites of the Nornalup Complex and granulite facies recrystallization occurred at 1300 Ma and the younger granites were emplaced between 1200 and 1100 Ma (Pidgeon 1990; David Nelson 1991–1992, personal communication).

CONCLUSIONS

The West Australian Craton and its adjacent orogens have a Precambrian history of repeated generation, dispersal, and aggregation of continental crust. They indicate the mobility of rafts of Precambrian continental crust and the operation of plate tectonic processes that were at least as active as those of the present. The same major lines of crustal weakness were repeatedly reactivated as zones of continental rifting, collision, and intra-cratonic deformation, controlling belts of uplift and subsidence and the location of sedimentary basins.

Major episodes of continental rifting are recorded by: the eruption of the late Archean flood basalts on the Pilbara Craton and east-west dikes cutting the Yilgarn Craton; the rifting and dispersal of portions of these cratons prior to early Proterozoic reassembly with different continental fragments; and the mid-Proterozoic rifting and dispersal of fragments of the West Australian Craton along the lines now marked by the Albany-Fraser, Paterson, and Pinjarra Orogens.

Major periods of collision and aggregation of crustal fragments occurred at: 2700–2600 Ma forming and amalgamating much of the Yilgarn Craton; 2000–1800 Ma joining the Pilbara and Yilgarn Cratons along the Capricorn Orogen to form the West Australian Craton; 1300–1100 Ma along the Albany-Fraser Orogen, Paterson Orogen, and possibly Pinjarra Orogen, combining the West Australian Craton with other crustal fragments to form the Austral-Antarctic Craton; and 700–600 Ma leading to crustal stacking along the Paterson and Pinjarra Orogens.

These episodes of continental aggregation coincide with similar activity seen in other Precambrian crust around the Earth. They may reflect the formation of Precambrian supercontinents.

ACKNOWLEDGMENTS

Thanks are expressed to David Nelson, Cees Swager, Alan Thorne, Ian Tyler, Alec Trendall, and Ian Williams for the opportunity to refer to their recent work, for discussion of the geology, and for comments on parts of the text. The review is published with permission of the director of the Geological Survey of Western Australia.

Literature Cited

Arndt, N. T., Nelson, D. R., Compston, W., Trendall, A. F., Thorne, A. M. 1991. The age of the Fortescue Group, Hamersley Basin, Western Australia, from ion microprobe zircon U-Pb results. *Aust. J. Earth Sci.* 38: 261–81

Awramik, S. M., Schopf, J. W., Walter, M. R. 1983. Filamentous fossil bacteria 3.5×10^9 years old from the Archean of Western Australia. *Precambrian Res.* 20: 357–74

Barley, M. E. 1987. The Archaean Whim Creek Belt, an ensialic fault-bounded basin in the Pilbara Block, Australia. *Precambrian Res.* 37: 199–215

Barley, M. E., McNaughton, N. J. 1988. The tectonic evolution of greenstone belts and setting of Archaean gold mineralization in Western Australia: geochronological constraints on conceptual models. *Geol. Dept. & Univ. Extension, Univ. West. Aust. Publ.* 12: 23–40

Barley, M. E., Eisenlohr, B. N., Groves, D. I., Perring, C. S., Vearncombe, J. R. 1989. Late Archean convergent margin tectonics and gold mineralization: a new look at the Norseman-Wiluna Belt, Western Australia. *Geology* 17: 826–29

Bickle, M. J., Bettenay, L. F., Boulter, C. A., Groves, D. I. Morant, P. 1980. Horizontal tectonic interaction of an Archean gneiss belt and greenstones, Pilbara Block, Western Australia. *Geology* 8: 525–29

Bickle, M. J., Bettenay, L. F., Barley, M. E., Chapman, H. J., Groves, D. I., et al. 1983. A 3500 Ma plutonic and volcanic calc-alkaline province in the Archaean east Pilbara Block. *Contrib. Mineral. Petrol.* 84: 25–35

Bickle, M. J., Bettenay, L. F., Chapman, H. J., Groves, D. I., McNaughton, N. J., et al. 1989. The age and origin of younger granitic plutons of the Shaw Batholith in the Archean Pilbara Block, Western Australia. *Contrib. Mineral. Petrol.* 101: 361–76

Bickle, M. J., Bettenay, L. F., Chapman, H. J., Groves, D. I., McNaughton, N. J., et al. 1993. Origin of the 3500 Ma–3300 Ma calc-alkaline rocks in the Pilbara Archaean: isotope and geochemical constraints. *Precambrian Res.* In press

Blake, T. S. 1993. Late Archaean crustal extension, flood basalt volcanism, and continental rifting: the lower Mount Bruce Supergroup, Western Australia. *Precambrian Res.* In press

Blake, T. S., Groves, D. I. 1987. Continental rifting and the Archaean-Proterozoic transition. *Geology* 15: 229–32

Blockley, J. G. 1980. Tin deposits of Western Australia with special reference to the associated granites. *Geol. Surv. West. Aust. Mineral Resources Bull.* 12

Bretan, P. G. 1985. *Deformation processes within mylonite zones associated with some fundamental faults.* PhD thesis. Univ. London, U.K.

Campbell, I. H., Hill, R. I. 1988. A two-stage model for the formation of the granite-greenstone terrains of the Kalgoorlie-Norseman area, Western Australia. *Earth Planet. Sci. Lett.* 90: 11–25

Chin, R. J., de Laeter, J. R. 1981. The relationship of new Rb-Sr isotopic dates from the Rudall Metamorphic Complex to the geology of the Paterson Province. *Geol. Surv. West. Aust. Annu. Rep.* 1980: 80–87

Chin, R. J., Williams, I. R., Williams, S. J., Crowe, R. W. A. 1980. *Rudall, W. A., Geol. Surv. West. Aust.* 1: 250 000 *Geol. Ser. Explan. Notes*

Clarke, G. L. 1991. Proterozoic tectonic reworking in the Rudall Complex, Western Australia. *Aust. J. Earth Sci.* 38: 31–44

Compston, W., Arriens, P. A. 1968. The Precambrian geochronology of Australia. *Can. J. Earth Sci.* 5: 561–83

Compston, W., Nesbitt, R. W. 1967. Isotopic age of the Tollu Volcanics, W. A. *Geol. Soc. Aust. J.* 14: 235–38

Compston, W., Pidgeon, R. T. 1986. Jack Hills, evidence of more very old detrital zircons in Western Australia. *Nature* 321: 766–69

Daniels, J. L. 1974. The geology of the Blackstone Region, Western Australia. *Geol. Surv. West. Aust. Bull.* 123

Daniels, J. L. 1975. Musgrave Block. In *The Geology of Western Australia, Geol. Surv. West. Aust. Mem.* 2: 194–205. Perth: Geol. Surv. West. Aust.

de Laeter, J. R., Hickman, A. H., Trendall, A. F., Lewis, J. D. 1977. Geochronological data concerning the eastern extent of the Pilbara Block. *Geol. Surv. West. Aust. Annu. Rep.* 1976: 56–62

Etheridge, M. A., Rutland, R. W. R., Wyborn, L. A. I. 1987. Orogenesis and tectonic processes in the early to middle Proterozoic of northern Australia. In *Precambrian Lithospheric Evolution. Am. Geophys. Union Geodyn. Ser.* 17, ed. A.

Kröner, pp. 131–47. Washington, DC: Am. Geophys. Union

Ewers, W. E., Morris, R. C. 1981. Studies of the Dales Gorge Member of the Brockman Iron Formation, Western Australia. *Econ. Geol.* 76: 1929–53

Fletcher, I. R., Wilde, S. A., Rosman, K. J. R. 1985. Sm-Nd model ages across the margins of the Archaean Yilgarn Block, Western Australia—III. The western margin. *Aust. J. Earth Sci.* 32: 73–82

Fletcher, I. R., Myers, J. S., Ahmat, A. L. 1991. Isotopic evidence on the age and origin of the Fraser Complex, Western Australia: a sample of mid-Proterozoic lower crust. *Chem. Geol.* 87: 197–216

Froude, D. O., Ireland, T. R., Kinny, P. D., Williams, I. S., Compston, W., et al. 1983. Ion microprobe identification of 4100–4200 Myr-old terrestrial zircons. *Nature* 304: 661–18

Gee, R. D. 1979. Structure and tectonic style of the Western Australian shield. *Tectonophysics* 58: 327–69

Gee, R. D., Baxter, J. L., Wilde, S. A., Williams, I. R. 1981. Crustal development in the Yilgarn Block. In *Archaean Geology, Geol. Soc. Aust. Spec. Publ.*, ed. J. E. Glover, D. I. Groves, pp. 43–56. Perth, Aust.: Geol. Soc. Aust.

Glikson, A. Y. 1979. Early Precambrian tonalite-trondhjemite sialic nucleii. *Earth Sci. Rev.* 15: 1–73

Goellnicht, N. M., Groves, D. I., McNaughton, N. J. 1991. Late Proterozoic fractionated granitoids of the mineralized Telfer area, Paterson Province, Western Australia. *Precambrian Res.* 51: 375–91

Goode, A. D. T. 1978. High temperature, high strain rate deformation in the lower crustal Kalka Intrusion, Central Australia. *Contrib. Mineral. Petrol.* 66: 137–48

Goode, A. D. T., Moore, A. C. 1975. High pressure crystallization of the Ewarara, Kalka and Gosse Pile intrusions, Giles Complex, central Australia. *Contrib. Mineral. Petrol.* 51: 77–97

Groves, D. I., Batt, W. D. 1984. Spatial and temporal variations of Archaean metallogenic associations in terms of evolution of granitoid-greenstone terrains with particular emphasis on the Western Australian Shield. In *Archaean Geochemistry*, ed. A. Kröner, G. N. Hansen, A. M. Goodwin, pp. 73–98. Berlin: Springer-Verlag

Groves, D. I., Archibald, N. J., Bettenay, L. F., Binns, R. A. 1978. Greenstone belts as ancient marginal basins or ensialic rift zones. *Nature* 273: 460–61

Groves, D. I. Phillips, G. N., Ho, S. E., Houstoun, S. M. 1985. The nature, genesis and regional controls of gold mineralization in Archaean greenstone belts of the Western Australian Shield: a brief review. *Geol. Soc. S. Afr. Trans.* 88: 135–48

Groves, D. I., Phillips, G. N., Ho, S. E., Houstoun, S. M., Standing, C. A. 1987. Craton-scale distribution of Archaean greenstone gold deposits: predictive capacity of the metamorphic model. *Econ. Geol.* 82: 2045–58

Hallberg, J. A. 1986. Archaean basin development and crustal extension in the northeastern Yilgarn Block, Western Australia. *Precambrian Res.* 31: 133–56

Hickman, A. H. 1983. Geology of the Pilbara Block and its environs. *Geol. Surv. West. Aust. Bull.* 127. 268 pp.

Hill, R. I., Campbell, I. H., Griffiths, R. W. 1991. Plume tectonics and the development of stable continental crust. *Explor. Geophys.* 22: 185–88

Hoffman, P. F. 1988. United plates of America, the birth of a craton: Early Proterozoic assembly and growth of Laurentia. *Annu. Rev. Earth Planet. Sci.* 16: 543–603

Horwitz, R. C. 1982. Geological history of the early Proterozoic Paraburdoo hinge zone, Western Australia. *Precambrian Res.* 19: 191–200

Horwitz, R. C., Smith, R. E. 1978. Bridging the Yilgarn and Pilbara Blocks, Western Australia. *Precambrian Res.* 6: 293–322

Hynes, A., Gee, R. D. 1986. Geological setting and petrochemistry of the Narracoota Volcanics, Capricorn Orogen, Western Australia. *Precambrian Res.* 31: 107–32

Idnurm, M., Giddings, J. W. 1988. Australian Precambrian polar wander: a review. *Precambrian Res.* 40–41: 61–88

Krapez, B., Barley, M. E. 1987. Archaean strike-slip faulting and related ensialic basins: evidence from the Pilbara Block, Australia. *Geol. Mag.* 124: 555–67

Lindsay, J. F., Korsch, R. J. 1991. The evolution of the Amadeus Basin, central Australia. In *Geological and Geophysical Studies in the Amadeus Basin, Central Australia. Bur. Mineral Resources, Geol. Geophys. Bull.* 236, ed. R. J. Korsch, J. M. Kennard, pp 7–32. Canberra: Bur. Mineral Resources

Lindsay, J. F., Korsch, R. J., Wilford, J. R. 1987. Timing and breakup of a Proterozoic supercontinent: evidence from Australian intracratonic basins. *Geology* 15: 1061–64

Lowe, D. R. 1980. Stromatolites, 3,400-Myr old from the Archaean of Western Australia. *Nature* 284: 441–43

Morris, R. C., Horwitz, R. C. 1983. The origin of the iron-formation-rich Hamersley Group of Western Australia—

deposition on a platform. *Precambrian Res.* 21: 273–97

Muhling, J. R. 1988. The nature of Proterozoic reworking of early Archaean gneisses. Mukalo Creek area, southern Gascoyne Province, Western Australia. *Precambrian Res.* 40–41: 341–62

Muhling, J. R. 1990. The Narryer Gneiss Complex of the Yilgarn Block, Western Australia: a segment of Archaen lower crust uplifted during Proterozoic orogeny. *J. Metamorphic Geol.* 8: 47–64

Muhling, P. C., Brakel, A. T. 1985. Geology of the Bangemall Group—the evolution of an intracratonic Proterozoic basin. *Geol. Surv. West. Aust. Bull.* 128

Myers, J. S. 1985. The Fraser Complex—a major layered intrusion in Western Australia. *Geol. Surv. West. Aust. Rep.* 14: 57–66

Myers, J. S. 1988. Early Archaean Narryer Gneiss Complex, Yilgarn Craton, Western Australia. *Precambrian Res.* 38: 297–308

Myers, J. S. 1989. Thrust sheets on the southern foreland of the Capricorn Orogen, Robinson Range, Western Australia. *Geol. Surv. West. Aust. Rep.* 26: 127–30

Myers, J. S. 1990a. Western Gneiss Terrane. In *Geology and Mineral Resources of Western Australia, Geol. Surv. West. Aust. Mem.* 3: 13–31. Perth: Geol. Surv. West. Aust.

Myers, J. S., compiler 1990b. Excursion 1: Narryer Gneiss Complex. In *Third Int. Archaean Symp., Perth*, 1990, *Excursion Guidebook*, ed. S. E. Ho, J. E. Glover, J. S. Myers, J. R. Muhling, pp. 61–95. Perth, Aust.: Geol. Dept., Univ. West. Aust.

Myers, J. S. 1990c. Precambrian tectonic evolution of part of Gondwana, southwestern Australia. *Geology* 18: 537–40

Myers, J. S. 1990d. Gascoyne Complex. In *Geology and Mineral Resources of Western Australia, Geol. Surv. West. Aust. Mem.* 3: 198–202. Perth: Geol. Surv. West. Aust.

Myers, J. S. 1990e. Anorthosite in the Leeuwin Complex of the Pinjarra Orogen, Western Australia. *Aust. J. Earth Sci.* 37: 241–45

Myers, J. S. 1990f. Albany-Fraser Orogen. In *Geology and Mineral Resources of Western Australia, Geol. Surv. West. Aust. Mem.* 3: 255–64. Perth: Geol. Surv. West. Aust.

Myers, J. S. 1993. Esperance, W. A. *Geol. Surv. West. Aust.*, 1: 1 000 000 Ser. Explan. Notes. In press

Myers, J. S., Watkins, K. P. 1985. Origin of granite-greenstone patterns, Yilgarn Block, Western Australia. *Geology* 13: 778–80

Nemchin, A. A., Pidgeon, R. T., Wilde, S.

A. 1993. Timing of the late Archaean granulite facies metamorphism and the evolution of the southwestern Yilgarn Craton of Western Australia. *Precambrian Res.* In press

Pidgeon, R. T. 1990. Timing of plutonism in the Proterozoic Albany Mobile Belt, southwestern Australia. *Precambrian Res.* 47: 157–67

Pidgeon, R. T., Horwitz, R. C. 1991. The origin of olistoliths in Proterozoic rocks of the Ashburton Trough, Western Australia, using zircon U-Pb isotopic characteristics. *Aust. J. Earth Sci.* 38: 55–63

Pidgeon, R. T., Wilde, S. A. 1990. The distribution of 3.0 Ga and 2.7 Ga volcanic episodes in the Yilgarn Craton of Western Australia. *Precambrian Res.* 48: 309–25

Richards, J. R., Blockley, J. G., de Laeter, J. R. 1985. Rb-Sr and Pb isotope data from the Northampton Block, Western Australia. *Aust. Inst. Min. Metall. Proc.* 290: 43–55

Rutland, R. W. R. 1973. Tectonic evolution of the continental crust of Australia. In *Implications of Continental Drift to the Earth Sciences*, vol. 2, ed. D. H. Tarling, S. K. Runcorn, pp 1011–33. London: Academic

Schopf, J. W., Packer, B. M. 1987. Early Archaean (3.3-billion to 3.5-billion-year-old) microfosils from Warrawoona Group, Australia. *Science* 237: 70–73

Swager, C., Witt, W. K., Griffin, T. J., Ahmat, A. L., Hunter, W. M., et al. 1990. A regional overview of the late Archaean granite-greenstones of the Kalgoorlie Terrane. In *Third Int. Archaean Symp., Perth, 1990, Excursion Guidebook*, ed S. E. Ho, J. E. Glover, J. S. Myers, J. R. Muhling, pp 205–20. Perth, Aust.: Geol. Dept., Univ. West. Aust.

Thorne, A. M., Seymour, D. B. 1991. Geology of the Ashburton Basin, Western Australia. *Geol. Surv. West. Aust. Bull.* 139. 141 pp.

Thorne, A. M., Trendall, A. F. 1994. Geology of the Fortescue Group, Western Australia. *Geol. Surv. West. Aust. Bull.* In press

Thorpe, R. I., Hickman, A. H., Davis, D. W., Mortensen, J. K., Trendall, A. F. 1992. U-Pb zircon geochronology of Archaean felsic units in the Marble Bar region, Pilbara Craton, Western Australia. *Precambrian Res.* 56: 169–90

Trendall, A. F. 1976. Striated and faceted boulders from the Turee Creek Formation—evidence for a possible Huronian glaciation on the Australian continent. *Geol. Surv. West. Aust. Annu. Rep.* 1975: 88–92

Trendall, A. F. 1979. A revision of the Mount Bruce Supergroup. *Geol. Surv. West. Aust. Annu. Rep.* 1978: 63–71

Trendall, A. F. 1983. The Hamersley Basin. In *Iron Formations—Facts and Problems*, ed. A. F. Trendall, R. C. Morris, pp. 69–129· Amsterdam: Elsevier

Trendall, A. F. 1990. Hamersley Basin. In *Geology and Mineral Resources of Western Australia, Geol. Surv. West. Aust. Mem.* 3: 163–89. Perth: Geol. Surv. West. Aust.

Trendall, A. F., Blockley, J. G. 1970. The iron formations of the Precambrian Hamersley Group, Western Australia, with special reference to the associated crocidolite. *Geol. Surv. West. Aust. Bull.* 119

Tyler, I. M. 1991. The geology of the Sylvania inlier and southeast Hamersley Basin. *Geol. Surv. West. Aust. Bull.* 138. 108 pp.

Tyler, I. M. Thorne, A. M. 1990. The northern margin of the Capricorn Orogen, Western Australia—an example of an Early Proterozoic collision zone. *J. Struct. Geol.* 12: 685–701

Tyler, I. M., Fletcher, I. R., de Laeter, J. R., Williams, I. R., Libby, W. G. 1992. Isotope and rare earth element evidence for a late Archaean terrane boundary in the southeastern Pilbara Craton, Western Australia. *Precambrian Res.* 54: 211–29

Veevers, J. J., McElhinny, M. W. 1976. The separation of Australia from other continents. *Earth Sci. Rev.* 12: 139–59

Walter, M. R., Buick, R., Dunlop, J. S. R. 1980. Stromatolites 3,400–3,500 Myr old from the North Pole area, Western Australia. *Nature* 284: 443–45

Watkins, K. P., Hickman, A. H. 1990. Geological evolution and mineralization of the Murchison Province, Western Australia. *Geol. Surv. West. Aust. Bull.* 137

Watkins, K. P., Fletcher, I. R., de Laeter, J. R. 1991. Crustal evolution of Archaean granitoids in the Murchison Province, Western Australia. *Precambrian Res.* 50: 311–36

Wells, A. T., Foreman, D. J., Ranford, L. C., Cook, P. J. 1970. Geology of the Amadeus Basin. *BMR Bull.* 100. Canberra: Bur. Mineral Resources

Wijbrans, J. R., McDougall, I. 1987. On the metamorphic history of an Archaean granitoid greenstone terrain, East Pilbara, Western Australia, using the ^{40}Ar/^{39}Ar age spectrum method. *Earth Planet. Sci. Lett.* 84: 226–42

Wilde, S. A., Murphy, D. M. K. 1990. The nature and origin of Late Proterozoic high-grade gneisses of the Leeuwin Block, Western Australia. *Precambrian Res.* 47: 251–70

Williams, S. J. 1986. Geology of the

Gascoyne Province, Western Australia. *Geol. Surv. West. Aust. Rep.* 15. 85 pp.

Williams, I. R. 1990a. Bangemall Basin. In *Geology and Mineral Resources of Western Australia, Geol. Surv. West. Aust. Mem.* 3: 308–28. Perth: Geol. Surv. West. Aust.

Williams, I. R. 1990b. Yeneena Basin. In *Geology and Mineral Resources of Western Australia, Geol. Surv. West. Aust. Mem.* 3: 277–282. Perth: Geol. Surv. West. Aust.

Williams, I. R. 1992. The geology of the Savory Basin, Western Australia. *Geol. Surv. West. Aust. Bull.* 141

Williams, I. S., Collins, W. J. 1990. Granite-greenstone terranes in the Pilbara Block, Australia, as coeval volcano-plutonic complexes; direct evidence from U-Pb zircon dating of the Mount Edgar Batholith. *Earth Planet. Sci. Lett.* 97: 41–53

Witt, W. K., Swager, C. P. 1989. Structural setting and geochemistry of Archaean I-type granites in the Bardoc-Coolgardie area of the Norseman-Wiluna belt, Western Australia. *Precambrian Res.* 44: 323–51

Annu. Rev. Earth Planet. Sci. 1993. 21:487–523
Copyright © 1993 by Annual Reviews Inc. All rights reserved

UNDERSTANDING PLANETARY RINGS

Larry W. Esposito

LASP, CB 392, University of Colorado, Boulder, Colorado 80309–0392

KEY WORDS: solar system origins, planets, satellites

1. INTRODUCTION

Planetary rings, which were until recently thought unique to the planet
Saturn, have now been observed around all the giant planets. These rings
are composed of many particles with a broad range in size. The observed
ring systems are quite diverse. Jupiter's ring is thin and composed of dust-
like small particles. Saturn's rings are broad, bright, and opaque. Uranus
has narrow, dark rings among broad lanes of dust which are invisible from
Earth. Neptune's rings include incomplete arcs restricted to a small range
of the circumference. All rings lie predominantly within their planet's
Roche limit, where tidal forces would destroy a self-gravitating body. They
are also within the planet's magnetosphere and in the case of Uranus they
are within the upper reaches of the planetary atmosphere.

The common occurrence of ring material around the planets is one of
the major scientific findings of the past 15 years. The discovery of the new
ring systems is the result of spacecraft and ground-based observations,
including magnetospheric studies, imaging, and stellar occultations. The
structure of the rings and their composition vary among the various
planets, and likewise within each ring system. The broadest set of rings
and the most identified processes are found in Saturn's system.

The rings display the following structural features: vertical thickness
considerably greater than the average particle size; dark lanes, gaps, and
other opacity variations; eccentric shapes and inclined orientations; sharp
edges; azimuthal brightness variations, arcs, and clumps; waves and wakes;
and incomplete, kinked, and apparently braided configurations. Some of

487

0084–6597/93/0515–0487$02.00

these features have been explained by a range of gravitational interactions with nearby moons, including orbital resonances.

Beyond the interactions with moons (many of which were likewise recently discovered by spacecraft), the ring particles interact with the magnetosphere via charging, plasma drag, and interaction with the ambient magnetic and electric fields. Electrostatic effects may lift small particles off the surface of the larger ring particles to create the dark, radial lanes seen in the *Voyager* Saturn pictures which were termed "spokes." Particles may feel a gas drag from the extended planetary atmosphere.

The size distribution of ring particles extends from sub-micron dust, through meter-sized particles, to small imbedded moons, including the recently discovered "Pan," about 10 km in radius. Perhaps 100–1000 moons bigger than 1 km orbit each of the giant planets, but were too small to be detected by spacecraft cameras . Theoretical expectations and some data support the idea that the particles in a ring will segregate in size, both radially and vertically.

The composition of ring particles is well known only for Saturn. Spectroscopic, occultation, and neutron measurements all imply that Saturn's rings particles are almost entirely water ice. For the other ring systems, the particles resemble the nearby small moons and probably contain significant silicate and in the case of Uranus and Neptune, possible carbonaceous material. Even in Saturn's rings, color variations may indicate compositional differences in different parts of the rings.

Radio occultations at two wavelengths have provided size information for Saturn's and Uranus' rings at a number of locations (unfortunately excluding Saturn's B ring because of its opacity) in the range of roughly 1 cm to 10 m. Information on smaller particles is from photometry and differential opacity in stellar occultations. The derived size distributions can be characterized as broad power-law distributions, with power-law exponents of 2.5 to 3.5.

We have a first order understanding of the dynamics and key processes in rings, much of it based on previous work in galactic and stellar dynamics. The rings are a kinetic system, where the deviations from perfect circular, equatorial motion can be considered as random velocities in a viscous fluid. Unfortunately, the models are often idealized (for example, treating all particles as hard spheres of the same size) and cannot yet predict many phenomena in the detail observed by spacecraft observations (for example, sharp edges).

The rings show many youthful features: Saturn's ice is bright, Uranus' rings are narrow, Neptune's arcs are constrained to a small range of longitude, and Jupiter's particles are so small that they will be dragged away in a thousand years or so. The momentum transferred between rings and the nearby moons should have caused them to spread much further apart than they are now. Further, the small moons discovered by *Voyager* also could not have survived the flux of interplanetary meteoroids for the age of the solar system. In much less time, these small moons would be shattered by an impacting object. This realization provides a potential solution to the problem of young rings: These impacts may not only destroy the moon, but regularly they can re-create the ring systems that are gradually spreading and being ground to dust. Thus, the moons not only sculpt the rings' structure, they may have provided the reservoirs for past and future ring systems.

We emphasize this hypothesis that most rings are likely much younger than the solar system and that new rings are episodically created by the destruction of small moons near the planets. This idea is one example of how the recent spacecraft observations have indicated a larger role for catastrophic events in the history of the solar system. The role of giant impacts seems essential in explaining the history of planetary rings. Thus, the study of rings connects with similar processes in the formation of the Earth's moon and the question of the demise of the dinosaurs.

The range of phenomena seen in planetary rings was unexpected and gives insight into the processes in other flattened astrophysical systems. An interesting parallel can be drawn between the processes observed now in planetary ring systems and those that occurred at the time of the origin of the planets. Clearly, the rings are not now accreting to form planets, as the original planetesimals did. However, many similar processes are occurring now in rings that resemble those in the solar nebula. Models of the present processes in rings can be compared in detail to the observations, allowing testing and refinement not possible for the early solar system.

This review briefly describes the rings in the solar system, and the mathematical and physical approaches to understanding them. I emphasize the variety of ring phenomena revealed in the past decade from space and Earth observations. I address in some detail the ideas for the recent history of rings and conclude with a discussion of the major open questions and future opportunities to learn about planetary rings. This paper benefits from the recent review by Nicholson & Dones (1991), and from previous reviews by Esposito et al (1984), Cuzzi et al (1984), Esposito (1986), and Esposito et al (1991), and the forthcoming review by Porco et al (1993).

2. DESCRIPTION OF PLANETARY RINGS

For the most part, planetary rings are composed of small particles orbiting inside the Roche limit of the giant planets. The Roche limit is the distance closer than which a fluid particle would be disrupted by tidal forces from the planet; at the limiting distance, the tides are just balanced by the self-gravity of the object (Weidenshilling et al 1984). An interesting exercise is to compare all the ring systems normalized to the equatorial radius of each planet [see Figure 1, from Nicholson & Dones (1991)]. The spacecraft observations have led to the discovery of numerous small moons (called

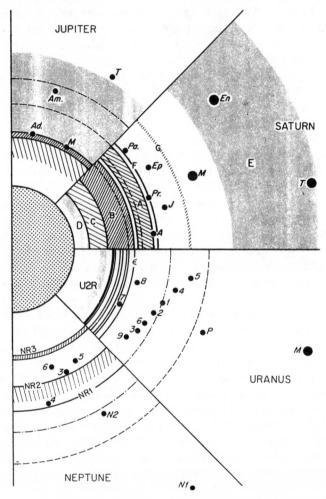

Figure 1 A comparison of the four planetary ring systems, including the nearby satellites, scaled to a common planetary equatorial radius. Density of cross-hatching indicates the relative optical depth of the different ring components. Synchronous orbit is indicated by a dashed line, the Roche limit for a density of 1 g cm⁻³ by a dot-dash line. For Uranus and Neptune, temporary IAU satellite nomenclature is used: 1986U1–9, 1989N1–6. (From Nicholson & Dones 1991.)

"ringmoons") near the rings. The physical properties of these moons have been reviewed by Thomas (1989). As we shall see, these moons are dynamically (and, likely, genetically) related to the nearby rings.

The properties of the particles in the rings are incompletely known. We have the best information for Saturn's rings, which are the brightest and have been studied from the ground since their discovery by Galileo in 1610 (e.g. van Helden 1984). Our current understanding is summarized in Table 1 from Nicholson & Dones (1991).

3. APPROACHES TO UNDERSTANDING RINGS

3.1 Particle Dynamics

The first modern step toward understanding planetary rings was made just before the discovery of Uranian rings in 1977 with theoretical studies of simple unperturbed collisional systems. The collisional dynamics of a differentially rotating disk of particles have been studied extensively; see, for example, Lynden-Bell & Pringle (1974), Brahic (1975, 1977), Goldreich & Tremaine (1978), and Lin & Papaloizou (1979). The main results can be summarized in the following way (Brahic 1977): After a very fast flattening within a time of the order of few tens of collisions per particle, the system reaches a quasi-equilibrium state in which the thickness of the newly formed disk is finite (i.e. the centers of the particles do not lie in the same plane) and in which collisions still occur. Under the combined effect of differential rotation and of inelastic collisions, the disk spreads very slowly and particles move both inwards and outwards carrying some angular momentum. In the absence of external confining forces, the spreading time is of the order of the time it takes particles to random walk a distance equal to the ring width. During a collision, part of the relative velocity arising from differential rotation is transformed into vertical and radial motion. A steady state is established between the energy received by the system and the energy drained away through inelastic collisions, so that the energy which is continually lost is obtained at the expense of the bodies moving inward and outward. Because angular momentum is conserved, the energy lost by the inward motion of a portion of the particles is larger than the energy gained by the remainder that are moving outwards: To spread, the disk must give up a small amount of its total energy.

3.2 Mechanisms for Ring Confinement

The first analytical studies solved the Boltzmann equation with a specified collision integral on the right hand side. In order to solve this equation, Goldreich & Tremaine (1978, 1982), Araki & Tremaine (1986), Shu &

Table 1 Rings and their particle properties

Planet	Ring	Large particles[a]				Small particles[b]				
		τ_{large}	r (cm)	P_{large}	Albedo	τ_{dust}	r (μm)	p_{dust}	σ (g/cm²)[c]	M (g)
Jupiter	main	3×10^{-6}	>1	[3]	[0.015]	2×10^{-6}	<0.3–16	2.5±0.5	$>5\times10^{-6}$	$>3\times10^{14}$
	halo	$<1\times10^{-6}$	—	—	—	2×10^{-6}	[0.3–16]	>2.5	—	—
	gossamer	$<1\times10^{-7}$	—	—	—	1×10^{-7}	[0.3–16]	[2.5]	—	—
Saturn	A, B	0.4–2.5	1–500	2.7–3.0	0.4–0.6	<0.01–0.05	0.6±0.2	—	20–100	2×10^{22}
	Cassini Div	0.05–0.15	1–750	2.8	0.2–0.4	—	—	—	5–20	4×10^{20}
	C	0.05–0.35	1–(100–500)	3.1	0.12–0.30	—	—	—	0.4–5	1×10^{20}
	F	0.02	≈1	3.3±0.2	0.6	0.1	[0.1–10]	4.1±1.1	—	—
	E	$<2\times10^{-7}$	>11	—	—	1.5×10^{-5}	1.0±0.3	—	—	—
	G	$<6\times10^{-7}$	—	—	—	1×10^{-6}	[0.1–10]	[3±1]	—	—
Uranus	ε	0.5–2.3	>70†	[3]	0.018	2×10^{-3}	[1]	[2.5]	25	5×10^{18}
	other 8 classical	0.2–0.6	>1	—	0.010–0.018	$(2$–$8)\times10^{-4}$	[1]	[2.5]	2–10	3×10^{17}
	λ = 1986U1R	<0.05	—	—	—	0.1–0.2‡	0.1?	—	—	—
	outer dust band	—	—	—	—	2×10^{-5}	[1]	[2.5]	—	—
Neptune	Adams = 1989N1R	0.01	—	—	[0.015]	0.005	—	[2.5]	—	—
	Leverrier = 1989N2R	0.005	—	—	[0.015]	0.01	—	[2.5]	—	—
	N1R arcs	0.03	—	—	[0.015]	0.04	—	[2.5]	—	—
	N3R, N4R	$(3$–$12)\times10^{-5}$	—	—	[0.015]	5×10^{-5}	—	[2.5]	—	—

Note: Quantities in brackets [...] are assumed, not derived. When a range of values is given, except for particle sizes, it indicates actual variation through broad regions of the ring, excluding narrow gaps or ringlets. Optical depths are at a wavelength of 0.5 mm. The large particle optical depths refer to the geometric cross section of the particles ($Q = 1$); the small particle optical depths include diffraction ($Q = 2$). (See Cuzzi 1985.)

[a] Large particles: backscattering, larger than ~1 mm. τ_{dust} = optical depth in large particles. r (cm) = size range in centimeters. P_{large} = power-law index of differential distribution. Albedo = spherical, or Bond, albedo.

[b] Small particles: forward scattering, smaller than ~1 mm. τ_{dust} = optical depth in small particles. r (μm) = size range in microns. p_{dust} = power-law index of differential size distribution, if applicable. (E ring and spokes ring have narrow size distributions.)

[c] σ (g/cm²) = surface mass density of ring in grams per square centimeter. M (g) = mass of ring in grams.

† Particle size derived from equality of optical depths at 3.6 and 13 cm in *Voyager* radio science experiment ($r > 70$ cm) appears inconsistent with prediction of self-gravity model.

‡ *Voyager* UVS and PPS occultations at same ring longitude measured $\tau \approx 0.2$ at $\lambda = 0.11$ μm, $\tau \approx 0.1$ at $\lambda = 0.28$ μm. Rapid fall of optical depth with increasing wavelength suggests subwavelength-sized particles.

Surface densities given for Uranian rings are predictions of self-gravity model.

References: Jupiter: Showalter et al 1987. Saturn: Showalter et al 1987. Saturn: Cuzzi et al 1984, Figure 10 (τ_{large}); Zebker et al 1985 (particle size distribution), Dones et al 1989, Doyle & Grün 1990, Cooke 1991 [albedo, τ_{dust}, r (μm)]; Pollack et al, preprint (F ring); Showalter 1989 (E and G rings); Rosen 1989 (σ); Showalter et al 1991 (E ring). Uranus: French et al 1986 (τ_{large}); Gresh et al 1989 (particle size distribution); Ockert et al 1987 (albedo, τ_{dust}, λ ring); Holberg et al 1987 (λ ring); Goldreich & Tremaine 1979 (σ). Neptune: Smith et al 1989, Nicholson et al 1990.

Stewart (1985), and Hammeen-Anttila (1978) have introduced simplifications to the complete Boltzmann equation based on several assumptions. An important question arises whether the "academic problem" then considered has anything to do with real rings. Rather than trying to reproduce any particular observed feature, these studies have served to investigate the important physical mechanisms.

A very significant mechanism has been quantitatively studied by Goldreich & Tremaine (1978, 1982). Through gravity, a nearby satellite will alter a ring particle's orbit, making it elliptical. This effect is especially pronounced where the period of the ring particle orbit is in the ratio of small integers to the period of the perturbing satellite. Such a location is termed a resonance and labeled by the ratio of periods, i.e. "the Mimas 2:1 resonance." The overall effect of the resonant perturbations on a population of particles in a disk is to increase in some places the density of particles and to decrease it in other places; this pattern is static in the satellite's frame, but moves through the disk and thereby generates spiral density waves, like those in spiral galaxies (cf Shu 1970). If the perturbing satellite is exterior to the ring, this wave moves outwards from the resonance, removing energy and angular momentum. An isolated ring particle would not suffer any systematic drift after the encounter with the satellite, but the rings are dense enough to have numerous collisions between particles, and the particles involved in such collisions gradually move inward towards the planet. In other words, encounters with a satellite increase radial motions while collisions circularize the orbits. A balance is achieved in which the particle orbits accommodate to the perturbing influence (Brophy et al 1993). The net effect is thus a repulsion of the orbits (see Greenberg 1983). The exterior satellite removes angular momentum from the ring, adds energy to nearby particles, and transfers the angular momentum from the ring to its orbit (e.g. Lin & Papaloizou 1979). If the torque carried off by this wave exceeds that carried by the particles diffusing across the resonance, the wave truncates the ring and an edge is formed (Goldreich & Tremaine 1978). An example is provided by the Uranus system, where known satellites (for the ε ring) as well as small undetected satellites (for the other rings) on each side of a narrow ring can thus constrain its edges and prevent ring spreading. This has been termed "shepherding." Elliptical and inclined rings can also be generated by such a confining mechanism (Goldreich & Tremaine 1982).

The precise mechanism for shepherding has not been completely determined (Borderies et al 1989, Brophy et al 1990), but can be qualitatively understood as follows. Satellites exert torques on the boundary of a ring located at a low order resonance. For the simplest case of a circular ring and a circular satellite orbit, the strongest torques occur where the ratio of the satellite orbit period to the ring particle period equals $m/(m+1)$ with m an integer. The torque is of the order of:

$$T_m = \pm m^2 (G^2 m_s^2 \sigma)/(\Omega^2 r^2) \tag{1}$$

(Goldreich & Tremaine 1980). G, Ω, m_s, r, and σ are, respectively, the gravitational constant, the orbital frequency, the satellite mass, the orbital radius, and the surface mass density evaluated at the resonance location. If the spacing between neighboring resonances from a nearby satellite is very small, the widths of individual resonances can be greater than the separations so that resonances overlap. Thus, it is useful to sum the discrete resonance torques and to define the total torque on a narrow ringlet of width Δr:

$$T \sim \pm (G^2 m_s^2 \sigma r \Delta r)/(\Omega^2 x^4) \tag{2}$$

where x is the separation between the satellite and the ringlet ($r \gg x \gg \Delta r$) (Goldreich & Tremaine 1980).

The presence of dissipation is essential since the torque would vanish without it. The exact nature of the dissipation does not affect the expression of the torque (Meyer-Vernet & Sicardy 1984, Greenberg 1983; although this standard expression for the torque has been questioned by Brophy et al 1993) as suggested by the fact that the torque expression does not contain explicitly any dependence on dissipation. Particle collisions are evidently the main source of dissipation. The physics involving many colliding particles sets the study of ring dynamics apart from classical celestial dynamics, in which no collisions are assumed, and fluid dynamics, in which more frequent collisions are known to occur than in planetary rings (where the mean free path of an individual particle is much longer than in the fluid case).

Note that the resonance torque would not act on isolated test particles. For example, due to the rarity of collisions in the asteroid belt, this mechanism is not responsible for the formation of the Kirkwood gaps.

The rate of transfer of angular momentum can also be calculated by a perturbative approach without reference to individual resonances. The gravitational interaction of a ring particle with a satellite occurs primarily close to encounter. At encounter, the tangential component of the relative velocity of the particle with respect to the satellite is reduced and thus angular momentum is exchanged with the net result that the ring experiences a torque (Lin & Papaloizou 1979). Furthermore, a particle initially moving on a circular orbit acquires a radial velocity and thereafter moves on a Keplerian ellipse. In a frame corotating with the perturbing satellite, all particles initially moving in circular orbits must follow similar paths after encounters. Thus, each perturbing satellite generates a standing wave.

In the inertial frame, each particle moves on an independent Keplerian ellipse, but the pericenters of these elliptical orbits and the phases of the particles create a sinusoidal wave which moves through the ring with the angular velocity of the perturbing satellite. For a more complete discussion, see Shu (1984). Numerous such *density waves* have been observed in Saturn's rings (see below) and perhaps also in Uranus' rings (Horn et al 1989). The damping of these waves by collisions can result in a net exchange of angular momentum between the satellite and the ring particles. This phenomenon is similar to the dynamical friction studied in stellar dynamics.

The discovery of ten small satellites inside the orbit of Miranda during the *Voyager 2* Uranus encounter has provided a remarkable opportunity to test theoretical predictions for the dynamics of Uranian rings. Porco & Goldreich (1987) have shown strong kinematical evidence to support the hypothesis that Cordelia (1986 U7) and Ophelia (1986 U8) are the inner and outer shepherds for the ε ring, and that Cordelia is an outer shepherd for the γ ring. Goldreich & Porco (1987) have explored whether the orbital resonance involving these satellites provide an adequate explanation for the confinement of the Uranian rings. They have demonstrated that these satellites are capable of confining the ε ring if the mass and the thickness of the ring are within a given range of values. The outer edges of the δ and the γ rings can also be confined by these satellites, but the drag due to the planet's extended neutral hydrogen exosphere poses a severe problem for the shepherding of the α and β rings (see below and French et al 1991).

4. RING PHENOMENA

4.1 *Saturn's Broad Rings*

Saturn with its rings remains one of the truly beautiful objects in the sky. It is still one of the most requested and observed targets for small telescopes at observatories and planetaria. The ringed planet is also a popular symbol for the wonders of space. The magnificent images from *Voyagers 1* and *2* insure that this will continue to be true for some time.

Some years ago the broad, general structure of Saturn's rings was characterized by ground-based observers. Figure 2 is a tracing of ring brightness by Dollfus (1970). Almost every feature in this brightness profile has a clear basis in the optical depth profile observed in the most recent studies; for comparison, Figure 3 shows the vertical optical depth from the *Voyager 2* Photopolarimeter (PPS) stellar occultation (Esposito et al 1983b).

One notable aspect from Figure 3 is the similarity of the inner edges of Saturn's B and A rings (at about 91,000 and 121,000 km). These similar features have not been explained by any known resonance with Saturn's

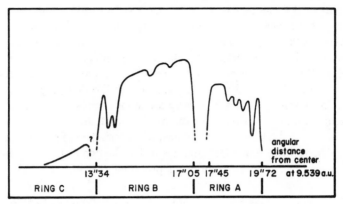

Figure 2 The overall structure of Saturn's rings, as determined by ground-based observations. (After Dollfus 1970.)

moons. One explanation is that they are formed from a balance between collisions among the ring particles and the production of debris from meteoroid collisions. The movement of ring material by impacts from meteoroids is termed "ballistic transport" (Durisen et al 1989, 1993). The fragments from a meteoroid collision have distinct orbits which re-intersect the rings. At a later ring crossing, the fragments collide with ring particles, imparting energy, momentum, and mass at that location. This transport modifies the ring evolution expected from slow spreading and resonant interactions, especially near the edges of opaque rings. This is another example of how impacts with meteoroids influence the history of the rings.

A number of features we see in Saturn's rings appear to have two components—one broad and one narrow. Figure 4 compares Saturn's F ring opacity to that on the Uranus' η ring. The Uranus' δ ring also shows a similar structure (Elliot et al 1985, French et al 1991). The phenomenon of narrow rings imbedded in broader ones is thus more general than its occurence in just Saturn's rings.

A detail we still do not have a good explanation for is the abundant microstructure in Saturn's ring. (See Figure 5.) The spacecraft data showed that the rings were not merely broad and homogeneous. Indications came first from the studies of the dark side of the rings and the ring passage observations analyzed by Lumme & Irvine (1979), and from the *Pioneer 11* data (Esposito et al 1980). Although these studies did not have the resolution that *Voyager* had, observations of the underside of the rings showed light leaking through that could not have penetrated through a homogeneous ring. Still, it was quite unexpected when *Voyager*

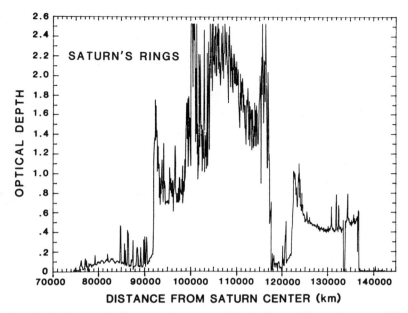

Figure 3 Optical depth of Saturn's rings measured by the *Voyager* Photopolarimeter (PPS) stellar occultation. The resolution has been degraded by averaging the original data in bins of 600 measurements. (From Esposito et al 1983b.)

approached Saturn for the first time (and even again, for the second time with the higher resolution provided by the stellar occultation) that we would continue to discover more structure in the rings. The structure is apparent at all ranges. The shape of the power spectrum is approximately $1/f$ down to the noise level: that is, we see structure at all scales in the rings, down to the best resolution we have (~ 100 m, Esposito et al 1983a).

The proposed explanation for this microstructure involves diffusional instabilities (e.g. Ward 1981, Lin & Bodenheimer 1981); however, Araki & Tremaine (1986) and Wisdom & Tremaine (1988) have shown that this type of instability would not actually occur. Furthermore, Brophy & Esposito (1989) have shown that the optical depth distribution is inconsistent with the model predictions. Therefore, we currently have no satisfactory model for creating the Saturnian ring microstructure.

4.2 *Waves*

The *Voyager* photopolarimeter stellar occultation and the radio occultation provided a very large set of waves to look at—approximately 50 density waves and a small number of bending waves (Esposito et al

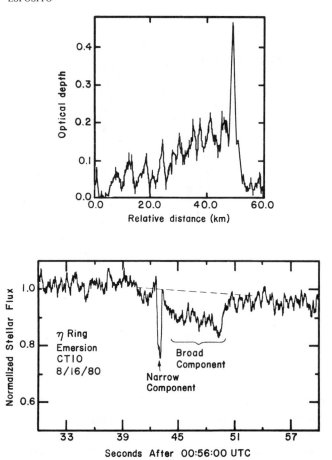

Figure 4   (*Top*) Optical depth of Saturn's F ring from the *Voyager* PPS occultation. (From Lane et al 1982.) (*Bottom*) Optical depth of Uranus η ring. (From Elliot et al 1981.) Optical depth increases upward in top, downward in bottom.

1983a,b; Esposito 1986; Rosen 1989; Rosen et al 1991a,b). Figure 6 shows the Saturn Mimas 5:3 and Janus 2:1 density waves propagating outward in the rings. Why are these waves of interest? We cannot yet get our hands on the rings, thus, we are in the same situation as the plasma physicist in the lab—we cannot place our measuring instrument into the medium that we want to measure. These waves are diagnostics in the same way that waves sent through a plasma are plasma diagnostics: We see the dynamic response of the medium to these waves and we are able to infer a number of physical properties—mass density, viscosity, vertical thickness. As an

Figure 5 Small scale structure in Saturn's B ring, as seen by the *Voyager* camera. The four panels were taken from photographs at four different longitudes. The dark area at the top is the Cassini division. Smallest features seen are about 20 km in width. (From Smith et al 1982.)

example, the total mass of Saturn's rings can be estimated to be about the same as its small moon, Mimas (Esposito et al 1983a).

The ability of the theory to reproduce the exact wave structure is more mixed. For the Mimas 5:3 density wave (Figure 6), the general shape can be well reproduced by theoretical models (e.g. Longaretti & Borderies 1986). On the other hand, for the Janus 2:1 wave in Figure 6 (the longest wave train and also the first wave seen in the *Voyager* photopolarimeter data), we are still a long way from a complete explanation. Part of the wave behavior is understood. For example, the theoretical prediction for the location of wave crests is well matched by the waves that we see. The

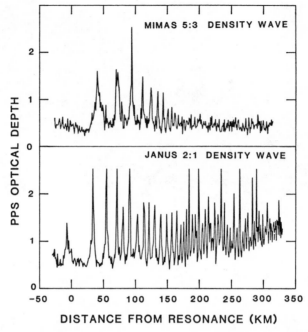

Figure 6 Two density waves in Saturn's rings. Note the amplitudes are not small compared to the background optical depth and there is little resemblance to a pure sine wave.

wave train includes 79 observed waves located as predicted by the linear theory (Shu 1984). From this wavelength behavior we estimate the ring mass density at the wave location to be about 70 g/cm^2 (Lane et al 1982, Holberg et al 1982, Esposito et al 1983a). The derived mass density of the rings does not seem to be changed very seriously by more detailed analyses which include nonlinear effects (see Shu et al 1985a,b). However, these same models have not reproduced the amplitudes of the wave crests (especially for the thicker parts of the ring), which remains an unsolved problem.

4.3   *Spokes*

Thirty-seven days before its closest approach to Saturn, *Voyager 1* discovered dark, nearly radial, wedge-shaped features in Saturn's B ring; these finger-like markings are called *spokes* (Smith et al 1982; see Figures 7 and 8). It was quickly determined from *Voyager* high phase angle pictures that these spokes contain a significant fraction of micron-sized particles.

Some recent results on the particle size are due to Doyle & Grün (1990) and Eplee & Smith (1987).

The small size of the particles explains why they appeared dark on *Voyager*'s approach and bright after it: Particles of about the same size as the light wavelength will preferentially scatter light forward. These small particles are also most easily moved by electric and magnetic forces, since their charge to mass ratio can be large. This consideration leads naturally to a possible connection with electromagnetic effects. The *Voyager* planetary radio astronomy experiment discovered a strong source of kilometric wavelength radiation (SKR). An interrelation between spokes and SKR seems plausible, since power spectral analysis of spoke activity (Porco 1983) revealed a significant peak with period 640.6 ± 3.5 min, consistent with the 639.4 min period of rotation in Saturn's magnetic field. Thus, the

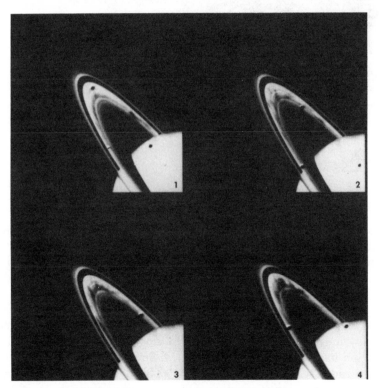

Figure 7   *Voyager* camera observations of "spokes" in Saturn's rings, showing a range of spoke activity. Analysis shows that the strongest activity occurs at one particular magnetic longitude.

Figure 8 A spoke seen in forward scattered light (i.e. backlit by the Sun) is now brighter than the surrounding ring: This and similar *Voyager* images indicate that the spoke particles have a size near the wavelength of visible light.

spokes are preferentially created at a peculiar longitude of Saturn which is also known to emit SKR. The most plausible explanation for this correlation would be the existence of a magnetic anomaly in Saturn's dipole field creating anomalous behavior of the magnetosphere near the rings.

One of the most puzzling aspects of spokes is that they extend across the B ring for many thousands of kilometers. Whether spoke particles actually move radially or are observed close to their points of initial elevation has not yet been demonstrated. Smith et al (1982) observed a narrow radial spoke grow to a length of 6000 km in < 5 min. If the spokes are the manifestation of a discharge which proceeds along the length of a spoke, then this formation time implies a minimum disturbance speed of 2×10^6 cm s^{-1}. This is much greater than any expected mechanical propagation speed for the rings; e.g. velocity dispersion of the larger ring particles is $\lesssim 0.5$ cm s^{-1}.

Voyager measurements have shed some light on this. By measuring the motions of two narrow-forming spokes (including the one reported in Smith et al 1982), Grün et al (1983) determined that inside the corotation point the trailing edges travel with the angular speed of the magnetic field, thereby remaining radial, while the leading edges move at Keplerian speeds. After formation, both edges tilt away from radial at the Keplerian rate. (Both edges of all other spokes measured in that study moved with Keplerian motion.) The angle of the wedge produced by the differential motion between edges is taken to be a measure of the time during which the spoke was active. Typical active times inferred from wedge-angle measurements of old spokes were 1 to 3 hr. Thus, they form in minutes and persist for hours. A possible cause is the impact of meteoroids on the rings, which create small puffs of ionized gas (Morfill et al 1983a,b). One possibility recently proposed is that they are the manifestation of magnetosonic waves in the partly ionized ring disk. This could explain their rapid propagation following the impact of a meteoroid (Tagger et al 1991). Cuzzi & Durisen (1990) propose that the spokes are more likely to occur near dawn (as observed) because meteoroid impacts are more energetic on the night side of the planet where the ring particle orbital velocity adds to Saturn's orbital velocity.

4.4   *Narrow and Eccentric Rings*

Until the discovery of the Uranian rings in 1977 (Elliot et al 1977) it seemed natural that all rings should be circular. Mutual collisions would cause eccentric motions to damp out, and the particle orbits to become circular. The rings of Uranus (Figure 9) instead turned out to be eccentric, inclined, narrow, and sharp edged! Most of these oddities can now be explained as due to the effects of nearby satellites (Goldreich & Tremaine 1982, Dermott 1984, Borderies et al 1984). To avoid the smearing out of an elliptical ring, Goldreich & Tremaine (1979) invoked the self-gravity of the ring particles themselves to create a precession of the ring that just counteracts the precession caused by the planet's gravity. This makes a very explicit prediction for the mass density of the rings of Uranus, which is unfortunately not confirmed by the spacecraft observations (Goldreich & Porco 1987, Esposito & Colwell 1989, Esposito et al 1991, French et al 1991). Unfortunately, no viable alternative has been proposed and explaining the uniform precession remains a major outstanding puzzle.

The Uranian rings have provided a prototype for understanding other narrow, eccentric features in Saturn's rings (e.g. Porco et al 1984a,b; Porco & Nicholson 1987; Porco 1990).

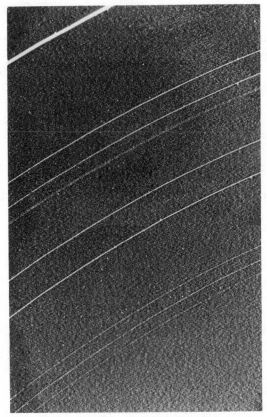

Figure 9 Uranian rings are seen in an image obtained by *Voyager 2* from a distance of
about a million kilometers as the spacecraft approached Uranus. The threadlike rings, which
are for the most part densely packed with particles, are only a few kilometers wide. They are
separated by hundreds of kilometers of virtually empty space. A new ring originally designated 1986U1R (now known as ring lambda) is barely visible between the outermost ring
(epsilon) and the next bright ring (delta). The rings reflect only 1% of the incident light.

4.5 *Dusty Rings*

Dust is found in all the planetary ring systems (Burns et al 1984, Showalter
et al 1987, Ockert et al 1987, Smith et al 1989, Esposito et al 1991,
Showalter et al 1991), generally in faint bands most easily visible when
backlit by the sun (indicating a particle size of about $1\,\mu$m or less). The
majority of the dust is concentrated near the equator. The dust in these
rings is likely derived from nearby satellites (Burns et al 1980; Colwell &
Esposito 1990a,b), and may also coat them. The dust dynamics is determined by the planet's gravity and magnetic field, and the local plasma and

gas densities. If dust becomes charged it has a tendency to corotate with the planetary magnetic field. Radiation from the sun can charge the dust grains and causes a relativistic drag force (Mignard 1984). Furthermore, the dust is ground by a continual flux of micrometeoroids (Burns et al 1980).

The premier "dusty" ring is that of Jupiter (Figure 10), whose light is dominated by scattering from small particles. The dust is derived from small satellites, has its orbits modified by orbital evolution due to plasma drag, and spreads vertically and horizontally. Just as the gravity of moons sculpt the broader rings of macroscopic particles, resonances with Jupiter's magnetic field cause boundaries and other features in the ring (Schaffer & Burns 1987) and perturb the particles into inclined orbits which explain the observed vertical thickness of the ring halo (Schaffer 1989).

Dust is prominent in Saturn's E, F, and G rings (Smith et al 1982), Neptune's rings (Smith et al 1989), and between the rings of Uranus, as evident from the backlit view of Figure 11.

4.6   Clumpy Rings

Prior to the spacecraft observations of the past 15 years it was possible to hold an idealized model of rings, which among other attributes assumed rings to possess azimuthal symmetry. Some azimuthal variations were known in Saturn's A ring (Camichel 1958, Lumme et al 1977). But the actual extent of longitudinal heterogeneity was mostly unexpected. We have now observed plentiful examples. Among this group, which includes Saturn's F ring and Uranus' η and λ rings, two cases are prominent: Saturn's Encke Gap ringlet (Figure 12) and Neptune's Adams' ring (Figure 13). The irregular, kinky structure of the Encke Gap ringlet was attributed to small nearby moons, and in fact, one has been discovered there in a reanalysis of the Voyager camera images (Showalter 1991). The Neptune ring structure is a set of 5 small arcs in a longitudinal range of only 25° (Porco 1991). The Voyager pictures show a very dim but continuous complete ring underlying these arcs, which was unfortunately too transparent to be detected from Earth-based occultations (Nicholson et al 1990). A model of Goldreich et al (1986) explained the arcs through corotation resonances between the ring particles and large, nearby inclined moons. The Voyager pictures did not show any moons as large as predicted, but the radial excursions of the ring particles clearly show the influence of Neptune's moon Galatea and the particles azimuthal distribution can be consistent with Goldreich's model (Porco, 1991). Unfortunately, collisions between the ring particles are likely to eject them from the weakly confining resonances and we must consider this model incomplete. A possible

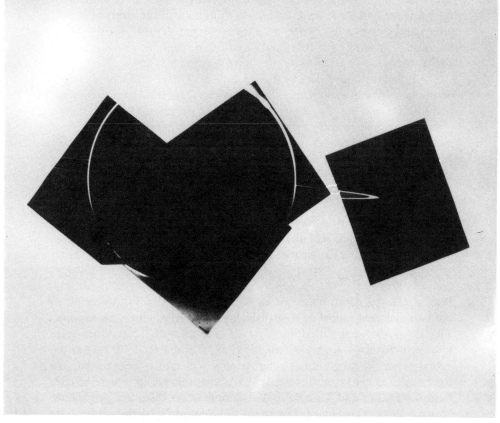

Figure 10 One of the most spectacular of the *Voyager 2* images obtained from inside the shadow of Jupiter. Looking back toward the planet and the rings with its wide-angle camera, *Voyager* took these photos on July 10, 1979 from a distance of 1.5 million kilometers. The ribbon-like nature of the rings is clearly shown. The planet is outlined by sunlight scattered from a haze layer high in the atmosphere. On each side, the arms of the ring curving back toward the spacecraft are cut off by the planet's shadow as they approach the brightly outlined disk.

improvement is that additional unseen moons can assist in this confinement (Sicardy & Lissauer 1992), similar to the moon Pan's effect on the Encke ringlet. The azimuthal structure in Saturn's F ring (often referred to as "braids," Figure 14) can also be attributed to unseen moons (Kolvoord et al 1990, Kolvoord & Burns 1992). In the model of ring creation by giant impact, these hypothesized nearby moons may be the largest fragments of a satellite that was shattered to form the ring (see below).

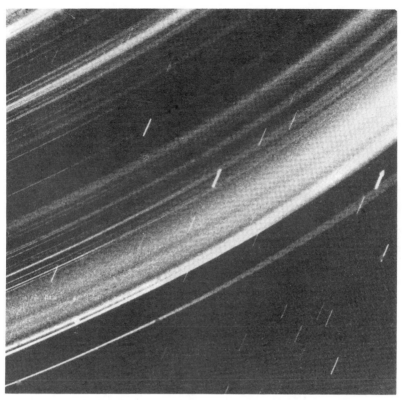

Figure 11   A backlit view of the Uranus rings shows numerous rings previously unseen from Earth, or by *Voyager* on its approach. This brightening is similar to that seen for Saturn's spokes (Figure 8) and Jupiter's ring (Figure 10), and shows that these new rings are mostly composed of micron-sized dust. The classical nine rings of Uranus can be found among the rings shown in this photograph but are undistinguished. Only the newly discovered lambda ring (see Figure 9) is especially bright.

5.   AGE AND EVOLUTION OF RINGS

5.1   *Introduction*

As described by Harris (1984), the rings of the planets likely result from the same process that created the regular satellites. Their orbits are prograde, equatorial, and nearly circular. A question that immediately arises is whether they are 1. the uncoagulated remnants of satellites which failed to form or 2. the result of a disruption of a pre-existing object. A related question highlighted by the apparent youth of rings is whether this latter process of ring creation by satellite destruction continues to the present time. This possibility thus mixes the *origin* of the rings with their sub-

Figure 12 Within the Keeler Gap lie at least two discontinuous ringlets. This *Voyager 2* figure, with a resolution of about 10 kilometers, shows one of these rings to be kinky (left). The kinks are spaced about 700 kilometers apart, approximately ten times more closely than the F ring kinks photographed by *Voyager 1*.

sequent *evolution*. Whatever their origin, the sculpted nature of the rings of Saturn, Uranus, and Neptune requires active processes to maintain their current state.

Because of the short time scales for viscous spreading, gas drag, particle coagulation, and transport of momentum to the forming planet, Harris (1984) argues that rings did not form contemporaneously with their primary planets, but were created later by disruption of satellites whose large size had made them less subject to the early destructive processes. The pieces of the disrupted satellite are within the Roche zone where tidal forces keep them from coagulating. This explains naturally the shepherd satellites and ring moons around the various giant planets as the largest pieces remaining after the destruction.

Conversely, both Lissauer et al (1988) and Ip (1988) have shown that it is very unlikely that a body large enough to create the Saturnian rings could be disrupted by the meteoroid flux recorded on the surfaces of the remaining Saturnian satellites. Dones (1991) argues that a comet may have

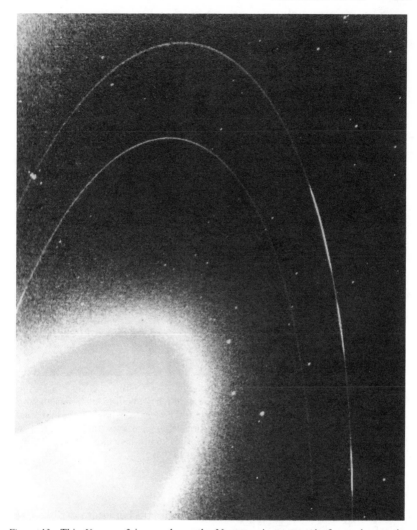

Figure 13   This *Voyager 2* image shows the Neptune ring system in forward scattering geometry (phase angle of 134°). Clearly seen are the three ring arcs and the Adams and Leverrier rings. The direction of motion is clockwise; the longest arc is trailing. The resolution in this image is about 160 km. (From Smith et al 1989.)

come too close to Saturn and been disrupted to form the rings. Harris (1984) notes that the satellite disruption hypothesis is particularly attractive for the Uranian rings; below, arguments are presented that processes currently active in the Uranus and Neptune systems may require some *recent* ring origins.

Figure 14   One of the most exciting photos of the *Voyager 1* Saturn encounter shows the F ring from a distance of 750,000 kilometers. At a resolution of about 15 kilometers, this outer ring suddenly revealed a complex braided structure. Two narrow bright rings that appear braided are visible, in addition to a broad diffuse component apparently separated from them by about 100 kilometers. Also visible is a kink, or knot, where the ring seems to depart dramatically from a smooth arc. (From Smith et al 1981.)

5.2   *Age of the Rings*

Estimates of the age of the rings can discriminate between possible scenarios for ring formation. A process of considerable importance is the meteoroid bombardment of the rings which darkens them (Doyle et al 1989) and grinds them to dust (Northrop & Connerney 1987). If lifetimes of some components are much less than the age of the solar system, those parts cannot have a primordial origin. This argument indicates a recent origin of the material observed in the Jovian ring (Burns et al 1980), Saturn's A ring (Esposito 1986), Saturn's F ring (Cuzzi & Burns 1988), Uranus' rings (Esposito & Colwell 1989, Colwell & Esposito 1990a), and Neptune's rings (Smith et al 1989; Colwell & Esposito 1990b, 1992a,b).

The narrowness of the observed rings of Uranus and Neptune raised the first concern about the age of these rings. Because the interparticle collision time is short [less than half the orbital period for optically thick rings (Stewart et al 1984)] the particles collide and interchange momentum which causes the ring to spread (see Section 3). Modeling a narrow ring as a fluid of single-size particles, one can define a kinematic viscosity

$$v \approx \frac{c^2}{2\Omega}\left(\frac{\tau}{1+\tau^2}\right), \qquad (3)$$

where c is the interparticle random velocity, Ω the mean motion, and τ the optical depth (Cook & Franklin 1964, Goldreich & Tremaine 1978). The minimum viscosity would be achieved if random motions are minimized, in which case the ring behaves as an incompressible liquid (Borderies et al 1985). This gives

$$v_{min} = \Omega\left(\frac{\sigma}{\rho}\right)^2, \qquad (4)$$

where σ is the surface mass density of the ring and ρ the density of the ring particles. Modeling the ring spreading as a diffusion process leads immediately to the lifetime of a narrow ring of width Δr. The lifetime is

$$\Delta t \approx \frac{\Delta r^2}{v}. \qquad (5)$$

For the rings of Uranus, if we use $\sigma_{min}/\rho = 80$ g/cm^2 (Gresh et al 1989), $\rho = 1.4$ g/cm^3 (average for the Uranian satellites, Tyler et al 1986), and $\Delta r = 100$ km (maximum width of the ε ring), we get $v_{min} > 1$ cm^2/sec and thus, $\Delta t \lesssim 10^6$ years. Clearly, these rings must be confined, most likely by shepherding satellites (Goldreich & Tremaine 1980, Section 3 above.)

A major success of the *Voyager* encounter was the discovery of the satellites shepherding the ε ring (Smith et al 1986, Porco & Goldreich 1987). However, the apparent lack of other satellites to shepherd the other Uranian rings is also significant. This may be because the other shepherds were too small to be seen by the *Voyager* cameras ($R < 10$ km, Smith et al 1986). The shepherds (seen and unseen) could then confine all the rings, avoiding the very short lifetimes associated with unconstrained spreading.

5.3 Age of Uranus' Rings

A strong argument for the youth of rings are related to the rates of momentum transfer between the objects in the Uranus ring system (Esposito & Colwell 1989). These rates were first considered in light of the *Voyager* data by Goldreich & Porco (1987). As discussed above, angular momentum flows outward through an unperturbed disk carrying the viscous torque (for particles in Keplerian orbits)

$$T_v = 3\pi\sigma v\Omega r^2.\tag{6}$$

We believe that the narrow rings of Uranus do not spread because an inner shepherd supplies this torque at the inner boundary, while an external shepherd carries it off at its outer edge. For the Uranus ε ring we know the sizes and locations of the shepherds and can estimate their masses from their size (Porco & Goldreich 1987). Lower limits to the mass of the ε ring can be inferred from the radio occultation results (Gresh et al 1989, Gresh 1990). For $\sigma > \sigma_{min} \equiv 80$ g/cm^2, the mass of the ε ring $M_\varepsilon > 10^{19}$ g. Since this is only a lower limit, it leaves open the possibility that the ε ring is more massive than the shepherds! If the ring is more massive than the shepherds, transferring momentum to them cannot significantly slow the spreading. However, the mass of the inner shepherd Cordelia is some three times the minimum mass of the ε ring (with an uncertainty of perhaps a factor of two, Porco & Goldreich 1987), so as long as M_ε is close to its minimum value this difficulty is avoided. Goldreich & Porco (1987) show in this case that the satellites are massive enough to supply and carry off the viscous torque Tv.

However, a lower limit to the viscous torque may be established by considering the drag on ring particles from the extended Uranian exosphere. This exospheric drag provides a lower limit to the torques transferred to the satellites shepherding the ε ring. The viscous torque must exceed the drag torque or otherwise the outer edge would be blown inward by the transfer of momentum as the ring particles plow into the exosphere. The excess torque, $T_v - T_D$, is carried off by the resonance coupling to Ophelia. This momentum transfer causes Ophelia to evolve outward, away from the ring, and the ring spreads, maintaining its sharp outer edge at the instantaneous location of the Ophelia resonance.

The minimum viscous torque is thus $T_v = T_D$. At the ε inner edge, a resonance transfers the momentum of the inwardly diffusing ring particles to Cordelia. This momentum transfer, $T_v + T_D > 2T_D$, causes Cordelia's semi-major axis to decrease. Since the mass of Cordelia is known, we can estimate how long it would require for Cordelia to evolve to its current separation, assuming the minimum torque and zero initial separation. The maximum duration of this shepherding is thus $t_{max} = \Delta L/(2T_D)$, where ΔL is the total change in angular momentum, and T_D is 9×10^{16} ergs (Goldreich & Porco 1987). This calculation gives $t_{max} = 6 \times 10^8$ years, considerably smaller than the age of the solar system (Esposito & Colwell 1989). Cordelia could not have transferred this momentum to a larger inner moon since none exists. Further, it could not have transferred it to one of the larger outer moons through a resonance that may have existed in the past since capture into a resonance libration with an outer moon is not possible (Peale 1986) given the tidal expansion of the orbit of the larger outer satellite and the shrinking of Cordelia's orbit. This age, t_{max}, is a strong upper limit because only minimum values for the torque T_v have been considered. In the minimum case where $T_v = T_D$, no torque is transferred to Ophelia while a torque $T_v + T_D = 2T_D$ is transferred to Cordelia, causing it to evolve inward. Even shorter ages are implied for possible shepherds of the α and β rings.

One proposed solution to the short lifetimes of the Uranus rings is to continually create the dust, rings, and small moons by disruption of larger objects (Esposito 1986, Smith et al 1986, Cuzzi & Esposito 1987, Esposito & Colwell 1989, Colwell & Esposito 1990a). Esposito & Colwell (1989) propose two families of objects, too small to be seen by *Voyager*: small satellites of $R \lesssim 10$ km, called "ring precursors," and belts of particles $R \sim 100$ m called "moonlet belts." These are similar to the material proposed to create Saturn's F ring by Cuzzi & Burns (1988). Meteoroid ejecta from rings, moons, and moonlet belts by gas drag leads to a continuous low-optical depth sheet of dust in the main ring system, with the highest optical depth near the ε ring where gas drag is slower. At the location of a moonlet belt, a dust band would be visible. Esposito & Colwell (1989) propose that these processes along with variations in the widths, optical depths, and size distributions of the moonlet belts account for the variety of forms of dust bands seen by *Voyager* at Uranus (see Figure 11).

5.4 Age of Neptune's Rings

The azimuthal variability of Neptune's rings is another strong indication of youth. Although Goldreich et al (1986) have shown how the arcs might be maintained and Porco (1991) has shown the *Voyager* observations to be consistent with this model (see above), nonetheless, interparticle collisions will allow particles or their fragments to escape the resonance. Particles in orbits outside the small range of the corotation resonance will rapidly circulate to fill the entire ring circumference. Thus, the extreme azimuthal structure observed in Neptune's Adams ring is unlikely to be a remnant of the formation of Neptune.

5.5 Satellite Disruption

Making rings by disrupting satellites naturally leads to consideration of the moons' evolution. The history of the small satellites of Uranus and Neptune has been studied in detail by Colwell & Esposito (1992a,b). Their stochastic simulations of the moons' collisional fragmentation [using the fragmentation results of Housen & Holsapple (1990)] confirm the conclusions of Smith et al (1986, 1989) that these moons are not primordial. Colwell & Esposito follow the process of satellite disruption from an initial distribution through successive disruptions, ignoring re-accretion. Two approaches are used: a Monte Carlo simulation follows the history of only the largest fragment after each disruption; a Markov chain calculation follows the stochastic evolution of all the fragments.

This raises the question of re-accretion of fragments of a disrupted satellite in or near the planet's Roche zone. Previous estimates by Burns et al (1984) and Stevenson (1986) are based on the calculation of Soter (1971). Soter computed the time scale for an ejected particle to collide with its source body. Canup & Esposito (1992) have used a matrix formulation of the integrodifferential coagulation equation and applied it to the ensemble of satellite fragments calculated by Colwell & Esposito (1992a) to improve on Soter's (1971) calculation. This method uses a particle-in-a-box calculation for the mutual collision probability. Re-accretion probabilities as a function of time are shown in Figure 15 for Neptune's moons Galatea, Larissa, and Proteus. These time scales are similar to Soter's and likewise do not consider the velocity distribution and its evolution, tidal

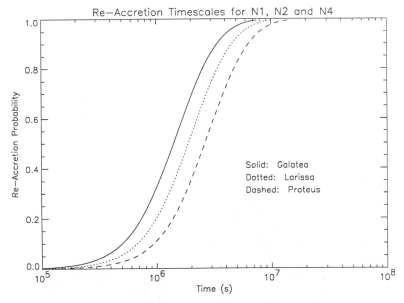

Figure 15 Probability of re-accretion of three Neptune satellites, following disruption by a comet impact. As can be seen, typical re-accretion times are about one year. (From Canup & Esposito 1992.)

effects, or fragmentation of re-accreting particles. Nonetheless, these results show that re-accretion should not be ignored: Realistic calculations in the Roche zone will form a critical part of our understanding of ring and moon evolution. Already we expect some limited re-accretion within rings as particles may temporarily stick together between higher velocity collisions. These temporary aggregations were termed dynamic ephemeral bodies (DEBs) by Weidenschilling et al (1984). Despite our present ignorance, the existence of rings surrounding the planets shows that accretion does not dominate near the planet and the present models' neglect of accretion in ring history calculations seems justified.

Once the original moon is destroyed for the first time, the "collisional cascade" to smaller size through successive disruptions occurs relatively quickly since the smaller moons are easier to destroy. Monte Carlo simulations of 300 separate histories of the moons of Neptune give the results shown in Figure 16. For an original moon of radius 40 km, the solid line shows the mean, asterisks the median, and squares the mode for the distribution of largest surviving fragment as a function of time. These results show that the evolution of moon populations from catastrophic fragmentation is more complex than can be described by a simple time scale. The speed of this cascade is dependent on the size distribution of the impactors, which is, unfortunately, a poorly known factor. For example, the difference between a power-law index of 2.5 and 3.5 is quite significant. Better constrained measurements of the impactor size distribution are strongly desired.

The Markov chain follows the evolution of the complete size distribution of moons and fragments. These simulations show that the more numerous smaller fragments can outlive the largest fragment followed in the Monte Carlo formulation and that the collisional debris from an initial complement of satellites will approximate a simple power-law distribution. Colwell & Esposito find a cumulative size power-law index of 2.5 for the collisionally evolved system. We note that after 4 billion years, the remnants of an original 100-km radius moon would include about one thousand 1–10 km "moonlets."

The calculations of Colwell & Esposito show that the small satellites of Neptune must have evolved through catastrophic fragmentation after the end of satellite and planet formation 4 billion years ago. The production of the currently observed smaller satellites is a natural consequence of the successive break-up of larger satellites.

5.6 *Ring Formation*

The size distribution and velocity distribution of the satellite fragments were further studied by Colwell & Esposito (1992b). They found that a

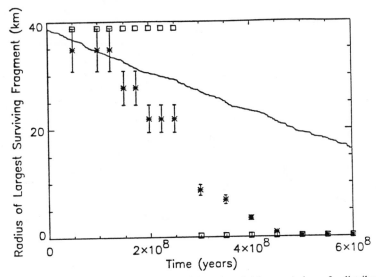

Figure 16 Mean (*solid line*), median (*asterisks*), and modal (*squares*) sizes of a distribution of 300 stochastic simulations of the collisional cascade for an initial 40 km radius moon, at the orbit of Thalassa. Error bars on the median connect the two size bins which fall on either side of the 50% cumulative probability.(From Colwell & Esposito 1992a.)

narrow ring is the natural outcome of the disruption, as the fragments' orbits quickly fill the entire circumference at the original orbital radius of the destroyed moon. In the production of this debris, a small number of large fragments are also created. Colwell & Esposito (1990a) argue that these larger fragments will naturally clear gaps. The discovery of Pan in Saturn's Encke Gap (Showalter 1991) may provide one realization of this. As the initial ring spreads, its edges will cross through the resonances with larger bodies in the system, allowing the edges of the rings to be shepherded and sharpened. The moons will then spread with the rings (albeit slower, because of the moons' larger mass) and possibly their evolving orbits will resonate with yet larger satellites. This can explain the "Cordelia Connection" found by Murray & Thompson (1990) for the Uranus rings, i.e. that several of the unseen moons which are needed to halt the radial spreading of the Uranian rings are exceptionally close to resonances with Cordelia. Furthermore, they note that Cordelia itself is very close to a resonance with Rosalind. This "linking-up" due to orbital resonances and associated transfer of angular momentum to more massive satellites will slow the initially rapid viscous spreading of the newly-formed ring.

The moons can also confine the ring material in azimuth: Neptune's arcs provide a vivid example. Like the radial lock-up, this confinement occurs quite quickly, otherwise differential rotation due to Kepler shear and differential precession would smear out the arcs on a time scale of years. Porco (1991) has shown how the azimuthal structure of the ring-arcs may be understood as material trapped in 7 of 86 possible Galatea corotation resonances (see Figure 17). All the ring-arcs span an azimuthal range of only 25°: Thus, the resonance locations are only thinly occupied.

We can estimate the probability that Neptune's particular arrangement arose by a set of chance events (Esposito & Colwell 1992). One possibility is that each of the resonance locations was filled by an independent event. Examples might be a collision between two fast moving particles in one of the resonance zones, or the destruction of a small moon at or near the resonance site. Each event would release material to be subsequently trapped in that corotation resonance.

Esposito & Colwell (1992) find that the random hypothesis has likelihood 10^{-8} (i.e. the probability of chance occurrence of all the ring arcs clustering together is 10^{-8}). The conclusion from this calculation is that we are most likely seeing the result of a single, recent event. This could have been the collision of a meteoroid with a small moon of Neptune,

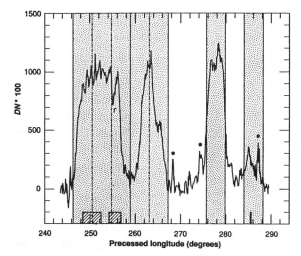

Figure 17   A radially averaged longitudinal scan of the arc region as seen in Figure 13 and precessed back to a common epoch using the arcs' mean motion, $n = 820.1185°$ per day. The asterisks indicate the positions of stars; the symbol r indicates an incompletely removed camera reseau. The stippled bars show the location of corotation resonances, each 4.186° wide, that might explain the azimuthal confinement of the observed arcs in the model of Goldreich et al (1986). As can be seen, not all resonance zones are occupied by arcs, and some of the observed arcs fill more than one resonance. (From Porco 1991.)

resulting in the destruction of the moon; the smaller fragments were trapped in 7 of the 10 nearby corotation sites. This destruction could also simultaneously leave several large fragments (as explained above) to assist in the azimuthal confinement as proposed in the model of Lissauer & Sicardy (1990).

5.7 *Summary*

Short lifetimes inferred from meteoroid bombardment imply interrelated histories for rings and nearby small moons in the planetary ring systems. Detailed simulations show the plausibility of a model where disrupted satellites provide the source of the Uranus and Neptune rings. The satellites confine the rings radially and azimuthally. The entire ring-moon system may lock up through resonances and evolve as a unit. Satellites excite the random motions with the rings to determine the size distribution through collisional equilibrium and particularly the fractions of dust observed by the *Voyager* cameras for the Uranus and Neptune rings.

6. CONCLUSIONS

Our present understanding of planetary rings highlights a number of important scientific questions. One question is to explain the differences between the planetary ring systems, although one proposal (Lane et al 1989) is that these differences represent different random outcomes of the same stochastic processes of ring creation and destruction. At the same time, our description of the various ring systems is still incomplete, especially knowledge of the complete size distribution and the composition of the ring particles (which in fact, may vary *within* each system). In addition, we need to measure accurately the locations of the rings and their three-dimensional structure. Present models of ring dynamics should be compared in detail with the best measurements for testing and refinement.

The questions of the ages of the rings, their recent origins, and history have been brought into sharp focus by the spacecraft observations of many youthful features and the calculations that the present rings could not have persisted for the age of the solar system. Perhaps the most important question is how our understanding of present processes in planetary rings can be fruitfully compared with those in the early solar nebula to explain the origin of the planets out of a flat disk of interacting particles, dust, and gas. We can also hope to apply this understanding to other flattened, rotating systems like galaxies and accretion disks.

Our consideration of current understanding of planetary rings and those key issues which arise leads to several major objectives for spacecraft investigations to enhance our understanding of planetary rings. The first is to measure radial, azimuthal, and vertical structure at high spatial resolution and multiple times. This involves not only high resolution imaging, radio, stellar, and solar occultations from planetary orbit—which for Jupiter's rings can be provided by the NASA *Galileo* mission and for Saturn's rings by the NASA/ESA *Cassini* Mission—but also full utilization of ground-based stellar occultations.

The second goal is to measure the complete size distribution and composition of the ring systems at multiple locations. This size distribution includes the smallest transient dust particles and embedded moons within and near the ring system. The composition and flux of bombarding meteoroids should be determined also. Space missions using radio occultation, spectroscopy, and photometric measurements contribute to this objective.

The third objective is to develop models of ring processes and history that are consistent with the best Earth and space observations and establish the rings' relation to moons, atmospheres, and magnetospheres in their planetary systems and to the early solar system. This development will depend on a better description from future spacecraft and Earth-based data and would provide a direct benefit to our understanding of the origin of the solar system. The most productive mode for these advances will involve close collaboration between the observers and those developing the models. The explosion of our understanding of planetary rings in the past decade shows the power of such a combined approach.

Acknowledgments

I appreciate helpful discussions with Glen Stewart, Tom Brophy, Joshua Colwell, and Andre Brahic, and the careful reading of my manuscript by Jeff Cuzzi and Luke Dones. This review was partly written while I was a visitor at the Max Planck Institute for Aeronomy. I thank the NASA planetary geology program, outer planets data analysis program, and the *Voyager* and *Cassini* projects for support of this research.

Literature Cited

Araki, S., Tremaine, S. D. 1986. The dynamics of dense particle disks. *Icarus* 65: 83–109

Borderies, N. P., Goldreich, P., Tremaine, S. D. 1984. Unsolved problems in planetary rings dynamics. See Greenberg & Brahic 1984, pp. 713–36

Borderies, N. P., Goldreich, P., Tremaine, S. D. 1985. A granular flow model for dense planetary rings. *Icarus* 63: 406–20

Borderies, N. P., Goldreich, P., Tremaine, S. D. 1989. The formation of sharp edges in planetary rings by nearby staellites. *Icarus* 80: 344–60

Brahic, A. 1975. A numerical study of a gravitating system of colliding particles: applications to the dynamics of Saturn's rings and to the formation of the Solar System. *Icarus* 25: 452–58

Brahic, A. 1977. Systems of colliding bodies

in a gravitational field: numerical simulation of the standard model. *Astron. Astrophys.* 54: 895–907

Brophy, T. G., Esposito, L. W. 1989. Simulation of collisional transport processes and the stability of planetary rings. *Icarus* 78: 181–205

Brophy, T. G., Stewart, G. R., Esposito, L. W. 1990. A phase-space fluid simulation of a two-component narrow planetary ring: particle size segregation, edge formation, and spreading rates. *Icarus* 83: 133–55

Brophy, T. G., Esposito, L. W., Stewart, G. R., Rosen, P. D. 1993. Numerical simulation of satellite-ring interactions: resonances and satellite-ring torques. *Icarus*. Submitted

Burns, J. A., Showalter, M. R., Cuzzi, J. N., Pollack, J. B. 1980. Physical processes in Jupiter's ring: clues to its origin by Jove! *Icarus* 44: 339–60

Burns, J. A., Showalter, M. R., Morfill, G. E. 1984. The ethereal rings of Jupiter and Saturn. See Greenberg & Brahic 1984, pp. 200–72

Canup, R. M., Esposito, L. W. 1992. *Reaccretion of disrupted planetary satellites.* Presented at the 1992 DPS Meet., Munich

Camichel, H. 1958. Mesures photométriques de Saturne et son anneau. *Ann. Astrophys.* 21: 231–42

Colwell, J. E., Esposito, L. W. 1990a. A numerical model of the Uranian dust rings. *Icarus* 86: 530–60

Colwell, J. E., Esposito, L. W. 1990b. A model of dust production in the Neptune ring system. *Geophys. Res. Lett.* 17: 1741–44

Colwell, J. E., Esposito, L. W. 1992a. Origins of the rings of Uranus and Neptune. 1. Statistics of satellite disruptions. *J. Geophys. Res.* 97: 10,227–41

Colwell, J. E., Esposito, L. W. 1992b. Origins of the rings of Uranus and Neptune. II. Initial distributions of disrupted satellite fragments. *J. Geophys. Res.* Submitted

Cook, A. F., Franklin, F. A. 1964. Rediscussion of Maxwell's Adams prize essay on the stability of Saturn's rings. *Astron. J.* 69: 173–200

Cooke, M. L. 1991. *Radial variation in the Keeler Gap and C ring photometry.* PhD dissertation. Cornell Univ.

Cuzzi, J. N. 1985. The rings of Uranus: not so thick, not so black. *Icarus* 63: 312–16

Cuzzi, J. N., Lissauer, J. J., Esposito, L. W., Holberg, J. B., Marouf, E. A., et al. 1984. Saturn's rings: properties and processes. See Greenberg & Brahic 1984, pp. 73–199

Cuzzi, J. N., Esposito, L. W. 1987. The rings of Uranus. *Sci. Am.* 237: 52–66

Cuzzi, J. N., Burns, J. A. 1988. Charged particle depletion surrounding Saturn's F ring: evidence for a moonlet belt? *Icarus* 74: 284–324

Cuzzi, J. N., Durisen, R. H. 1990. Bombardment of planetary rings by meteoroids: general formulation and effects of Oort cloud projectiles. *Icarus* 84: 467–501

Dermott, S., 1984. Dynamics of narrow rings. See Greenberg & Brahic 1984, pp. 589–640

Dollfus, A. 1970. Optical reflectance polarimetry of Saturn's globe and rings. *Icarus* 40: 171–79

Dones, L. R. 1987. *Dynamical and photometric studies of Saturn's rings.* PhD dissertation. Univ. Calif., Berkeley

Dones, L. R. 1991. A recent cometary origin for Saturn's rings? *Icarus* 92: 194–203

Dones, L. R., Showalter, M. R., Cuzzi, J. N. 1989. Simulations of light scattering in planetary rings. In *Dynamics of Astrophysical Disks*, ed. J. Sellwood, pp. 25–26. Cambridge: Cambridge Univ. Press

Doyle, L. R., Dones, L., Cuzzi, J. C. 1989. Radiative transfer modelling of Saturn's outer B ring. *Icarus* 80: 104–35

Doyle, L. R., Grün, E., 1990. Radiative transfer modelling constraints on the size of spoke particles in Saturn's rings. *Icarus* 85: 168–90

Durisen, R. H., Cramer, N. L., Murphy, B. W., Cuzzi, J. N., Mullikin, T. L., et al. 1989. Ballistic transport in planetary ring systems due to particle erosion mechanisms. I. Theory, numerical methods and illustrative examples. *Icarus* 80: 136–66

Durisen, R. H., Bode, P. W., Cuzzi, J. N., Cederblom, S. E., Murphy, B. W. 1993. Ballistic transport in planetary ring systems due to particle erosion mechanisms. II. Theoretical models for Saturn's A and B ring inner edges. *Icarus*. Submitted

Elliot, J. L., Dunham, E. W., Mink, D. J. 1977. The rings of Uranus. *Nature* 267: 328–30

Elliot, J. L., French, R. G., Frogel, J. A., Elias, J. H., Mink, D. J. et al. 1981. Orbits of the nine Uranian rings. *Astron. J.* 86: 444–55

Elliot, J. L., Baron, R. L., Dunham, E. D., French, R. G., Meech, K. J., et al. 1985. The 1983 June 15 occultation by Neptune. I. Limits on a possible ring system. *Astron. J.* 90: 2615–23

Eplee, R. E., Smith, B. A. 1987. A dynamical constraint on particulate sizes for Saturn's B ring spokes. *Icarus* 69: 575–77

Esposito, L. W. 1986. Structure and evolution of Saturn's rings. *Icarus* 67: 345–57

Esposito, L. W., Dilley, J. P., Fountain, J. W. 1980. Photometry and polarimetry of Saturn's rings from Pioneer 11. *J. Geophys. Res.* 85: 5948–56

Esposito, L. W., O'Callaghan, M., West, R.

A. 1983a. The structure of Saturn's rings: implications from the Voyager stellar occultation. *Icarus* 56: 439–52

Esposito, L. W., O'Callaghan, M., Simmons, K. E., Hord, C. W., West, R. A., et al. 1983b. Voyager photopolarimeter stellar occultation of Saturn's rings. *J. Geophys. Res.* 88: 8643–49

Esposito, L. W., Cuzzi, J. N., Holberg, J. B., Marouf, E. A., Tyler, G. L., Porco, C. C. 1984. Saturn's Rings: structure, dynamics and particle properties. In *Saturn*, ed. T. Gehrels, M. S. Matthews, pp. 463–545. Tucson: Univ. Ariz. Press

Esposito, L. W., Colwell, J. E. 1989. Creation of the Uranus rings and dust bands. *Nature* 339: 637–740

Esposito, L. W., Brahic, A., Burns, J. A., Marouf, E. A. 1991. Particle properties and processes in Uranus' rings. In *Uranus*, ed. J. T. Bergstralh, E. D. Miner, M. S. Matthews, pp. 410–68. Tucson: Univ. Ariz. Press

Esposito, L. W., Colwell, J. E. 1992. *Neptune's rings and satellite system: collisional origin and evolution.* Presented at Spring AGU Meet., Montreal, Canada

French, R. G., Elliot, J. L., Levine, S. E. 1986. Structure of the Uranian rings. II. Ring orbits and widths. *Icarus* 67: 134–63

French, R. G., Nicholson, P. D., Porco, C. C., Marouf, E. A. 1991. Dynamics and structure of the Uranian rings. In *Uranus*, ed. J. T. Bergstralh, E. D. Miner, M. S. Matthews, pp. 327–409. Tucson: Univ. Ariz. Press.

Goldreich, P., Tremaine, S. D. 1978. The velocity dispersion in Saturn's rings. *Icarus* 34: 227–39

Goldreich, P., Tremaine, S. D. 1979. Towards a theory for the Uranian rings. *Nature* 277: 97–99

Goldreich, P., Tremaine, S. D. 1980. Disk-satellite interactions. *Astrophys. J.* 241: 425–41

Goldreich, P., Tremaine, S. D. 1982. The dynamics of planetary rings. *Annu. Rev. Astron. Astrophys.* 20: 249–83

Goldreich, P., Tremaine, S., Borderies, N. 1986. Toward a theory for Neptune's arc rings. *Astron. J.* 92: 195–98

Goldreich, P., Porco, C. C. 1987. The shepherding of the Uranus rings. II. Dynamics. *Astron J.* 93: 730–37

Greenberg, R. 1983. The role of dissipation in the shepherding of ring particles. *Icarus* 53: 207–18

Greenberg, R., Brahic, A., eds. 1984. *Planetary Rings.* Tucson: Univ. Ariz. Press. 784 pp.

Gresh, D. L. 1990. *Voyager radio occultation by the Uranian rings: structure, dynamics and particle size.* PhD dissertation. Stanford Univ.

Gresh, D. L., Marouf, E. A., Tyler, G. L., Rosen, P. A., Simpson, R. A. 1989. Voyager radio occultation by Uranus's rings. I. Observational results. *Icarus* 78: 131–68

Grün, E. Morfill, G. E., Terrile, R. J., Johnson, T. V., Schwehm, G. J. 1983. The evolution of spokes in Saturn's B ring. *Icarus* 54: 227–52

Hammeen-Anttila, K. A. 1978. An improved and generalized theory for the collisional evolution of Keplerian systems. *Astrophys. Space Sci.* 58: 477–519

Harris, A. W. 1984. The origin and evolution of planetary rings. See Greenberg & Brahic 1984, pp. 641–59.

Holberg, J. B., Forester, W., Lissauer, J. J. 1982. Identification of resonance features within the rings of Saturn. *Nature* 297: 115–20

Holberg, J. B., Nicholson, P. D., French, R. G., Elliot, J. L. 1987. Stellar occultation probes of the Uranian Rings: a comparison of Voyager UVS and Earth-based results. *Astron. J.* 94: 178–88

Horn, L. J., Yanamandra-Fisher, P. A., Esposito, L. W., Lane, A. L. 1989. Physical properties of the Uranian Delta ring from a possible density wave. *Icarus* 76: 485–92

Housen, K. R., Holsapple, K. A. 1990. On the fragmentation of asteroids and planetary satellites. *Icarus* 84: 226–53

Ip, W-H 1988. An evaluation of a catastrophic fragmentation origin of the Saturnian ring system. *Astron Astrophys.* 199: 340–42

Kolvoord, R. A., Burns, J. A., Showalter, M. R. 1990. Periodic features in Saturn's F ring. *Nature* 345: 675–77

Kolvoord, R. A., Burns, J. A. 1992. Three-dimensional perturbations of particles in a narrow planetary ring. *Icarus* 95: 253–64

Lane, A. L., Hord, C. W., West, R. A., Esposito, L. W., Coffeen, D. L., et al. 1982. Photopolarimetry from Voyager 2: preliminary results on Saturn, Titan, and the rings. *Science* 215: 537–43

Lin, D. N. C., Papaloizou, J. 1979. Tidal torques on accretion disks in binary systems with extreme mass ratios. *MNRAS* 186: 799–812

Lin, D. N. C., Bodenheimer, P. 1981. On the stability of Saturn's rings. *Astrophys. J.* 248: L83–86

Lissauer, J. J., Squyres, S. W., Hartmann, W. K. 1988. Bombardment history of the Saturn system. *J. Geophys. Res.* 93: 13,776–804

Lissauer, J. J., Sicardy, B. 1990. *Models of arcs in Neptune's 63K ring.* Paper presented at 23rd COSPAR Meet., the Hague, Netherlands

522 ESPOSITO

Longaretti, P.-Y., Borderies, N. 1986. Non-linear study of the Mimas 5: 3 density wave. *Icarus* 67: 211–33

Lumme, K. A., Esposito, L. W., Irvine, W. M., Baum, W. A. 1977. Azimuthal brightness variations of Saturn's rings. II. Observations at an intermediate tilt angle. *Astrophys. J.* 216: L123–26

Lumme, K. A., Irvine, W. M. 1979. A model for the azimuthal brightness variations in Saturn's rings. *Nature* 282: 695–96

Lynden-Bell, D., Pringle, J. E. 1974. The evolution of viscous discs and the origin of the nebular variables. *MNRAS* 168: 603–37

Meyer-Vernet, N., Sicardy, B. 1984. On the physics of resonant disk-satellite interaction. *Icarus* 69: 157–75

Mignard, F. 1984. Effects of radiation forces on dust particles. See Greenberg & Brahic 1984, pp. 333–66

Morfill, G. E., Fechtig, H., Grün, E., Goertz, C. K. 1983a. Some consequences of meteoroid impacts on Saturn's ring. *Icarus* 55: 439–47

Morfill, G. E., Grün, E., Johnson, T. V., Goertz, C. K. 1983b. On the evolution of Saturn's spokes: theory. *Icarus* 53: 230–35

Murray, C. D., Thompson, R. P. 1990. Orbits of shepherd satellites deduced from the structure of the rings of Uranus. *Nature* 348: 499–502

Nicholson, P. D., Cooke, M. L., Matthews, K., Elias, J., Gilmore, G. 1990. Five stellar occultations by Neptune: further observations of ring arcs. *Icarus* 87: 1–39

Nicholson, P. D., Dones, L. 1991. Planetary rings. *Rev. Geophys.* 29: 313–27 (Suppl.)

Northrop, T. B., Connerney, J. E. 1987. A meteorite erosion model and the age of Saturn's rings. *Icarus* 70: 124–37

Ockert, M. E., Cuzzi, J. N., Porco, C. C., Johnson, T. V. 1987. Uranian ring photometry: results from Voyager 2. *J. Geophys. Res.* 92: 14,969–79

Peale, S. J. 1986. Orbital resonances, unusual configurations and exotic rotation states among the planetary satellites. In *Satellites*, ed. J. A. Burns, M. S. Matthews, pp. 159–223. Tucson: Univ. Ariz. Press.

Porco, C. C. 1983. *Voyager Observations of Saturn's rings*. PhD dissertation. Calif. Inst. Technol., Pasadena

Porco, C. C. 1990. Narrow rings: observation and theory. *Adv. Space Res.* 10(1): 221–29

Porco, C. C. 1991. An explanation for Neptune's ring arcs. *Science* 253: 995–1001

Porco, C. C., Nicholson, P. D., Borderies, N., Danielson, G. E., Goldreich, P., et al. 1984a. The eccentric Saturnian ringlets at 1.29 R_S and 1.45 R_S. *Icarus* 60: 1–16

Porco, C. C., Danielson, G. E., Goldreich,

P., Holberg, J. B., Lane, A. L. 1984b. Saturn's non-axisymmetric ring edges at 1.95 R_S and 2.27 R_S. *Icarus* 60: 17–28

Porco, C. C., Goldreich, P. 1987. Shepherding to the Uranian rings. I. Kinematics. *Astron. J.* 93: 724–29

Porco, C. C., Nicholson, P. D. 1987. Eccentric features in Saturn's outer C ring. *Icarus* 72: 227–52

Porco, C. C., Cuzzi, J. N., Esposito, L. W., Lissauer, J. J., Nicholson, P. D. 1993. Neptune's rings. In *Neptune*, ed. D. P. Cruikshank. Tucson: Univ. Ariz. Press. In press

Rosen, P. A. 1989. *Waves in Saturn's rings probed by radio occultation*. PhD dissertation. Stanford Univ.

Rosen, P. A., Tyler, G. L., Marouf, E. A. 1991a. Resonance structures in Saturn's rings probed by radio occultation. I. Methods and examples. *Icarus* 93: 3–24

Rosen, P. A., Tyler, G. L., Maouf, E. A., Lissauer, J. J. 1991b. Resonance structures in Saturn's rings probed by radio occultation. II. Results and interpretation. *Icarus* 93: 25–44

Schaffer, L. E. 1989. *The dynamics of dust in planetary magnetospheres*. PhD dissertation. Cornell Univ.

Schaffer, L. E., Burns, J. A. 1987. The dynamics of weakly charged dust: motion through Jupiter's gravitational and magnetic field. *J. Geophys. Res.* 92: 2264–80

Showalter, M. R. 1989. Anticipated time variations in (our understanding of) Jupiter's ring system. In *Time-Variable Phenomena in the Jovian System*, ed. M. J. S Belton, R. A. West, J. Rahe. NASA SP–494

Showalter, M. R. 1991. The visual detection of 1981S13 and its role in the Encke Gap. *Nature* 351: 709–13

Showalter, M. R., Burns, J. A., Cuzzi, J. N., Pollack, J. B. 1987. Jupiter's ring system: new results on structure and particle properties. *Icarus* 69: 458–98

Showalter, M. R., Cuzzi, J. N., Larson, S. M. 1991. Structure and properties of Saturn's E ring. *Icarus* 94: 451–73

Shu, F. H. 1970. On the density wave theory of galactic spirals. *Astrophys. J.* 160: 99–112

Shu, F. H. 1984. Waves in planetary rings. See Greenberg & Brahic 1984, pp 513–61

Shu, F. H., Stewart, G. R. 1985. The collisional dynamics of particulate disks. *Icarus* 62: 360–83

Shu, F. H., Yuan, C., Lissauer, J. J. 1985a. Nonlinear spiral density waves: an inviscid theory. *Astrophys J.* 291: 356–76

Shu, F. H., Dones, L., Lissauer, J. J., Yuan, C., Cuzzi, J. N. 1985b. Nonlinear spiral density waves: viscous damping. *Astrophys. J.* 299: 542–73

Sicardy, B., Lissauer, J. J. 1992. Dynamical models of the arcs in Neptune's 63K ring (1989N1R). *Adv. Space Res.* 12(11): 11(97)–11(111)

Smith, B. A., Soderblom, L. A., Batson, R., Bridges, P., Inge, J., et al. 1982. A new look at the Saturn system: the Voyager 2 images. *Science* 215: 504–37

Smith, B. A., Soderblom, L. A., Beebe, R., Bliss, D., Boyce, J. M., et al. 1986. Voyager 2 in the Uranian system: imaging science results. *Science* 233: 43–64

Smith, B. A., Soderblom, L. A., Banfield, D., Barnet, C. Basilevsky, A. T., et al. 1989. Voyager 2 at Neptune: imaging science results. *Science* 246: 1422–49

Soter, S. 1971. The dust belts of Mars. *CRSR Rep. No.* 462, Cornell Univ.

Stevenson, D. J., Harris, A. W., Lunine, J. I. 1986. Origins of Satellites. In *Satellites*, ed. J. A. Burns, M. S. Matthews, pp. 39–88. Tucson: Univ. Arizona Press

Stewart, G. R., Lin, D. N. C., Bodenheimer, P. 1984. Collision-induced transport processes in planetary rings. See Greenberg & Brahic 1984, pp. 447–512

Tagger, M., Henriksen, R. N., Pellat, R. 1991. On the nature of spokes in Saturn's rings. *Icarus* 91: 297–314

Thomas, P. C. 1989. The shapes of small satellites. *Icarus* 77: 248–74

Tyler, G. L., Sweetnam, D. N., Anderson, J. D., Campbell, J. K., Eshleman, V. R., et al. 1986. Voyager 2 radio science observations of the Uranian system: atmospheres, rings, satellites. *Science* 233: 79–84

van Helden, A. 1984. Rings in astronomy and cosmology 1600–1900. See Greenberg & Brahic 1984, pp. 12–24

Ward, W. R. 1981. On the radial structure of Saturn's rings. *Geophys. Res. Lett.* 8: 641–43

Warwick, J. W., Evans, D. R., Romig, J. H., Sawyer, C. B., Desch, M. D., et al. 1986. Voyager 2 radio observations of Uranus. *Science* 233: 102–6

Weidenschilling, S. J., Chapman, C. R., Davis, D. R., Greenberg, R. 1984. Ring particles: collisional interactions and physical nature. In *Planetary Rings*, ed. R. Greenberg, A. Brahic, pp. 367–415. Tucson: Univ. Ariz. Press

Wisdom, J., Tremaine, S. 1988. Local simulations of planetary rings. *Astron. J.* 95: 925–40

Zebker, H. A., Marouf, E. A., Tyler, G. L. 1985. Saturn's rings: particle size distributions for thin layer models. *Icarus* 64: 531–48

Annu. Rev. Earth Planet. Sci. 1993. 21:525–55

IMPACT EROSION OF TERRESTRIAL PLANETARY ATMOSPHERES

Thomas J. Ahrens

Lindhurst Laboratory of Experimental Geophysics, Seismological Laboratory 252-21, California Institute of Technology, Pasadena, California 91125

KEY WORDS: atmospheric escape, atmospheric loss, earth accretion, early atmosphere

1. INTRODUCTION

The idea that planetary atmospheres can erode as a result of impact, and thus lose mass along with solid and molten high velocity ejecta during accretional infall of planetesimals follows from such early thoughtful works as that of Arrhenius et al (1974), Benlow & Meadows (1977), Ringwood (1979), and Cameron (1983). Ahrens et al (1989) describe how planetary impact accretion (and impact erosion) concepts lead naturally, from the idea that atmospheres form and erode during planetary growth.

   The theory of planetary system formation from a disc of gas, and later, gas and dust, corotating around a proto-sun, which evolves into increasingly larger planetesimals, is described by Safronov (1969), Wetherill (1980, 1990), and Kaula (1979) (Figure 1). In this model, planets grow as a result of mutual attraction and collision of planetesimals within a solar nebula which evolved from the primordial disc. An important step in the wide acceptance of this model is the demonstration that the requirement of planetesimals with diameters > 10 m will form from a dust and gas mixture before the nebular gas is removed from the terrestrial planet region in $\sim 10^6$ years (Weidenschilling 1988, 1989). Recently, very strong support for the early phase of the Safronov-Wetherill-Kaula scenario of planetary growth from a gaseous and possibly dusty disc of planetesimals, has come

525

0084–6597/93/0515–0525$02.00

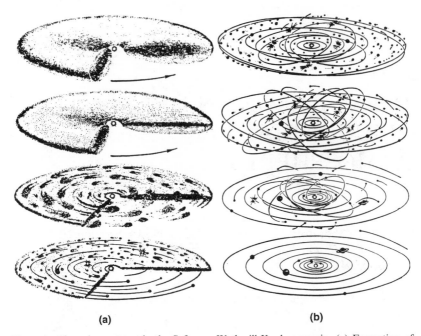

Figure 1 The solar system via the Safronov-Wetherill-Kaula scenario. (*a*) Formation of asteroid-size intermediate bodies form the dust component of the solar nebula. (*b*) Runaway accretion to intermediate-sized bodies. Impact accretion of intermediate-sized bodies into planets. Accretion of gas by giant planets is not shown. The initially flat system of intermediate bodies thickens due to their mutual gravitational perturbations. (After Levin 1972.)

(a)   (b)

from discovery using infrared, optical, and radio imagery of disc-shaped circumstellar gas and dust clouds around T-Tauri stars (e.g. HL Tauri and R Monocerotis) (Beckwith et al 1986, Sargent & Beckwith 1987) and main sequence stars (e.g. β-Pictoris) (Smith & Terrile 1984) (Figure 2). Although the gaseous rotating discs around T-Tauri stars have been imaged via microwave radio interferometry and infrared techniques, it has not yet been demonstrated that these discs contain accreting planetesimals. Recent observations of the ^{13}CO emission spectra and the blackbody thermal emission from the circumstellar gas disc around HL Tauri indicate that this disc extends out to a radius of 2000 AU, but is less than 380 AU thick (where 1 AU is the Earth-Sun distance $= 1.5 \times 10^8$ km). Moreover, the spectra are consistent with the gas and dust moving in bound orbits around this star (Sargent & Beckwith 1992).

Here I review current ideas about the nature of the planetesimals—composition, size distribution, and the planetary encounter velocity. Pre-

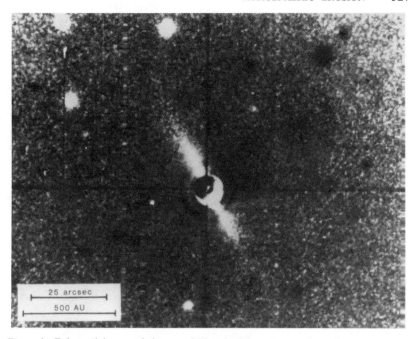

Figure 2 Enhanced image of the star β-Pictoris demonstrates what appears to be the beginnings of another solar system. The disc of material surrounding β-Pictoris extends 60×10^9 km from the star, which is located behind a circular occulting mask in the center of the image. The disc material is probably composed of gases and grains of ices, carbonaceous chondrite-like organic substances, and silicates. These are the materials from which the comets, asteroids, and planets of our own solar system are thought to have formed. (After Smith & Terrile 1984.)

vious papers on accretion and erosion of planetary atmospheres as a result of multiple impacts are also reviewed. Finally, the effects of blowing off a substantial fraction of the atmosphere from a terrestrial planet due to a single giant body impact are discussed.

2. PLANET FORMING MATERIALS

The planets and minor objects in the solar system appear to have accreted from the following three components:

1. Planetesimals similar to meteorites.

    The constitution of the terrestrial planets suggests that they accreted largely from planetesimals with a range of composition including primitive objects, such as C1 chondrites, as well as objects similar to differentiated metal and silicate meteorites. The linear relation in Figure 3 indicates how similar the major element composition of the chondritic meteorites are to the Sun. It is presumed that these objects are similar in bulk chemistry to some of the planet-forming planetesimals. This concept is reinforced by Figure 4, which illustrates that the noble gas abundance patterns of terrestrial planets are similar to each other and to primitive meteorites such as C1 chondrites. Meteorites are often taken to be typical of the planetesimals existing within the inner zone of the solar nebula from which the terrestrial planets accreted. The vestiges of the planetesimals of the inner solar system are believed to be the asteroids. We presumably sample these objects via meteorites that fall on the Earth. Some of the planetesimals that formed the terrestrial planets were also probably similar in composition to the

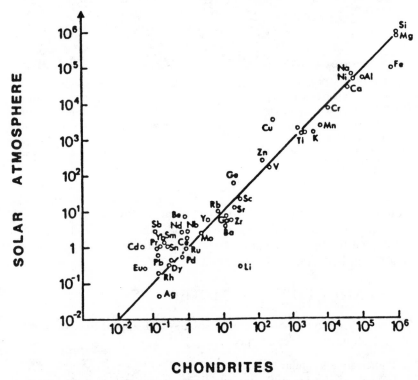

Figure 3 Atomic abundance of the elements in the solar photosphere vs the abundance in chondritic meteorites. Plot is normalized with respect to 10^6 atoms of Si. (After Allègre 1982.)

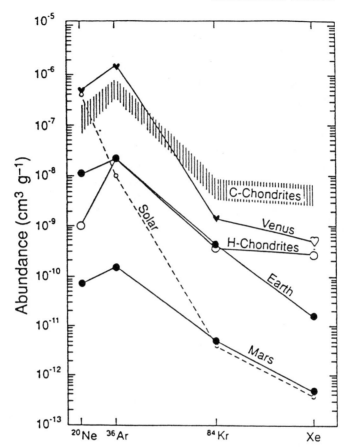

Figure 4 Abundances of noble gases in planetary atmospheres and chondritic meteorites given as cubic centimeters per gram of rock. [After E. Anders, personal communication in Owen et al (1992).]

present comets. Largely because of the gravitational perturbation from passing stars in the Galaxy, comets are perturbed from their orbits in the Oort cloud (which is spherically concentric with the Sun) at a radius of 10^4 to 10^5 AU, and possibly via the Kuiper belt (Duncan et al 1988) can achieve highly eccentric orbits which, near perihelion, result in collision with the terrestrial planets. The large masses of Jupiter and Saturn also play an important role in perturbation of cometary orbits, once these objects enter the planetary zone. Shoemaker et al (1990) estimated that some 30% of the recent impactor flux on the Earth may be cometary. How much of this flux has provided the volatile budgets of the terrestrial planets is presently unclear (e.g. Grinspoon & Lewis 1988 and Donahue & Hodges 1992) because of the similarity of the cometary nonvolatile major element inventory with that of the C1 chondrites (Jessberger et al 1989).

2. Cometary planetesimals.

In contrast to the silicate and iron-rich planetesimals, the planetesimals that made up the cores of the giant outer planets—Jupiter and Saturn, and to a lesser degree Neptune and Uranus, and possibly Pluto—were ice-rich. In analogy to the terrestrial planets whose planetesimals are probably related to the main belt asteroids, the remnants of the planetesimal swarm that formed the icy and silicate cores of the giant planets are now associated with comets. The generally smaller size (< 10 km diameter) of the comets (relative to asteroids) suggests that in the outer solar system the density of matter in the solar nebula was never great enough for the ice-rich planetesimals to experience substantial mutual gravitational attraction, resulting in impact accretion.

3. Solar nebulae gas.

In contrast to terrestrial planets, Jupiter and Saturn, and to a lesser degree Uranus and Neptune, appear to have gravitationally captured large quantities of solar nebular gases, after building up their initial planetary core from ice, silicate, and carbon-rich objects. The large planets thus retain a large solar-like reservoir of H_2 and He. Pollack & Bodenheimer (1989) suggested that the ratio of carbon to hydrogen in the atmospheres (which dramatically increases in the order: Jupiter, Saturn, Uranus, and Neptune) directly reflects a decreasing budget of accreted solar nebular hydrogen.

3.   TERRESTRIAL PLANETARY VOLATILES

On the basis of the general similarity of the noble gas abundances of the terrestrial planet atmospheres to the noble gas component that was processed within planetesimals before their accretion, I infer that terrestrial planet atmospheres originated from planetesimals similar to primitive meteorites. Cometary-like planetesimals may have also contributed to the terrestrial planet volatile inventory. Notably, the thermal and gravitational evolution of the planets are distinctly different; however, their relative noble gas inventories are similar. The solid phases containing noble gases presumably are still present in the asteroids and demonstratively occur in meteorites. Small meteorites impact at sufficiently slow terminal velocity that the noble gases are not released upon impact. They still contain the complement of noble gases present 4.6 Gyr ago prior to the planet-forming epoch (Figure 1). The planetesimal origin of noble gases in planetary atmospheres is also indicated by the observation that the two key mass-selective, gas-loss mechanisms (e.g. Hunten et al 1989) for atmospheric escape—Jeans loss and hydrodynamic escape—predict abundance patterns, starting from a solar (noble gas) pattern, that are quite different from those in the atmospheres of Mars, Earth, and Venus.

4.   SIZE DISTRIBUTION OF PLANETESIMALS

In Safronov's (1969) theory of planetary accretion, a simple power law distribution of the differential number of objects, $n(m)dm$, which occurs in a mass range dm is assumed:

$$n(m) = [N_0 \exp(-m/m_0)]/m_0, \tag{1}$$

where $n(m)dm$ is the number density of objects with mass between m and $m+dm$, N_0 is the initial number density of objects, and m_0 is the mean mass of the body distribution. Wetherill (1990) showed that in previous "smooth" growth descriptions of accreting objects (e.g. Safronov 1969), the increase of mass, dM/dt, could be described by

$$\frac{dM}{dt} = \pi R^2 \bar{V}_{rel} \rho \left(1 + \frac{8}{3} \frac{\pi \rho_p R^2}{\bar{V}_{rel}^2}\right), \tag{2}$$

where $\bar{V}_{rel}$ is the average relative velocity between the large (accreting) objects (of density ρ_p and radius R) and the nearby small objects (to be accreted), and ρ is the small body mass density in space. Equation 2 can be written as

$$dM/dt = kR^q \tag{3}$$

where k is a constant. When $\bar{V}_{rel}$ is large compared to the escape velocity from the large accreting object (V_e), $q \sim 2$. When $\bar{V}_{rel} \ll V_e$, $q \sim 4$. Initially smooth accretion models (Nakagawa et al 1983, Safronov 1969) used a constant intermediate value of $q = 3$. Moreover, recent modeling (Stewart & Kaula 1980, Stewart & Wetherill 1988) demonstrated that in swarms of unequal mass objects in an accreting disc, the mutual gravitational perturbations which result increase the mean velocities (essentially, $\bar{V}_{rel}$) of the smaller objects, such that the effective value of q in Equation 3 increases from ~ 2 to ~ 4. As a result, a phenomenon termed "runaway growth" occurs as a result of the marginally larger objects growing rapidly at the expense of their smaller neighbors. At 1 AU, smooth growth from

10^{20} gram objects to 10^{24} gram objects occurs in $\sim 10^4$ years, then a runaway growth occurs in an $\sim 10^5$ year period. Much of the proto-solar system mass ends up in 10^{26} gram (lunar to Mars size) proto-planets. Accretion of such large proto-planets gives rise to large body impacts during the latter stages of accretion.

5.   PLANETESIMAL IMPACT VELOCITIES

As shown in Figure 1, the major element of motion of the planetesimals in ideal circular Keplerian circular orbits around the Sun is velocity (v_s), which, in the solar reference frame, is given by

$$v_s = \sqrt{GM_\odot/r}, \tag{4}$$

where G is the gravitational constant (6.67×10^{-8} dyn-cm^2/g^2 or 6.7×10^{-11} nm^2/kg^2), $M_\odot$ is the solar mass, and r is the distance of the planetesimal from the Sun's center. Even for perfectly circular planar orbits, two objects at slightly different solar distances, initially will differ in velocity. Hence, the object closest to the Sun, m_1, will move in at a slightly higher speed, passing an object, m_2, which is further from the Sun. Gravitational interaction occurs in this two-body encounter, such that the object m_1 will experience a radial velocity increase given by

$$\delta v = m_2 v_s/(m_1+m_2)E. \tag{5}$$

Here, E is an encounter parameter which depends on the geometry and the relative masses of the objects. Continual gravitational interaction of adjacent objects gives rise to increasing orbital eccentricities, as well as inclination of the orbits. These in turn give rise to collisions which tend to damp out the velocity and orbital perturbations such as described by Equation 5. On average, for planetesimals relative to one another in orbit in a corotating disc of particles, their encounter velocities will be v_∞. Safronov pointed out that the largest particles with an escape velocity, v_e, will have encounter velocities:

$$v_\infty = v_e/\sqrt{2\theta}. \tag{6}$$

Here the local escape velocity is

$$v_e = \sqrt{2Gm_p/R_p} = \sqrt{2R_p g}, \tag{7}$$

where m_p is the mass of the largest planetesimal in a region, R_p is the largest planetesimal radius, and g is its gravitational acceleration. Here, θ, the Safronov parameter, is usually taken to be about 4 or 5. The result depends critically on $q < 2$. Otherwise, v_∞ will be drastically reduced by multiple encounters with the smallest particles.

The impact velocity of a planetesimal is therefore

$$v_i = \sqrt{(v_e^2 + v_\infty^2)}. \tag{8}$$

Equation 8 implies that as planets grow by accretion, the planetesimal impact velocity is always somewhat greater than the planetary escape velocity. Moreover, Equation 2 indicates that as planets grow, so does the mass of the planetesimals which impact their surface. Both runaway growth and velocity of impact considerations have led to efforts to understand the essential physics of large-body impacts on the terrestrial planets by Benz et al (1989, 1986, 1987, 1988) and Kipp & Melosh (1986).

6. COACCRETION OF PLANETARY ATMOSPHERES

Although Lange & Ahrens (1982b) suggested that the impact-induced dehydration of water-bearing minerals in planetesimals such as serpentine $Mg_3Si_2O_5(OH)_4$ would produce a largely water-rich atmosphere on the growing planets, it was Abe & Matsui (1985) who first suggested the possibility that water in this atmosphere, and possibly the dust produced by planetesimal impact, could drastically alter the thermal regime on the surface of growing planets (Figure 5). They assumed that serpentine in planetesimals brought the Earth and the other terrestrial planets their

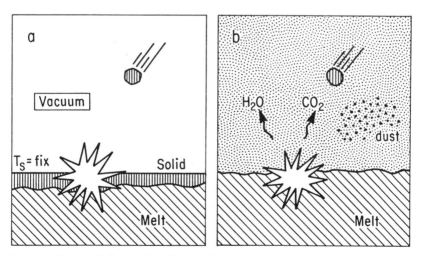

Figure 5 Cartoon indicating the difference between the thermal regime of accretion via (a) previous studies and (b) "thermal blanketing." [Figure from Abe & Matsui (1985).]

water inventory during their accretion. Once impact pressures of infalling planetesimals exceeded $P \simeq 23$ GPa (for a porous regolith), the supply of water to the atmosphere was assumed to begin. Abe & Matsui (1985) estimated the peak shock pressure by

$$P = \rho_0[C_0 + (K' + 1)v_i/8]v_i/2, \tag{9}$$

where ρ_0 and C_0 are impacting planetesimal initial density and zero-pressure bulk sound velocity, respectively, and K' is the partial derivative of the bulk modulus with respect to pressure. In these atmospheric accretion models the shock pressure is assumed to act on the entire impacting planetesimal and H_2O is released, as well as lesser amounts of CO_2, NH_3, SO_2, and other volatiles present in primitive meteorites (e.g. C1 carbonaceous chondrites). Previous calculations have also included contributions to the atmosphere induced by shock-loading of volatiles already present in the material of the planetary surface layer (e.g. Lange & Ahrens 1982a).

Using the experimental value of 23 GPa as the shock pressure required to induce complete water loss for serpentine, Lange & Ahrens (1982b) and Abe & Matsui (1986) concluded that once the radius of the Earth reached between 0.2 to 0.4 of the present value ($R_\oplus$), thermal blanketing of the Earth caused by a dense water atmosphere occurred. Matsui & Abe (1986a) point out that thermal blanketing is a more severe condition than the greenhouse effect. In a greenhouse effect, sunlight penetrates through the atmosphere (in the visible) but the thermal energy to be reradiated by the planetary surface in the infrared is trapped because of the infrared opacity of the planetary atmosphere resulting from abundant CO_2 and H_2O. Thermal blanketing is more severe, because solar radiation incident on the top of the atmosphere is completely scattered by the fine aerosols and impact ejecta and all the thermal energy is absorbed by the greenhouse gases, H_2O and CO_2. Impact cratering calculations (Ahrens et al 1989, O'Keefe & Ahrens 1977a) demonstrate that for large impactors, 60 to 90% of the energy of the impact is delivered as internal energy of the near surface material. The major effect of the proto-atmosphere then is to provide an insulating blanket to the flux of heat due to impacts on the surface of the growing planet. These processes are described by the equation

$$(1 + 1/2\theta)\frac{Gm_p}{R_p}\dot{m}dt = 4\pi R^2(F_{atm} - F_i)dt + C_p\dot{m}_p(T_s - T_p)dt + C_p m_s\dot{T}_s dt, \tag{10}$$

where the left-hand side is the rate of kinetic energy supplied to the surface provided by the impacting planetesimals. The first term on the right is the balance of heat flux of the atmosphere, where F_{atm} is the energy flux escaping from the surface to interplanetary space and F_i is the energy flux from the interior to the surface layer. The second and third terms are the heat sinks to the planet as a result of heating a larger planetary material of mass m_s from planetesimal temperature T_p to the higher surface temperature T_s. For a graybody radiative equilibrium atmosphere

$$F_{atm} = 2(\sigma T_s^4 - S_0/4)/[(3kp_0/2g) + 2], \qquad (11)$$

where σ is the Stefan-Boltzmann constant, S_0 is the solar flux, k is the absorption coefficient in the atmosphere, and p_0 is the surface pressure. Thermal blanketing as described by Equations 10 and 11, with the reasonable values of the constants chosen by Abe & Matsui (1985, 1986), quickly leads to temperatures above the solidus of crustal (basaltic) rocks (Figure 6a). Moreover, as more planetesimals impact the planet, the additional water provided begins to dissolve in what is the start of a magma ocean. Abe & Matsui showed that the surface temperature should be buffered by the solidus of hydrous basalt, ~ 1500 K, and the mass of H_2O in the atmosphere is nearly constant at 10^{24} g (essentially the present surface water budget) (Figure 6b). They showed that the mass of H_2O in the atmosphere remains nearly constant (due to a negative feedback effect). This effect can be demonstrated for a small surface temperature increase. This causes an increase in the fraction of molten basalt. That in turn induces additional water to dissolve in the molten silicate. Loss of water from the atmosphere then decreases the effectiveness of atmospheric blanketing and the result is that the small increase in temperature is nullified by the system's negative feedback.

The effectiveness of thermal blanketing of impact energy by the massive proto-atmosphere, as well as the feedback effect of water solubility in molten silicates, has been independently verified in a study by Zahnle et al (1988).

The termination of the coaccretion of an atmosphere and planet, which was modeled to occur on Earth, Venus, and Mars, can occur via three different mechanisms.

1. Abe & Matsui (1988) suggested that as the accretion rate decreased, the impact energy flux at the base of the atmosphere decreased and gradually solar heating dominated over impact heating. The surface temperature then declined below the melting point of hydrous basalt. With decreasing temperature, the water condensed and formed terrestrial oceans. Oceans may have formed on Venus (Matsui & Abe 1986b)—however, the larger solar ultraviolet flux gives rise to an enhanced photodisassociation to hydrogen and oxygen in the upper

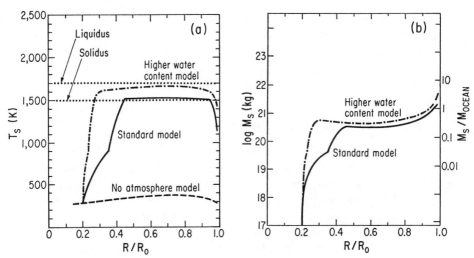

Figure 6   (*a*) The evolution of surface temperature during accretion of a model Earth from planetesimals containing 0.1% H_2O. The radius R is normalized by the final value, R_0. The dashed curve gives the calculated surface temperature without an impact-generated atmosphere (accretion period is 5×10^7 yr). The model surface temperature is affected by an atmosphere (which begins to greatly increase its mass once the impact velocity exceeds a critical value). The rapid rise in the surface temperature of the "standard model" which occurs after the Earth grows to $\sim 0.3R_0$, is due to an increase in the total mass of the atmosphere because of the initiation of the complete dehydration reaction of the surface layer. Once the surface temperature reaches the melting temperature, it remains nearly constant. [Figure after Matsui & Abe (1986a).] (*b*) The total mass of the impact-generated H_2O atmosphere is plotted against the normalized radius for the standard planetesimal models. Note that the total atmospheric mass, M_s, remains nearly constant after the Earth grows to $0.4R_0$ and is very close to the present mass of the Earth's oceans (1.4×10^{21} kg). [Figure after Matsui & Abe (1986a).]

atmosphere and subsequent Jeans escape of a large fraction of the planet's hydrogen inventory gives rise to the presently observed enhancement of the D/H ratio of Venus relative to the Earth of $\sim 10^2$.

2. Atmospheric loss occurs via multiple impact erosion. This is discussed in the next section.

3. Sudden partial or complete atmospheric loss occurs as a result of a large body impact. This is discussed in Section 8.

7.   ATMOSPHERIC EROSION BY IMPACT CRATERING

In addition to bringing volatiles to accreting atmospheres, the infall of planetesimals can erode planets and their oceans and atmospheres. For atmosphere-free solid and molten silicate planets, because the mechanical impedance of the impacting planetesimals is, in general, similar to that of a planet, the amount of ejecta which can escape from a planet with a given surface escape velocity depends only on impact velocity (and hence energy per unit mass) (O'Keefe & Ahrens 1977b), whereas the net gain or loss of a planetary atmosphere depends on total impact energy. O'Keefe & Ahrens (1982) calculated the energy partitioning into an atmosphere overlying a planet. They found that upon impact of a planet with projectiles of radii less than the atmospheric scale height, where the ejecta was primarily solid or molten, the amount of energy imparted to the atmosphere by direct passage through the atmosphere was only a few percent. Moreover, very little of the atmosphere achieved upward velocities in excess of the escape velocity. Walker (1986) showed that a very small portion of the atmosphere, shocked by the meteoroid, achieved sufficient enthalpy density to expand to greater than escape velocity. Using a numerical explosion model, Jones & Kodis (1982) showed that for the Earth, atmospheric explosion energies $> 5 \times 10^{26}$ ergs induced significant atmospheric blow-off.

Subsequently, Ahrens & O'Keefe (1987) and Ahrens et al (1989) employed a model in which they assumed all the energy of the impactor is delivered to the planetary surface and applied a theory [developed by Zel'dovich & Raizer (1966, Chapter 12) and Bach et al (1975)] for the shock acceleration of the atmosphere by an explosion (Figure 7). In this model, the time for atmospheric escape is related to atmospheric density near the Earth's surface, ρ_{00}, explosion energy, E, and atmospheric scale height, H, by

$$t = C_1(\rho_{00}H^5/E)^{1/2}, \tag{12}$$

where the constant, C_1, is approximately equal to 25. Moreover, the initial atmospheric shock velocity for atmospheric escape is

$$D = \alpha H/t, \tag{13}$$

where $\alpha \cong 6$. Assuming the Earth impactor has a velocity $v_e = 11$ km/sec and a strong shock condition exists such that $D \cong 11$ km/sec, Equation 13 yields $t = 4.4$ sec and the minimum impact energy calculated from Equation 12 is 1.9×10^{27} ergs. A projectile carrying this energy, if composed of silicate, will have a radius of ~ 0.5 km. The energy, 1.9×10^{27} ergs, is somewhat greater (by a factor of 20) than that from the numerical calculations of Jones & Kodis (1982).

For more energetic impacts (larger impactors), Melosh & Vickery (1989) and Vickery & Melosh (1990) developed a simple atmospheric cratering model applicable, for example, on the Earth in the 4.5×10^{27} to 9.9×10^{30}

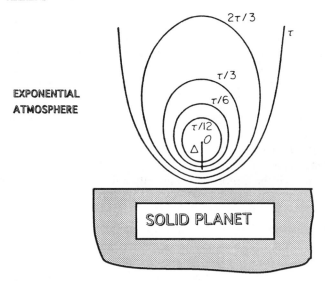

Figure 7 Shock front at successive instants of time for a strong explosion at high altitude. Sections shown are formed by passing a vertical plane through the origin of the explosion. The density of the atmosphere changes by a factor of e over the atmospheric scale height, Δ. Note that $\tau = (\rho_c \Delta^5/E)^{1/2}$, where ρ_c is the density at the altitude of the explosion and E is the energy. (After Zel'dovich & Raizer 1966.)

erg energy range (Figure 8). For 20 km/sec, 2.7 g/cm^3 impactors, these energies correspond to impactor radii of 0.6 to 7.6 km. They pointed out that for high-velocity planetary impacts, which penetrate the atmosphere, the projectile and a proportional mass of target become vaporized. For simplicity they assumed that the mass vaporized is equal to twice the mass of the impactor. This assumption is consistent with detailed computer simulation of impact on planetary surfaces (O'Keefe & Ahrens 1977a). The resulting gas plume then expands at a speed greater than the planetary escape velocity and carries with it the overlying planetary atmosphere. A conservative model of the plume expansion gives the mean (mass-averaged) velocity of expansion as

$$v_{\text{exp}} = [2(e - h_{\text{vap}})]^{1/2}, \tag{14}$$

where e is the internal energy per unit mass of the impactor and h_{vap} is the total enthalpy per unit mass, starting at the ambient temperature required to vaporize the projectile or target material. For silicate and ice, h_{vap} is 13×10^{13} and 3×10^{13} ergs/g, respectively. The energy per unit mass imparted by the impact-induced shock wave for like materials is $e \approx v^2/8$,

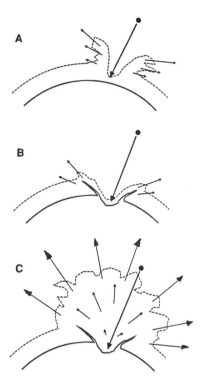

Figure 8 Cartoons of atmospheric impact indicating how projectile momentum can be transferred to a planetary atmosphere. (*A*) The projectile enters the atmosphere, heating, compressing, and accelerating the atmospheric gases ahead of it. (*B*) Solid ejecta from a growing crater pass through the atmosphere, transferring some or all of their momentum to the atmosphere by drag. Only a small quantity of atmosphere is ejected via the mechanisms of *A* and *B*. (*C*) The impact-generated vapor plume expands upward and outward. (After Vickery & Melosh 1990.)

where we assume the shock particle velocity u is $v/2$. It is easy to show that the minimum impact velocity required for the vapor plume to exceed the escape velocity is

$$v_{\mathrm{m}} = \sqrt{8[(v_{\mathrm{e}}^2/2) + h_{\mathrm{vap}}]}.$$ (15)

Atmospheric erosion occurs if the impact-induced momentum of this shock-induced gas when combined with the mass of the overlying atmosphere has sufficient velocity to escape the planet. By assuming a self-similar velocity profile in the total expanding gas cloud proposed by Zel'dovich & Raizer (1966, p. 104) of the form

$$\rho(r) = A(1 - r^2/R^2)^\alpha/R^3,$$ (16)

where R is the radius of the front of the gas cloud and r is the radius to a point within the gas cloud, A and α are determined by assuming conservation of mass and energy above the impact site. Vickery & Melosh (1990) used a value of $\gamma = 9/7$ to infer values of $\alpha = 11$ and $A = 15.4M_{\mathrm{t}}$, where γ is the gas cloud's polytropic exponent, and M_t is the total mass of the vapor cloud. When the projectile velocity exceeds the minimum impact velocity for atmospheric escape (Equation 15), escape occurs in atmosphere directly above the impact site. As the impact energy is increased, a cone with increasing angle θ is ejected (Figure 9). As impacts become more energetic, the maximum energy in the Vickery-Melosh model corresponds to ejection of an air mass above the tangent plane (Figure 9) of 3×10^{18} g (for the case of the Earth) or 6×10^{-4} of the total atmospheric budget. Vickery & Melosh showed that, for the case of the above approximations, when the mass of the projectile exceeds the mass of the atmosphere above the horizontal tangent, m_c, all of the atmosphere above a tangent plane to Earth is ejected. Thus when

$$m \geq m_c \equiv H m_a / 2 R_p \qquad (17)$$

Equation 17 indicates that for smaller m_a, the mass of the atmosphere gives rise to a smaller mass m_c, which can erode an atmosphere. Table 1 gives m_c values for the terrestrial planets. Since smaller projectiles are more numerous and thinner atmospheres erode rapidly, Equation 17 indicates that once an atmosphere starts eroding, erosion is accelerated until the planet is stripped. Figure 10 shows the maximum atmospheric mass that can be expelled by a spectrum of impactors for three different planets for

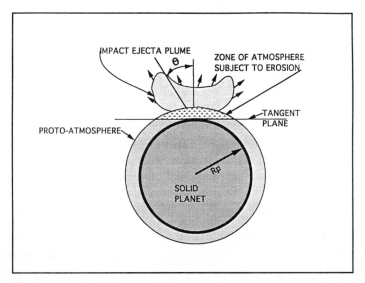

Figure 9 The impact-generated gas is assumed to interact only with the atmosphere lying above the plane tangent to the Earth at the center of impact. (After Vickery & Melosh 1990.)

Table 1 Characteristics of impact erosion models

	Vickery-Melosh model	
Planet	Tangent mass (g)	Tangent mass, impact energy (ergs)
Earth	2.6×10^{18}	9.9×10^{30}
Venus	5.9×10^{20}	1.6×10^{33}
Mars	4.4×10^{16}	4.5×10^{28}

	Total blow-off model			
	Free-surface velocity for atmospheric escape (km/sec)		Impact energy for total blow-off (ergs)	
Planet	$\gamma = 1.1$	$\gamma = 1.3$	$\gamma = 1.1$	$\gamma = 1.3$
Earth	1.60	2.75	6.20×10^{37}	1.84×10^{38}
Venus	1.00	2.45	2.10×10^{36}	1.20×10^{37}
Mars	1.03	1.20	2.90×10^{34}	5.70×10^{35}

a given accretion (veneer) mass being added to the planet. Melosh & Vickery (1989) demonstrated how Mars' atmosphere could have been eroded from an initial surface pressure of 0.7 bars to the present 7×10^{-3} bars in the first 1.4 Gyr of solar system history.

In conclusion, note that in the case of the Earth, impact erosion is important only for impacts more energetic than $\sim 10^{27}$ ergs. Thus the Abe-Matsui scenario, which only deals with atmospheric accretion, is relevant if the projectiles impacting the Earth's surface are much smaller in radius than the atmospheric scale height. For projectiles with radii in the km range, the research summarized in this section indicates that impact erosion needs to be taken into account in accretion models of the terrestrial planets.

8.   ATMOSPHERIC BLOW-OFF BY GIANT IMPACTS

As discussed in Section 3, as the planets accreted according to the Safronov-Wetherill-Kaula scenario, the planetesimal impactors also grew in size and it appears likely that some planetesimals grew to radii in the 2000 to 3000 km range—comparable to the size of the smaller planets (e.g. Mercury and Mars). These considerations have motivated the numerical modeling of large body impacts by Benz et al and Kipp & Melosh cited in Section 5.

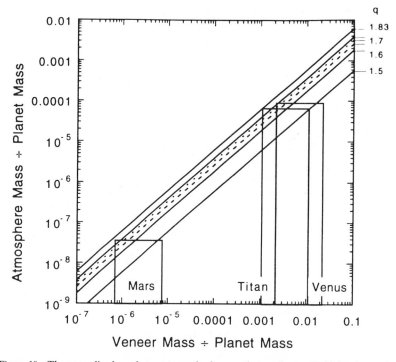

Figure 10 The normalized maximum atmospheric mass that can be expelled by an impacting veneer of normalized mass for three values of q, the power law exponent describing the (differential) mass spectrum of impactors. Mars (*dots*) and Venus (*dashes*) are quantitatively similar (both shown for $q = 1.7$). Rectangles indicate veneer masses (width corresponds to plausible q values) needed to remove present atmospheres of Titan, Mars, and Venus. (After Zahnle et al 1992.)

Previously, the effects of large body impact on the Earth's atmosphere have only been briefly described by Ahrens (1990). To calculate the energy, and hence, approximate planetesimal size, such that upon impact the entire planetary atmosphere is blown off, I employ a different approach than previous efforts and consider a shock wave that is entirely propagated within a terrestrial planet as sketched in Figure 11.

For a large impact on a terrestrial planet, where the impactor dimensions are greater than the atmospheric scale height (Table 2), the direct shock wave strips the atmosphere in the vicinity of the impactor (Figure 9). The air shock is also refracted around the entire planet. However, most of the projectile energy will be delivered to the solid planet (O'Keefe et al 1982). The effect on the overlying atmosphere of a great impact on the solid planet is considered below.

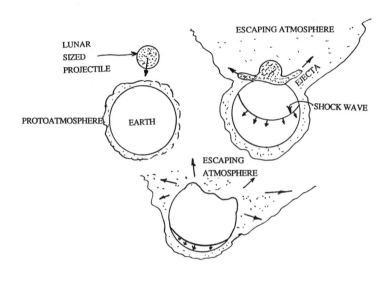

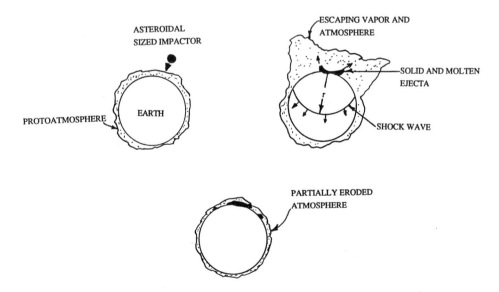

Figure 11 (*Top*) Sketch of lunar-sized planetesimal impacting the Earth. The proto-atmosphere is blown away by the shock wave-induced motion of the solid or molten planet. (*Bottom*) Impact of asteroidal-sized impactor and resultant partial eroded atmosphere.

Table 2   Terrestrial planet and atmosphere properties

| | Solid planet | | | |
Planet	Radius (km)	Core radius (km)	Surface gravity (cm/sec^2)	Surface escape velocity (km/sec)
Earth	6352	2769	981	11.18
Venus	6052	2827	887	10.37
Mars	3396	1146	372	5.03

| | Atmosphere | | | |
Planet	Molecular weight (daltons)	Surface pressure (bars)	Scale height (km)	Surface temperature (K)	Surface density (g/cm^3)
Earth	29	1.01	8	288	1.29×10^{-3}
Venus	44	92	15	735	6.49×10^{-2}
Mars	44	0.007	11	215	1.20×10^{-5}

The key calculation is to relate the particle velocity of the solid planet-atmosphere interface, u_{fs}, antipodal of a major impact to the atmospheric free-surface velocity, v_e. The velocity, v_e, is achieved as a result of the atmosphere being shocked first by the solid planet moving at velocity v_{fs}, and then becoming further accelerated to velocity, v_e, as a result of the upward propagating shock wave reflecting (isentropically) at the effective top of the atmosphere (Figure 12). This reflection provides an additional velocity increment, u_r. Gas speeds greater than the escape velocity are thus achieved. This is a conservative calculation since I use the density and pressure of the atmosphere at its base. Moreover, the atmosphere covering the planet closer to the impact than the antipode is expected to achieve yet higher velocity because it is shocked by the decaying air wave, and also shocked to higher pressures by the solid planet. Note that, in general, as a shock wave is propagated upward in an exponential atmosphere, because the density encountered by the traveling shock is decreasing, the shock velocity and particle velocity increase with altitude as discussed in Section 6 and by Zel'dovich & Raizer (1966). Thus, one can safely neglect shock attenuation in the atmosphere, and assume the particle velocity at the solid planet-atmosphere interface (the independent variable) and calculate the shock pressure induced in the gas by the outward surface of the Earth. The solid Earth therefore acts like a piston with velocity, u_{fs}, pushing on the atmosphere.

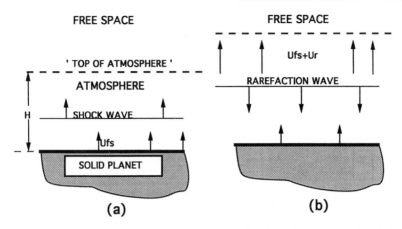

Figure 12 Sketch of shock- and rarefaction wave-induced motion of atmosphere being driven by planetary free-surface velocity, u_{fs}. Atmosphere geometry is approximated by a layer of scale height, H. (*a*) Shock wave driven particle velocity, u_{fs}. (*b*) Increase of gas velocity to $u_{fs}+u_r$ upon "reflection" of shock at the "top of atmosphere."

The pressure behind the shock wave, p_1, for different outward rock velocities, u_{fs}, can be determined by solving for p_1 in the following set of equations (e.g. Equations 1.78 and 1.79 of Zel'dovich & Raizer 1966):

$$|u_{fs}| = u_0 - u_1, \tag{18}$$

where

$$u_0 = \{V_0[(\gamma-1)p_0+(\gamma+1)p_1]/2\}^{1/2} \tag{19}$$

is the particle velocity of the unshocked gas with respect to the shock wave in the atmosphere and

$$u_1 = \left\{\frac{V_0[(\gamma+1)p_0+(\gamma-1)p_1]^2}{2[(\gamma-1)p_0+(\gamma+1)p_1]}\right\}^{1/2}, \tag{20}$$

where u_1 is the particle velocity of the shocked atmosphere relative to the shock front. Surface values of p_0 and $V_0 = 1/\rho_{00}$ for the planetary atmospheres used to calculate p_1 are given in Table 2. As shown in Figure 7, the upward propagation of the shock wave in a decreasing atmosphere gives rise to a great increase in both shock and particle velocity which we approximate by imagining that this shock "reflects" at the "top of the atmosphere" at a scale height, H. This is sketched in Figure 12*b*. This approximation should be valid if the duration of the impact induced particle velocity, u_{fs} ($\sim 10^2$ sec, see Figure 14), is long compared to the travel time (~ 20 sec) of a sound wave in the atmosphere up to a scale height.

Upon reflection of the shock wave at the top of the atmosphere, isentropic release from p_1 to zero pressure gives rise to an additional large increase in particle velocity which is described by the Riemann integral:

$$u_r = \int_0^{p_1} (-dV/dp)^{1/2}\, dp. \tag{21}$$

Upon substituting for an ideal polytropic gas, with a polytropic exponent, γ, Equation 21 yields

$$u_r = (p_0^{1/\gamma} V_0/\gamma)^{1/2} p_1^{1/2 - 1/(2\gamma)}/[1/2 - 1/(2\gamma)]. \tag{22}$$

The shock-induced outward atmospheric velocity, u_e, is given by

$$u_e = |u_{fs}| + u_r. \tag{23}$$

We assume that when

$$|u_e \geq v_e| \tag{24}$$

atmospheric blow-off occurs. When $u_e = v_e$, the corresponding value of u_{fs} is denoted by u_{fse}. The outward rock velocity versus outward atmospheric velocity for the Earth, Venus, and Mars, is shown in Figure 13 for values of the polytropic exponent in the range from 1.1 to 1.3. This range encompasses the effective likely range of γ which is expected to decrease from 1.3 to 1.1 with increasing gas ionization. Equation 24 is satisfied for $\gamma = 1.1$ to 1.3 for outward rock velocities of 1.60 to 2.25 km/sec for Earth, 1.00 to 2.45 km/sec, for Venus, and 0.27 to 1.2 km/sec for Mars. What impact energies will produce these outward rock velocities for the terrestrial planets?

Fortunately, the strength of the shock-wave induced compressional wave that results upon propagation completely through planets with varying iron core sizes overlain by silicate mantles has been recently calculated for objects that have core to planetary radius ratios of 0.333 and 0.466 (Watts et al 1991). In these calculations, the energy of the surface source was $E_w = 3.1 \times 10^{34}$ ergs. Notably, the calculation of Watts et al (1991), when scaled as discussed below, agrees closely with those of Hughes et al (1977) upon which earlier estimates of the energy required to blow-off the Earth's atmosphere were based (Ahrens 1990).

I scaled the results, which are given as peak compressional wave stresses experienced by material directly beneath the antipode of the impact point for a core to planetary radius ratio of 0.333 and a planet radius $R_w = 1500$ km, to that for Mars, which has a core to radius ratio of ~ 0.34 and a planetary radius of $R_p = 3396$ km (Table 2). Similarly, I used the Watts et al result for a core to planetary radius ratio of 0.47 to provide estimates of the peak stress beneath the antipode for an impact on the Earth and Venus for which the actual core to planetary radius ratios are 0.44 and 0.47, respectively. Since planetary gravity was not included in the calculations, we employ cube scaling (Melosh 1989, p. 112) to adjust the results for planetary size.

The energy of the equivalent surface source, E_p, assumed for an impact on the actual planet of interest is

$$E_p = E_w(R_p/R_w)^3. \tag{25}$$

To relate the peak pressure experienced by the cell beneath the antipodes for the 0.33 and 0.47 core to planetary radius ratios, peak shock pressures of $P_1 = 2.03$ and $P_1 = 2.10$ GPa were used. To convert these values to shock particle velocity, u_1, I assumed a surface density of $\rho_0 = 2.72$ g/cm^3 and a shock velocity $U_s = 5$ km/sec in the momentum equation:

$$u_1 = P_1/(\rho_0 U_s) \tag{26}$$

and then made the common approximation that the outward rock (free-surface) velocity u_{fsw} corresponding to the Watts et al calculation is

$$u_{fsw} = 2u_1. \tag{27}$$

The energy E_{fs}, required of an impactor to obtain the upper and lower bounds of u_{fs} necessary to launch the atmosphere to escape velocity in the calculations of Figure 13, can then be calculated from

$$E_{fs} = E_p(u_{fs}/u_{fsw})^2. \tag{28}$$

Thus, for complete atmospheric blow-off, values for E_{fs} of 6.2×10^{37} to 1.8×10^{38} ergs are inferred for the compressional wave induced motion for the Earth. This compares to $\sim 10^{37}$ ergs previously calculated for the compressional wave by Ahrens (1990). For Venus, E_{fs} varies from 2.1×10^{36} to 1.2×10^{37} ergs, whereas for Mars values of 2.9×10^{34} to 5.7×10^{35} ergs are needed for complete blow-off (Table 1).

The above impact energies may be somewhat of an overestimate as the later arriving antipodal surface (Rayleigh) wave is expected to have a greater vertical amplitude, and hence, higher free-surface velocity.

Although calculation of antipodal compressional wave amplitude via finite difference methods, such as employed by Hughes et al and Watts et al are straightforward, obtaining surface wave forms requires more computational effort. Recently, H. Kanamori (private communication, 1992) has calculated the antipodal surface wave displacement for the Earth for a point force step pulse of 10^{16} dynes (Figure 14) using the method of

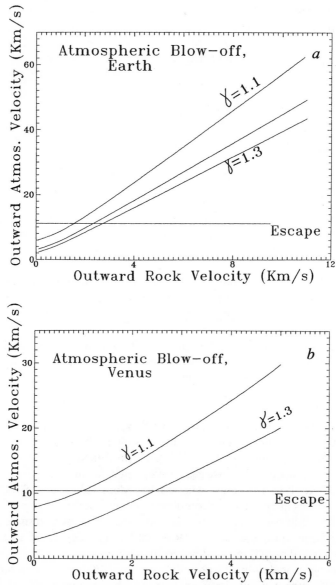

Figure 13 Relationship of outward free-surface velocity (u_{fs}) to outward atmospheric velocity (v_e) for a polytropic atmosphere with various values of γ. (*a*) Earth, (*b*) Venus, (*c*) Mars.

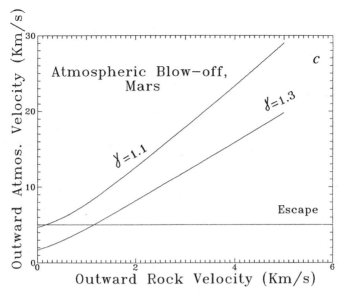

Figure 13 (*continued*).

Båth (1966). Because the elastic displacement at the source is singular, it is convenient to estimate the energy in the surface wave as it passes over the equator (relative to the source region) from the equation (Kanamori & Hauksson 1992):

$$E \cong 2\pi(2\pi R_p)\bar{h}U \int_0^\infty \dot{x}(t)dt,$$ (29)

where R_p is the Earth's radius, $\bar{h} = 3 \times 10^7$ cm is approximately 1/3 the wavelength of the fundamental Rayleigh wave Airy phase, $U = 3.6 \times 10^5$ cm/sec is the group velocity, and $\rho = 2.7$ g/cm^3. We assume that the peak to peak displacement of Figure 14b occurs over a period of ~ 200 sec. Moreover, the integral is approximated by a 400 sec time interval. Thus Equation 29 yields an energy of $E = 2.6 \times 10^{11}$ ergs. Application of the same scaling as used in Equation 28, for the Earth, yields a seismic energy of 8 to 16×10^{34} ergs for $\gamma = 1.3$ and 1.1, respectively. Although this is a factor of 10^3 to 10^4 lower than is given for the energy inferred from body wave amplitudes (after Equation 28), the inefficiency of inducing a seismic surface wave from a surface hypervelocity impact needs to be taken into account. This factor is poorly known. Schultz & Gault (1975) estimate only 10^{-3} to 10^{-4} of the impact energy is carried from the impact region away as seismic energy. Their estimate may be too low for a giant impact.

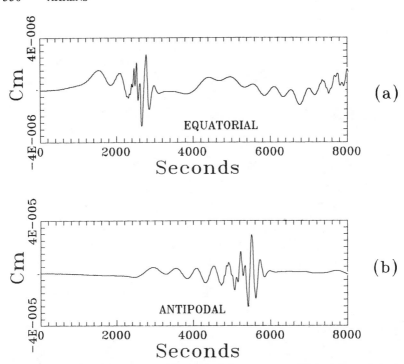

Figure 14 Theoretical free-surface elastic displacement of the Earth versus time upon sudden application of a surface (point) source of 10^{16} dynes. (*a*) Displacement history at 90° (equatorial) from source. (*b*) At 180° (antipodal) from the source. [After H. Kanamori (private communication, 1992).]

More research is obviously needed. Nevertheless, with our present knowledge of cratering mechanics, it appears difficult to estimate the expected difference between body and surface wave induced atmospheric blow-off, and the energy estimates given on the basis of body waves should be tentatively accepted as the best order-of-magnitude estimate. Clearly, estimating antipodal surface wave amplitudes for large impacts on the other terrestrial planets, is even more uncertain and is not attempted here.

Finally, it is useful to estimate the energy and mass fraction of the planetary atmosphere blown off in going from the Melosh-Vickery model of tangential blow-off (Table 1) to the condition of complete blow-off. For energies less than those required to eject the entire atmosphere, I assume a simple power law for the decrease of particle velocity with radius, r, from the impact point (Figure 11*b*) and assume that as the decaying stress wave interacts with the free-surface of a spherical planet, atmospheric blow-off occurs if the particle velocity is greater or equal to 0.5 u_{fse}—which is the value of u_{fs} required to blow off the atmosphere for the antipodal case. Thus, the amount of atmosphere blown off is related to the area of a sphere subtended by an arc of radius r at the point where the particle velocity has decayed to 0.5 u_{fse} (Figure 11b).

To determine an empirical relation for the attenuation of particle velocity, I assume the form (Melosh 1989, p. 62):

$$u = u_0/r^n. \tag{30}$$

I first calculate the radius of a hemisphere, r_1, enclosing a unit mass

$$r_1 = [3/(2\pi\rho_0)]^{1/3}. \tag{31}$$

Using the values of $E_{fs\,min}$ and $E_{fs\,max}$ calculated from Equation 28 to designate the minimum and maximum energies obtained from Figure 13 for $\gamma = 1.1$ and $\gamma = 1.3$, respectively, the shock particle velocity associated with each energy is given by

$$u_{1\,min} = (E_{fs\,min})^{1/2} \tag{32a}$$

$$u_{1\,max} = (E_{fs\,max})^{1/2}. \tag{32b}$$

Denoting $|u_{fs}|$ calculated from Equation 18 using $\gamma = 1.1$ and $\gamma = 1.3$ as $u_{fs\,min}$ and $u_{fs\,max}$ I can obtain from Equations 27 and 30, expressions for the particle velocity decay parameters, n_1 and n_2:

$$n_1 = \log(2u_{1\,min}/u_{fs\,min})/\log(2R_p/r_1) \tag{33a}$$

$$n_2 = \log(2u_{1\,max}/u_{fs\,max})/\log(2R_p/r_1). \tag{33b}$$

For a stress wave particle velocity which decays to a value of $u_{fsw}/2$ at a radius r_e from the impact point, the mass of atmosphere blown off is

$$m_e = (\rho_0/g)A, \tag{34}$$

where the term in parentheses is the atmospheric mass per unit area and A is the area of a sphere subtended by an arc of length r (Figure 11b).

From geometrical arguments it can be shown that the area of the planet subtended by an arc of length r is

$$A = \pi r^2. \tag{35}$$

The energies (minimum and maximum) associated with the radius r_e, for $\gamma = 1.3$ and $\gamma = 1.1$ values are:

$$E_{min} = [u_{fs\,min}(r_e/r_1)^{n_1}/2]^2 \tag{36a}$$

$$E_{max} = [u_{fs\,max}(r_e/r_1)^{n_1}/2]^2. \tag{36b}$$

The normalized mass of the atmosphere blown off, m_e/m_a, versus both E_{min} and E_{max} are shown in Figure 15.

Table 1 shows that the energy for atmospheric loss above a tangent plane is a small fraction ($\sim 10^{-8}$ to 10^{-4}) of the energy required to drive off the entire atmosphere. For the Earth, this total loss energy is $\sim 10^{38}$ ergs and would be achieved via an impact of a lunar-sized object at 20 km/sec. In the case of Venus, the impact of a smaller ~ 800 km radius object, at ~ 20 km/sec, will drive off the atmosphere. For Mars, the impact of a 160 km radius object at 20 km/sec will drive off the atmosphere. It may be, as suggested by Cameron (1983), that the terrestrial planets all suffered several giant impacts and their present atmospheres may reflect accretion and outgassing since the last great impact event.

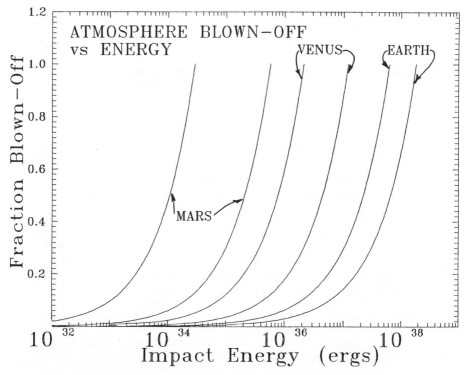

Figure 15   Calculated fraction of atmosphere blown off versus impactor energy for Earth, Venus, and Mars. Lower and higher energy curves for each planet correspond to assumed polytropic exponent of ideal gas of $\gamma = 1.1$ and 1.3, respectively.

ACKNOWLEDGMENTS

I appreciate receiving helpful comments on this paper from G. W. Wetherill, A. Vickery, A. W. Harris, and L. R. Rowan, as well as private communications from G. Chen and J. Melosh. Special thanks go to H. Kanamori who computed the results shown in Figure 14. Many of the ideas developed in this review have come from collaborations with J. D. O'Keefe, M. A. Lange, and J. A. Tyburczy. This research is supported by NASA and is Contribution #5198, Division of Geological and Planetary Sciences, California Institute of Technology.

Literature Cited

Abe, Y., Matsui, T. 1985. The formation of an impact-generated H_2O atmosphere and its implications for the early thermal history of the Earth. *Proc. Lunar Planet. Sci. Conf. 15th, Part 2, J. Geophys. Res.* 90: C545–59

Abe, Y., Matsui, T. 1986. Early evolution of the Earth: accretion, atmosphere formation and thermal history. *Proc. XVII Lunar Planet. Sci. Conf., I, J. Geophys. Res.* 91: E291–302

Abe, Y., Matsui, T. 1988. Evolution of an impact-generated H_2O-CO_2 atmosphere and formation of a hot proto-ocean on earth. *J. Atmos. Sci.* 45: 3081–3101

Ahrens, T. J. 1990. Earth accretion. In *Origin of the Earth*, ed. J. Jones, H. Newsom, pp. 211–27. Houston: Oxford Univ. Press

Ahrens, T. J., O'Keefe, J. D. 1987. Impact on the Earth, ocean, and atmosphere. *Proc. Hypervelocity Impact Symp., Int. J. Impact Eng.* 5: 13–32

Ahrens, T. J., O'Keefe, J. D., Lange, M. A. 1989. Formation of atmospheres during accretion of the terrestrial planets. In *Origin and Evolution of Planetary and Satellite Atmospheres*, ed. S. K. Atreya, J. B. Pollack, M. S. Matthews, pp. 328–85. Tucson: Univ. Ariz. Press

Allègre, C. J. 1982. Cosmochemistry and primitive evolution of planets. In *Formation of Planetary Systems*, ed. A. Brahic, pp. 283–364. Paris: Centre Natl. d'Etudes Spatiales

Arrhenius, G., De, B. R., Alfven, H. 1974. Origin of the ocean. In *The Sea*, ed. E. D. Goldberg, pp. 839–61. New York: Wiley

Bach, G. G., Kuhl, A. L., Oppenheim, A. K. 1975. On blast waves in exponential atmospheres. *J. Fluid Mech.* 71: 105–22

Båth, M. 1966. Earthquake energy and magnitude. *Phys. Chem. Earth* 7: 115–65

Beckwith, S., Sargent, A. I., Scoville, N. Z.,

Masson, C. R., Zuckerman, B., et al. 1986. Small-scale structure of the circumstellar gas of HL Tauri and R Monocerotis. *Astrophys. J.* 309: 755–61

Benlow, A., Meadows, A. J. 1977. The formation of the atmospheres of the terrestrial planets by impact. *Astrophys. Space Sci.* 46: 293–300

Benz, W., Cameron, A. G. W., Melosh, H. J. 1989. The origin of the moon and the single impact hypothesis, III. *Icarus* 81: 113–31

Benz, W., Slattery, W. L., Cameron, A. G. W. 1986. The origin of the moon and the single impact hypothesis, I. *Icarus* 66: 515–35

Benz, W., Slattery, W. L., Cameron, A. G. W. 1987. Origin of the moon and the single impact hypothesis, II. *Icarus* 71: 30–45

Benz, W., Slattery, W. L., Cameron, A. G. W. 1988. Collisional stripping of Mercury mantle. *Icarus* 74: 516–28

Cameron, A. G. W. 1983. Origin of the atmospheres of the terrestrial planets. *Icarus* 56: 195–201

Donahue, J. M., Hodges, R. R. Jr. 1992. Past and present water budget of Venus. *J. Geophys. Res.* 97: 6083–92

Duncan, M., Quinn, T., Tremaine, S. 1988. The origin of short-period comets. *Astrophys. J.* 328: 69–73

Goldreich, P., Ward, W. R. 1973. The formation of planetesimals. *Astrophys. J.* 183: 1051–61

Grinspoon, D. H., Lewis, J. S. 1988. Cometary water on Venus: implications for stochastic impacts. *Icarus* 74: 21–35

Hughes, H. G., App, F. N., McGetchen, T. R. 1977. Global effects of basin-forming impacts. *Phys. Earth Planet. Inter.* 15: 251–63

Hunten, D. M., Pepin, R. O., Owen, T. C. 1989. Planetary atmospheres. In *Meteor-*

ites and the Early Solar System, ed. J. F. Kerridge, M. S. Matthews, pp. 565–94. Tucson: Univ. Ariz. Press

Jessberger, E. K., Kissel, J., Rahe, J. 1989. The composition of comets. In *Origin and Evolution of Planetary Satellite Atmospheres*, ed. S. K. Atreya, J. B. Pollack, M. S. Matthews, pp. 167–91. Tucson: Univ. Ariz. Press

Jones, E. M., Kodis, J. W. 1982. Atmospheric effects of large body impacts; The first few minutes. In *Geological Implications of Impacts of Large Asteroids and Comets on the Earth*, ed. L. T. Silver, P. H. Schultz, pp. 175–86. Geol. Soc. Am. Spec. Pap.

Kanamori, H., Hauksson, E. 1992. A slow earthquake in the Santa Maria Basin, California. *Bull. Seismol. Soc. Am.* 82: 2087–96

Kaula, W. M. 1979. Thermal evolution of Earth and Moon growing by planetesimal impacts. *J. Geophys. Res.* 84: 999–1008

Kipp, M. E., Melosh, H. J. 1986. Short note: A preliminary study of colliding planets. In *Origin of the Moon*, ed. W. K. Hartmann, R. J. Phillips, G. T. Taylor, pp. 643–48. Houston: Lunar Planet. Inst.

Lange, M. A., Ahrens, T. J. 1982a. The evolution of an impact-generated atmosphere. *Icarus* 51: 96–120

Lange, M. A., Ahrens, T. J. 1982b. Impact-induced dehydration of serpentine and the evolution of planetary atmospheres. *Proc. Lunar Planet. Sci. Conf. 13th, Part 1, J. Geophys. Res., Suppl.* 87: A451–56

Levin, B. J. 1972. Origin of the earth. In *The Upper Mantle*, ed. A. R. Ritsema, pp. 7–30. Amsterdam: Elsevier

Matsui, T., Abe, Y. 1986a. Evolution of an impact-induced atmosphere and magma ocean on the accreting Earth. *Nature* 319: 303–5

Matsui, T., Abe, Y. 1986b. Impact induced atmospheres and oceans on Earth and Venus. *Nature* 322: 526–28

Melosh, H. J. 1989. *Impact Cratering, A Geologic Process.* New York: Oxford Univ. Press. 245 pp.

Melosh, H. J., Vickery, A. M. 1989. Impact erosion of the primordial Martian atmosphere. *Nature* 338: 487–89

Nakagawa, Y., Hayashi, C., Nakazawa, K. 1983. Accumulation of planetesimals in the solar nebula. *Icarus* 54: 361–76

O'Keefe, J. D., Ahrens, T. J. 1977a. Impact induced energy partitioning, melting, and vaporization on terrestrial planets. *Proc. Lunar Sci. Conf., 8th*, Vol. 3, *Geochim. Cosmochim. Acta*, Suppl. 8: 3357–74

O'Keefe, J. D., Ahrens, T. J. 1977b. Meteor-ite impact ejecta: dependence on mass and energy lost on planetary escape velocity. *Science* 198: 1249–51

O'Keefe, J. D., Ahrens, T. J. 1982. The interaction of the Cretaceous-Tertiary extinction bolide with the atmosphere, ocean, and solid earth. *Geol. Soc. Am. Spec. Pap.* 190: 103–20

Owen, T., Bar-Nunn, A., Kleinfeld, I. 1992. Possible cometary origin of heavy noble gases in the atmospheres of Venus, Earth, and Mars. *Nature* 358: 43–45

Pollack, J. B., Bodenheimer, P. 1989. Theories of the origin and evolution of the giant planets. In *Origin and Evolution of Planetary and Satellite Atmospheres*, ed. S. K. Atreya, J. B. Pollack, M. S. Matthews, pp. 564–602. Tucson: Univ. Ariz. Press

Ringwood, A. E. 1979. *Origin of the Earth and Moon.* Berlin, New York: Springer-Verlag. 295 pp.

Safronov, V. S. 1969. *Evolution of the Proto-Planetary Cloud and Formation of the Earth and Planets.* Moscow: Nauka. Transl. for NASA and NSF by Isr. Prog. Sci. Transl. as NASA-TT-F677

Sargent, A. I., Beckwith, S. 1987. Kinematics of the circumstellar gas of HL Tauri and R Monocerotis. *Astrophys. J.* 323: 294–305

Sargent, A. I., Beckwith, S. V. 1991. The molecular structure around HL Tauri. *Astrophys. J. Lett.* 382: L31–35

Schultz, P. H., Gault, D. E. 1975. Seismically induced modification of lunar surface features. *Proc. Lunar Sci. Conf. 6th* 3: 2845–62

Shoemaker, W. M., Wolfe, R. F., Shoemaker, C. S. 1990. Asteroid and comet flux in the neighborhood of Earth. In *Global Catastrophes in Earth History*, ed. V. L. Sharpton, P. D. Ward, pp. 155–70. Geol. Soc. Am. Spec. Pap.

Smith, B. A., Terrile, R. J. 1984. A circumstellar disk around β-Pictoris. *Science* 226: 1421–24

Stewart, G. R., Kaula, W. M. 1980. Gravitational kinetic theory for planetesimals. *Icarus* 44: 154–71

Stewart, G. R., Wetherill, G. W. 1988. Evolution of planetesimal velocities. *Icarus* 74: 542–53

Vickery, A. M., Melosh, H. J. 1990. Atmospheric erosion and impactor retention in large impacts with application to mass extinctions. In *Global Catastrophes in Earth History*, ed. V. L. Sharpton, P. O. Ward, pp. 289–300. Geol. Soc. Am. Spec. Pap.

Walker, J. C. G. 1986. Impact erosion of planetary atmospheres. *Icarus* 68: 87–98

Watts, A. W., Greeley, R., Melosh, H. J. 1991. The formation of terrains antipodal to major impacts, *Icarus* 93: 159–68

Weidenschilling, S. J. 1988. Formation processes and time scales for meteorite parent bodies. In *Meteorites and the Early Solar System*, ed. J. F. Kerridge, M. S. Matthews, pp. 348–71. Tucson: Univ. Ariz. Press

Weidenschilling, S. J., Donn, B., Meakin, P. 1989. Physics of planetesimal formation. In *The Formation and Evolution of Planetary Systems*, ed. H. A. Weaver, L. Danly, pp. 131–50. Cambridge: Cambridge Univ. Press

Wetherill, G. W. 1980. Formation of the terrestrial planets. *Annu. Rev. Astron. Astrophys.* 18: 77–113

Wetherill, G. W. 1990. Formation of the Earth. *Annu. Rev. Earth Planet. Sci.* 18: 205–56

Zahnle, K. J., Kasting, J. F., Pollack, J. B. 1988. Evolution of a steam atmosphere during Earth's accretion. *Icarus* 74: 62–97

Zahnle, K., Pollack, J. B., Grinspoon, D. 1992. Impact-generated atmospheres over Titan, Ganymede, and Callisto. *Icarus* 95: 1–23

Zel'dovich, Y. B., Raizer, Y. P. 1966. *Physics of Shock Waves and High-Temperature Hydrodynamic Phenomena*. Vol. 112. New York: Academic. 916 pp.

SUBJECT INDEX

and atmospheric CO_2, 245–47
and deep water formation, 245
Glacial position
of the North Atlantic Current,
244
Glasgow
air quality in, 152
Glengarry Group
in the Capricorn Orogen, 466,
469
Global Circulation Models
and global methane mixing ratios, 415
Global Volcanism Network
(GVN), 428
Gneiss
in the Albany-Fraser Orogen,
478, 479, 480
in the Capricorn Orogen, 470
in the Murchison Terrane, 459
in the Narryer Terrane, 458
in the Paterson Orogen, 473,
475
in the Pinjarra Orogen, 477
in the Yilgarn Craton, 456,
457, 458
Gondwana supercontinent, 454
Gorge Creek Group
in the Pilbara Craton, 461–62
Gorham, Eville, 152–53
Granite
in the Albany-Fraser Orogen,
478, 480
and Archean cratons, 454
in the Capricorn Orogen, 466,
468, 470
in the Kalgoorlie Terrane,
460, 461
in the Murchison Terrane,
459, 460
in the Narryer Terrane, 458,
459
in the Paterson Orogen, 474,
475
in the Pilbara Craton, 461,
463
in the Pinjarra Orogen, 477
in the Yeneena Group, 476
in the Yilgarn Craton, 456,
457, 458
Graphite
and carbonaceous chondrite
meteorites, 290
Gravitational attraction
in the outer solar system, 530
Gravitational forces
and crater ejecta, 352
Gravitational impacts
and catastrophic impacts, 370
Gravity
and crater formation, 358–61,
363, 371
and scaling impact processes,
338, 339, 340, 341, 352
Gravity flows
and sediment formation, 89–
109
Great Basin
ichnofabric index in carbonate
strata of, 218–20
Great Britain
and acid deposition, 169

see also England
Great Salinity Anomaly, 241–
43, 245
Greenhouse effect, 534
by trace gases, 421, 422
Greenhouse gases, 407
Greenland ice cores
and atmospheric methane mixing ratios, 412
Greenland ice shelf
and deep water formation, 233
Greenland-Iceland-Norwegian
(GIN) Seas, 236, 250
in a global circulation
scheme, 248
Greenland Sea
convective mixing in, 238–40
deep water formation in, 242
and open ocean convection,
237
Greenland Sea Deep Water
(GSDW), 236
Greenstone(s)
and Archean cratons, 454
in the Kalgoorlie Terrane,
460, 461
in the Murchison Terrane, 459
in the Pilbara Craton, 461,
463
in the Yilgarn Craton, 456,
457, 458
Greywackes
in the Capricorn Orogen, 469
Groundwaters
and acid deposition, 164–65,
170
chemistry of, 162
Gulf of Suez
and deep water formation, 233
Gulf of the Lions
and deep water formation, 233
Gypsum
and carbonaceous chondrite
meteorites, 273
crusts on carbonate stone, 167
Gyrocompass, 8, 16

H

Halogens
and carbonaceous chondrite
meteorites, 264
Hamersley Group
in the Pilbara Craton, 462,
463, 464–65
Hardey Formation
in the Pilbara Craton, 463–64
Harzburgite, 38
Hawaii
and hotspot volcanism, 433
Hawaiian hotspot, 441
Heat flow
and accretionary wedges, 323
in the Barbados wedge, 324
continental, 14
and the décollement, 309–11
oil field, 12–13
at sea, 12
terrestrial, 1, 13–15, 17
Heat production
radiogenic and non radio-
genic, 13–14

Heat transport
and the ocean and atmo-
sphere, 228, 229
in the oceans, 229, 230
Heaviside function
and seismic waves, 377
Helium (He)
and the giant outer planets,
530
isotopes of, 16
of the world ocean, 13
Helminthoida
in the Troll Field, 223
Heywood Fault
in the Albany-Fraser Orogen,
479, 480
High field strength element
(HFSE)
and arc magmas, 176, 177,
191, 193, 194, 197, 199
and the subduction zone component, 178
High-resolution transmission
electron microscopy
(HRTEM)
and carbonaceous chondrite
matrices, 257, 288
and the Orgueil meteorite, 271
Holocene
and atmospheric methane mixing ratios, 412
Holocene volcanoes, 438
volcanic eruptions in, 430
magnitude and frequency
of, 443–44
Hotspot volcanism, 16, 433, 442
Howardites, 138, 139
Huckleberry Ridge Tuff, 444
Hydraulics
of turbidity currents, 90–98
Hydrogen (H)
and carbonaceous chondrite
meteorites, 270, 296, 297
and the giant outer planets,
530
isotopes
and carbonaceous chondrite
meteorites, 266–67,
270
in meteorite material, 266–
67, 270
ions
and acid groundwaters,
165
and acidification of surface
waters, 164
alteration of materials by,
167
concentrations in the U.S.,
158
deposition in the U.S., 158
and watershed processes,
162
on Venus, 535–36
Hydrogen peroxide
and acid rain, 156
Hydroxide
and carbonaceous chondrite
meteorites, 275
Hydroxyl
and carbonaceous chondrite
meteorites, 293

CUMULATIVE INDEXES

CONTRIBUTING AUTHORS, VOLUMES 1–21

CHAPTER TITLES, VOLUMES 1–21

ANNUAL REVIEWS INC.

a nonprofit scientific publisher
4139 El Camino Way
P. O. Box 10139
Palo Alto, CA 94303-0897 • USA

ORDER FORM

ORDER TOLL FREE
1-800-523-8635
from USA and Canada

FAX: 415-855-9815

Annual Reviews Inc. publications may be ordered directly from our office; through booksellers and subscription agents, worldwide; and through participating professional societies. **Prices are subject to change without notice.** California Corp. #161041 • ARI Federal I.D. #94-1156476

- **Individual Buyers:** Prepayment required on new accounts by check or money order (in U.S. dollars, check drawn on U.S. bank) or charge to MasterCard, VISA, or American Express.
- **Institutional Buyers:** Please include purchase order.
- **Students/Recent Graduates:** $10.00 discount from retail price, per volume. Discount does not apply to Special Publications, standing orders, or institutional buyers. **Requirements:** [1] be a degree candidate at, or a graduate within the past three years from, an accredited institution; [2] present proof of status (photocopy of your student I.D. or proof of date of graduation); [3] Order direct from Annual Reviews; [4] prepay.
- **Professional Society Members:** Societies that have a contractual arrangement with Annual Reviews offer our books to members at reduced rates. Check your society for information.
- **California orders** must add applicable sales tax.
- **Canadian orders** must add 7% General Sales Tax. GST Registration #R 121 449-029. Now you can also telephone orders Toll Free from anywhere in Canada (see below).
- **Telephone orders,** paid by credit card, welcomed. **Call Toll Free 1-800-523-8635** from anywhere in USA or Canada. From elsewhere call 415-493-4400, Ext. 1 (not toll free). Monday – Friday, 8:00 am – 4:00 pm, Pacific Time. Students or recent graduates ordering by telephone must supply (by FAX or mail) proof of status if current proof is not on file at Annual Reviews. Written confirmation required on purchase orders from universities before shipment.
- **FAX: 415-855-9815** – 24 hours a day.
- **Postage paid** by Annual Reviews (4th class bookrate). UPS ground service (within continental U.S.) available at $2.00 extra per book. UPS air service or Airmail also available at cost. UPS requires a street address. P.O. Box, APO, FPO, not acceptable.
- **Regular Orders:** Please list below the volumes you wish to order by volume number.
- **Standing Orders:** New volume in series is sent automatically each year upon publication. Please indicate volume number to begin the standing order. Each year you can save 10% by prepayment of standing-order invoices sent 90 days prior to the publication date. Cancellation may be made at any time.
- **Prepublication Orders:** Volumes not yet published will be shipped in month and year indicated
- **We do not ship on approval.**

ANNUAL REVIEWS SERIES *Volumes not listed are no longer in print*	Prices, postpaid, per volume. USA / other countries (incl. Canada)	Regular Order Please send Volume(s):	Standing Order Begin with Volume:
Annual Review of **ANTHROPOLOGY**			
Vols. 1-20 (1972-1991)............................$41.00/$46.00			
Vol. 21 (1992).......................................$44.00/$49.00			
Vol. 22 (avail. Oct. 1993)...................$44.00/$49.00		Vol(s). _____	Vol._____
Annual Review of **ASTRONOMY AND ASTROPHYSICS**			
Vols. 1, 5-14 (1963, 1967-1976)			
16-29 (1978-1991)............................$53.00/$58.00			
Vol. 30 (1992).......................................$57.00/$62.00			
Vol. 31 (avail. Sept. 1993)...................$57.00/$62.00		Vol(s). _____	Vol._____
Annual Review of **BIOCHEMISTRY**			
Vols. 30-34, 36-60 (1961-1965, 1967-1991) .... $41.00/$47.00			
Vol. 61 (1992) ....$46.00/$52.00			
Vol. 62 (avail. July 1993) ....$46.00/$52.00		Vol(s). _____	Vol._____

Annual Review of **BIOPHYSICS AND BIOMOLECULAR STRUCTURE**
Vols. 1-20 (1972-1991)............................\$55.00/\$60.00
Vol. 21 (1992)....................................\$59.00/\$64.00
Vol. 22 (avail. June 1993)...................\$59.00/\$64.00 Vol(s). _____ Vol._____

Annual Review of **CELL BIOLOGY**
Vols. 1-7 (1985-1991)............................\$41.00/\$46.00
Vol. 8 (1992)....................................\$46.00/\$51.00
Vol. 9 (avail. Nov. 1993)...................\$46.00/\$51.00 Vol(s). _____ Vol._____

Annual Review of **COMPUTER SCIENCE**
Vols. 1-2 (1986-1987)............................\$41.00/\$46.00
Vols. 3-4 (1998-1989/1990)...................\$47.00/\$52.00 Vol(s). _____ Vol._____
Series suspended until further notice. Purchase the complete set for the special promotional price of \$100.00 USA / \$115.00 other countries, when all four volumes are ordered at the same time. Orders at the special price must be prepaid.

Annual Review of **EARTH AND PLANETARY SCIENCES**
Vols. 1-19 (1973-1991)............................\$55.00/\$60.00
Vol. 20 (1992)....................................\$59.00/\$64.00
Vol. 21 (avail. May 1993)...................\$59.00/\$64.00 Vol(s). _____ Vol._____

Annual Review of **ECOLOGY AND SYSTEMATICS**
Vols. 2-12, 14-22 (1971-1981, 1983-1991).........\$40.00/\$45.00
Vol. 23 (1992)....................................\$44.00/\$49.00
Vol. 24 (avail. Nov. 1993)...................\$44.00/\$49.00 Vol(s). _____ Vol._____

Annual Review of **ENERGY AND THE ENVIRONMENT**
Vols. 1-16 (1976-1991)............................\$64.00/\$69.00
Vol. 17 (1992)....................................\$68.00/\$73.00
Vol. 18 (avail. Oct. 1993)...................\$68.00/\$73.00 Vol(s). _____ Vol._____

Annual Review of **ENTOMOLOGY**
Vols. 10-16, 18 (1965-1971, 1973)
20-36 (1975-1991)............................\$40.00/\$45.00
Vol. 37 (1992) ....\$44.00/\$49.00
Vol. 38 (avail. Jan. 1993) ....\$44.00/\$49.00 Vol(s). _____ Vol._____

Annual Review of **FLUID MECHANICS**
Vols. 2-4, 7, 9-11 (1970-1972, 1975, 1977-1979)
14-23 (1982-1991) ....\$40.00/\$45.00
Vol. 24 (1992) ....\$44.00/\$49.00
Vol. 25 (avail. Jan. 1993) ....\$44.00/\$49.00 Vol(s). _____ Vol._____

Annual Review of **GENETICS**
Vols. 1-12, 14-25 (1967-1978, 1980-1991) ....\$40.00/\$45.00
Vol. 26 (1992)....................................\$44.00/\$49.00
Vol. 27 (avail. Dec. 1993)...................\$44.00/\$49.00 Vol(s). _____ Vol._____

Annual Review of **IMMUNOLOGY**
Vols. 1-9 (1983-1991) ....\$41.00/\$46.00
Vol. 10 (1992) ....\$45.00/\$50.00
Vol. 11 (avail. April 1993) ....\$45.00/\$50.00 Vol(s). _____ Vol._____

Annual Review of **MATERIALS SCIENCE**
Vols. 1, 3-19 (1971, 1973-1989)..................\$68.00/\$73.00
Vols. 20-22 (1990-1992) ....\$72.00/\$77.00
Vol. 23 (avail. Aug. 1993) ....\$72.00/\$77.00 Vol(s). _____ Vol._____

From:

Name _____

Address _____

_____

_____ Zip Code _____

Place
Stamp
Here

ANNUAL REVIEWS INC.
4139 EL CAMINO WAY
P. O. BOX 10139
PALO ALTO CA 94303-0897

ANNUAL REVIEWS SERIES	Prices, postpaid, per volume. USA / other countries (incl. Canada)	Regular Order Please send Volume(s):	Standing Order Begin with Volume:
Volumes not listed are no longer in print			

Annual Review of PUBLIC HEALTH
Vols. 1-12	(1980-1991) .... $45.00/$50.00		
Vol. 13	(1992) .... $49.00/$54.00		
Vol. 14	(avail. May 1993) .... $49.00/$54.00	Vol(s). _____	Vol._____

Annual Review of SOCIOLOGY
Vols. 1-17	(1975-1991) .... $45.00/$50.00		
Vol. 18	(1992) .... $49.00/$54.00		
Vol. 19	(avail. Aug. 1993) .... $49.00/$54.00	Vol(s). _____	Vol._____

NEW! Comprehensive Multiyear Index to Annual Review publications on computer disks. Available in the fall of 1992. Price to be announced.

❏ Please send complete information when available.   ❏ DOS   ❏ MAC

SPECIAL PUBLICATIONS	Prices, postpaid, per volume. USA / other countries (incl. Canada)	Regular Order Please send:

The Excitement and Fascination of Science
Volume 1	(1965 softcover) .... $25.00/$29.00	_____	Copy(ies).
Volume 2	(1978 softcover)...................... $25.00/$29.00	_____	Copy(ies).
Volume 3	(1990 hardcover).................... $90.00/$95.00	_____	Copy(ies).

(Volume 3 is published in two parts with complete indexes for Volume 1, 2, and both parts of Volume 3. **Sold as a two-part set only.**)

Intelligence and Affectivity:
Their Relationship During Child Development

	(1981 hardcover).................... $8.00/$9.00	_____ Copy(ies).

Send To: **ANNUAL REVIEWS INC., a nonprofit scientific publisher**
4139 El Camino Way   •   P. O. Box 10139
Palo Alto, CA 94303-0897   USA

❏ Please enter my order for publications indicated above. Prices are subject to change without notice.

Date of Order _____   ❏ Proof of student status enclosed

Institutional Purchase Order No. _____   ❏ California order, must add applicable sales
Individuals: Prepayment is required in U.S.   tax
funds or charge to bank card listed below.   ❏ Canadian order must add 7% GST.

❏ Amount of remittance enclosed: _____   ❏ Optional UPS shipping (domestic ground
_Or_ charge my   ❏ VISA   service except to AK or HI), add $2.00 per
                                                  volume. UPS requires a street address. No
❏ MasterCard   ❏ American Express   P.O. Box, APO or FPO.

Account Number _____ Exp. Date ____ / ____

Signature_____

Name _____
        please print
Address _____
        please print
_____ Zip Code _____

_____ Send free copy of current _Prospectus_ ❏
Area(s) of interest   Calif. Corp. No. 161041   ARI Federal I.D. No. 94-1156476